THEORY AND
DESIGN OF
STEEL STRUCTURES

EDITOR'S FOREWORD

Dramatic innovations and developments have occurred in civil and structural engineering in recent years. Difficulties of analysis which appeared insurmountable only twenty years ago have largely disappeared with the advent of the mainframe computer and the finite elements method; new generation microcomputers now increasingly provide such analyses with great convenience and economy. The engineer today has more time to devise new forms of construction, to improve design details, and to allow for phenomena and data which were previously overlooked or approximated. Much of this new expertise has been used to improve the design of ships and aircraft, offshore platforms, subway systems, high-rise towers and buildings, and many other forms of construction previously designed by rules-of-thumb and simple codes of practice. There is now much more internationalism in engineering too, with design methods and codes becoming more standardized, and large computers providing technical literature and patent information from all over the world. There is a need for these advances to be presented to an international audience by leading engineers of international repute; this is the purpose of the new Civil Engineering Series by Chapman and Hall.

The first of the new series is by Professors G. Ballio and F. Mazzolani of Milan and Naples respectively. Both authors have imposing lists of publications in this area, both are seasoned in theoretical and experimental research, both are eminent authorities in Italy, in Europe and on the international scene. They have the problems of the steelwork designer at heart and are dedicated to helping him by explaining and improving current practice. Their approach is fundamental and general, with instances of many varied examples of systems, joints and components. Complete design examples, however, are not included and should be sought elsewhere in professional journals and periodicals. Such examples, at their best, demonstrate conceptual thinking and ingenious application, but all of them need to be sound and economical in their detailed anatomy, a feature which this book sets out to establish and improve. I believe that every structural steel designer with pride in his own competence ought to find time to read and absorb this substantial contribution to his understanding and expertise.

E. Lightfoot
Oxford

FORTHCOMING TITLES

Probabilistic Methods in Structural Engineering
G. Augusti, A. Baratta and F. Casciati

Underground Structures
P. S. Bulson

THEORY AND DESIGN OF STEEL STRUCTURES

Giulio Ballio

Dipartimento di Ingegneria Strutturale
Politecnico di Milano, Italy

Federico M. Mazzolani

Istituto di Tecnica delle Costruzioni
Università di Napoli, Italy

LONDON   NEW YORK
Chapman and Hall

First published 1979 in Italian by
ISEDI
Revised English edition published 1983 by
Chapman and Hall Ltd
11 New Fetter Lane, London EC4P 4EE
Published in the USA by
Chapman and Hall
733 Third Avenue, New York NY 10017

© 1983 G. Ballio and F. M. Mazzolani

Printed in Great Britain by
J. W. Arrowsmith Ltd., Bristol

ISBN 0 412 23660 5

British Library Cataloguing in Publication Data

Ballio, Giulio
  Theory and design of steel structures.
  1. Steel, Structural   2. Structural design
  I. Title   II. Mazzolani, Federico M.
  III. Strutture in acciaio. *English*
  624.1'821   TA684
  ISBN 0-412-23660-5

To our wives
Delia and Silvana

F O R E W O R D

by Professor T.V. Galambos

Department of Civil and Mineral Engineering
University of Minnesota,USA

This book *Theory and Design of Steel Structures* by Professors Ballio
and Mazzolani contains most of what structural steel design engineers
need to know in order to do their work properly. The book contains
not only everyday solutions but also the treatment of many unusual
and vexing problems which do not nicely fit into a convenient category.
Many practical solutions are provided for such complex problems as
frame stability and local buckling. The treatment of connections and
stability is almost as thorough as texts devoted solely to these
special topics.

This book is, however, not just a lexicon of unusual problems where
one goes when the more familiar avenues have failed to provide an
answer; the treatment of the material is firmly based on the funda-
mentals of structural mechanics, and the experimental basis of the
problem is thoroughly explained.

There are few other books in the English language which match the
breadth and the thoroughness of this text; however, such books exist.
What makes this book unique is the authors' attempt to summarize and
compare the design specifications of the major codes throughout the
world. To be sure, the bias of the book is European, but North
American users will find that what is given here is impossible to
find without excessive and frustrating searching of codes and reports
in a half dozen languages. In a world where trade is international
and design engineers must understand the rules and the requirements
in many parts of the world, the information provided by Professors
Ballio and Mazzolani is priceless.

CONTENTS

x

PREFACE

When we began to write this book in 1978, we had two aims. First of all, we wanted it to be useful to different groups of readers, such as civil engineering students, engineers beginning to design metal structures and experienced steel designers. Secondly, it was our intention to explain the bases of the design methods suggested by the various national and international codes.

When the theory of structures is being discussed, it is not always possible to neglect local situations, habits, experience, in other words the engineering practice of each country. Nor is it possible to avoid any reference to current recommendations when one is introducing calculation methods, the validity of which are founded more on experience than on incontrovertible theoretical bases. For this reason, it is difficult for a book on the theory of structures completely to satisfy the needs of designers in different countries.

On the other hand, nowadays designers work more for foreign clients than for their own domestic market and, therefore, from an educational point of view, there is an advantage in a book that tries to overcome the cultural differences between engineers of different countries and help them look for a common language.

However, these intentions notwithstanding, our book is certainly influenced by our Latin origins, which lead us to be less pragmatic, less simple ... sometimes less certain than our English-speaking counterparts. On the other hand, we maybe have a greater disposition to criticism and discussion, a greater readiness to question the statements of codes and thus search for generalizations. This is, perhaps, the advantage of this book: it is not pragmatic, it raises problems, discusses them, sometimes offers a number of different solutions, and always tries to point out the origins and the range of validity of each method.

The plan of this book follows the logic of any structural design, which requires that the effects of external loads acting on the structural model must be lower than those of structural resistance. In order to evaluate the effects of the loads, the actual structure must be reduced to simplified models, according to rules which are never rigorous nor precisely expressed in mathematical terms. These models may be worked out by considering the expected collapse mechanism and they are often based on the intuition and experience of the designer. The designer must then define a design method suitable for evaluating the degree of reliability required for the structure. These problems are analysed in the first three chapters.

The following three chapters deal with material strength without referring to any particular structural types. Steel sections, material properties and imperfections, and assembly by means of welding and bolting are analysed.

The last three chapters deal with the behaviour of connections, the strength and the stability of the various structural components. The semi probabilistic limit state method is always used in calculations. It is, however, very easy to utilize such results in the allowable stress method, if we divide the design strength by a given safety factor.

Two major topics have not been covered in the book: resistance under fatigue and plastic design. Historically speaking, fatigue has been the particular concern of mechanical engineers. Nowadays civil engineers are beginning to appreciate the need for knowledge of this problem; the time has certainly come when co-operation with mechanical engineers is necessary, to single out and solve the problems of civil engineering structures requiring fatigue resistance.

Plastic design, on the other hand, is traditionally linked to civil engineering and has given excellent results as far as the systematic organization of the various structural analysis methods is concerned. Unfortunately, however, our knowledge in this field is not yet sufficiently advanced as to allow a generalized application. In fact, general computer programs for plastic design are incomplete or too expensive: the existing ones mostly concern the analysis of bent elements, when buckling phenomena and interaction between axial forces and bending are absent. It is to be hoped that in the next few years the gap between research and the needs of designers will be reduced.

We have also omitted specific references to the design of cold formed profiles and of composite steel and concrete structures. This is essentially due to the approach adopted in this book, which is purposely directed more to structural analysis methods than to the calculation of any specific type as in steel construction. Where reference has been made to structures, as in the first chapter, it is simply as a means of giving examples of the approach to the structural problem.

We hope that we have succeeded in avoiding the criticism which is often raised by those who operate in the field of designing and building structures, namely that those who study, experiment and sometimes draw up specifications complicate design methods and recommendations, thus creating a useless division between theory and practice.

Traditionally, in the field of steel structures, there has always been a fruitful exchange of information between researchers and designers. We believe that, to a large extent, this has been possible thanks to the continual activity of international and national organizations, such as IABSE (International Association for Bridge and Structural Engineering), ECCS (European Convention for

Constructional Steelwork), SSRC (Structural Stability Research Council) and the various national steelwork associations.

We sincerely hope that this exchange of ideas will continue and increase. This book can be considered as our contribution to the debate.

We wish to thank all those who have made this English edition possible: the international revisers and publishers who encouraged us, Dr Lightfoot who gave much precious advice and, particularly, Mr Leech who revised the text, playing a major part in rendering our work more acceptable for English speaking readers. Our warmest thanks to him.

We thank Claudia Weiss who translated the book from the Italian, a task often rendered difficult due to the existence of so many technical terms.

Guido Capra, Giorgio Sforzini, Gennaro Vedova, Giancarlo Zaupa contributed to the graphic work, Flavia Bellini, Teresa Bianchi, Giuliana Comini and Isabella Peroni typed the text, Francesco Ballio and Giovanni Novati glued the figures.

Giulio Ballio
Milano   1983

Federico M. Mazzolani
Napoli

1 Lower case letter symbols

a distance, geometrical dimension; depth of the throat section of a fillet weld

b width; side of a fillet weld

b_d width of a diagonal tension band in a sheet panel

b^{eff} effective width

$b^{eff,n}$ effective width for n bolts

b^f width of flange

b^{nom} nominal value of the flange width

b_s width of stiffener

c geometrical dimension; length of the zone where a concentrated load is applied; specific heat

c_p pressure coefficient

$c^{p,e}$ external pressure coefficient

$c^{p,f}$ total pressure coefficient

$c^{p,i}$ internal pressure coefficient

$c_{p,l}$ local pressure coefficient

d diameter; diameter of the shank of a bolt; static depth of the reinforced concrete section; distance between the centroids of the flanges in an I beam; distance between the centroids of the cords in a truss

d_h diameter of a hole in a web

d^m average diameter of a bolt

d^n diameter of the core of the shank of a bolt

d_{res} diameter of the stress section of the thread

e eccentricity

e_* fictitious eccentricity

e_o distance between centroid and shear centre

e^w web eccentricity

e_x (e_y) components of the eccentricity in direction $x(y)$

f strength

f_b bearing strength (for bolts); bond strength (for concrete)

$f_{b,d}$ design value of bearing (bond) strength

f_c	compressive strength
$f_{c(c),k}$	characteristic value of concrete strength measured on cube
$f_{c,k}$	characteristic value of compressive strength
$f_{c,m}$	mean value of compressive strength
f_d	design value of strength
$f_{d,V}$	design value of one bolt-shear strength
$f_{d,N}$	design value of one bolt-tensile strength
$f_{d,w}$	design value of a butt weld strength
f_e	elastic limit strength
f_k	characteristic strength
$f_{k,N}$	characteristic value of one bolt-tensile strength
$f_{k,V}$	characteristic value of one bolt-shear strength
f_{lim}	limit strength
f_m	mean value of strength
f_o	strength at the limit of proportionality
$f_{0,2}$	conventional strength at 0.2%
f_t	ultimate tensile strength
$f_{t,k}$	characteristic value of ultimate tensile strength
f_u	ultimate strength
$f_{u,w}$	ultimate strength of a weld
f_y	yield strength
$f_{y,f}$	yield strength of a flange
$f_{y,k}$	characteristic value of yield strength
$f_{y,m}$	mean value of yield strength
$f_{y,nom}$	nominal value of yield strength
$f_{y,N}$	yield strength of a bolt in tension
$f_{y,w}$	yield strength of a web
f^w	strength of a weld
f_∞	strength for $n = \infty$ fatigue cycles
g	distributed permanent load (for length or surface unit)
g_k	characteristic value of the permanent load
g_m	mean value of the permanent load
h	depth
h_b	depth of a beam
h_c	depth of a column
h_h	depth of a rectangular hole in a web
h_{nom}	nominal value of the depth
h_s	snow depth
$h_{s,d}$	design value of the snow depth
h_w	depth of the web of a beam
i	radius of gyration
i_c	radius of gyration of a batten plate
i_{eq}	equivalent radius of gyration
$i_{min}, (i_{max})$	minimum (maximum) radius of gyration
i_x, i_y	radius of gyration around $x(y)$ axis
i_*	fictitious radius of gyration
k	coefficient; ratio; coefficient for reducing the ultimate moment because of an axial force; ratio between ultimate stresses in shear and in tension; coefficient for combining stress in fillet welds; thermal conductivity; stiffness of an axial or rotational spring

k_b — flexural stiffness of a beam

k_c — flexural stiffness of a column

k_{min} — minimum stiffness of a spring

$k_{\sigma,x}$ ($k_{\sigma,y}$) — buckling, coefficient for a rectangular plate submitted to normal stress in $x(y)$ direction applied on both opposite sides

k_τ — buckling coefficient for a rectangular plate

l — *see L*

m — distributed moment for unit length

m_{lim} — limit moment

m_{res} — resisting moment for unit length

m_x (m_y) — moment acting about the $x(y)$ axis

n — number of numerical exponent

n_{tot} — total number

n_v — number of the shear stress section in bolted joint

n_f — number of friction surfaces in a bolted joint

$n_\perp$ — perpendicular stress in a fillet weld (referred to the turned throat section)

p — probability; pitch of the bolt thread; pitch of the bolts in joints; distance between the joints of a truss; pressure

p (f) — probability density function

p_{adm} — allowable probability for not overcoming a given limit state

p_k — probability corresponding to the characteristic value of a random variables

p_u — probability of collapse

p_{ij} (P) — stress tensor at the point P

p_{cr} — dynamic pressure corresponding to the critical speed

p_{ref} — reference value of the pressure

q — distributed live load (for length or surface unit)

q_k — characteristic value of the live load

q_m — mean value of the live load

q_n — perpendicular component of the live load

q_o — reference value of the live load

$q_{s,o}$ — reference value of the snow load

$q_{s,m}$ — averange value of the snow load

$q_{s,k}$ — characteristic value of the snow load

q_t — tangential component of the load

q_u — value of the load causing plastic collapse of the structure

r — radius

r_{eff} — effective or equivalent radius

r_t — centroid and tangent to the middle line distance in a hollow section

s — standard deviation; curvilinear coordinate; distance; snow distributed load

t — time; distributed torque for unit length; thickness

t_f	flange thickness
$t_{f,nom}$	nominal value of the flange thickness
t_o	initial thickness
t_{min}	minimum thickness
t_s	thickness of a stiffener
t_w	web thickness
$t_{w,nom}$	nominal value of the web thickness
$t_\perp ; t_{\parallel}$	shear stresses in a fillet weld (referred to the turned throat section)
u	displacement component in the x direction; perimeter
u_o	initial middle span deflection in the x direction
v	displacement component in the y direction; speed
v_o	initial out of straightness; middle span deflection in a simply supported beam
v_m	mean value of the deflection; mean value of speed
v_{max}	maximum beam deflection
v_M	deflection due to bending moment
v_r	residual deflection
v_V	deflection due to shear
v_{cr}	critical speed
v_d	design value of speed
$v_{m}, \Delta t_o$	mean value of speed with a given return period Δt_o
v_{nom}	nominal value of speed
v_{ref}	reference value of speed
v_w	web deviation
w	dispacement component in the z direction; wind load (for length or surface unit)
w_L	wind load (parallel to the wind direction)
w_t	wind load (perpendicular to the wind direction)
x	random variable
x_k^+	upper fractile
x_k^-	lower fractile
x_k	characteristic value
x_m	mean value
x, y, z	coordinates
y	distance of a point from the centroid of a section
z	lever arm; depth from the ground level

2 Capital letter symbols

A	area; bearing stress area
A_c	area of a batten plate
A_d	area of a diagonal bar
A_{eff}	effective area
A_f	area of flange
A_g	gross area
A_m	area of the moment diagram
A_n	net area

A_{nom}	nominal value of an area
A_o	measured area of a test specimen
A_{red}	reduced area
A_{res}	stress resistant area
A_s	area of a stiffener
$A_{s,L}$ $(A_{s,t})$	area of a longitudinal (transverse) stiffener
$A_{s,min}$	minimum area of a stiffener
A_t	area of a chord
A_u	area of the minimum section of the test specimen after rupture
A_w	area of a web; area exposed to wind
$A_{w,n}$	net area of the web
C	integration constant
D	flexural stiffness of a plate
E	Young's modulus; seismic force
E_m	mean value of the elastic modulus
E_{red}	reduced modulus
E_t	instantaneous tangent modulus
$E_{t,m}$	mean value of tangent modulus
F	force, external action, load
F_c	ultimate compressive force
F_{cr}	Euler's critical value of the force in elastic range
F_d	design value of a load
F_e	value of the load causing the achievement of the elastic limit in a point of the structure
F_f	friction load
F_h	horizontal component of a force
F_k	characteristic value of a force
$F_{k,n}$	characteristic value of a force having the $1 - n$ probability to be overcome
F_L	longitudinal component of a force
F_{lim}	limit value of a force
F_m	mean value of a force
F_n	perpendicular compnent of a force
F_N	force causing axial stress
F_{red}	reduced value of a force
F_{serv}	serviceability value of a force
F_t	transversel or tangential component of a force; ultimate value of tensile force
F_u	ultimate value of a force (collapse load)
F_v	vertical component of a force
F_V	force causing shear stress (in bolted joints)
$F_{V,u}$	ultimate value of a force causing shear stress
$F_{V,f}$	value of the slipping force in a bolted joint
F_w	ultimate value of the web force
F_y	squash load
$F^{\Delta t_o}$	value of a force corresponding to a return period Δt_o
F	dimensionless value of a force
$F(\ )$	function

G	shear modulus; permanent load
G_k	characteristic value of the permanent load
H	horizontal component of a force
H_L	longitudinal component of an horizontal force
H_t	transversel component of an horizontal force
H_x (H_y)	horizontal component in $x(y)$ direction
H^*	fictitious horizontal force
I	turbulence intensity; moment of inertia of a plane area
I_b (I_c)	moment of inertia of a beam (of a column)
I_c	moment of inertia of a batten plate
I_d	moment of inertia of a diagonal bar
I_{eq}	equivalent moment of inertia
I_f	moment of inertia of a flange
I_m	average moment of inertia
I_{min} (I_{max})	minimum (maximum) moment of inertia
I_n	moment of inertia of the net section
I_{nom}	nominal value of the moment of inertia of a section
I_{red}	reduced moment of inertia
I_s	moment of inertia of a stiffener
$I_{s,L}$ $(I_{s,t})$	moment of inertia of a longitudinal (transverse) stiffener
I_t	moment of inertia of a chord
I_T	twisting moment of inertia
I_w	moment of inertia of a web
I_x (I_y)	moment of inertia related to the $x(y)$ axis
I_ω	warping moment of inertia
L	length
L	length of wave propagation
L_b	effective length of a bolt
L_c	reference length of a test specimen; effective length of a compression member
$L_{c,h}$	effective length in the horizontal plane
$L_{c,v}$	effective length in the vertical plane
L_d	length of a diagonal bar
L_h	length of a rectangular hole in a web
L_o	initial distance between the reference points in a specimen
L_t	length of a chord
L_{tot}	total length
L_u	final distance beween the reference points in a specimen after rupture
M	moment, bending moment
M_a (M_b)	bending moment at the edge a(b)
M_{adm}	allowable bending moment
M_{cr}	Euler's critical moment for flexural-torsional buckling in elastic range
M_d	design value of the bending moment
$M_{d,pl}$	design value of the plastic bending moment
M_D	moment corresponding to the flexural-torsional buckling of a beam

M_e	elastic moment (corresponding to the achievement of the elastic limit at a point of the section)
M_{eq}	equivalent bending moment
$M_{e,x}$ $(M_{e,y})$	elastic moment for bending about the $x(y)$ axis
M_f	bending moment acting on flanges only
$M_{f,pl}$	plastic moment evaluated considering the yielding of the flanges only
M_i	bending moment on the member i
M^m	average bending moment
M^{max}	maximum bending moment
M^o	bending moment in the middle span
M_{pl}	plastic moment (corresponding to the plastic collapse of the section)
$M_{pl,x}$ $(M_{pl,y})$	plastic moment for bending about the $x(y)$ axis
M^{res}	resistance moment
$\bar{M}$	ultimate value of bending moment
M^u	bending moment acting on the web
M^w	warping moment
M^ω	dimensionless value of the bending moment
N	axial force
N_c	maximum load bearing capacity of a compression members; compression axial force in a strut
$N_{c,x}$ $(N_{c,y})$	maximum load bearing capacity of a compression members in plane buckling with bending about the $x(y)$ axis
$N_{c,xy}$	maximum axial force for a compression member for flexural torsional buckling
$N_{c,\theta}$	maximum axial force for a compression member in torsional buckling
N_{cr}	Euler's critical load
$N_{cr,x}$ $(N_{cr,y})$	Euler's critical load in plane buckling with bending about $x(y)$ axis
$N_{cr,xy}$	Euler's critical load in flexural-torsional buckling
$N_{cr,\theta}$	Euler's critical load in torsional buckling
N_d	axial force in a diagonal bar
$N_{d,o}$	design force for one bolt in tension only
N_e	elastic axial force (corresponding to the achievement of the elastic limit in the section)
N_f	axial force in a flange
N^{lim}	limit axial force
N^m	average axial force
N_p	axial force corresponding to the loss of prestress in a preloaded bolt connection
N_{pl}	plastic axial force (corresponding to the plastic collapse of the section)
N^s	prestressing due to the tightening of bolts
N^t	tensile axial force in a chord or in a tie
$\bar{N}$	ultimate value of the axial force
N^u	dimensionless value of the axial force
Q	prying force; variable load, live load
Q_k	characteristic value of the variable load
$Q_{k,i}$	characteristic value of the variable load number i

Q_m	mean value of the variable load
Q_{max}	maximum value of the variable load
Q_{nom}	nominal value of the variable load
R	force resultant; reaction of a support; generic resistance
R_b	resistance of a bolted joint
R_t^b	ultimate tensile resistance
$R(\ )$	resistance depending on a given stress
$R(f_d)$	resistance evaluated on the basis of design strength
$R(f_y)$	resistance evaluated on the basis of yield strength
R_e	Reynold's number
S	first moment of a plane area; internal action; action effect; stress resultant; shear resultant acting at the connection between the web and the flange of a beam; total load due to snow
$S_x\ (S_y)$	first moment referred to the $x(y)$ axis
S_ω	warping first moment
S_{lim}	limit internal action
S_{red}	reduced load due to snow
S_t	Strouhal's number
T	twisting moment; torsional moment; temperature; foundamental period
T_{pl}	plastic twisting moment (corresponding to the plastic collapse of the section)
T^s	torque (per bolt)
T^T	uniform torsional moment
T^u	ultimate torsional moment
T^ω	warping torsional moment
T_o	initial temperature
T^*	transition temperature
V	vertical component of a force; shear force; volume
V^a	steel volume
V^c	concrete volume
V^d	design shear force
$V^{d,o}$	design force for one bolt in shear only
$V^{d,N}$	reduced design force for one bolt in tension and shear
$V^{d,b}$	design force for one bolt in bearing
$V^{d,t}$	design strength for the scratching of the plate
V^f	slipping force in a bolted joint
$V^{f,m}$	average slipping force in a bolted joint
$V^{f,o}$	friction force for one bolt in shear only
$V^{f,N}$	reduced friction force for one bolt in shear and tension
V^m	average value of the shear force
V_{pl}	plastic shear force (corresponding to plastic collapse of the section)
V^u	ultimate shear force
$V^{u,f}$	ultimate shear force for flanges only
$V^{u,w}$	ultimate shear force for web only
V^V	shear force for buckling in the web
$V_{w,pl}$	plastic shear force for web

W	total wind load; elastic modulus (in bending)
W_{nom}	nominal value of the elastic modulus
W_{red}	reduced value of the elastic modulus
W_x (W_y)	elastic modulus for bending about the $x(y)$ axis
X	random quantity; force parallel to the x axis
Y	unknown force; force parallel to the y axis
Z	force parallel to the z axis; plastic modulus (in bending)
Z_{nom}	nominal value of the plastic modulus

3 Greek letter symbols

α	angle, ratio, exponent, coefficient; ratio between both sides of a rectangular panel; section shape factor; deflection limit $(= v_{max}/L)$; coefficient of thermal expansion
α_{cr}	critical load multiplier
α_u	collapse load multiplier
α_w	fillet weld coefficient
α_x (α_y)	section shape factor for bending about the $x(y)$ axis
α_c	critical load correction factor (for stiffened plates)
α_c^*	limit value of the correction factor α_c
$\alpha_{c,red}^*$	reduced value of the correction factor α_c^*
β	angle, ratio, exponent, coefficient; buckling factor (for evaluating the buckling length)
β_{eff}	reduced value of the buckling factor obtained by correcting β
β_w	efficiency coefficient for fillet welds
γ	generic coefficient, dimensionless ratio
γ_F	partial load factor
γ_G	partial load factor for permanent load
γ_Q	partial load factor for variable loads
γ_a	reduction coefficient of the steel strength
γ_c	reduction coefficient of the concrete strength
γ_f	reduction coefficient of the friction strength in bolted joint
γ_m	reduction coefficient of the material strength
$\gamma_{m,pl}$	reduction coefficient of the material strength in plastic design
γ_N	reduction coefficient of the axial strength of a bolt
γ_V	coefficient of reduction of the shear strength of a bolt
γ_w	coefficient of reduction of the strength of a weld
γ	relative flexural rigidity of a stiffener
γ_L	relative flexural rigidity of a longitudinal stiffener
γ_o	relative flexural rigidity of a flange
γ_t	relative flexural rigidity of a transverse stiffener
γ^*	optimum relative flexural rigidity of a stiffener
γ^{**}	actual relative flexural rigidity of a stiffener
γ_L^*	optimum relative flexural rigidity of a longitudinal stiffener

γ_t^*	optimum relative flexural rigidity of a transverse stiffener
γ	specific weight
γ_a	specific weight of the steel
γ_c	specific weight of the concrete
γ_s	specific weight of the snow
δ	logarithmic decrement; variation coefficient; relative axial rigidity of a stiffener
δ_L	relative axial rigidity of a longitudinal stiffener
δ_t	relative axial rigidity of a transverse stiffener
δ_o	relative axial rigidity of a flange
δ^*	optimum relative axial rigidity of a stiffener
ϵ	strain
ϵ_c	compressive strain
ϵ_e	elastic limit strain
ϵ_r	residual strain
ϵ_t	strain at failure
$\epsilon_{z,\omega}$	strain due to warping
ζ	dimensionless coordinate in z direction; dimensionless ratio
η	dimensionless coordinate in y direction; coefficient; dimensionless factor; parameter of imperfection
η_r	reduction factor for flexural-torsional buckling
η_w	efficiency coefficient of a butt weld
θ	angle, rotation
θ_d	slope of the diagonal field in a web panel
θ_o	initial rotation
θ_u	ultimate rotation of a section (corresponding to the collapse of the structure)
λ	slenderness
λ_c	conventional slenderness of proportionality
λ_e	slenderness of proportionality
λ_{eq}	equivalent slenderness
λ_{red}	reduced slenderness
$\lambda_x(\lambda_y)$	slenderness for bending about the $x(y)$ axis
λ_w	conventional slenderness of the web
λ^*	fictitious slenderness
$\bar{\lambda}$	dimensionless slenderness
$\bar{\lambda}_M$	normalized slenderness for flexural torsional buckling
$\bar{\lambda}_N$	normalized slenderness for compression and bending
μ	coefficient, dimensionless ratio; friction coefficient; ratio between empty and full; ratio between Euler's critical load and axial load; percentage of reinforcement
ν	Poisson's ratio; safety factor; kinematic viscosity of the air
ν_c	safety factor against elastic-plastic instability
ν_{cr}	safety factor against elastic Euler's instability

ξ	dimensionless coordinate in x direction; dimensionless ratio, dimensionless coefficient
ρ	coefficient; dimensionless ratio; efficiency; ratio between residual stress and yield stress; density
σ	normal stress
$\sigma_{(a)}$	normal stress of the steel
σ^{adm}	allowable stress
$\sigma^{adm,N}$	allowable tensile stress for one bolt
$\sigma^{adm,V}$	allowable shear stress for one bolt
σ_b	normal stress in a bolt
σ_c	strength corresponding to the force that causes failure of a compression member
$\sigma_{(c)}$	normal stress in concrete
$\sigma^{c,adm}$	allowable stress in concrete
σ^{cr}	Euler's critical stress (for compression members)
$\sigma^{cr,D}$	Euler's critical stress (for members in bending)
$\sigma^{cr,o}$	Euler's critical stress (for an infinite plate)
$\sigma^{cr,o}_{cr,x}$ $(\sigma_{cr,y})$	Euler's critical stress in $x(y)$ direction for a rectangular plate loaded by axial forces simultaneously acting in x and y directions
$\sigma_{cr,x,o}$ $\sigma_{cr,y,o}$	Euler's critical stress for a rectangular plate loaded by axial forces acting in $x(y)$ direction only
σ_D	maximum stress acting in a member subjected to flexural buckling in elastic-plastic field
σ_{id}	ideal stress
σ_{lim}	limit stress
σ_m	average stress
σ_M	stress due to bending
σ_{min}, σ_{max}	extreme values of stress in fatigue
σ_{max}	maximum value of stress
σ_N	stress due to axial load
σ_r	residual stress
$\sigma_{r,c}$ $(\sigma_{r,t})$	compressive (tensile) residual stress
σ^{ref}	reference stress
$\sigma_{r\perp}$, $\sigma_{r\parallel}$	residual stress perpendicular (parallel) to the axis of the welded joint
$\sigma_{cr,red}$	reduced critical stress
σ_t	tensile stress
$\sigma^{w,adm}$	allowable stress in a weld
σ_x, σ_y, σ_z	components of normal stress
$\sigma_{z,\omega}$	warping normal stress
σ_1, σ_2, σ_3	principal stresses
$\sigma_\perp$, $\sigma_\parallel$	normal stress in a weld (referred to the throat section)
τ	shear stress
τ_{adm}	allowable shear stress
$\tau_{adm,V}$	allowable shear stress for one bolt
τ_b	shear stress in a bolt
$\tau_{c,adm}$	allowable shear stress for concrete
τ_{cr}	Euler's critical shear stress for a rectangular plate

$\tau_{cr,o}$ Euler's critical shear stress for a rectangular plate in shear only

τ_m average shear stress

τ_T shear stress due to uniform torsion

τ_u ultimate value of shear stress

τ_ω shear stress due to warping

$\tau_{xy}, \tau_{xz}, \tau_{yz}$ components of shear stress

$\tau_\perp, \tau_\parallel$ shear stress in a weld (referred to the throat section)

Φ dimensionless coefficient; limit value of the angle of friction

χ dimensionless coefficient; coupling coefficient in flexural-torsion buckling; reduction coefficient of the restraint coefficient; shear factor; curvature

ψ dimensionless coefficient; coefficient taking into account partial stress redistribution in bending; combination loading factor of variable loads

ω sectional area

Δ component of relative displacement

Δ_c displacement at collapse load

Δh variation of height

ΔL variation of length

Δx increment of the magnitude x

Δv instanteneous speed of a wind gust

$\Delta \sigma$ amplitude of variation of the stress

Δt period

ΔT return period

ΔT_s^o expected life of the structure

ϕ diameter of a hole or of a bar

ϕ energy

ϕ_D energy for variation of shape

ϕ_V energy for variation of volume

4 Abbreviations

AASHTO	American Association of State Highway and Transportation Officials
ACAI	Associazione fra i Costruttori in Acciaio Italiani
ACI	American Concrete Institute
AFNORM	Association Française de Normalisation
AISC	American Institute for Steel Construction
AISI	American Iron and Steel Institute
ASTM	American National Standard
ASCE	American Society of Civil Engineers
ASME	American Society of Mechanical Engineers
AWS	American Welding Society
BCSA	British Constructional Steelwork Association
BSI	British Standards Institution
CEB	Comité Européen du Béton
CECM-ECCS-EKS	European Convention for Constructional Steelwork

CIB	Conseil International du Bâtiment
CNR	Consiglio Nazionale delle Ricerche
CRC	Column Research Committee of Japan
DIN	Deutsche Industrie Normen
DSV	Deutschen Stahlbau Verband
FIP	Fédération International de la Précontraint
IABSE-AIPC	International Association for Bridge and Structural Engineering
JSSC	Japan Society of Steel Construction
NBN	Institut Belge de Normalisation
NEN	Nederlands Normalisatie Instituut
RILEM	Réunion International Laboratoires d'Essais sur Materiaux
SIA	Società Svizzera degli Ingegneri e Architetti
SSRC	Structural Stability Research Council
UNI	Ente Nazionale Italiano d'Unificazione
VDE	Verein Deutscher Eisenhuttenleute
WES	Welding Engineering Society (Giappone)
WRC	Welding Research Council

C H A P T E R   O N E

Structural Systems

1.1 INTRODUCTION

Structural steel is used for such a variety of types of structures,
ranging from residential buildings to industrial plants, from mill
buildings to cranes, from transmission towers to silos, etc, that no
single approach could possibly be expected to cover all applications.
In any instance the structural scheme will largely depend on the
purpose of the construction. It should be useful, however, to try to
find some general features which are common to all steel structures
and to provide guide lines for structural design criteria. In the
following, therefore, some matters which can help in the structural
choice are considered in relation to the main problems involved in the
design of two of the most common types of structure: the multi-storey
building and the single storey industrial building.

Obviously, what is said concerning these structures can be extended
to other systems: the choice of these two types can be considered as
a pretext for illustrating certain logical patterns which are
sufficiently general to be useful for a proper design conception of
any kind of steel structure .

Some outstanding text books dealing with the structural analysis of
steel systems are given in references |1-20| at the end of the
chapter. Design rules and constructional details may be found in
|21-28|. An international survey of the present state of the art on
steel structures |29| also contains a most extensive bibliography on
different aspects.

1.1.1 *Kinematics of structures*

A steel structure is the result of assembling the various parts which
are prefabricated, generally from sections or profiles and sheets,
away from the building site.

In order to explain this definition better, the normal sequence of
events leading to the completed structure are:

(a) Production of profiles and sheets in steelworks
(b) Transformation of profiles and sheets into pre-assembled
structural elements or structural sets in the fabrication shop

(c) Transportation and erection of the structural elements or sets on site

The form of construction of a steel structure is widely influenced by this sequence, which can be considered the converse of that which characterizes a non-prefabricated concrete structure.

The latter is entirely produced on site by casting the concrete into the forms, so that a monolithic and therefore extremely hyperstatic structure is created. Consider for example a typical beam to column connection of a concrete structure (Fig. 1.1). The constraint of the members composing the structure is such that the joint can be considered as built in (Fig. 1.1a). If a greater freedom of relative movement between the members is required, special technological or constructional devices must be applied (Fig. 1.1b).

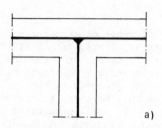 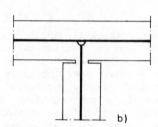

Fig. 1.1

The steel structure, on the contrary, produced by assembling various parts, has a fairly low degree of mutual constraint between the structural members and special technological or constructional devices must be applied to increase redundancy of the structure.

Fig. 1.2a schematically shows one of the simplest and most economical connections linking a beam to a column. If a greater rigidity of the connection is required, more complicated and expensive procedure must be applied (Fig. 1.2b).

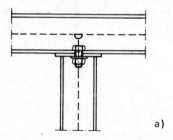 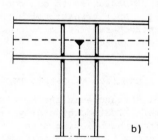

Fig. 1.2

The economic need of minimizing the working processes and simplifying the erection operations therefore requires a low degree of mutual constraint between the structural members. In designing a

2

structure and its joints in this way, due consideration must be given
to possible instability, which must be overcome by introducing
stabilizing elements. For example, consider a succession of beam to
column connections of the type shown in Fig. 1.2a. They can belong to
a structural system as shown in Fig. 1.3a. It can be seen that it is
necessary to introduce bracings (Fig. 1.3b) to stabilize the structure
and thus enable it to resist external load, whatever their direction
might be.

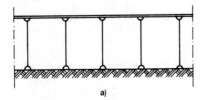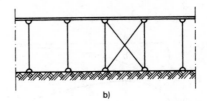

a)   b)

Fig. 1.3

1.1.2 *Stability of the structural elements*

Steel is a material whose stress strain-relationship is practically
symmetrical in both tension and compression. For structural purposes,
it can be schematically represented as in Fig. 1.4. A steel
structural element can behave asymmetrically due to buckling phenomena
of compressed parts. Such phenomena affect columns, beams, web panels,
etc. They produce a lowering effect on the load carrying capacity of
the elements which often influences the structural choices and the
design.

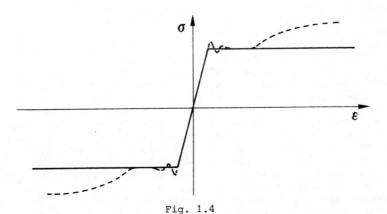

Fig. 1.4

These problems are examined in Chapter 9. It is, however, worth-
while to consider here the behaviour of a compression member, just
from the qualitative point of view, in order to deduce some useful

conclusions for structural design purposes.

The behaviour of the compression member of Fig. 1.5 can be represented by the load/displacement ($N - \Delta$) curve, Δ being the relative shortening of its two ends.

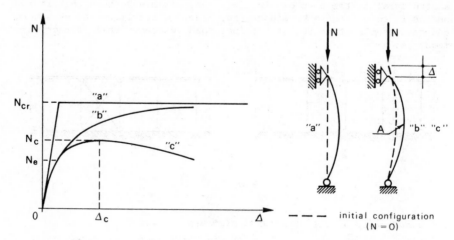

Fig. 1.5

If the bar is initially straight and subjected to axial compression only, its behaviour is represented by curve 'a' of Fig. 1.5, which is valid in the perfectly elastic field for displacements which are small in relation to its length. The bar remains straight as long as the value of the load N is lower than the Euler critical load $N_{cr} = \pi^2 EA/\lambda^2$, where E is Young's modulus of the material, A the cross sectional area and λ the slenderness ratio, L/i (length/radius of gyration). If $N < N_{cr}$, the shortening of the bar depends only on the axial deformability of the material and the $\Delta - N$ relationship is linear and identical (except for the sign) to the one in tension.

If $N = N_{cr}$, any configuration of the bar represents an equilibrium configuration, and therefore Δ is indeterminate.

Actual structural members made from industrial bars are not perfectly straight. They always possess some geometrical imperfections (e.g. out-of-straightness of their axes, out-of-plumb due to erection tolerances). If one assumes once more that the bar is made of perfectly material, its behaviour is defined by curve 'b' of Fig. 1.5. As the load increases, the shortening of the bar increases more than linearly. Because of the load eccentricity due to the geometrical imperfections, bending arises in addition to compression. The Euler critical load N_{cr}, however, represents the asymptotical value of the load the compression member can bear.

But the bar is not made of perfectly elastic material. Let the value of the applied load be N_e when the elastic limit of the material is reached in the most stressed fibre of the section (the inside one at mid-height, point A of Fig. 1.5). For $N > N_e$ the bar performs elasto-plastically and its shortening is greater than that foreseen in

4

the hypothesis of a perfectly elastic material. It is possible to evaluate the load N_c defining the maximum load carrying capacity of the bar.

For $N > N_c$, a state of equilibrium is impossible and for $\Delta > \Delta_c$ the bar can be in equilibrium under a load $N < N_c$ (decrescendant or unstable branch).

In conclusion, a simple compression member is a structural element having an asymmetrical N - Δ relationship (Fig. 1.6). Tension members can be considered, in good approximation, as behaving in a perfectly elasto-plastic way, with a limiting load of $N_{pl} = f_y A$, where A is the cross sectional area and f_y is the yield stress of the material. Under compression the N - Δ law is curvilinear and the limiting load for a given slenderness depends on the geometrical and structural imperfections of the industrial bar.

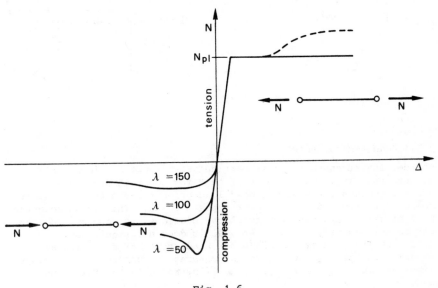

Fig. 1.6

1.1.3 *Space behaviour of structures*

The two foregoing sections immediately suggest a third: in order to check the overall rigid-body stability of a structure and to evaluate correctly the buckling phenomena of its structural elements, one must always conceive the structure in three-dimensional space, even when it can be considered and analysed as composed of various substructures behaving in one plane.

Consider for example a structure consisting of a set of identical frames (Fig. 1.7a), the upper joints of which cannot translate transversely. This can be examined as a number of single frame substructures each of which can be represented and analysed as a

5

planar structure (Fig. 1.7b) in order to evaluate the stresses
resulting from the applied loads F. The buckling effects must be
considered not only in the plane, but also outside the plane of the
frame (Fig. 1.7c), to find the most unfavourable conditions for the
structure.

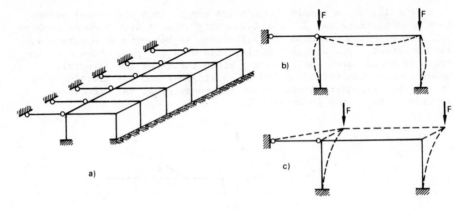

Fig. 1.7

1.1.4 *Dead load effects*

The density strength ratio is particularly favourable for steel
structures. It is only 20-30% of that for concrete structures. In
steel structures, therefore, dead load effects are often negligible
compared to those due to live or imposed loads. This is an advantage
most of all during the structural calculations: structure dead load
can be assumed very approximately as it does not significantly
influence the results. On the other hand, some effects which are
usually neglected in the design of concrete structures often become
important for steel structures, as illustrated in the following
examples:

(a) A plane steel roof weighs about $0.15 - 0.30$ kN m^{-2}, while a
concrete structure weighs $2 - 3$ kN m^{-2}. A snow load of 0.75 kN m^{-2} is
therefore equal to about 70-85% of the total amount for a steel
structure and to about 20-30% of the total amount for a concrete
structure. Consequently, although the small reduction in permanent
load of a steel structure is certainly an advantage as far as
seismic resistance is concerned, a snowfall is more dangerous for
a steel structure than for a concrete one

(b) On the same roof structure, wind can create a suction of
$0.30 - 0.50$ kN m^{-2} (Fig. 1.8). Usually, this is not taken into
consideration in the case of a concrete structure because it is on
the safe side as it reduces dead load effects. In the case of a
steel structure, the forces in the members can change their sign:
e.g., if the roof beam is a truss, the bottom bar, which normally

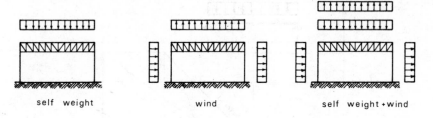

self weight   wind   self weight+wind

Fig. 1.8

is in tension, might become compressed and therefore be subject to buckling. If slenderness is great, a slight compression can be more decisive than tension in determining the size of member.

1.1.5 *Deformability of structural elements*

In steel structures, the ratio between maximum stresses in the members due to service load and material elastic modulus is 3-4 times higher than in the case of concrete structures. Furthermore, as pointed out in 1.1.4, the member carrying capacity is often totally available for serviceability if dead load effects are negligible. As a consequence, the deflection of a steel member under service load is often rather high and may sometimes be unacceptable. Therefore deflection controls in service conditions are of great importance. Consider for example a simply supported I-beam under a uniformly distributed load q:

$$\sigma = \frac{1}{8} \frac{qL^2}{W} \quad ; \quad v_{max} = \frac{5}{384} \frac{qL^4}{EI} \quad ; \quad I = Wh/2$$

Hence,

$$v_{max}/L = \frac{5}{24} \frac{\sigma_{max}}{E} \frac{L}{h} \tag{1.1}$$

where v_{max}/L is the ratio of mid-span deflection to length; L/h the length to depth ratio; E the material elastic modulus; W the elastic modulus of the beam section and σ_{max} the maximum stress produced by the bending moment due to the service loads on the structure.

Fig. 1.9 shows the lines corresponding to Equation (1.1) for various values of σ_{max} and the field where the usual values of L/h lie. It can be seen that displacement limitations within the range of $L/350$ - $L/500$ can be more restrictive than those required by the material strength. They therefore strongly influence the design, particularly when the steel structure is used together with other constructional elements requiring sufficiently rigid structures for their satisfactory performance: e.g. brick walls for façade or partitions supported by steel decks or beams.

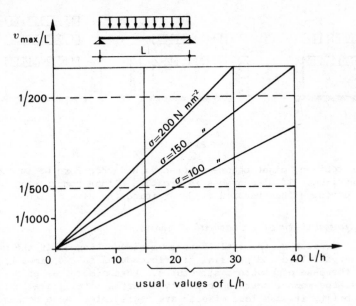

Fig. 1.9

1.1.6 *Interaction between bending moment and axial load*

In many common types of structures used for buildings the vertical
elements are compressed by the vertical loads and bent by horizontal
forces due to, for example, wind or earthquakes.

In steel structures, due to buckling phenomena, interaction between
bending moment and axial action is always unfavourable and, where
slenderness is great, extremely lowering. The situation is different
in the case of concrete structures, in which column slenderness is
usually rather low and furthermore axial force/bending moment
interaction can also be favourable.

Fig. 1.10 illustrates such a fact: a steel section (Fig. 1.10a)
is always less resistant in compression than in tension ($M = 0$) and
the presence of a bending moment in any case reduces its load carrying

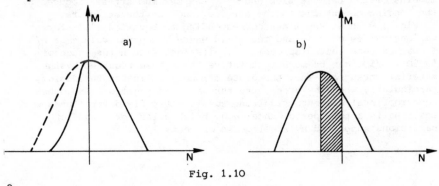

Fig. 1.10

capacity. A concrete section (Fig. 1.10b) obviously resists better in compression than in tension and with low values of axial loading will not be adversely affected by bending until the bending tensile stress exceeds the axial compression.

Bending/axial action interaction is typical of framed structures. Therefore, it is not surprising that it is not particularly economical to design a steel structure to resist horizontal forces. These can be dealt with in other ways.

1.2 MULTI-STOREY BUILDINGS

A multi-storey building can be represented, in its simplest form, by a series of floor systems bearing upon columns (Fig. 1.11). It is not intended to define the various structural possibilities nor describe different types of structure. The aim here is to illustrate a possible approach to the structural problem with a sufficient degree of generality to allow the identification of possible design schemes and calculation models. A possible approach can consist in analysing the effects of external loads on the structure and then classifying them according to the type of internal action that the load itself creates in the various parts of the structure.

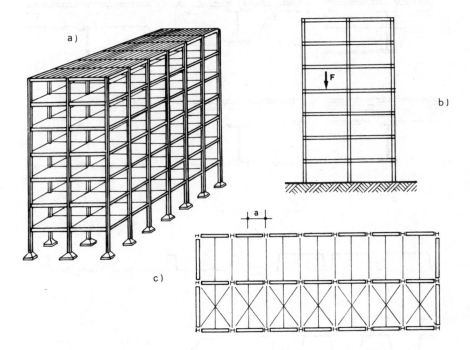

Fig. 1.11

1.2.1 *Effects of vertical loads*

Consider a vertical load F concentrated at a given point on any floor of the building (Fig. 1.11b). By means of appropriate structural elements, its effects must be transmitted to the foundation system. These structural elements and the main functions they are to fulfil must be determined.

By means of elements capable of covering small spans, load F can be brought to bear upon beams that are disposed at a maximum spacing compatible with the resistance of these elements (Fig. 1.11c). These (secondary) beams are as long as one of the column spacings and bear upon other principal beams which discharge the effects of load F into the columns.

Among other solutions, the roofing slab elements can be:

Panels made of normal or prestressed concrete or of concrete mixed with bricks, cast in situ (Figs 1.12a,b) or prefabricated (Figs 1.12c,d)

Corrugated sheets filled with inert material (Fig. 1.12e)

Corrugated sheets filled with concrete (Fig. 1.12f)

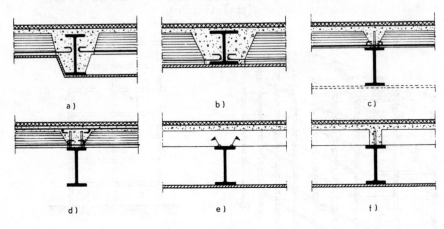

Fig. 1.12

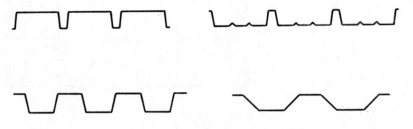

Fig. 1.13

Fig. 1.13 illustrates a few examples of corrugated sheet shapes. An example of steel corrugated sheets with reinforced concrete is illustrated in Fig. 1.14.

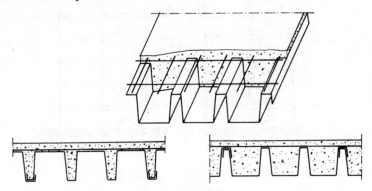

Fig. 1.14

It is thus possible to recognize the main effects created by a vertical load on the members of a multi-storey building.

In accordance with equilibrium conditions, the horizontal elements are mainly bent in the vertical plane and the vertical elements are compressed. Possible axial load in the beams and bending in the columns depend on the compatibility of the rotation of the members at their junction.

1.2.2 *Effects of horizontal loads*

Consider a horizontal load F concentrated in any point of the building (Fig. 1.15a). By means of a cladding element (façade) it is transmitted to two continuous floors (Fig. 1.15b). There is no loss in generalization if one considers a load H acting in the floor slab and applied at any point. Its effects must still reach the foundation system: the supporting elements of each floor slab will distribute the effects of H to the column joints and thence to the foundation (Fig. 1.15c).

From the foregoing it is clear that the horizontal elements are axially loaded and bent in their own plane, while the vertical elements are mainly bent.

1.2.3 *Structural types*

If the effects of horizontal and of vertical loads are now combined, it can be seen that it is possible to conceive the structure as a system of members capable of resisting axial, bending and shearing actions. Such behaviour is however possible only if there are rigid connections between the vertical members (columns) and the horizontal ones (beams), creating a framed structure with a high degree of hyperstatics.

As a consequence:

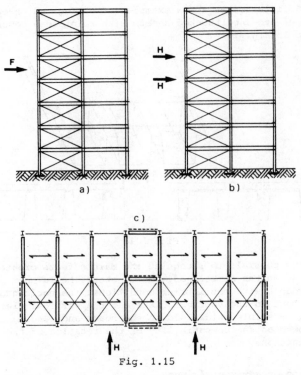

Fig. 1.15

The joints between the various elements are complicated
Interaction between axial loads and bending moments is determinant
in column design
Overall structure deformability can be great as it depends on the
moment of inertia of the columns only

Fig. 1.16 shows some constructional details of joints between
horizontal and vertical members which are designed to resist bending
moments and therefore are typical of rigid framed structures. Fig.
1.16a shows a joint which can transfer limited bending moments as the
web of the column undergoes local bending. The joint shown in Fig.
1.16b is obviously more resistant: the stiffeners recreate the section

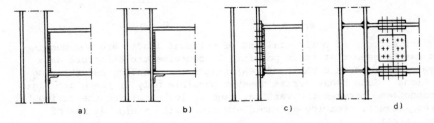

Fig. 1.16

of the beam and the column web must bear the shearing force only. These site welded joints are expensive and cause delay in erection. This is why bolted joints with end plate (Fig. 1.16c) or cover plates (Fig. 1.16d) offer suitable alternatives.

To avoid the problems of rigid frame construction it is however possible to conceive a structure capable of resisting the effects of vertical and horizontal loads by distributing the carrying function among the various members. To this end a hypostatic (simple) structure as shown in Fig. 1.17a is capable of transferring the effects of vertical loads to the foundation. In such a structure the beams are bent in the vertical plane, columns are simply compressed and the hinged joints between beams and columns must absorb only shearing forces.

To resist horizontal forces and to transfer their effects to the foundation, it is sufficient to provide a cantilever fixed to the ground (Fig. 1.17b).

By combining these two elementary structures, the structure shown in Fig. 1.17c is obtained: it is isostatic and therefore capable of carrying the effects of loads in any direction. In it:

Joints are simple (hinges)
Deformability is related to the behaviour of the bracing system (cantilever)
Interaction between axial load and bending moment in the column is virtually absent

Bracings, however, do present some complications. Their foundation system, for instance, has to resist the horizontal loads where they are concentrated instead of distributing them over the whole building area.

In such a structural scheme, the columns, which are simply compressed by the vertical loads, work mainly as pin-ended bars and may be either continuous or interrupted at the various floors, according to whether the main beams are interrupted or continuous. Therefore, by means of columns and beams, it is possible to absorb any

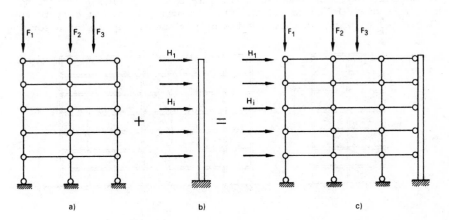

Fig. 1.17

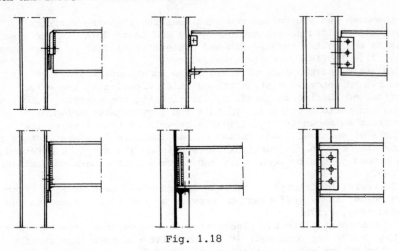

Fig. 1.18

vertical load and to transmit it to the foundations. Fig. 1.18
illustrates some constructional details of joints connecting
horizontal and vertical members, which are capable of transmitting
only axial and shearing forces. They are therefore suitable for
pin-ended structures only and not for rigid joint structures.

Bracing structures generally take one of two forms. They can be
reinforced concrete cores, usually located around stairways, or they
can be of steel forming vertical trusses with diagonals between the
beam/column joints. In this case, the most common systems are the St.
Andrew's cross (Fig. 1.19a) and the K bracing (Fig. 1.19b).

Bracing structures are mainly loaded in shear and bending and their
deformability must be checked under serviceability conditions in order
to limit displacements of the whole building.

In an actual structure the design of all members necessary for the
overall equilibrium requires to take into account the space behaviour
of the building. The generic horizontal load H in Fig. 1.15 must no
longer be distributed to all the floor/column joints, but only to the
floor/bracing ones, as they are the only elements capable of absorbing

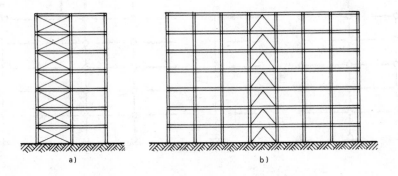

a)   b)

Fig. 1.19

its effects and of transmitting them to the foundation. Furthermore,
there must be a sufficient number of bracing structures to allow any
horizontal load however directed to be balanced.

It follows that:

(a) It must be possible to consider any floor system as a plane
structure, restrained by the vertical bracings
(b) Bracings are external restraints of the floor system: they
therefore must provide a system of at least three degrees of
restraint
(c) The floor system must be capable of resisting the internal
forces due to the applied horizontal loads

To fulfil requirement (a), diagonal bracings must be introduced in
the plane of the floor, thus transforming the floor system itself into
a horizontal truss. As an alternative, the slab or prefabricated
concrete elements of the floor system can be assumed to resist the
horizontal forces as a plane plate structure, the deformability of
which is normally negligible. In this case, however, the erection of
the steel skeleton requires particular care because it is unstable
until the floor elements are placed. Temporary bracing therefore must
be used in this situation.

To fulfil requirement (b), reinforced concrete bracing systems can
be considered as having one, two or three degree of restraint,
depending upon their resistance to bending in only one plane, bi-axial
bending or bi-axial bending and torsion respectively.

Steel truss type bracings, acting only in their own plane, are
simple restraints with reference to the floor system.

Finally, requirement (c) is fulfilled by evaluating internal forces
in the floor system elements due to the horizontal loads, depending on
the location of the vertical bracings.

These considerations can be illustrated by using some classic
examples. Fig. 1.20 shows a three-dimensional structure for a multi-
storey building with steel bracing. Every point of the floor system
is fixed in two directions. In particular, the diagonals reaching
points A and B restrain all the points in line '1' also in the 'x'
direction. The floor bracing is able to receive external forces from
both in direction 'x' and 'y' and to transmit them to the vertical
bracings as follows. Under the action of distributed load q on the
long façade,

$$R_1 = R_2 = qL/2 \quad ; \quad R_3 = 0$$

Due to the loads applied to the transverse façade, one has:

$$R_1 = - R_2 = (F_1 2a + F_2 a)/L \quad ; \quad R_3 = F_1 + F_2 + F_3$$

The spatial structure can be reduced to plane substructures whose
static schemes are shown in Fig. 1.21. The façade along axis '3' is
directly braced in its plane as well as by the transverse bracings of
axes 'a' and 'b'. Also all the joints of the intermediate transverse
frames and the longitudinal frames on axes '1' and '2' are prevented
from any displacement because they are all connected to the vertical
bracings through the floor bracings.

15

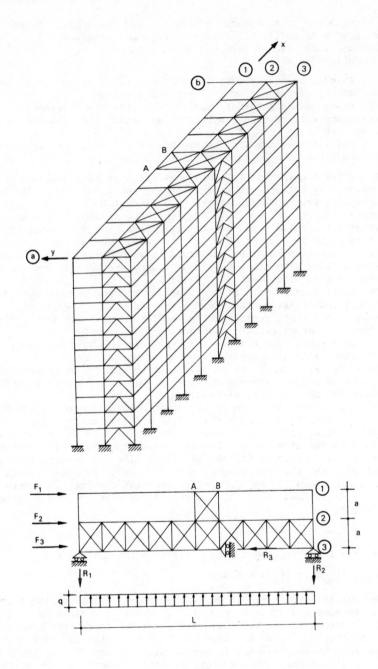

Fig. 1.20

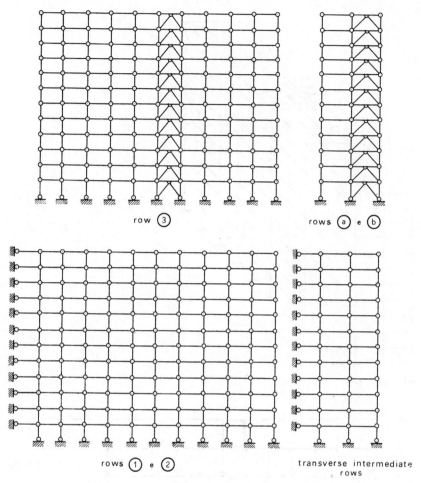

row ③

rows ⓐ e ⓑ

rows ① e ②

transverse intermediate rows

Fig. 1.21

Fig. 1.22 represents the spatial structural scheme of a multi-storey building with a reinforced concrete bracing core. The floor bracing solution illustrated in Fig. 1.23a is correct if the four walls of the staircase core are efficient.

If only three sides of the staircase core are efficient, it is necessary to distribute the forces acting in the longitudinal direction to the wall which is able to receive them (Fig. 1.23b), using additional floor diagonals.

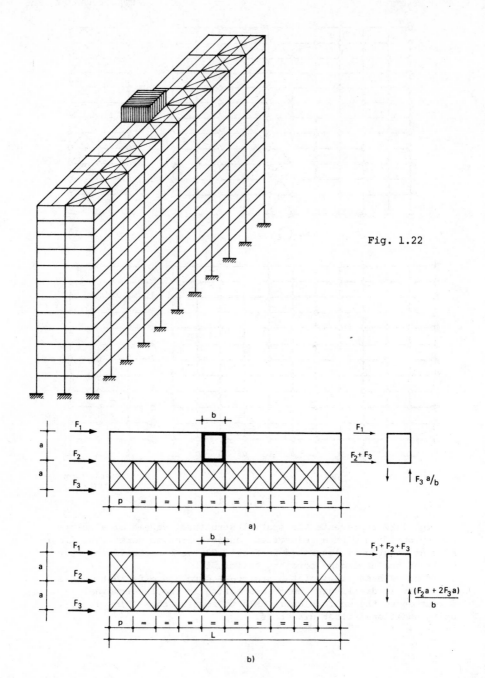

Fig. 1.22

a)

b)

Fig. 1.23

1.3 SINGLE STOREY BUILDINGS

Consider a generic single storey industrial building covered by a roof having mainly a protection function against the weather.
Columns are at the corners of rectangular or square bays having sides $L \times a$ (Fig. 1.24), in which at least one dimension is not inconsiderable ($L > 15$ m). In the building there may be one or more systems of rails for cranes. The following analysis will consider the simplest case (Fig. 1.25) of a building with a plane roof with two

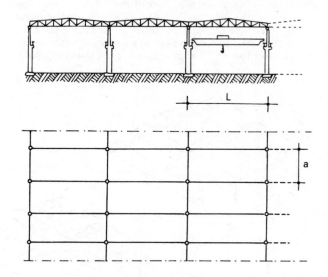

Fig. 1.24

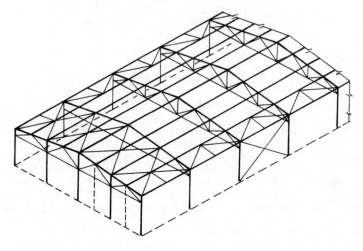

Fig. 1.25

19

slopes for the shedding of water and snow. In order to identify
possible structural systems the same kind of approach as described for
multi-storey buildings in Section 1.2 will be followed.

1.3.1 *Effects of vertical loads*

A generic vertical load F acting on the roof (Fig. 1.26) can be
transmitted to the principal beams or trusses by means of the purlins
(Fig. 1.26a). If the principal beams are more than say 6 m apart,
secondary beams or rafters must be provided to keep the span of the
purlins reasonable. The spacing of the purlins will depend to some
extent on the form of roof sheeting used. Profiled steel sheeting is
now available which can safely span 4 m or more although for roofs it
is more usual to use one of the shallower profiles on spans of about
2 m.

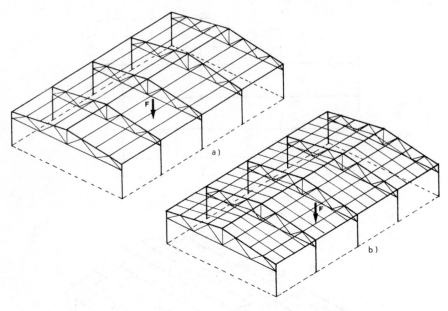

Fig. 1.26

Purlins are mainly subjected to bi-axial bending due to the roof
slope. The vertical load F must therefore be resolved into the two
components $F_n = F \cos \alpha$, perpendicular to the slope, and $F_t = F \sin \alpha$,
tangential to the slope (Fig. 1.27). As the flexural behaviour of the
profiles normally used for purlins is weaker in the direction of the
slope it is often necessary to provide them with additional support in
the form of threaded bars (sag bars) at the middle or third points of
the span (Fig. 1.28). Obviously, they must be supported at the top if
their purpose is to be fulfilled.

In the slope plane the purlin can thus be considered as a continuous
beam. Since the maximum commercial length of structural steel purlin

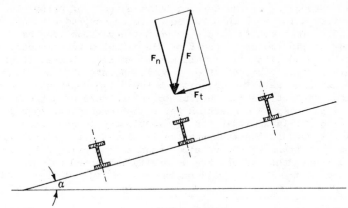

Fig. 1.27

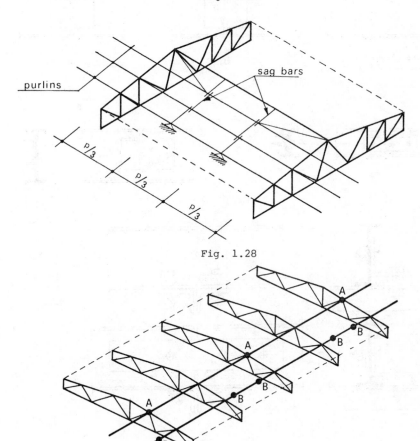

purlins

sag bars

p/3

p/3

p/3

Fig. 1.28

A

A

A

B

B

B

B

B

Fig. 1.29

21

profiles is about 12 m, the spacing between the roof principles does
not generally allow more than two spans to be covered with one bar
(Fig. 1.29). To simplify joints between the various lengths of one
continuous purlin (points A and B of Fig. 1.29) a web connection only
is generally used (Fig. 1.30). It may be entirely bolted (Figs. 1.30a,
b) or only partially bolted (Fig. 1.30c). Joints of Figs. 1.30d,e are
designed to allow relative rotations between two contiguous lengths.
Two typical connections for purlins to the rafters of a truss are
shown in Fig. 1.31. A connection for an unbroken continuous purlin is
shown in Fig. 1.31a and that for two contiguous lengths whose
separation occurs at the truss rafter is illustrated in Fig. 1.31b.

The purlins transfer the vertical load to the principal beam. If
this is a truss, the rafters are generally subjected to pure
compression if the purlins fall on the joints (Fig. 1.32a). If they

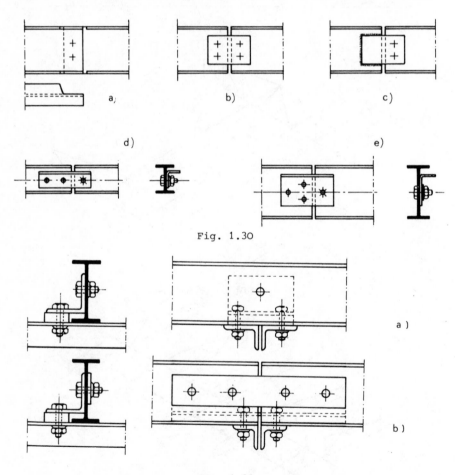

Fig. 1.30

Fig. 1.31

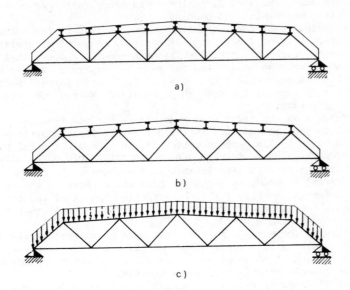

a)

b)

c)

Fig. 1.32

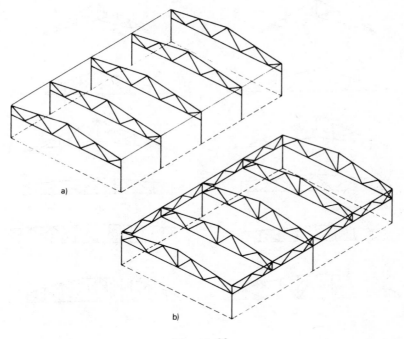

a)

b)

Fig. 1.33

are located away from the joints (Fig. 1.32b) or if the roofing element transmits the load directly to the truss rafter (Fig. 1.32c), they undergo compression and bending.

The principal beam can in turn transmit its load to the columns either directly (Fig. 1.33a) or indirectly by means of a lateral beam (Fig. 1.33b), which will therefore be bent under a concentrated load.

Shapes and schemes of trusses can vary as required. Some classical types are shown on the left hand side of Fig. 1.34: (a) Scissors; (b) Fink or Polonceau; (c) English or Howe; (d) Warren; (e) Pratt or Mohnié; (f) Bowstring.

The schemes on the right hand side can be derived from the classical ones by introducing additional bars. Their positions are chosen in order to limit the buckling length of the principal bars in compression and also to receive the concentrated load of a purlin.

The buckling or effective lengths of the compression members can be easily defined in order to evaluate their slenderness in the truss plane. Referring to Fig. 1.35, the effective length of the top boom elements in the plane of the truss $L_{c,v}$, can be assumed equal to the distance between the nodes.

Compression members can also be subjected to lateral buckling, i.e. they can buckle outside their plane. It is therefore necessary to evaluate the out of plane effective length $L_{c,h}$, which depends on the three-dimensional behaviour of the structure.

The roof scheme of Fig. 1.35 shows a structure whose behaviour out

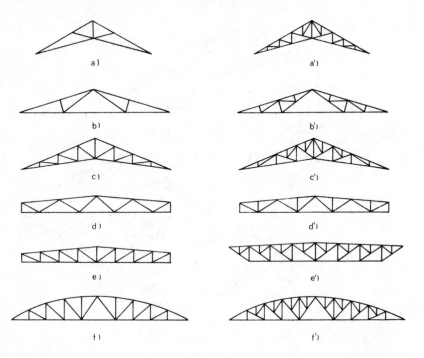

a)   a')

b)   b')

c)   c')

d)   d')

e)   e')

f)   f')

Fig. 1.34

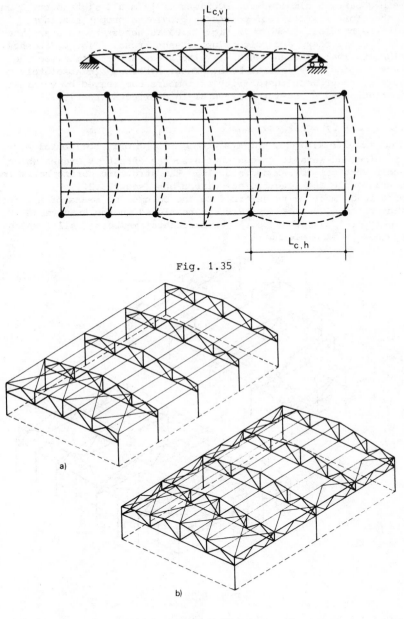

Fig. 1.35

Fig. 1.36

of plane depends entirely on the flexural rigidity of the trusses, which often is negligible. The trusses are in fact connected by the purlins or the roofing element, but if one considers them as simple

25

pin-ended bars, a simultaneous buckling of them all might occur. One
must therefore fix all roof points by providing proper bracings.
Transverse bracing as shown in Fig. 1.36a is necessary to stabilise
the trusses if there are columns under each truss. Fig. 1.36b shows
the longitudinal bracing, which is also necessary to stabilize the
trusses bearing on a lateral beam. These bracings are absolutely
compulsory to prevent instability of members compressed by vertical
load effects and to resist horizontal load effects (see 1.3.2).

1.3.2 *Effects of horizontal loads*

In industrial buildings, one must consider not only horizontal effects
due to wind and seismic forces, but also the effect of cranes which
transmit vertical forces and also load the structure horizontally in
both transverse and longitudinal directions (see 3.3.2).

Wind loads must be transferred to the columns by means of a proper
framed system. Vertical members can be placed in the transverse
façade (Fig. 1.37), to support the horizontal members (rails) which
carry the cladding elements.

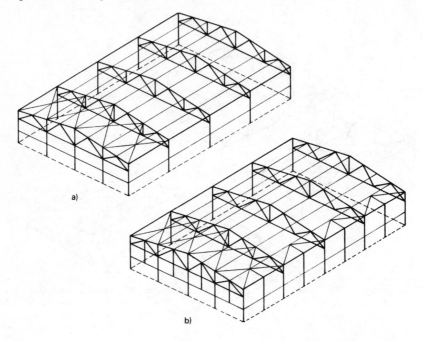

Fig. 1.37

Vertical members are generally subject to bending in one direction.
Horizontal ones are bent by wind forces in the horizontal plane and by
the façade weight in the vertical plane. The rails therefore undergo
bi-axial bending and it may be necessary to reduce their span by using
tie-rods (Fig. 1.37b). The longitudinal façade can be framed as the

transverse one (Fig. 1.37a) using the columns as vertical elements. The columns are therefore bent by wind effects.

If the column spacing is too great, intermediate posts can be introduced to limit the span of the horizontal members. In this case, the load acting on such vertical members must be transferred to the columns by means of longitudinal bracing in the plane of the roof (Fig. 1.37b).

Summing up, due to wind effects on the façade elements:

Horizontal bars are bent in the horizontal plane and furthermore bear the façade weight
Vertical bars can be considered as bent beams supported at the ground and at the roof (and at the rail level, if any)
Roof bracings transmit forces from the vertical bars to the roof fixed support points.

Horizontal transverse (Fig. 1.38a) and longitudinal (Fig. 1.38b) crane loads must be transmitted to the columns together with the vertical loads. Crane girders can be of various designs depending on the type and load carrying capacity of the crane. The size and type

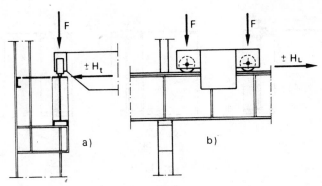

Fig. 1.38

of crane rail fixed to the top of the crane girder will also depend on the crane capacity.

In the case of modest cranes, it might be convenient to use an I-beam and to increase its performance in the horizontal plane by welding one or two angle bars (Figs.1.39a,b) or a channel section (Fig. 1.39c) to the upper flange.

For heavier cranes, a fabricated I-beam with proper horizontal bracing is used. Such bracing can be provided by means of a truss where the upper flange of the beam becomes a truss member (Fig. 1.39d), or it can be set between two coupled girders (Fig. 1.39e). If vertical and lower horizontal bracings are provided in addition to the upper horizontal bracing, a torsionally efficient box section is obtained (Fig. 1.39f).

If an inspection platform, parallel to the crane girder, is available, the platform deck can be incorporated in and form part of the upper bracing (Figs 1.39g,h,i).

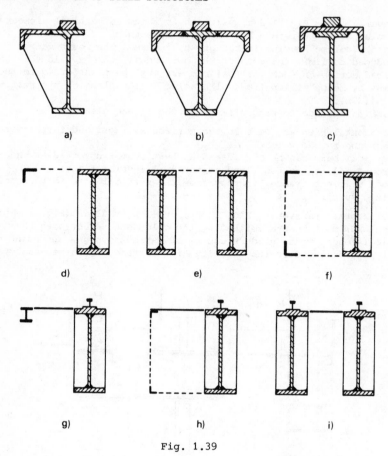

Fig. 1.39

1.3.3 *Structural types*

The foregoing brief remarks on the effects of vertical and horizontal
loads lead to the following considerations:

   (a) Columns (Fig. 1.40) are compressed by the vertical loads
transmitted by the roof (V_1), and by the crane (V_2), and by the dead
load (V_3) and cladding weight (V_4). They are also bent by the wind
forces transmitted at the roof (W_1) and at possible supports of the
horizontal façade members (W_2). Also the horizontal forces due to
the crane (H_t) produce bending in the columns
   (b) In order to transfer the horizontal forces to the fixed points
and to stabilize the structure auxiliary structures are needed.
They must be able to render the roofing structure isostatic at
least in its own plane. There are two possible solutions to this
problem. One can consider the entire roof as just one bent element,
the web of which consists of the roof panels (Fig. 1.41a). In this

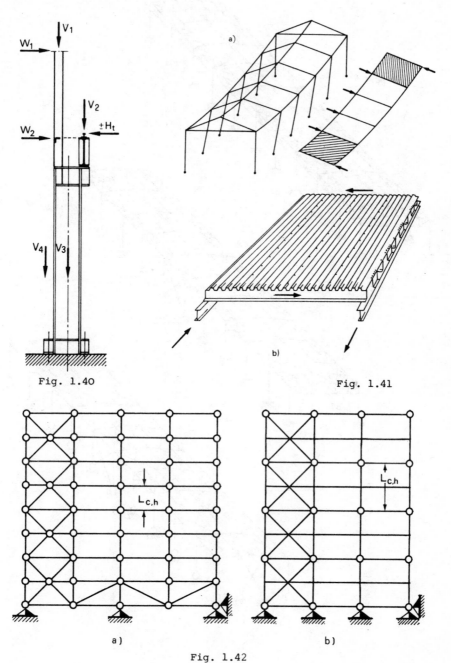

Fig. 1.40

Fig. 1.41

Fig. 1.42

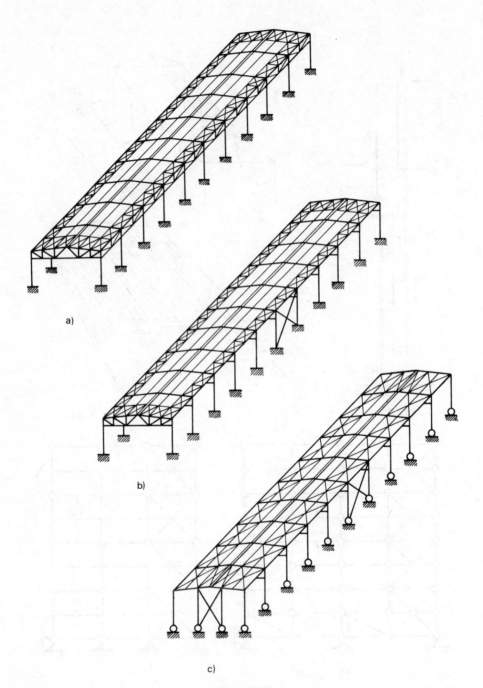

a)

b)

c)

Fig. 1.43

30

case, the panels must be of corrugated sheets and must resist shear forces (stressed skin design) |30|. They must be joined together in a statically efficient way and along both directions to the other roofing elements (Fig. 1.41b)

A more conventional approach is to provide a sufficient number of diagonals so as to render the roofing structure isostatic in its own plane, once the outer restraints consisting of the vertical resistant elements have been introduced. Fig. 1.42a represents a bracing system which makes isostatic a roof composed of purlins, principal beams and lateral beams. The longitudinal bracing prevents the displacement of the mid-point of the lateral beams and consequently determines free bending length of the compression members. Transverse bracing is provided by diagonals which meet it and join the middle purlin. The effective length of the upper members in the main truss is therefore $L_{c,h}$. The roof shown in Fig. 1.42b differs from the previous one for two reasons. All the trusses are supported by the columns and therefore a longitudinal bracing is no longer necessary at their intersection. The trusses are not connected to the intermediate purlins, so the effective length of the compression member is twice the previous one.

The next matter to be considered is how to provide external restraints to prevent roof structure movements. Fig. 1.43 shows three typical ways of providing adequate vertical structures to acheive this. They consist of:

(a) Frames in both directions (Fig. 1.43a)
(b) Frames in the transverse direction and braced pin-ended structures in the longitudinal direction (Fig. 1.43b)
(c) Pin-ended structure with bracing in both directions (Fig. 1.43c)

The solution (a) is not suitable for I-shaped columns having a preferential direction in bending.

The solution (c) optimizes the column behaviour by minimizing the bending in it. On the other hand, it must take advantage of roof beam-like behaviour and therefore it requires solutions based on the shear diaphragm action in the sheeting or leading to a considerable increase in bracing. This solution therefore implies a greater complication of constructional details and limits the possibility of future extension.

The solution (b) with frames in the transverse direction and pin-ended braced structures in the longitudinal direction is the most common one (Fig. 1.43b).

The longitudinal bracings absorb the horizontal forces in the longitudinal direction, such as wind against the façade and starting up and stopping forces of cranes. Wind loads (Fig. 1.44) must be transmitted from the façade to the vertical bracing by means of the transverse roof bracing and the purlins or beams which are placed along the edges. To reduce the slenderness of these elements which are in compression, it is useful to introduce longitudinal bracing in the roof, even though it is not needed to guarantee equilibrium of the roof in its own plane.

Vertical longitudinal bracings can be a problem in respect of the use of the building. For this reason, besides the St Andrew's Cross

31

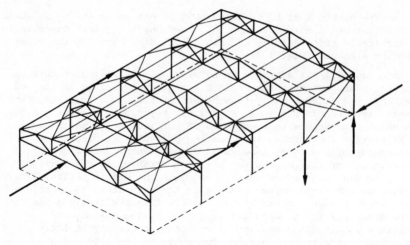

Fig. 1.44

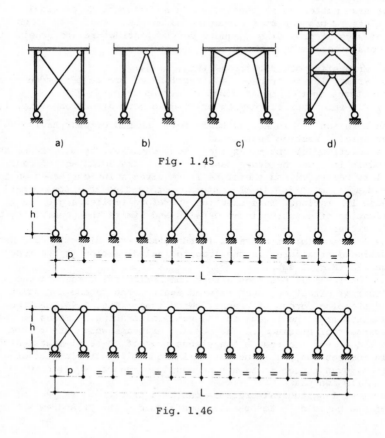

a)   b)   c)   d)

Fig. 1.45

Fig. 1.46

32

(Fig. 1.45a), other solutions which permit easier passage (Figs 1.45b, c,d) are often adopted.

The bracings are preferably located at the middle of the side of the building (Fig. 1.46a). Placing them at the ends (Fig. 1.46b) is best avoided as thermal variations can create a state of constraint leading to compression in the longitudinal members and additional forces in the bracings.

On the other hand, the solution with only one bracing can entail excessive expansion, specially if the length is considerable (L = 60 - 100 m). It is therefore necessary to introduce expansion joints, which can be provided by slotting the bolt connection holes in the longitudinal structural elements.

Transverse frames vary considerably according to whether there are crane girders or not. Those illustrated in Figs 1.47a,b are typical solutions without cranes or with limited capacity cranes. The solutions illustrated in Figs 1.47c,d are typical of mill buildings with heavy cranes. They include stepped columns formed by a lower part and an upper part with reduced inertia set between the rail level and the roof.

The solution shown in Fig. 1.47e is aesthetic, but expensive: its height allows space for the horizontal rail bracing.

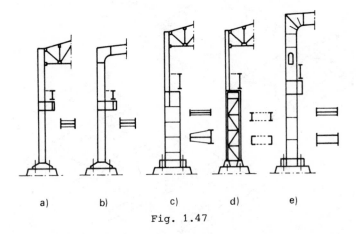

a)   b)   c)   d)   e)

Fig. 1.47

Basically, all solutions can be summed up in the three typical schemes illustrated in Fig. 1.48.

If the stiffness of the transverse member of the frame is comparable to the column stiffness (Fig. 1.48a), the scheme behaves as a classical frame. If transverse beam stiffness is great compared to column stiffness, the beam can be considered undeformable (Fig. 1.48b). In this case, each column is subjected to a bi-triangular bending moment distribution, which vanishes at the mid-height ($h/2$) and reaches its maximum value ($M = Hh/4$) at the ends. Consequently, if the beam is a truss, its top and bottom members are loaded by tension and compression forces having a value of $N = M/d = Hh/4d$, which could become dangerous as they also tend to

33

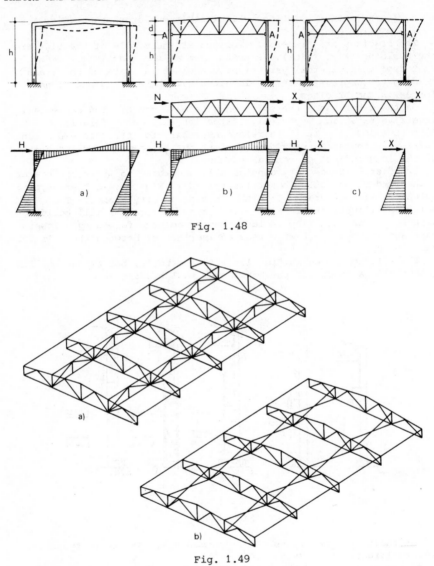

Fig. 1.48

Fig. 1.49

compress certain lower members near the columns. As the lower part of
the truss is not usually braced, even a modest compression might play
an important rôle due to the high slenderness involved.

Finally, one can consider the truss as a pin-ended member, which is
useful only for the purpose of distributing forces H to the continuous
columns. In order to obtain this structural scheme, it is sufficient
to make the joints at points A of Fig. 1.48b flexible, e.g. by
slotting the holes if the joints are bolted. In this way, the diagram
shown in Fig. 1.48c becomes possible. This solution leads to the

34

maximum bending moment in the column, but is has the advantage of not creating axial compression in the roof girder elements, in addition to that already existing due to vertical loads, for which a slenderness limitation is in any case necessary. This solution is particularly advisable when the column has a great inertia as in Figs 1.47c,d.

As a consequence of the compressive forces induced in the lower bars of the roof truss in the scheme illustrated in Fig. 1.48b, it may be necessary to restrain the truss against lateral buckling. This is achieved by introducing longitudinal bracings on the nodes (Fig. 1.49a), or by vertical cross bracing between lower and upper nodes, one member of which will be in tension whichever direction the buckling tends to take place (Fig. 1.49b).

1.4 CALCULATION MODELS

The foregoing considerations relate to an ideal structure having perfect constraints. On the other hand, the corresponding constructional details have shown that the connections between the various members which comprise the structure are considerably different from the assumed idealizations. It is therefore important to point out that any approach to structural design must be based on simplifying hypotheses and schemes which make the transition between actual and model structure possible. Only the model can be studied by the methods of structural analysis. The corresponding results will be as much closer to the actual behaviour of the structure as the model more nearly interprets the structure itself.

An obvious question to ask is whether the introduction of simplifying hypotheses leads to define a model whose behaviour is on the safe side or not. In other words, it is necessary to check whether the results obtained on the model - and, first of all, its ultimate load carrying capacity at collapse - are safe or optimistic and therefore erring against safety.

In answering this question it may be helpful to apply the plastic design static theorem. In a structure undergoing a system of external forces F_j, $\alpha_u F_j$ are the values of the loads that, if applied, would produce the collapse of the structure, α_u being the actual collapse factor. If, for a generic load αF_j, it is possible to find a distribution of internal forces which balance the external forces and when the structure also complies everywhere with a given plasticity criterion,

$$\alpha \leq \alpha_u$$

This theorem is valid if the following hypotheses are satisfied:

   Effects of local buckling are absent
   Second order effects have no influence
   Strain values at each point of the structure are lower than those
   corresponding to material failure

A calculation model will be therefore as much closer to actual behaviour as the compatibility conditions are more strictly satisfied. Any solution is however on the safe side, even though compatibility is not complied with, provided that:

It respects the equilibrium between internal and external forces
It respects material strength
The structure guarantees enough ductility, which is necessary to
avoid localized fractures for load values below those for local or
overall structure collapse.

Obviously, once the calculation model has been defined, stability
of members must be checked and, in the case of highly deformable
structures, the influence of second order effects on vertical loads
must be assessed. Some typical examples of steel structure
calculation models follow.

1.4.1 *The pin-ended structure*

The model of a generic pin-ended structure (Fig. 1.50a) can be studied
with reference to various positions of the ideal hinges. They can be
located, for example, in any one of the three positions shown in Fig.
1.50b: results will be on the safe side provided the dimensions of the
various structural elements comply with the assumed model. From the
three cases of Fig. 1.50 the following criteria can be deduced for
calculating columns, beams and joint sections $X - X$ and $Y - Y$.

Scheme 1

Columns A and B are simply compressed
L is the span for calculating the beam
The joint section $X - X$ must resist not only a shear force $V = R_1$,
but also a moment $M = R_1 a$
The joint section $Y - Y$ must resist not only a shear force $V = R_1$,
but also a moment $M = R_1 (a + e)$

Scheme 2

Column B is compressed ($N = R_{1B} + R_{2B}$) and bent by a moment
$M = a (R_{1B} - R_{2B})$ concentrated at the node
Column A is compressed ($N = R_1$) and bent by a moment $M = R_1 a$
concentrate at the node
$L - 2a$ is the span for calculating the beam
The joint section $X - X$ must only a shear force $V = R_1$
The joint section $Y - Y$ must resist not only a shear force $V = R_1$,
but also a moment $M = R_1 e$

Scheme 3

Column B is compressed ($N = R_{1B} + R_{2B}$) and bent by a moment
$M = (R_{1B} - R_{2B}) (a + e)$ concentrated at the node
Column A is compressed and bent by a moment $M = R_1 (a + e)$
concentrated at the node
$L - 2(a + e)$ is the span for calculating the beam
The joint section $X - X$ must resist a shear force $V = R_1$ and a
moment $M = R_1 e$
The joint section $Y - Y$ must resist a shear force $V = R_1$ only

Each of these three models is on the safe side and can therefore be
assumed for calculation. The choice will be made considering the
structural element or the joint that is the weakest part of the

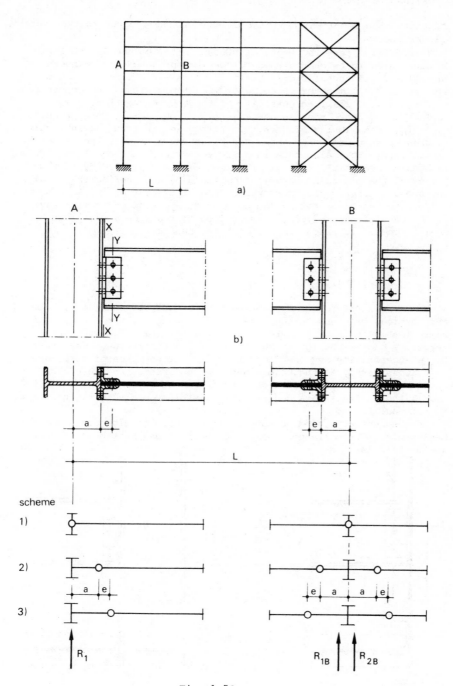

Fig. 1.50

structure: the model which minimizes the internal forces in that part will be chosen.

In the first case the state of stress in the column is the lowest. In can therefore be chosen when columns are oriented according to their lower stiffness (Fig. 1.51a). Bending effects in the columns are, in fact, eliminated in spite of slight moments in the joints due to a relatively small eccentricity of the bolt connection.

The second case is often conservative if the columns are oriented according to maximum stiffness. In this case, in fact, eccentricity is greater than half the column depth and could necessitate larger connections. This also entails a greater stress in the columns, due to bending moments. Their distribution can be evaluated by setting hinges at the mid-point between floors and by considering the columns supported by the bracing structures (Fig. 1.52a).

So, each vertical row can be considered by means of the isostatic scheme shown in Fig. 1.52b. Its horizontal reaction H_i is given by equilibrium at rotation around the hinge number i:

$$H_i = \frac{2M_i}{h_i} - \sum_1^{i-1} H_k$$

The effects of forces H_i, for each floor and for each column, loads the vertical bracings through the floor bracing. Their intensity is approximatively $\Delta R \, e/h$, wherein ΔR is the difference between the reactions of two contiguous beams and e/h is the ratio between the hinge eccentricity and the floor height. In the types of structure being considered, as the beam spans are comparable, ΔR depends only on any unbalanced accidental loads. Furthermore, as e/h is essentially small, these effects are generally negligible compared with those due to external loads.

On the contrary, bending moment effects on the columns are not negligible. The corresponding increase in stress must be considered in the calculations.

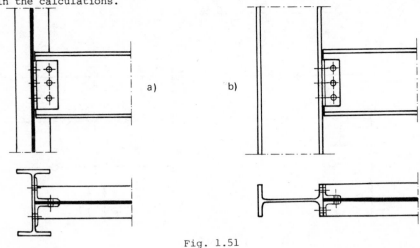

a)   b)

Fig. 1.51

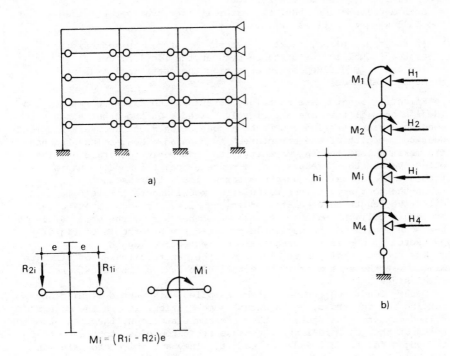

a)

b)

$M_i = (R_{1i} - R_{2i})e$

Fig. 1.52

1.4.2 *Truss and lattice girders*

Trusses and lattice girders, whether their joints are bolted (Fig. 1.53a) or welded (Fig. 1.53b), are usually analysed assuming each bar hinged at its ends. This calculation model thus does not take account of the bending moments due to joint rigidity, but considers the bars

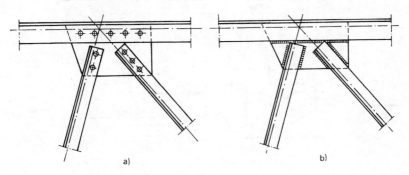

a)   b)

Fig. 1.53

39

under simple tension or compression.

Such a calculation model is certainly conservative, provided the two following conditions are satisfied:

In the truss plane, buckling lengths L_c of compressed bars are assumed equal to the distance between the ideal hinges
The truss is defined by intersecting the centroidal axes of its members at nodes

Effective lengths are certainly less than the distance between two ideal hinges if the bars are rigidly joined. On the other hand, bending moments arising from such a connection might decrease the load bearing capacity of the structure. As soon as a beam-column reaches its maximum load bearing capacity, its ends begin to rotate and its effective length increases. As a consequence, the model behaviour tends to the actual one as the external loads increase.

On the contrary, if the calculation model underestimates the effective length in ultimate conditions (i.e. for ends rotating) a premature collapse will take place. In other words, the moments due to beam continuity can be neglected provided the hypothesis of rigidly connected joints is not used to reduce effective length.

The second condition requires that the centroidal axes of all bars meeting at the same node intersect to give what can be considered the ideal hinge.

Such a condition, which must be complied with when designing and setting out the truss, could entail some difficulties in the case of bolted members. In fact, if they are angle sections, they cannot be holed at the centroidal axis, because it is too close to the root of the angle for the nut and bolt to fit. They are therefore holed on the bolt line axis (Fig. 1.54a).

If the centroidal axes meet in one point, the gusset plate is loaded by a force N, whose application line coincides with the

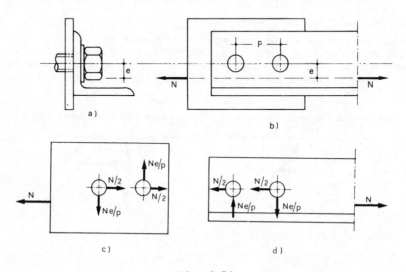

Fig. 1.54

centroidal axis of the bar (Fig. 1.54b). The bolt axis is eccentric to this, so the bolts have to resist an additional moment $M = Ne$.
This moment acts both in the plate (Fig. 1.54c) and in the bar (Fig. 1.54d). In addition to the axial force in the bar it is loaded in bending by this additional moment which also increases the stress in the bolt. The use of the centroidal axes makes the plate design and marking difficult, because the centre lines of the holes do not meet in the same point (Fig. 1.55a). This is why one often prefers setting out the truss on the bolt line axes (Fig. 1.55b).

Against this technological advantage one must consider the static requirements: the centroidal axes no longer meet in one point, but they intersect variously in points A, B and C in Fig. 1.55b. The resulting additional moment $M = N_4 e$ must be distributed among the bars meeting at the node.

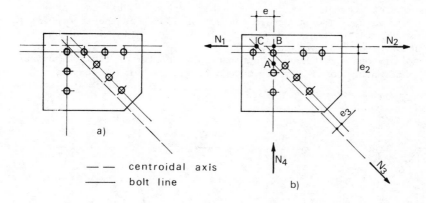

— — centroidal axis
——— bolt line

Fig. 1.55

As this effect is repeated at each point, in each bar there is a bending moment which should not be neglected in principle. On the other hand, such additional effects are generally much greater as one approaches the supports (axial force N_4 is in fact proportional to the shear force). If the upper and lower bars are loaded axially and the forces decrease towards the supports, the additional bending moments do not produce a dangerous weakening of the structure, even if it has been designed taking into account the axial forces alone. For these reasons, recommendations accept marking on the bolt line axes for truss structures with angle bars, provided important fatigue phenomena are absent and the bolts are designed to resist an additional moment equivalent to the axial load existing in each bar multiplied by the distance between bolt line and centroidal axis. The bolt system in each bar must therefore be at least equal to that required for a structure which has been set out on the centroidal axes.

1.4.3 *Truss bracings*

Forces acting on bracing structures, such as the effects of wind, earthquakes and geometric imperfections, do not act in a given direction. Therefore, the geometry of some bracing systems can be designed and dimensioned in different ways according to whether a hyperstatic or isostatic behavioural model is considered.

Referring to the simple truss bracing shown in Fig. 1.56a, consider the analysis of a single diagonal system (Fig. 1.56b).

The structure is hyperstatic and its solution, at least as far as slight displacements are concerned, is determined by the compatibility condition (Fig. 1.56c) which imposes equality $\Delta_{AB} = \Delta_{CD}$ between the elongation of the tension diagonal AB and the shortening of the compression diagonal CD. If the relationship $N - \Delta$ between axial load N and length variation Δ (Fig. 1.57a) is equal in both tension and compression, then the axial load in both diagonals has the same absolute value. The structure can be considered as the superposition of two isostatic structures working in parallel (Fig. 1.57b) and its solution becomes immediate.

The diagonals can however substantially differ in their behaviour: the compressed bar CD may not have a linear behaviour because,

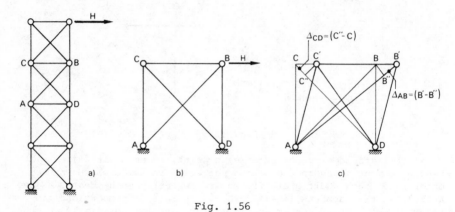

Fig. 1.56

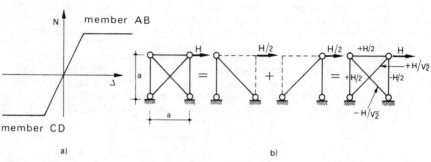

Fig. 1.57

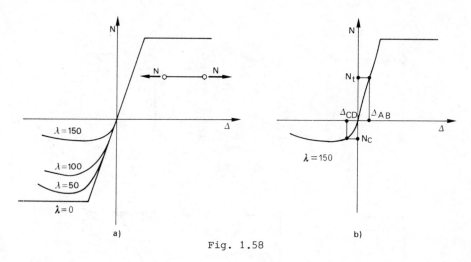

Fig. 1.58

although it remains elastic, it is subjected to buckling and the variation from linear behaviour increases as strut slenderness λ increases (Fig. 1.58a). In the case of high slenderness (Fig. 1.58b), the geometric condition $\Delta_{AB} = \Delta_{CD}$ requires an axial load N_c in the strut which is substantially lower than the axial load N_t in the tie.

There are thus two ways of dealing with bracing. It can be dimensioned so that both diagonals can resist both tension and compression. This requires a low slenderness ($\lambda \leq 100$), so that the behavioural difference between tension and compression bars is negligible. This solution is illustrated in Fig. 1.57b: both diagonals cooperate in resisting shear forces. Alternatively, the bracing can be dimensioned by considering the tension diagonal alone. This requires high slenderness ($\lambda \geq 200$) in order to ensure that when the stress reverses and the diagonal becomes a strut, it will remain elastic even if it buckles. Under this condition the bar in compression is redundant and the forces are wholly resisted by the tension bar. Bracings designed in this way are generally more economical, but structure deformability is greater. Furthermore, the possibility of buckling of compression diagonals makes this solution inadvisable whenever the bracing is located in the plane of façades or partition walls.

The above considerations are applicable also to other types of bracings.

The bracing shown in Fig. 1.59a for example consists of two inclined bars connected to a horizontal beam capable of resisting bending. It can be calculated by the method indicated in Fig. 1.59b or by that of Fig. 1.59c, according to whether the compression bar is taken into consideration or not. The bracing in Fig. 1.59b, corresponds to members of a truss bearing axial loads only. Sloped bars are one in tension and one in compression. As both bars are made identical, one must check that they can satisfactory resist the compressive load. In Fig. 1.59c only the tension bar is considered operative. Consequently the horizontal beam must also resist bending

43

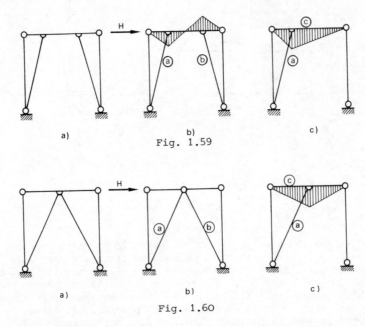

Fig. 1.59

Fig. 1.60

due to the external force H. In this case also the bracing can be economical, provided the compression bar is sufficiently slender to buckle while remaining elastic.

The same approach can be followed for the bracing systems shown in Fig. 1.60a. In Fig. 1.60b the bracing bars are designed to act in both tension and compression and this minimizes bending in the horizontal beam, whereas in Fig. 1.60c the bracing is designed to take tension only, the member in compression being ignored. This increases bending in the horizontal beam.

1.4.4 Roof bracings

Roofs consisting of a succession of parallel trusses or lattice girders (those for industrial buildings are typical) must be provided with bracings to prevent lateral buckling of their compression members (Fig. 1.61a). Due to initial imperfections (Fig. 1.61b) forces H acting against lateral buckling of the trusses are induced at the purlin connections which are fixed in position. These forces are balanced by reactions R applied to the truss supports.

In each joint of each of the n trusses, threfore, a force H is transmitted to the purlin. The bracing is thus loaded by forces having a value nH, where n is the number of trusses to be stabilized, which produce reactions nR (Fig. 1.61c).

If trusses buckle in the direction shown in Fig. 1.61b, purlins are in tension due to a maximum axial force nH, whereas the lateral beam is compressed.

44

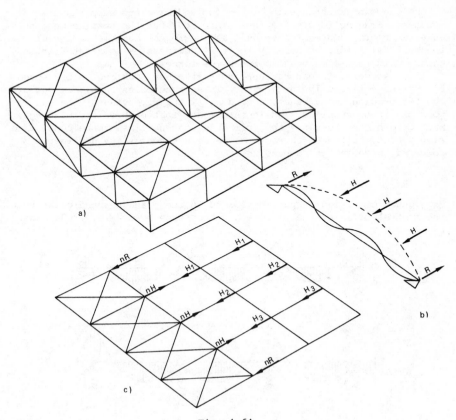

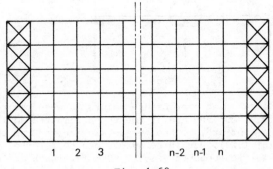

Fig. 1.61

Fig. 1.62

The upper members of trusses must however be stabilized in both directions and there are two possible methods for dealing with the problem, illustrated in Fig. 1.62.

One can suppose that each of the two end bracings supports $n/2$ trusses. Bracing forces are thus minimized, but the purlins must be able to absorb compression as well as the forces due to bending under the vertical loads. Generally their slenderness is so great that they cannot withstand even small compressive forces.

Alternatively, it can be supposed that truss buckling is prevented by purlins acting only in tension. This presupposes that only one of the two bracings must react, according to the direction of the possible truss buckling. While this means that each of the bracings must be designed to carry the full load, the purlins can be considered simply bent, as the effects of the tensile stresses are negligible compared to those due to bending.

1.4.5 *Roof purlins*

One method of designing purlins loaded in the plane perpendicular to the roof slope is by the so-called Gerber-type beam (Fig. 1.63a) where

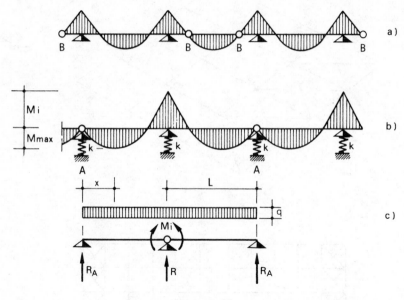

Fig. 1.63

the joints are located at appropriate points between the roof girders (points B of Fig. 1.29) and are of the types illustrated in Figs 1.30d,e. On the contrary, if the joints are located at the supports (points A of Fig. 1.29), purlin behaviour can be referred to that of a beam on three elastic supports (Fig. 1.63b) for which spring stiffness k depends on geometrical and mechanical characteristics of the principal beams and varies according to the mutual position of purlins and beam supports. Furthermore, in the case of bolted truss principals, supports can no longer be simulated with springs alone. It is also necessary to consider the possibility of inelastic settlement due to slip of bolts in the joints of the roof principals. Hence, even the most sophisticated elastic analysis of the problem cannot determine with absolute certainty what is the actual bending moment distribution in the purlins.

It is therefore preferable to seek an equilibrated solution, independent of the compatibility conditions between deflections of the principal beams and purlin supports.

The problem can be schematically represented as in Fig. 1.63c. Let the generic bending moment at the intermediate support be $M_i = \alpha qL^2$. The reactions associated with M_i can be expressed as follows:

$$R_A = qL/2 - M_i/L \quad ; \qquad R = qL + M_i/L$$

For maximum bending moment within the span the shear $V = R_A - qx$ is equal to zero, whence $x = L(0.5 - \alpha)$ and:

$$M_{max} = qL^2\{ 0.125 - 0.5\,\alpha(1 - \alpha)\}$$

Therefore, any equilibrated solution for the purlin will be on the safe side, provided that the underlying trusses are calculated for the value of load R corresponding to the moment distribution assumed for the purlin. The table (Fig. 1.64) indicates, for different values of α, the maximum span and support moment values and the consequent

α	M_{max}	M_i	R
0	$qL^2/8$	0	qL
0.05	$qL^2/9.88$	$qL^2/20$	$1.1\ qL$
0.0625	$qL^2/10.44$	$qL^2/16$	$1.125qL$
0.08579	$qL^2/11.66$	$qL^2/11.66$	$1.171qL$

Fig. 1.64

values of reaction R. For $\alpha = 0$ the load on the beam is reduced, but the stress in the purlin increases. For $\alpha = 0.08579$, the purlin is optimized and moments are equally distributed between span and support, but the principal beam is penalized.

To avoid penalizing the beam and to relieve the purlins from the maximum stress condition arising when $\alpha = 0$, joints must be staggered in plan. Thus one ensures that the same principal beam alternately bears loads having values R and $2R_A$, which on average are mutually balanced. For the purlins one can therefore assume $\alpha = 0.08579$ and the underlying beam is loaded by forces which are equivalent to those due to a distributed load qL. By this means material saving is optimized but standardization in fabrication and in connection design suffers.

1.4.6 *Distribution of horizontal forces in vertical bracings*

It has already been shown that structures with one or more floors
often require vertical bracing systems to discharge the effects of
horizontal loads to the foundation. One must therefore solve the
problem of distributing the horizontal forces acting at each floor
among the individual bracing elements. Such distribution is immediate
whenever the bracing system is isostatic i.e. when it produces three
degrees of restraint for each floor (Fig. 1.65a). It is sufficient to
evaluate the mutual reactions between slabs and individual bracing
elements in order to give a complete solution to the problem. In
particular, one must know the value of the forces acting on each
bracing element at each floor and the bending and shear forces in the
slab, which are needed for the calculation of possible horizontal
floor bracings.

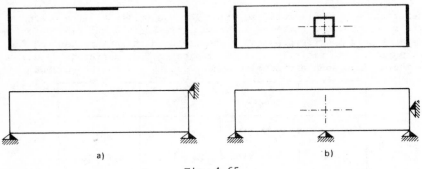

Fig. 1.65

The problem is different if the bracing system offers more than
three degrees of restraint. In this case, each slab is restrained in
a hyperstatic way (Fig. 1.65b) and the solution depends on the
stiffnesses of slab horizontal elements and of the individual elements
of the bracing system.

In this case it is impossible to define the solution exactly. The
stiffness of the individual elements is in fact largely influenced by
the behaviour of the connections. For example, the possibility of
slip in bolted joints due to clearance holes can significantly modify
the calculation results.

One generally prefers to distribute horizontal forces according to
equilibrium diagrams, which may be derived from a consideration of the
requirements for displacement compatibility in the elastic range. In
this respect the horizontal elements are usually assumed to be rigid
so that the solution depends only on the stiffness of the vertical
elements.

The structure can therefore be considered as a three-dimensional
frame for which each floor has three displacement components.

Referring to Fig. 1.66 and applying the deformation method to a
building having n floors, one has $3n$ unknown quantities: u, v, θ at
each floor. Horizontal loads applied to the building between its top

48

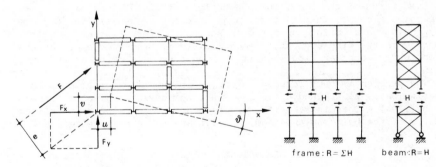

Fig. 1.66

and a generic floor produce three components F_x, F_y, $M = e\sqrt{F_x^2 + F_y^2}$.
For each frame and for each bracing beam, the resulting reaction R of
horizontal forces H (with components R_x and R_y in the assumed
reference system) and their torsion moment T can be expressed, at each
floor, as functions of the $3n$ unknowns.

At each floor one has the 3 equilibrium equations:

$$F_x - \Sigma R_x = 0 \quad ; \quad F_y - \Sigma R_y = 0 \quad ; \quad M - T - \Sigma R_x y - \Sigma R_y x = 0$$

A linear system of $3n$ equations in the $3n$ unknowns u, v, θ is thus
obtained.

Such a method is complicated and its application might be justified
in cases which are complex either due to the building plan or due to
bracing system heterogeneity. This method could lead to interesting
results if torsionally efficient bracing elements were present in
addition to plane elements, but without these conditions it is
advisable to distribute the horizontal forces in a simpler manner
neglecting the interconnection between the various floors of the
building. In this way, it is possible to distribute external forces

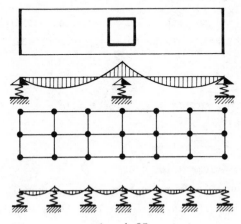

Fig. 1.67

applied to the i-th floor, independently of the behaviour of the
remaining floors. In this case, a generic floor (usually the top one)
is considered as being supported on springs having stiffness $k = 1/u$,
where u is the displacement of each individual bracing element at the
floor under consideration when it is loaded by unit force at the same
point.

The problem is thus reduced to one of the equilibrium of a rigid
slab which is restrained by means of springs having a given stiffness
(Fig. 1.67). The distribution so obtained is then extended to all the
floors of the building and used to calculate the bracing elements.

1.5 DIMENSIONAL TOLERANCES

Steel structures consist of prefabricated elements which must be
assembled together. Dimensional tolerances during fabrication and
erection are therefore quite important |31,32|.

Experience has shown that generally a structure which has been
produced and erected in accordance with common accuracy rules performs
satisfactorily, even with respect to the attachment of prefabricated
finishing elements. Limitations to dimensional tolerances, however,
must also be set for statical requirements. Internal stresses due to
forcing operations during erection, for example, which have not been
taken into consideration in calculations might alter structural
behaviour or render incorrect the methods and hypotheses adopted as
the basis for design.

The maximum values for dimensional tolerances given below are for
the steel structure loaded only by its own weight.

Greater deviations than those listed are acceptable provided that
both dimensional controls of such imperfections and calculation of the
corresponding effects on the structural members are performed. The
limitations are valid for private residential buildings or public
buildings (schools, hospitals, etc.) having one or more floors, but
they can also be assumed for industrial buildings. They can be
considered as attainable through normal fabrication and erection
procedures and therefore they do not imply additional costs.

Referring to Fig. 1.68, it is advisable that the following
limitations be complied with for the assembled building |33|:

Dimensional deviation of the building (Fig. 1.68a)

Length: $\Delta L \leq$ $\begin{cases} 20 \text{ mm} & \text{for } L \leq 30 \text{ m} \\ 20 + 0.25(L - 30) & \text{for } L > 30 \text{ m} \end{cases}$

Height: $\Delta h \leq$ $\begin{cases} 20 \text{ mm} & \text{for } h \leq 30 \text{ m} \\ 20 + 0.25(h - 30) & \text{for } h > 30 \text{ m} \end{cases}$

Column spacing: $\Delta L_i \leq 15$ mm

Floor height: $\Delta h_i \leq 5$ mm

Inclination of column (Fig. 1.68b)

Between adjacent floors (height h_i): $e_i \leq 0.0035 \, h_i$

Maximum deviation for the vertical line of floor number n through

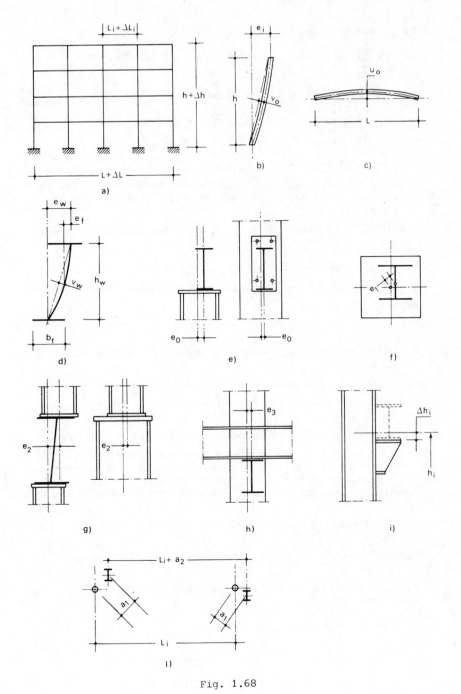

Fig. 1.68

51

the column base: $e_h \leq 0.0035 \frac{3}{n+2} \sum_1^n h_i$

Out of straightness of columns between floors (height h_i) (Fig. 1.68b): $v_0 = 0.0015 \, h_i$

Lateral deflection of girder (span L_i) (Fig. 1.68c): $u_0 \leq \frac{0.0015 \, L_i}{40 \text{ mm}}$

Defect in assembly of fabricated beams (Fig. 1.68d):

$e_w \leq h_w/150$; $\qquad v_w \leq h_w/75$; $\qquad e_f \leq \frac{b_f/40}{10 \text{ mm}}$

Unintentional eccentricity of girder bearing (Fig. 1.68e): $e \leq 5$ mm

In order to make compliance with these overall limitations easy, the individual parts before assembling should comply with the following limitations:

Dimensional deviation of elements to be assembled (Fig. 1.68a):

Beams: $\Delta L_i \leq 0$

Columns: -5 mm $\leq \Delta h_i \leq 0$ mm

Out-of-straightness of elements to be assembled (Figs. 1.68b,c):

Beams: $u_0 \leq L_i/1000$

Columns: $v_0 \leq h_i/1000$

Deviation from theoretical axis in connecting plates (Fig. 1.68f): $e_1 \leq 5$ mm

Deviation between column axes (Fig. 1.68g): $e_2 \leq 5$ mm

Deviation in level of bearing surfaces (Fig. 1.68i): -10 mm $\leq \Delta h_i \leq 0$

Deviation in position of bearing surfaces (Fig. 1.68h): $e_3 \leq 10$ mm

In order to facilitate compliance with the dimensional limitations for the finished building, it is also essential to check the accuracy of the setting out of the columns on their foundations.

Fig. 1.68l shows the following dimensional tolerances which should be met:

Maximum distance between theoretical axis and actual axis of a column: $a_1 \leq 15$ mm

Maximum variation of spacing between two adjacent columns: $a_2 \leq 10$ mm

Maximum variation of distance L between the first and last columns in the same row:

$a_3 \leq \begin{array}{ll} 20 \text{ mm} & \text{for } L \leq 30 \text{ m} \\ 20 + 0.25(L - 30) & \text{for } L > 30 \text{ m.} \end{array}$

REFERENCES

1. Baker, J. F. (1960) *The Steel Skeleton*, Vol. I, *Elastic behaviour and Design*, University Press, Cambridge
2. Baker, J. F., Horne, M. R. and Heyman, J. (1960) *The Steel Skeleton*, Vol. 2, *Plastic behaviour and Design*, University Press, Cambridge
3. Baldacci, R., Ceradini, G. and Giangreco, E. (1971) *Plasticità*, CISIA, Milan
4. Baldacci, R., Ceradini, G. and Giangreco, E. (1971) *Dinamica e stabilità*, CISIA, Milan
5. Bresler, B., Lin, T. Y. and Scalzi, J. B. (1968) *Design of Steel Structures*, 2nd Edn, Wiley, New York
6. Danieli, D. and De Miranda, F. (1970) *Strutture in acciaio per l'edilizia civile e industriale*, CISIA, Milan
7. Daussy, R. (1965) *Guide pratique de charpente métallique*, Lahure, Paris
8. Finzi, L. and Nova E. (1969) *Elementi strutturali*, CISIA, Milan
9. Gaylord, E. H. and Gaylord, C. N. (1972) *Design of Steel Structures*, Mc Graw Hill, New York
10. Grinter, L. E. (1962) *Theory of Modern Steel Structures*, 3rd edn., The Mac Millan Company, New York
11. Mc Guire, W. (1968) *Steel Structures*, Prentice Hall, New Jersey
12. Johnston, B. G. (1976) *Guide to Stability Design Criteria for Metal Structures*, 3rd Edn, Wiley, New York
13. Johnston, B. G., Lin, F. J. and Galambos, T. U. (1980) *Basic Steel Design*, Prentice Hall, New Jersey
14. Lorin, P. (1968) *Construction métallique*, Dunod, Paris
15. Lothers, J. E. (1972) *Design in Structural Steel*, 3rd edn., Prentice Hall, New Jersey
16. Masi, F. (1955) *Costruzioni metalliche*, Hoepli, Milan
17. Tall, L., Beedle, L. S. and Galambos, T. V. (1964) *Structural Steel Design*, The Ronald Press Company, New York
18. Trahair, N. S. (1977) *The Behaviour and Design of Steel Structures*, Chapman and Hall, London
19. Yu, W. W. (1973) *Cold-Formed Steel Structures*, Mc Graw Hill, New York
20. Zignoli, V. (1968) *Costruzioni metalliche*, UTET, Turin
21. AISC (1980) *Manual of Steel Construction*, 8th edn., AISC, Chicago
22. AISC (1968) *Plastic Design of Braced Multistorey Steel Frames*, AISC, Chicago
23. AISC (1959) *Plastic Design in Steel*, AISC, Chicago
24. AISC (1968) *Iron and Steel Beams*, AISC, Chicago
25. AISC (1966) *Structural Steel Detailing*, AISC, Chicago
26. BCSA and CONSTRADO (1978) *Handbook on Structural Steelwork*, BCSA, London and Constrado, Croydon
27. Merritt, F. S. (1972) *Structural Steel Designers' Handbook*, Mc Graw Hill, New York
28. VDE (1953) *Stahl im Hochbau*, Stahleisen, Düsseldorf
29. Council on Tall Buildings, Group SB (1979) *Structural Design of Tall Steel Buildings*, Volume SB of Monograph on Planning and Design of Tall Buildings, ASCE, New York

30. CECM-ECCS (1977) *European Recommendations for the Stressed Skin Design of Steel Structures*, Constrado, Croydon
31. Dal Pont, E. and Nascé, V. (1975) *Tecniche di montaggio*, CISIA, Milan
32. Magenta, G. (1970) *Lavorazioni in officina*, CISIA, Milan
33. CECM-ECCS (1978) *European Recommendations for Steel Construction*, ECCS, Brussels

C H A P T E R   T W O

Reliability of Steel Structures

2.1 CRITICAL COSIDERATIONS CONCERNING ALLOWABLE STRESS DESIGN

In order to check a structure or a structural element, one must assess whether a dangerous situation might be reached due to extreme events. Consider for example an element in tension carrying a static load F. To check whether it is safe means checking that stress $\sigma(F)$ due to F does not exceed a given limit f_{lim}, beyond which inadmissible conditions or damage could arise. The following operations are thus needed:

Define the load F and calculate the stress $\sigma(F)$

Define the limit f_{lim} correspondig to an assumed damage level

Compare σ to f_{lim}.

Let the damage level be defined as the yielding of the material. The limit f_{lim} thus coincides with the yield strength f_e. If however one assumes $\sigma(F) \leq f_e$ as a control condition, it is not possible to guarantee that yielding is never reached in the element. This is due to the possibility that:

Actual load is greater than that foreseen

Yield strength is lower than assumed

Fabrication or erection has produced residual stress in the structure.

It is therefore possible to define allowable stress as $\sigma_{adm} = f_{lim} / \nu$, equal to limit stress f_{lim} divided by a safety coefficient $\nu \geq 1$. The control according to the allowable stress method thus becomes: $\sigma(F) \leq \sigma_{adm}$.

This method is widely applied in structrual design, but it has some faults, specially when used to check safety concerning limit states other than the elastic one.

In this original version, in fact, the allowable stress method can be applied only to check whether the material resists elastically to the effects due to the so-called service loads. It is an extrapolation to use it to determine a safety index for the structures.

Let the allowable stress method be used for the design of the structure shown in Fig. 2.1a.

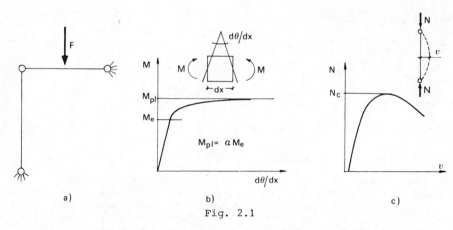

Fig. 2.1

Consider the bending of the most stressed section of the beam and column stability under the service load F. For the beam the maximum bending moment must be $M(F) \leq M_e/\nu$, $M_e = f_e W$ being the moment corresponding to the elastic limit of the section. For the column one must have $N_c \leq N_c/\nu$, N_c being the value of the load carrying capacity of the column according to its slenderness. If however one analyses the behaviour of the two structural elements until collapse, one notices that for a beam (Fig. 2.1b) collapse can be reached only when the bending moment $M_{pl} > M_e$ whereas the load N_c (Fig.2.1c) corresponds to the ultimate capacity of the column: the degree of safety (or, rather, the reliability) of the two structural elements is therefore different. Even though the foregoing example might seem absurd in as much as two different limit states are considered for the two structural elements, (strength for the beam, stability for the column) the applied approach is based on the allowable stress method according to actual recommendations.

The allowable stress method is open to another criticism: it is impossible to operate correctly when two or more independent loads must be combined.

Consider a column of a building: it can be bent by the wind and compressed by permanent and imposed loads. Bending moment and axial force therefore depend on independent causes.

Let a serviceability situation be characterized by values N and Fh of axial and bending effects. They intersect at point P (Fig. 2.2a), which must lie within the section resistance region.

The allowable stress method guarantees that $OA/OP \geq \nu$, i.e. it considers only a simultaneous increase in axial load and in bending moment. For increases of a different kind, safety margins are greater: if only axial load increases, then $CB/CP > \nu$ and if only bending moment increases, then $RS/PR > \nu$.

Now consider the case of a reinforced concrete column: its

56

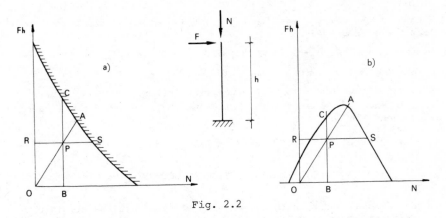

Fig. 2.2

resistance region is of the type shown in Fig. 2.2b. According to the allowable stress design $OA/OP > \nu$, but for an increase in the bending moment alone, it could be $BC/\overline{BP} < \nu$.

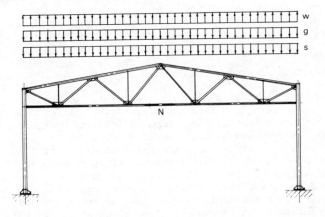

Fig. 2.3

Another typical example can be found considering roof structures of single storey buildings in areas where snow loads are small (Fig. 2.3): the truss lower chord is in tension because of self weight g and imposed load s due to snow; wind might create a suction on the roof, having a value w.

Suppose the values of self weight, snow and wind axial forces in the chord are respectively:

$$N(g) = 100 \text{ kN} ; \; N(s) = 200 \text{ kN} ; \; N(w) = 120 \text{ kN}$$

The out of plane lower chord slenderness is usually very high: thus the ultimate capacity in compression N_C will be low, compared to its ultimate strength in tension N_t. Assume:

$$N_t = 450 \text{ kN} \; ; \; N_c = -50 \text{ kN}$$

If a constant safety coefficient $\nu = 1.5$ is chosen in compliance with allowable stress design:

Max tension: $N(g) + N(s) = 100 + 200 = 300 \leq 450/1.5$

Max compression: $N(g) + N(w) = 100 - 120 = -20 \geq -50/1.5$

It is sufficient however, that wind effect be only 1.25 times the foreseen value for the collapse limit state due to chord buckling to be reached. In fact:

$$N(g) + 1.25N(w) = 100 - 1.25 \times 120 = -50 = N_c.$$

2.2 RELIABILITY OF STRUCTURES

A more rational approach to the problem of structural safety is the probabilitstic one |1,2,3,4|. According to this, in order to assess the reliability of a building, one must consider as random variables all the parameters whose uncertainty can influence the results. For a given structure, some critical sections can be chosen and internal stress resultants (axial and shear forces, bending and twisting moments) $S(F)$ can be determined.

It is possible to define the resistance region of the same sections, in terms of a function $R(S_{lim})$ mutually correlating limit values S_{lim} of the internal stress resultants. Such a function depends on material resistance f: $R = R(S_{lim} (f)) = R(f)$. The probability of $S(F)$ being comprised within the corresponding region defined by $R(f)$ can be assumed as the measure of structural reliability.

If such a probability value were independent of the type of structure and were only a function of material strength and load randomness, one could obtain a common reliability assessment for all structures, independent of the type of material.

Unfortunately, this is not the case: resistance and load distribution being equal, collapse probability depends on the model and on the variables used in defining the structure.

Before introducing some simplified control methods, the main characteristics of the definition of resistance and of load randomness will be examined (even though only from the qualitative point of view).

2.2.1 *Resistance randomness*

The resistance of a material can be defined by a reference stress (yielding for steel, resistance to compression for concrete).

Consider the results of tension tests on a sample of n_{tot} elements of structural carbon steel |5,6|. Represent by ordinates (Fig. 2.4) the number of times n when a value between f_i and $f_i + \Delta f_i$ occurs (in the figure, $f_i = 10 \text{ N mm}^{-2}$). A histogram is thus obtained, which can be approximated by a suitable probability density function $p(f)$ (P.D.F.). In general, if the probability that the value X of random variable x is inside the interval x and $x + \Delta x$, the probability density function $p(x)$ is:

58

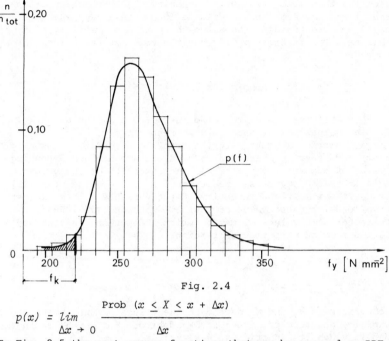

Fig. 2.4

$$p(x) = \lim_{\Delta x \to 0} \frac{\text{Prob } (x \leq X \leq x + \Delta x)}{\Delta x}$$

In Fig. 2.5 the most common functions that may be assumed as PDF are shown. The normal or Gaussian law (Fig. 2.5a) may be defined by just two parameters:

the mean value $\qquad x_m = \displaystyle\int_{-\infty}^{+\infty} xp(x)dx$

the standard deviation $\qquad s = \displaystyle\int_{-\infty}^{+\infty} p(x)(x - x_m)^2 dx$

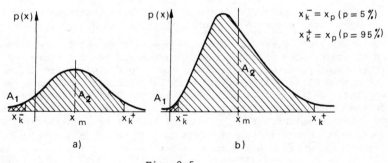

Fig. 2.5

Under the some assumptions usually adopted when applying probability theory to structural mechanics |7|, the extrems (type 1, maxima) law (Fig. 2.5b) may also be defined by the same parameters. The fractile x_p is defined as the value that has a probability p of being diminished. The fractile x_p is thus the value of variable x for which it is true:

$$p = \int_{-\infty}^{x_p} p(x)\,dx = Prob(x \leq x_p)$$

The probability p is represented in Fig. 2.5 by the areas A_1 and A_2 for the values $p = 5\%$ and $p = 9.5\%$ respectively.

Characteristic values x_p are the fractile of order p. Usually upper x_k^+ and lower x_k^- characteristic values are defined depending on the value of probability $p = k$. If the intensity of the variable x acts in the safety sense (resistance) a little value of p will be adopted (2%; 5%). If the intensity of the variable x acts against safety (loads) a great value of p will be adopted (95%; 98%).

Characteristic values x_p may be written depending on standard deviation s or on variation coefficient $\delta = s/x_m$:

$$x_k^+ = x_m + k_1 s = x_m(1 + k_1\delta)$$
$$x_k^- = x_m - k_2 s = x_m(1 - k_2\delta)$$

The coefficients k_1 and k_2 depend on the values of the chosen probability and from the type of PDF adopted. For $p = .5\%$ and normal probability density function $k_1 = k_2 = 1.64$.

2.2.2 Load randomness

Loads can in general be divided into two classes: permanent loads (which are constant throughout the life of the building) and variable loads. The former include structural self-weights, while the latter can be subdivided |5| into:

Loads due to fabrication, erection and testing

Loads due to the use of the structure (live loads, thermal effects, lateral thrust of materials against walls, etc.)

Loads due to natural phenomena (snow, wind, earthquake) or to unaccountable events (impacts, blast, accidents).

Permanent loads depend also on the specific weight of the elements and on the tolerances of geometric dimensions of the structure, of floor finishes, of permanent partitions, etc. To define their randomness, the method already illustrated with reference to resistance can be applied. Define for example, the randomness of the permanent load due to the finishing of slabs in residential buildings. It is possible to estimate weight per m^2 by measuring thickness and checking material quality on a number n_{tot} of slabs to plot the number of times a value between F_i and $F_i + \Delta F_i$ occurs. A histogram will thus be obtained, which can be interpolated by a probability density function $p(F)$ (Fig. 2.6). It is possible furthermore to define:

60

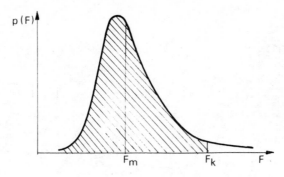

Fig. 2.6

Mean value F_m

Variation coefficient δ

Characteristic value $F_k = F_m(1 + k\delta)$ defined as the value having a given probability p_k of being diminished (95%) or $(1 - p_k)$ being exceeded (5%).

Loads due to fabrication, erection and testing can quite reasonably be generally considered deterministic. They are in fact chosen for each case as a function of the structural building type. On the contrary residual stresses due to a succession of operations which remain in the structure or in the structural element even after erection (shrinkage, imposed deformations, prestressing, etc.) must be considered as random parameter loads.

Variable live loads due to structure cannot be defined by means of the same type of study as the one used for parameter loads: observations must be repeated many times for each structure. In fact, throughout the life of the structure its purpose might change and therefore the survey of the necessary data to construct the histogram must take into account a period of time which is at least of the same order of magnitude as the expected life of the structure Δt_s. Consider, for example, a room and divide it into areas near the walls and a central area |5,8| (Fig. 2.7a). Depending on the position of furniture and therefore on observation instant t, varying values are found for maximum load F per surface in the two areas (Fig. 2.7b)

One can however consider that the parameters defining load randomness (mean value and standard deviation) do not depend on time: all the measured values can thus contribute to defining a histogram and therefore a probability density function curve (Fig. 2.7c). From this it is possible to define mean load F_m, variation coefficient δ and characteristic value F_k, following an analogous procedure to that used for permanent loads.

Loads due to natural phenomena or to unaccountable events are described according to a different model. They cannot in fact be foreseen a priori as they are independent of user's will. They differ from those described above because it is impossible to define, even qualitatively, a limit to their maximum value. For a crane, e.g., it is possible to

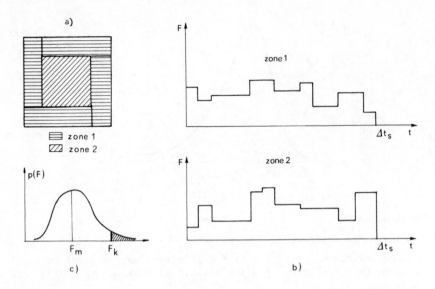

Fig. 2.7

define maximum lifting load. Even though such a value could be exceeded due to improper use of the machine, it can however be under human control and it is possible to introduce limitations in the use of the machine. Such a control cannot take place in the case of natural phenomena or accidents.

Such loads can therefore be defined only by using a statistical analysis in time and it is impossible to establish their distribution independently from the time parameter. Their values must therefore be related to return period Δt of the value itself. Once the value of such a period has been established, value F must be defined: it is on the average exceeded only once in the given time interval Δt. Let, for example, wind speed at a given site be defined. By means of an anemometer it is possible to record wind speed during a sufficiently long period (e.g. 20 years).

It is then possible to trace, on the basis of a number n_{tot} of data, a histogram for each of the following:

Daily   maximum speed n_{tot} = 365 x 20

Monthly maximum speed n_{tot} = 12 x 20

Yearly maximum speed n_{tot} =   20

Probability density functions derived from these histograms can be related together themselves applying statistical techniques and therefore, once the return period Δt is chosen, it is possible to define fairly closely the required value $F_{\Delta t}$, although only one probability function is known (Fig. 2.8a). A mean value and a characteristic value can also be determined, in the same way that was applied in the case of permanent and live loads. One can then define a mean value of the distribution of daily, annual of 50-year maximum.

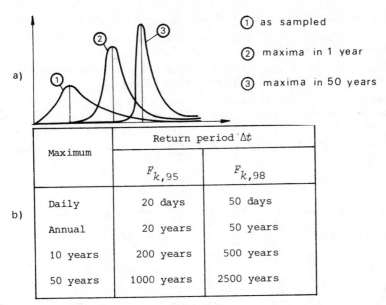

a)

① as sampled

② maxima in 1 year

③ maxima in 50 years

b)

Maximum	Return period Δt	
	$F_{k,95}$	$F_{k,98}$
Daily	20 days	50 days
Annual	20 years	50 years
10 years	200 years	500 years
50 years	1000 years	2500 years

Fig. 2.8

If one assumes, for example, as characteristic value that
corresponding to a 5% probability of being exceeded (i.e. that which,
on the average, is exceeded five times in a hundred), then the
characteristic value of the daily maximum distribution is given by a
wind speed which is exceeded on the average once every 20 days and
therefore has a return period Δt = 20 days. The table of Fig. 2.8b
relates (for the various maxima) the return period Δt to the
characteristic values $F_{k,95}$ or $F_{k,98}$, respectively defined as those
having a 5% or 2% probability of being exceeded.

2.2.3 *Limit states*

It is possible to assumes the reliability of a structure only if loads
and materials are defined and the characteristics of the limit state
one wishes to avoid are specified. The definition of the limit state
determines the calculation method which should be used to assess
effects S of applied loads F and of resistance $R(f)$, which is a
function of material property f.

Limit states can be defined $|5,9|$ as conditions in which the
structure or a part thereof ceases to fulfil one of its functions
or infrenges one of the criteria governing the performance which
it was designed. They can be divided into two categories:

Ultimate limit states corresponding to the maximum carrying
capacity of the structure or part

Serviceability limit states related to criteria governing the
normal use of the structure

63

Ultimate limit states of a structure include:

Loss of equilibrium of a part or of the whole of the structure considered as a rigid body (Fig. 2.9a)

Transformation of the structure or of one of its parts into a mechanism due to plastic hinges being formed (Fig. 2.9b)

Rupture of critical sections of a structure or excessive deformation before a mechanism is formed due to lack of ductility (Fig. 2.9c)

General or local instability due to second order effects (Fig. 2.9d)

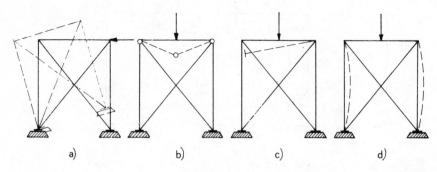

Fig. 2.9

Among ultimate limit states, mention must also be made of collapse due to fatigue. It must however be considered from a different point of view from those listed above, because the values of the loads that must be referred to it are those due to normal use of the structure. At the same time, fatigue phenomena are relevant only for certain structural types; load spectra and behaviour under fatigue of the material and joints have to be analysed. For all these reasons fatigue limit state will not be examined in the following.

Serviceability limit states |9| include:

Deformation of the structure or of any part of it which could adversely affect the appearance or efficiency of the structure

Local damage such as plasticizations, local luckling, bolted joint slipping, cracks in welded joints, which might entail excessive maintenance or lead to corrosion

Variations due to wind or to machinery that can render the structure unusable or cause an increase in loading state due to resonance phenomena or may lead to disconfort, alarm or impairment of its proper function

2.2.4 *Safety analysis*

Once the limit state one wishes to avoid has been defined and probability density functions of material strength and of acting loads are known, it is possible, at least theoretically, to determine a probability that load effects will not exceed resistance in one or more sections. Therefore it is possible to access structure reliability with reference to the given limit state.

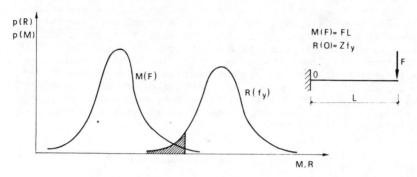

Fig. 2.10

Consider the cantilever of Fig. 2.10: external moment $M = FL$ has a probability density law which can be determined according to that for load F, if distance L is considered deterministic. Assuming the limit state to be structural collapse, the built-up joint section is critical.

Its resistance is defined by the ultimate moment of the section $M_{pl} = Zf_y$, where Z is the plastic modulus and f_y the material yield strength. If the section dimensions are assumed to be deterministic, resistance $R(f)$ has a probability density according to that of strength f_y. The probability that the external moment $M(f)$ is greater that the ultimate moment $R(f_y)$ determines the probability of attaining the considered limit state under consideration.

Checking the safety of a structure thus means checking that, once a distribution law is given for loads and resistance, probability p of exceeding a given limit state is not greater than the chosen value p_u which theoretically depends on the seriousness of the consequences of reaching that limit state |5,10|; therefore p_u has to be chosen according to the type of limit state and the type of structure being checked. On the other hand, it is not easy to choose an appropriate method for estabishing the value of the required probability p_u.

Value p_u could be defined by an economical optimization process of costs and benefits. In this case, a choice must be made on the basis of a comparison between higher building cost and possible economical damage ensuing from loss of use and collapse of the building. Such an approach would lead to the unacceptable conclusion that the economical value of human life should be calculated.

A definition of p_u can be sought by choosing social investment priorities on the basis of the cost of each life which is spared. This means calculating the cost of each human life saved as a function of the various related probability levels.

A value p_u can be established on the basis of the economical resources which are available to the community and according to expense priorities that must be given to protecting the lives of the members of the community itself. In this case risks due to structural collapse, have to be compared with protection against other dangers, such as illnesses,road accidents, pollution, etc. This method requires economical policy decisions that so far are obviously idealistic.

One must therefore try to define the value of p_u on the basis of past experience, assessing the probability of collapse for the various structural types of existing buildings that have proved reliable, at least from the qualitative point of view. Remarkable efforts have been made in the last few years in this direction and the recommendations of the various countries are beginning to show the results of these studies. Such results cannot be considered final but, at least for the time being they can be regarded as an attempt at finding rational control methods, which are useful to homogenize the various design criteria |11|.

2.3. THE SEMI-PROBABILISTIC METHOD

2.3.1 *Deduction of the method*

Once the limit state is given and the distribution probability functions defined for loads and strength, the statement that $p \leq p_u$ implies that the effects of the load characteristic value amplified by an appropriate coefficient, are lower than the structure resistance assessed on the basis of the limit strength characteristic value, possibly reduced by an appropriate coefficient. Briefly, the statement the $p \leq p_u$ means:

$$S(\gamma_F\, F_k) \leq R(f_k/\gamma_m)$$

where F_k and f_k are the characeristic values of F and f and coefficients γ_F and γ_m depend on:

Probability density function of F and f

The fractile according to which characteristic values F_k and f_k are defined

The accepted probability level p_u

If there are two or more loads, condition $p \leq p_u$ implies that the effect of one or more appropriate linear combinations of load characteristic values is lower than structural resistance:

$$S(\gamma_{F,1}\, F_1 + \gamma_{F,2}\, F_2 + \cdots \gamma_{F,n}\, F_n) \leq R(f_k/\gamma_m) \qquad (2.1)$$

If one defines:

Design combination F_d as the combination of load characteristic values multiplied by partial coefficients $\gamma_{F,i}$

Design strengh f_d as resistance characteristic value f_k divided by coefficient γ_m;

then the control formula (2.1) becomes:

$$S(F_d) \leq R(f_d) \tag{2.2}$$

If one lets:

G_k = characteristic value of permanent loads

$Q_{k,i}$ = characteristic values of variable loads

γ_G = partial coefficient of permanent loads

γ_Q = partial coefficient of the variable load having predominant effect

$\psi_i \gamma_Q$ = partial load coefficient of the other variable loads ($\psi_i \leq 1$)

then load F_d is generally expressed by the formula:

$$F_d = \gamma_d \, G_k + \gamma_Q (Q_{k,1} + \sum_2^n \psi_i \, Q_{k,2}) \tag{2.3}$$

In this equation the variable load $Q_{k,1}$, considered in its entirety, is called the basic or fundamental combination load. The $n-1$ variable loads $Q_{k,i}$ are called accompanying loads and coefficient ψ_i depends on the probability of the variable loads acting simultaneously among themselves.

If there are n variable loads, there are n combinations of the type (2.3). Each one assumes the i-th load as fundamental. In practice, many of the n design combinations can be discarded a priori as they are definitely on the safe side.

Coefficients γ_G, γ_Q, ψ_i referred to the loads and γ_F referred to the strength depend on the shape of the probability density function assumed for loads and strength, on possible load simultaneity, and on the chosen probability level. They also depend on the type of structure and on the quantitative influence of the single loads on the limit state for which reliability control is being performed. The above can be easily illustrated by two examples, respectively proving dependence of factors γ_F on structural type and the necessity of using varying factors $\gamma_{F,i}$ (or ψ_i) according to load value.

1st example. Consider the three beams shown in Fig.2.11. Assume they have the same transversal section the performance of which is characterized by limit moment value $M_{lim}(f_k)$. Following the semi-probabilistic method, if $L_3 = 2L_1$, and $L_2 = 1.5L_1$, the three beams can carry the same design load $Q_d = \gamma_Q \, Q_k = 4M_{lim}/L$. For the sake of simplicity assume that the resistances of the various beam sections are mutually independent and that their proability density law is of the normal type having variation coefficients δ_f.

For load Q assume an extreme probability density law of the second type having variation coefficient $\delta_Q = 20\%$. In the case of a coefficient $\gamma_Q = 1.75$, collapse probabilities of the three beams are

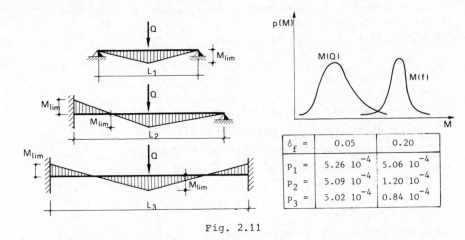

δ_f =	0.05	0.20
p_1 =	$5.26 \ 10^{-4}$	$5.06 \ 10^{-4}$
p_2 =	$5.09 \ 10^{-4}$	$1.20 \ 10^{-4}$
p_3 =	$5.02 \ 10^{-4}$	$0.84 \ 10^{-4}$

Fig. 2.11

given in the table of Fig. 2.11 for two different values of resistance variation coefficient δ_f (the two values are indicative, but they closely represent distributions of basic material strength and of welded or bolted connections. If variation coefficient δ_f is small, resistance may be considered deterministic and has practically no influence on results, which therefore depend only on load distribution shape. If variation coefficient δ_f is great, i.e. if resistance values are more scattered, structures (resistance being equal) are the safer the greater the number of critical sections and therefore the degree of hyperstaticity. Finally, for the same resistance characteristic value, the structure is safer the more scattered resistance distribution is.

2nd example. Consider two beams designed to carry the same variable load, made of different materials (e.g. one of steel and the other of concrete). Suppose the variable load characteristic value is 1.00 kNm^{-1}, whereas the permanent loads due to self weight are respectively 1.00 and 3.00 kNm^{-1} Applying the semi-probabilistic method and assuming $\gamma_G = 1.33$, $\gamma_Q = 1.5$, the following limiting resistance moments are necessary in the most stressed section:

For beam 1 $M_{lim} = (1.33 \times 1.0 + 1.5 \times 1.0)L^2/8$

For beam 2 $M_{lim} = (1.33 \times 3.0 + 1.5 \times 3.0)L^2/8$

If a probability density law of the normal type is assumed having variation coefficient of 5% and 10% respectively for the permanent load and for resistance, together with a variable load extreme distribution of the 2nd type having parameter $\delta = 20\%$ the following values are found for the probability of reaching the given limit state:

For beam 1: $p = 1.91 \times 10^{-4}$

For beam 2: $p = 1.13 \times 10^{-5}$

Beam 2 is therefore more reliable: in other words, in order to obtain the same reliability, the values of γ_F must also depend on the values of the ratios of the various loads applied to the structure.

From the above examples it is evident how difficult it is to find a simple and at the same time sufficiently general safety analysis.

Therefore the semi-probabilistic method cannot be considered in practice as a method guaranteeing uniform structural safety. It is more realistic to consider it as a rationalization of the allowable stress method: instead of applying a constant safety coefficient, it is possible to weight the coefficient, so as to reach results which in many cases are more consistent. Only more sophisticated approaches (level II and III methods) |5,7,12,13| may guarantee uniform structural safety.

2.3.2 *Applying the method*

The semi-probabilistic method can be of practical use if one defines:

The limit states one wishes to avoid and therefore calculation methods by which both the effects of load design combination (in particular internal stress resultants) and the strength of the structure (or of its elements) can be evalued.

Load characteristic values

Strength characteristic values

The values of γ_m, γ_F and therefore of γ_G, γ_Q and ψ_i, which are functions of the way in which characteristic values (2.2) and (2.3) are defined, of the limit state under consideration and of the risks of attaining it

Calculation methods for controlling whether a given limit state is reached or not depend on the character of the limit state itself.

A structure must be able to attain a serviceability limit state many times during its lifetime, without losing any of its functionality and thus without accumulating residual deformations.

In order to determine whether load effects may lead to a serviceability limit state, calculation methods based on elastic analysis have to be used.

Ultimate limit states are, of course, related to structural collapse. Strictly speaking, therefore, the methods which can be applied to assess the possibility of their attainment are based on plastic design. For such an approach, it is necessary to:

Define load design combinations

Evaluate the ultimate strength of critical sections, admitting elastic-plastic behaviour and stress redistribution in the section

Calculate the factors by which load design combination values must be multiplied to cause structural collapse.

Check that such factors are greater than unity so that the actual load design combinations do not cause the attainment of the ultimate limit state under consideration.

Such an approach is obviously difficult and in some cases impossible. Plastic design, in fact, often requires sophisticated numerical analyses based on hypotheses that are often disregarded in steel structures; buckling phenomena and brittle failure of connections often do not allow the formation of a sufficient number of plastic hinges and thus the reduction of the structure into a mechanism.

Methods based on plastic design, while of great help for certain types of structure, cannot therefore be recommended for all steel structures.

In order to keep the calculations for ultimate limit states simple and practical, it is therefore necessary to adopt a simpler approach. For this purpose one can define conventional ultimate limit states as those corresponding to the attainment of the elastic limit in the various sections of the structure |14|. In this case elastic methods of structural analysis are still valid and the procedure is as follows:

Load design combinations are defined

A reasonable calculation model of the structure is formumated on the basis of plastic design criteria (see 1.4), assuming the formation of plastic hinges in areas of the structure that are not subject to buckling phenomena or brittle failure of connections.

On such a model, distribution of internal actions is found by elastic analysis

Internal actions have to comply with criteria and not cause buckling. This requires that elastic strength or critical stresses for struts must not be exceeded at any point

Such a calculation lies on the side of safety and has the following advantages:

It is possible to carry out an easy structural analysis mostly based on elastic theory

Redistribution of stresses in the section are not accepted

Possible partial redistribution of internal actions are accepted only in those parts of the structure in which it is known a priori that buckling or brittle failure cannot prejudice the section behaviour beyond the elastic limit.

Characteristic values of actions are in many cases unknown. In fact, little is known about snow, earthquakes and lateral thrusts due to lifting equipment. In many cases codes fix nominal values for actions which, so long as statistically acceptable data are lacking, can be assumed as characteristic on the basis of experience.

Chapter 3 refers to some data that could possibly lead to designing a structure according to its own particular load conditions instead of those currently specified in the codes.

Characteristic values of yield strength are known, as well as, within close limits, the characteristic strength of simply compressed columns hinged at their ends and made of normal structural steel (yield between 240 and 400 N/mm^2). In other cases, actual recommendations
70

either suggest nominal values to be assumed as characteristic, or directly indicate design values f_d. They are defined on the basis of past experience. Chapters 4 to 9 indicate, at least for the most common cases, methods which have been recognized as reliable for evaluating design strengths of basic material, connections and members subjected to stability problems.

Load factors γ_F are indicated in codes according to the type of limit state considered. In |14| the following load factors are proposed:

For serviceability limit states, $\gamma_G = \gamma_Q = 1$

For ultimate limit states, $\gamma_G = 1.35$; $\gamma_Q = 1.5$

For ultimate limit states concerning erection when a greater probability of reaching them is accepted, it is possible to reduce by 10% and about 15% the factors for permanent and live loads respectively, i.e. $\gamma_G = 1.20$; $\gamma_Q = 1.30$

Combination coefficients ψ_i depend on load nature, structural type and the particular limit state. They have not yet been properly defined and there is still a need for studies for their rational assessment. Presently, the various codes indicate values of coefficients ψ_i defined by empirical and qualitative considerations.

Material coefficients γ_m essentially depend on the type of collapse (brittle, ductile) and on the degree of knowledge or the phenomenon (especially for buckling phenomena).
    There is, therefore, not much point in defining a value of γ_m for serviceability limit states: they are concerned rather with structure deformability than with strength. The only exception is that of slipping of friction grip bolted joints |15,16|. At present the relation between slipping force and the data determining it (tightening of the bolts, friction coefficient) are not statistically well known. For this reason many codes prefer to refer calculations to a mean value divided by a safety coefficient based on experience.
    For ultimate limit states, many actual codes give values of γ_m depending on the type of limit state and on calculation methods. Overall or local stability limit states are studied by calculation methods which are always on the safe side and without statistical significance (except for compressed columns). For such limit states, conventionally $\gamma_m = 1$.
    Ultimate limit states concerning brittle failure such as those of many joints, must be considered with particular caution. Statistical distributions of strength are generally non-existent; therefore $\gamma_m \geq 1$ is adopted depending on the type of connection (bolted or welded) and on the types of control. For ductile ultimate limit states, two different values of γ_m are usually adopted, according to the two calculation methods above. If plastic design is used and therefore redistribution in the structure and in the section are taken into consideration, $\gamma_m = 1.10$ is assumed. If ultimate state control is performed conventionally, applying elastic design methods and therefore neither section nor structure strength is fully used, it is justified to assume $\gamma_m = 1.0$. Analogous criteria for stating load and

material factors are assumed in US practice |17|. On the contrary a
different approach is followed by the Nordic Committee for Building
regulations |18|: load factors are much lower than those stated in
|14| but material factors are higher.

REFERENCES

1. AIPC-IABSE (1969) *Symposium on Concepts of Safety of Structures
    and Method of Design*, Preliminary & Final Report, London
2. ASCE (1972) Structural safety: a literature review, *Proceedings
    ASCE, Structural Division*, 98, ST 4, 845-884, New York
3. ASCE (1972) *Report of the Conference on Safety and Reliability of
    Metal Structures*, ASCE, New York
4. Ferry Borges, J. and Castanheta, M. (1971) *Structural Safety*,
    Laboratorio National de Engenharia Civil, Lisbon
5. ASCE (1979) *Tall Building Criteria and Loading*, ASCE, New York
6. Alpsten, G. A. (1972) *Variations in Mechanical and Cross Sectional
    Properties of Steel*, Proceedings of International Conference on
    Planning and Design of Tall Buildings, Lehigh University, Vol. 1b,
    755-807, Bethlehem, Pennsylvania
7. Beniamin, J. F. and Cornell, C. A. (1979) *Probability Statistics
    and Decisions for Civil Engineers*, Mc Graw Hill, New York
8. Green, W. E. (1972) *Stochastic Models and Live Loads Surweys*,
    Proceedings of International Conference on Planning and Design of
    Tall Buildings, Lehigh University, Vol. 1b, 35-58, Bethelhem,
    Pennsylvania
9. JCCS (1976) *Common Unified Rules for Different Types of
    Construction and Material*, CEB FIP (1978) International
    Recommendations, 3rd Edition, 5-49, Paris, CECM ECCS (1978)
    Recommendations, 248-291, Brussels
10. Lin, N. C. and Basler, E. (1972) *Safety Level Decisions*,
    Proceedings of International Conference on Planning and Design of
    Tall Buildings, Lehigh University, Vol. 1b, 961-972, Bethlehem,
    Pennsylvania
11. Trezos, C. (1977) Approche probabiliste de la sécurité et remarque
    sur les codes actuelles, *Construction Métallique*, 2, 25-59
12. Esteva, L. (1972) *Summary Report: Structural Safety and
    Probabilistic Methods*, Proceedings of International Conference on
    Planning and Design of Tall Buildings, Lehigh University, Vol. 1b,
    1043-1066, Bethlehem, Pennsylvania
13. Leporati, E. (1979) *The Assessment of Structural Safety*, Research
    Studies Press, Letchworth
14. CECM ECCS (1978) *European Recommendations for Steel Constructions*,
    ECCS, Brussels
15. Fisher, J. W., Kato, B., Woodward, M. M. and Frank, K. M. (1979)
    *Field Installation of High Strength Bolts in North America and
    Japan*, IABSE Surveys S8/78, Zurich
16. Fischer, J. W. and Struik, J. H. A. (1974) *Guide to Design Criteria
    for Bolted and Riveted Joints*, Wiley, New York
17. Galambos, T. V. (1981) Load and resistance factor design, *AISC
    Engineering Journal*, 3, 74-82
18. NKB (1978) *Recommendations for Loading and Safety Regulations*,
    Nordic Committee on Building, Report No. 6

C H A P T E R   T H R E E

Loads

3.1 INTRODUCTION

The various codes define load values to be used when designing
according to the semi-probabilistic method.

The following information concerning the present state of
probabilistic load definition can be of help in understanding
recommendations dealing with load and might guide the designer in
deciding whether it would be advisable to vary the relevant loads to
be assumed in the design according to the purpose of the structure.

Most of the relevant load values have been deduced from the studies
of the joint committee for structural safety CEB-CECM-FIP-CIB-IABSE-
RILEM |1,2| who synthetized the results obtained by a number of
authors in various European countries: for the time being, such data
should be considered indicative, although they might offer a starting
point for bringing the various codes up to date.

For the determination of seismic forces, because of the complexity
of the phenomenon and dependence of stress on building type (which in
many cases requires a dynamic analysis), the reader is referred to
specialized works |3-5|.

3.2 PERMANENT LOADS

The calculation of the permanent weight of a structure is
theoretically easy, but for many reasons it might raise considerable
difficulties. The value of such a load can often be small compared to
other vertical loads acting in service and this can influence the
determination of overturning controls to resist lateral forces.

The elements that generally raise the greatest doubts are:

Use of different materials than those originally planned for
partition walls, external walls, roofs and secondary elements.
Scarce knowledge in the design phase of the weight of finishing
materials that may be chosen according to the builder's
requirements
Inaccurate knowledge of material specific weights and variations in
building element dimensions

73

For the self weight of rolled profiles, existing codes in all countries recommend dimensional tolerances that usually limit variation in mass to 2%. The self weight of structural steel may therefore be considered as deterministic (specific weight γ_a = 77 kNm^{-3}).

In the case of a probabilistic analysis, such a value can be assumed as the mean value, with a variation coefficient δ = 2%.

For concrete members upper g_k^+ and lower g_k^- characteristic values are significant, depending on whether the loading due to self weight is or is not in favour of safety with respect to building stability. Let it be assumed |2| that:

$f_{c,k}$ characteristic strength at 28 days, measured on cylinders and expressed in N/mm^2 (equivalent to approx. 0.83 of that measured on a cube)

V_a/V_c volumetric ratio of reinforcement referred to all directions

γ_c specific weight; 25 kNm^{-3} is normally assumed

A_c area of transversal section determined according to design dimensions and disregarding the existence of steel bars, expressed in mm^2

η a coefficient equal to 1 for in situ concrete and 0.75 for precast structures

u perimeter of concrete section, including the internal surfaces for hollow sections, expressed in mm

r_{eff} = 2 A /u equivalent radius of the section expressed in mm
η = 0.90 + 0.002 $f_{c,k}$ + 2.25 V_a/V_c

If $f_{c,k}$ lies between 15 and 60 Nmm^{-2} and V_a/V_c < 0.04, the following formulae can express mean values and standard deviations:

$$g_m = \xi \, \gamma_c \, A_c \quad ; \quad s = \delta g_m = \eta \left(\frac{6}{r_{eff}} + 0.02\right) g_m$$

Upper and lower characteristic values can be expressed by the formula:

$$g_k = g_m \, (1 \pm k\delta) = g_m \, \alpha \quad \text{with: } \alpha = 1 + k\eta \left(\frac{6}{r_{eff}} + 0.02\right)$$

As the fractile of probability applied to define characteristic values varies, the values α may be obtained from Fig. 3.1. For example, if one assumes as upper and lower characteristic values respectively fractiles 0.95 and 0.05, for r_{eff} = 100 mm and for in situ concrete one gets:

$$g_k^+ = 1.132 \, g_m \quad ; \quad g_k^- = 0.868 \, g_m$$

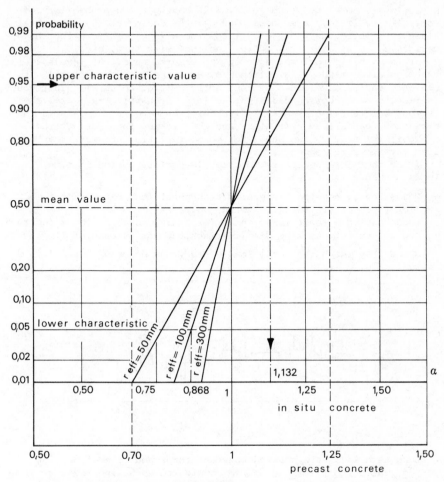

Fig. 3.1

3.3 IMPOSED LOADS

Almost all codes specify design loads for the most common types of
room use; load distribution to produce a reasonable maximum effect in
the resisting structure is, however, the designer's responsibility.
The values given in codes must be considered as static loads but
adequately increased to take into account any slight dynamic effects.
Significant dynamic effects or, in any case, those acting on particular
types of structure, have to be determined, taking into account
structural response, both to find the maximum loads due to vibration
and to allow for possible fatigue phenomena.

Thanks to the probabilistic study of imposed loads, it is possible
to obtain various assessments of actions according to the purpose of
each individual calculation: this is generally very important in any

75

design phase. One must find load values having a low probability of
coming true, if they are to be used for the design of the structural
element on which they act (e.g. load from office furniture for the
design of the slab, snow load for the roof structure, etc.).

It is rather pointless, however, to use the same load levels,
assuming all actions as simultaneous, when designing structural
elements indirectly loaded by such actions (e.g. columns). That would,
in fact, mean accepting different collapse probabilities p_o for the
various structural elements. It is therefore necessary to assess the
situation when a load acts alone and when taking into account the
probability that during the lifetime of the structure more than one
load may act simultaneously.

It is therefore necessary to study load distribution in time.
Loads can be classed as:

Loads having constant intensity throughout long periods of time
(load due to furnishings, stored goods, etc.)
Loads having a short duration (a few hours), such as loads due to a
big concentration of people (meetings, crowds in case of fire, etc.)

A model by which the total load history may be described |6| is
shown in Fig. 3.2.

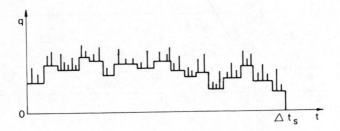

Fig. 3.2

Long-term loads are assumed constant throughout the occupation
period, while short-term loads are considered as having instant values.
In this case, it is rather complicated to determine the load value
having a given probability (e.g. 5%) of being exceeded: only a
stochastic model may take into account the possibility of short-term
and long-term loads acting simultaneously.

3.3.1 *Load intensity of floors*

The stochastic approach to the problem is complex and a final solution
has not yet been found. As far as imposed loads on floors are
concerned, a conventional load situation is generally chosen, taking
into consideration the fact that it can be rather rare for all the
slabs to be simultaneously loaded by the maximum foreseeable load. In
particular, for column design in multistorey buildings loads are
assumed to be uniformly distributed but, dependent on floor location,
most codes permit reductions in order somehow to allow for the
randomness of their distribution |7,8|.

To simplify matters the Joint Committee for Structural Safety CEB-CECM-FIP-CIB-IABSE-RILEM |2| collected the results of probabilistic analyses for the most common types of load intensity on floors, classing them (according to their use) into residential, office, sales, parking, floors.

3.3.1.1 Residential floors

In the case of loads foreseen for domestic rooms, it is appropriate to define the distribution of maximum values as sampled (useful for serviceability limit state) and that for a reference period of 50 years (useful for ultimate limit state, if one assumes the conventional life of the building to be 50 years).

Load intensity in floors $(kN\ m^{-2})$	As sampled		50 years	
Room destination	q_m	q_k	q_m	q_k
Living-rooms	0.30	0.49	0.45	0.64
Bedrooms	0.25	0.44	0.40	0.59
Kitchens and halls	0.10	0.29	0.25	0.44

Fig. 3.3

The table of Fig. 3.3 expresses in $kN m^{-2}$ mean values g_m and characteristic values q_k (fractile 95%) of load intensity due to furnishings, both as sampled and for 50-years periods, for which one can assume $s = 0.19\ kN m^{-2}$, independent of room purpose.

Total load values in 50 years due to furnishings and concentration of people can be considered independent of use, as follows:

Mean value	$1.00\ kNm^{-2}$
Standard deviation	$0.50\ kNm^{-2}$
Characteristic value (fractile 95%)	$1.75\ kNm^{-2}$

To allow for non-uniformity of loading, the resulting effects should be determined by multiplying the uniformly distributed loads listed in the table by appropriate concentration factors as given in 3.3.1.2.

Load intensity against probability is plotted in Fig. 3.4. Probability distributions of loads due to furnishings only as sampled are given in Fig. 3.4a; those for a 50-year period are given in Fig. 3.4b which also shows total load distribution (furnishings plus concentrations of people) independent of room use, for the 50-year period. These diagrams clearly lead to the values of Fig. 3.3.

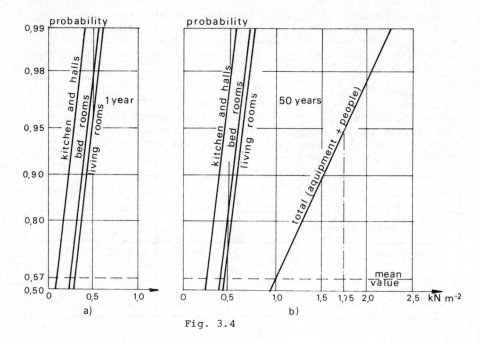

Fig. 3.4

3.3.1.2 Office building floors

In this case, loads are referred to a typical floor, which is neither the ground-floor nor a basement. Imposed load depends on the area of the surface on which it is applied. In fact it is more probable for a 4.00 kNm^{-2} load to be applied to 1 m^2 of surface (result: F = 4.00 kN) than to 10 m^2 (result: F = 40.00 kN). Probability distribution of load intensity must therefore be defined with reference to the area on which it is supposed to act (catchment or tributary area).

The designer has to choose the most unfavourable application area, i.e. the area deemed to be supported by the structural element under consideration. Fig. 3.5 shows in kNm^{-2} the probability distribution of load intensity as sampled and for a 50-year period.

In Fig. 3.6 mean values, standard deviation and characteristic values are quoted. They depend on A (in m^2) where A = 10 m^2. Imposed load values for ground and basement floors can be obtained by multiplying the values for other floors by 1.3 and 1.5 respectively.

For small areas, it is essential to evaluate the resulting effects taking into account any non-uniformity of loading. For this purpose one can use equivalent uniformly distributed loads obtained by multiplying the load intensities of Fig. 3.6 by factors as follows:

For bending moment in slabs supported on two sides: ξ = 1.10
For bending moment and shear in beams bearing slabs supported on two sides: ξ = 1.25
For bending moment in slabs supported on four sides: ξ = 1.35
For bending moment and shear in beams carrying slabs supported on four sides: ξ = 1.35

78

For axial forces in columns: $\xi = 1.15$.

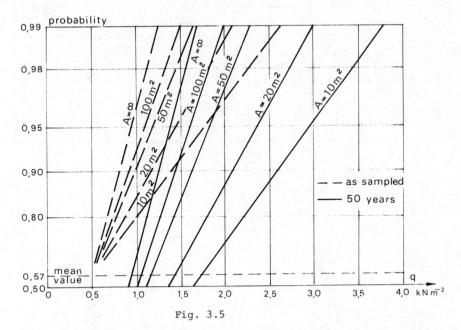

Fig. 3.5

Imposed loads on office floors (kN m^{-2})	As sampled	50 years
Mean value q_m	0.50	$\dfrac{12.80}{A + 6} + 0.95$
Standard deviation s	$\dfrac{6.80}{A + 6} + 0.25$	$\dfrac{6.80}{A + 6} + 0.25$
Characteristic value q_k (fractile 95%)	$\dfrac{12.70}{A + 6} + 0.97$	$\dfrac{25.00}{A + 6} + 1.42$

Fig. 3.6

3.3.1.3 Shop live loads

Shops can be divided into two areas: one open to the public (sales zone) and the other used for goods storage (non-sales zone). Imposed loads can be expressed by the formula $q' = q\,\alpha$, where α is a coefficient taking into account the type of occupation. Values of q expressed in kN m^{-2} are indicated in the table of Fig. 3.7 for the two areas considered and for $A \geq 10$ m^2.

The variation of load intensity as a function of probability is indicated in Fig. 3.8 for sales zones and in Fig. 3.9 for storage. In the latter case it depends on the catchment area and includes effects due to people.

79

Shop imposed loads (kN m^{-2})		As sampled		50 years	
Zone		Sales	Non sales	Sales	Non sales
Mean value	q_m	0.70	0.70	1.70	$\dfrac{15.00}{A+7}+1.95$
Standard deviation	s	0.54	$\dfrac{10.00}{A+7}+0.70$	0.54	$\dfrac{10.00}{A+7}+0.70$
Characteristic value (fractile 95%)	q_k	1.70	$\dfrac{18.00}{A+7}+2.00$	2.70	$\dfrac{35.00}{A+7}+3.25$

Fig. 3.7

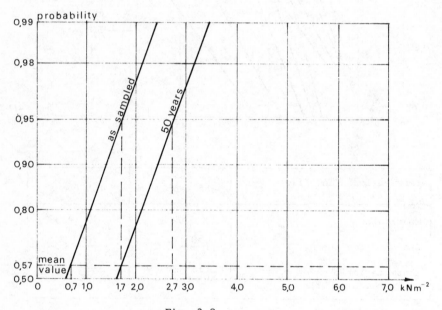

Fig. 3.8

Coefficient α depends on the type of occupancy and may have the following values both in sales and non-sales zones:

Furniture stores	$\alpha = 0.5$
Chemists and clothing stores	$\alpha = 0.7$
Department and general stores	$\alpha = 1.0$
Ironmongers and booksellers	$\alpha = 1.2$
Grocery stores	$\alpha = 1.4$

To allow for the non-uniformity of loading, load effects may be determined using the factors adopted in the case of office loads (see 3.3.1.2).

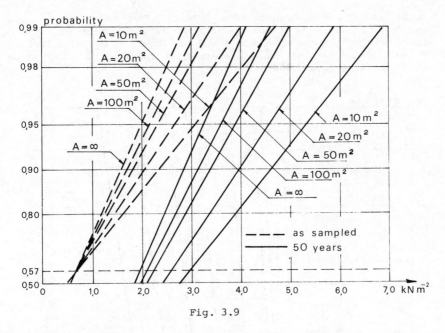

Fig. 3.9

3.3.1.4 Car parking

Parking areas can be divided into two classes, according to whether their disposition is orthogonal or oblique. The probability distribution of load intensity as sampled is indicated in Fig. 3.10.

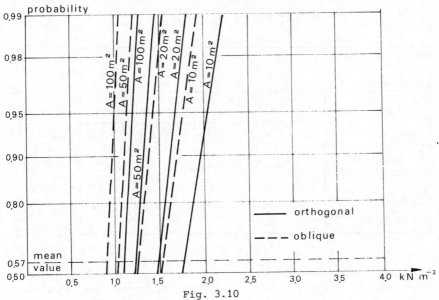

Fig. 3.10

Fig. 3.11 shows the distribution in 50 years. Mean values q_m, characteristic values q_k (fractile 95%) and standard deviation s as sampled and 50-year maximum loads are given in kNm^{-2} in Fig. 3.12 for the two types of parking. Such data are appropriate for tributary areas included between 10 and 100 m^2.

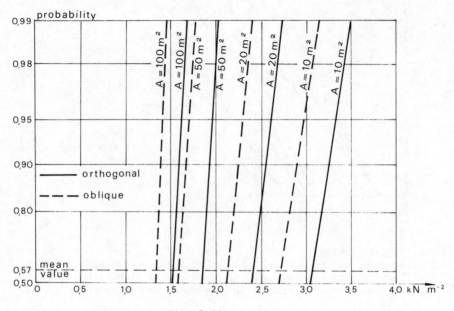

Fig. 3.11

Live imposed loads for parking (kN m^{-2})	As sampled		50 years	
Tipe	Orthogonal	Oblique	Orthogonal	Oblique
Mean value q_m	$0.83+3.00/\sqrt{A}$	$0.65+2.08/\sqrt{A}$	$0.83+7.00/\sqrt{A}$	$0.65+6.50/\sqrt{A}$
Standard deviation s	$0.46/\sqrt{A}$	$0.41/\sqrt{A}$	$0.46/\sqrt{A}$	$0.41/\sqrt{A}$
Characteristic value q_k (95%)	$0.83+3.86/\sqrt{A}$	$0.65+3.56/\sqrt{A}$	$0.83+7.86/\sqrt{A}$	$0.65+7.26/\sqrt{A}$

Fig. 3.12

To allow for non-uniformity of loading on small areas and to take account of the effects of load concentration, one may take equivalent

distributed load intensities having the same values as those above
multiplied by a concentration factor ξ defined thus:

$$\xi = 1 + \frac{1}{2}\left\{\frac{1}{\sqrt{A}} + \frac{1}{\sqrt[3]{A}}\right\}$$

where A is in m^2.

3.3.2 *Cranes*

When designing industrial buildings, it is absolutely necessary to
consider all significant actions due to the production process for
which the building is intended. The structure might in fact undergo
both impact and loads vaying in time; dynamic effects can be considered,
at least in the most common cases, by assuming static load values,
appropriately increased.

For structures supporting lifting equipment, it is advisable to
carry out a detailed analysis for each service condition: a typical
example is given by a travelling crane (Fig. 3.13), the hook of which
has great freedom of movement and can be displaced from a position in
any direction with a given acceleration. The crane can therefore
create dynamic reactions on the various structural elements of the
building, independent of its position.

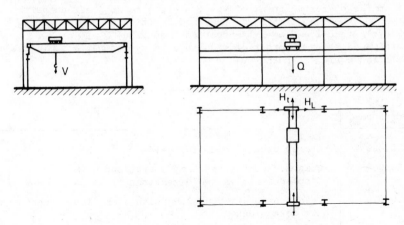

Fig. 3.13

Generally, loads that should be taken into consideration in the design
are caused by:

Weight of lifted loads
Crane self weight
Wind forces
Inertial forces and impacts
Thermal variations

The knowledge of the value of the loads and of their time history
is necessary in order to give a correct probabilistic description. At

83

present time this statement is obviously theoretical. Therefore, the various recommendations deterministically indicate conventional load combinations for structural design.

From a qualitative point of view, in addition to vertical component V a longitudinal horizontal component H_L and a transverse one H_t could be transmitted to the structure. The longitudinal component H_L can be created by inertial forces due to crane braking or starting up and to friction forces between wheel and rail.

The transverse component H_t is due to braking and starting up forces of the crab and to side thrust between wheel edge and rail. Such thrusts occur as the result of the rail not being straight and of the oblique traction loading due to the different dragging forces at the crane wheels.

In is practically impossible to carry out an analytical study of the individual effects and therefore it is preferable to establish the values of components H_L and H_t as a percentage of the vertical load V, including dynamic effects, at least for the most common types of lifting equipment. Load values recommended in the various codes are thus conventional ones largely based on experience.

The following analysis may be of help in understanding the operation characteristics which are common to the various types of plant. Results of some experimental surveys of vertical loads |9| are also quoted for the purpose of providing some qualitative indications.

From the operative point of view, a cycle can be defined as the process of hooking, lifting, conveying, positioning and unhooking. Each cycle is defined by a transported load value Q_i.

Fig. 3.14 shows the order of magnitude in cycles per annum, related to lifting equipment purpose.

Cycles per annum	Examples
20.000	Cranes used in building yards
40.000	Cranes used for seasonal operations with hooks displacements in two directions
100.000	Cranes used all year long with hook displacements in two directions
200.000	Cranes for metallurgical plants operating in three shifts all year long

Fig. 3.14

A number of cycles having individual durations t_i together form an operation having total duration t_o. An operation is therefore defined (Fig. 3.15) by transported load maximum value Q_{max} and mean value Q_m.

If the time scale is extended, it is interesting to define operation continuity, i.e. its duration with reference to a greater time interval than t_o. One can distinguish between long term continuous operations (Fig. 3.16d). The table of Fig. 3.17 refers the type of operation to

84

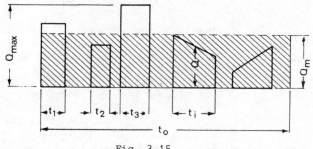

Fig. 3.15

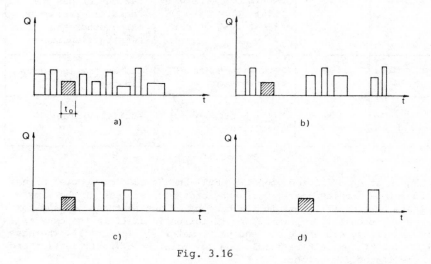

Fig. 3.16

the average number of service hours per year, service conditions and examples of utilization of their lifting capacity.

Finally, one can define average lifting performance q of the equipment as a mean value of transported load per time unit. In the case of loads varying from time to time in weight, average lifting performance is defined by

$$q = \sum_{1}^{n} {}_i Q_i \quad |kN/\text{hour}|$$

where n is the number of cycles per hour, whereas in the case of loads all of the same weight one has:

$$q = Q n \quad |kN/\text{hour}|$$

A correct approach to the problem of defining the vertical load for the design of lifting equipment is to consider magnitude and duration of the various operations as random quantities, so as to define the lifting power probability density law.

85

Type of operation	Average number of hours per annum	Operating conditions	Examples
Continuous for long period	6000	Continuous displacement without any day of rest	Cranes in steel fabrication shops or for concrete pouring in big prefabrication plants
Continuous	4500	Continuous work with interruption due to day of rest or rain	Harbour cranes for docks or cranes for large concrete construction sites
Periodical	2500	Work with seasonal interruption	Cranes for industrial plants
Short periods	500	Max 2 hours/day operation	Auxiliary cranes

Fig. 3.17

The scatter of lifted loads is small in the case of special cranes used for particular operations with loads more or less of the same weight. Scatter is larger in the case of cranes that can be used for various purposes. In particular, variation is small in the case of liquids (ladle cranes, etc.) and pasty materials (concrete), whereas it is greater for bulk materials due to variation of particle size and of their specific weight.

With reference to the type of load hooking equipment, minimum scatter is to be found in the case of cranes provided with an automatic grip unit, intended for the handling of materials of constant dimension.

In the case of electro-magnets, the force they develop depends also on load dimensions. Whenever the part to be handled is supported by a steel surface, considerable overloading can take place.

Maximum dispersion is obviously to be found in the case of manual load attachment, as the operator can suspend loads of any size.

Available experimental data are limited; some interesting data taken from |9| are given below:

If ratio $\psi_i = Q_i/Q_{nom}$ is defined as the ratio between the load existing throughout the generic cycle and the nominal one Q_{nom} it is possible to represent on a diagram a probability density law for parameter ψ (Fig. 3.18).

Curve 1 is representative of auxiliary cranes handling single pieces in steel fabrication shops: its mean value is 0.66 and its dispersion is not great, due to volume and specific weight of the material carried being constant. Also frequency of possible overloading ($\psi > 1$)

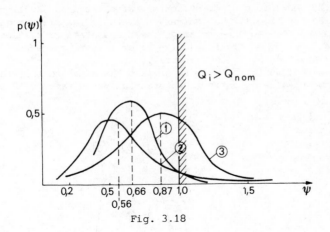

Fig. 3.18

is limited.

Curve 2 has a mean value ψ_m = 0.56, but it has a long tail and consequently the characteristic value is in excess of one.

Curve 3 has a high mean value (ψ_m = 0.87) and an high overloading frequency and therefore is typical of a machine used above its capacity.

Fig. 3.19 shows a typical load probability density function for a crane equipped with electro-magnets. Mean load value is near lifting capacity, while many overloads take place. This can be explained by excessive electro-magnetic power, which however is needed to maintain an adequate reserve against possible supply voltage drop and varying magnetic properties of handled materials.

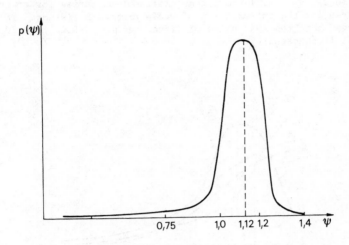

Fig. 3.19

Fig. 3.20 shows load distribution on various cranes handling bulk material. Shapes of distribution curves are quite similar among themselves and seem to undergo displacement parallel to the abscissa axis, as a function of greater (curve 1) or lesser (curve 6) difficulty in scooping up the material.

In all cases, deviation is more or less the same and the standard deviation/mean value ψ_m ratio is substantially constant at 12 to 15%.

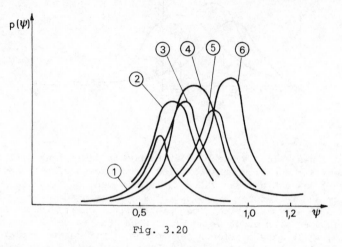

Fig. 3.20

Cranes used for prefabricated buildings, shipyards and heavy engineering plants, where monolithic pieces must be lifted, present low use of lifting power and great variation in the weight of the various loads. Fig. 3.21 shows some distribution curves for tower cranes used for the construction of prefabricated buildings. It may be observed that the mean value of lifted weight is 0.20 to 0.30 of maximum lifting capacity.

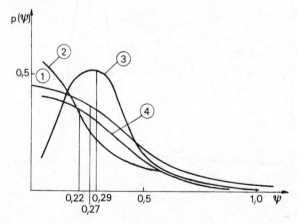

Fig. 3.21

In the case of cranes used for industrial construction, maximum lifting power is used to an even lower degree. This also applies to overhead travelling cranes in industrial plants producing small units. In the latter case, in fact, installed crane lifting power is not related to the loads to be lifted initially, but is chosen to provide a margin to allow for future production development. Fig. 3.22 shows a typical distribution for travelling cranes of various lifting capacities (100-2500 kN) for heavy engineering plants: the mean lifted load is only 9% of lifting capacity.

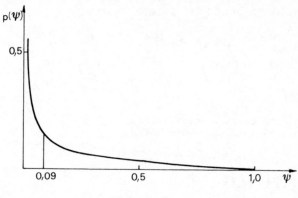

Fig. 3.22

Summing up, the ratio ψ_m of mean lifted load Q_m to nominal load Q_{nom} can be expected to reach values nearing those given in the table of Fig. 3.23, according to machine purpose. Obviously, such values are correct in the case of properly used machines and might be used in fatigue controls.

$\psi_m = \dfrac{Q_m}{Q_{nom}}$	Equipment purpose
0.05-0.15	Cranes used for auxiliary operations, emergency, repair work
0.20-0.30	Cranes used for erection of industrial structures and in mechanical plants
0.40-0.60	Cranes used for prefabricated blocks handling
0.70-0.90	All purpose cranes
1.0	Charging and ladle cranes; magnetical cranes Crabbin cranes designed for single kind of bulk material

Fig. 3.23

89

3.4 CLIMATIC LOADS

3.4.1 *Snow*

The roof of any building and its supporting structure must be able to bear accidental loads that might occur either during erection (e.g. the 1 kN concentrated load due to the weight of a person) or in service. Apart from special cases in which the roof is used for parking, pavement, roof garden, etc., snow load most heavily influences, in temperate and cold areas, roof dimensions. Snow load intensity q_s is given in terms of unit horizontal surface area. Referring to the three diagrams of Fig. 3.24, it follows that:

$$q_t = q_s \sin\alpha \cos\alpha \; ; \; q_n = q_s \cos^2\alpha \; ; \; M_{max} = \frac{1}{8} q_n L^2 = \frac{1}{8} q_s L_1^2$$

The value of q to be assumed can be expressed as:

$$q = \mu \, q_{s,o}$$

where:

  μ is a roof exposure coefficient

  $q_{s,o}$ is a snow load reference value.

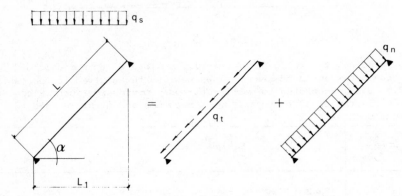

Fig. 3.24

According to the proposals of the Joint Committee CEB-CEM-CIB-FIP-IABSE-RILEM |2|, Fig. 3.25 shows the probability distribution of snow load intensity for an individual snowfall. Maximum yearly values are a function of the mean reference value $q_{s,m}$ depending on the site. One can assume a standard deviation $s = 0.50 \, q_{s,m}$ and a characteristic value (fractile 95%) $q_{s,k} = 1.93 \, q_{s,m}$.
Maximum snow values over 50 years are given by:

$$q_{s,m,50} = 2.53 \, q_{s,m} \; ; \; s_{50} = 0.50 \, q_{s,m} \; ; \; q_{s,k,50} = 3.46 \, q_{s,m}$$

For evaluating the effects of a possible snowfall together with other actions, one must apply a model able to describe the phenomenon as faithfully as possible. The natural phenomenon can be described as

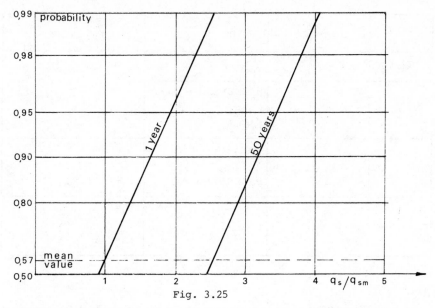

Fig. 3.25

a stochastic process based on random occurrences of random amplitude, duration and shape (Fig. 3.26).

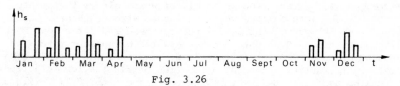

Fig. 3.26

The exposure coefficient μ essentially depends on the horizontal inclination α of the slope of the roof under consideration and the geometry of adjoining ones. Its value is not uniformly defined in the various national codes. For the most common cases of Fig. 3.27a, some indicative realistic values of μ are shown in Fig. 3.27b.

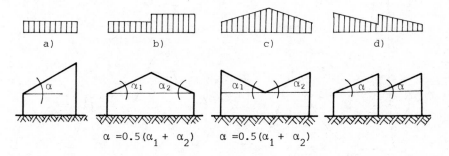

Fig. 3.27a

91

Roof slope	Case a	Case b	Case c, d
$0 \leq \alpha \leq 15°$	$\mu_1=0.8$	$\mu_1=\mu_2=0.8$	$\mu_1=0.8(30+\alpha)/30$
$15<\alpha \leq 30°$	$\mu_1=0.8$	$\mu_1=0.8+0.4(\alpha-15)/15$ $\mu_2=0.8$	$\mu_2=0.8$
$30°<\alpha<60°$	$\mu_1=0.8(60-\alpha)/30$	$\mu_1=1.2(60-\alpha)/30$ $\mu_2=0.8(60-\alpha)/30$	$\mu_1=1.6$ $\mu_2=0.8(60-\alpha)/30$
$\alpha \leq 60°$	$\mu_1=0$	$\mu_1=\mu_2=0$	$\mu_1=1.6$ $\mu_2=0$

Fig. 3.27b

3.4.2 *Wind*

Wind is an air displacement tending to balance pressure differences of
the various atmospheric strata due to uneven heating of the air. It
is therefore influenced by geographic position and by ground profile
the roughness, or variation in level, of which alters the stratum
depth and causes turbulence which increases with roughness. At any
given point in space, wind is defined by speed and direction, both in
relation to time. Such data can be measured using anemometers or
recorded by anemographs. Wind speed fluctuations can therefore be
developed in Fourier series as a function of fluctuation in period T;
it is also useful to define the wind power spectrum: $S(T)/T$ (Fig.3.28).
A study of the spectrum shows that energy production due to wind takes
place mainly in certain frequency bands, grouped around the four
following periods: 1 year, 4 days, 1/2 day and 1 minute |10,11|.

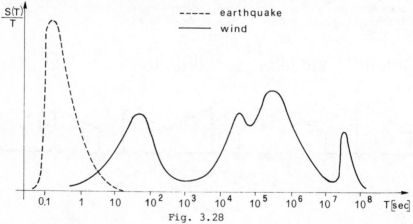

Fig. 3.28

The first is linked to a seasonal cycle, the second to metereological variations, the third to air temperature variations due to direct sun heating. Finally, the fourth corresponds to so called gusts, i.e. to sudden alterations in wind speed. It can also be seen that energy variation in the frequency band defined by periods between 10 minutes $(6 \times 10^2$ sec) and 2 hours $(7 \times 10^3$ sec) is negligible. Therefore two distinctly separate types of wind speed fluctuation can be defined, one slow and the other due to gusting fast. At any instant t, wind speed can therefore be expressed by the formula:

$$v(t) = v_m(t) + \Delta v(t)$$

where v_m is the mean wind speed corresponding to slow fluctuations and Δv instantaneous gust speed. If mean speeds in a given interval (including of course that between 10 minutes and 2 hours) are measured, a reliable value for v_m is found. On the basis of a series of such surveys, carried out over a period of at least 20 years, i.e. throughout an interval approximately corresponding to twice that characterizing the cycle of sun spots, which also influence wind phenomenon, it is possible to produce probability density functions of annual maximum wind speeds and thence to define characteristic values, or rather those having a fixed return period.

Statistical analysis of records has proved that an extreme distribution law of the first type can be assumed: it is characterized by values of the two parameters u and α allowing, at least in Europe, a practically constant product $|12|$ $(1/\alpha u = 0.13)$. Therefore, once a speed value $v_{m,50}$ having a return period $\Delta t_o = 50$ years is known, speed $v_{m,\Delta t_o}$ corresponding to a return period Δt_o is given by:

$$v_{m,\Delta t_o} = \alpha_3 \, v_{m,50} = v_{m,50} \left\{ 1 - 0.13 \log \left| -\log \left(1 - \frac{1}{\Delta t_o} \right) \right| \right\} / 1.507$$

Value α_3 is shown in Fig. 3.29.

Δt_o (years)	2	5	10	20	50	100	200	500	1000
α_3	0.70	0.79	0.86	0.92	1	1.06	1.12	1.20	1.26

Fig. 3.29

Also speed Δv representing wind speed fluctuations about the mean value v_m can be considered as a random variable. For the purposes of structural design, it can be characterized by its spectrum: it is thus possible to relate the energy produced by gusts to mean wind speed through turbulence intensity I, which in turn is a function of site ground roughness and of height above the ground.

3.4.2.1 Wind effects
The nature of the air displacement affecting a body can be qualitatively characterized by the two following factors:

93

Inertia forces acting on the air particles resisting change in direction
Viscosity forces between the various air strata in mutual relative movement.

If air speed is high and if body dimensions are considerable, inertia forces become relatively important and air movement becomes turbulent. The relation between inertia forces and air viscosity is characterized by Reynolds' number $R_e = v_m b/\nu$:

v_m is mean wind speed
b is smallest solid dimension in the plane perpendicular to the wind direction
ν is the kinematic viscosity of air, which can be assumed equal to 1.46×10^{-5} m^2 s^{-1}.

When wind blows |13| on a building (Fig. 3.30), the current separates and a high pressure cushion is formed on the front face. The wind flow is deflected and accelerated around the end walls and rejoins at some distance from the wall.

This entails the formation of opposite vortices with vertical axes and produces suction on the walls. Also the effect of the ground (Fig. 3.30b) influences the phenomenon by creating vortices with horizontal axes which alter the pressure regime caused by the vertical axes vortices.

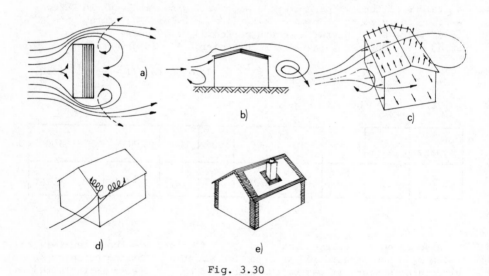

Fig. 3.30

The flow stream can produce a laminar or turbulent layer on the roof (Fig. 3.30c), depending on its slope. Vortices are in any case formed on the leeward slopes. Finally (Fig. 3.30d), the effects of corners or building discontinuities produce localized turbulent effects which might be the determining factor when calculating roof cladding. Fig. 3.30e schematically indicates the areas in which such effects could be significant.

Summing up, from a qualitative point of view, it is apparent that pressure due to wind at instant t on a point of a building depends on:

The energy related to wind mean speed v_m
The energy transferred to the building by a gust, which is superimposed on the mean wind stream
The effect of vortices having dimensions of the same order as those of the building itself, which enclose the building and create a simultaneous pressure variation on the entire building
The effect of vortices having smaller dimensions than those of the building, which produce significant variations of mean pressure on limited areas of the building. These have generally negligible effect on the resulting wind forces on the building but may have considerable effect with reference to control of secondary elements of the building

Usually dynamic effects due to gust fluctuations are negligible. Structure periods generally lie between 0.5 and 2 seconds: they are therefore (Fig. 3.28) sensitive to seismic effects, but fall within a wind spectrum area which is in favour of safety. Furthermore, gusts are not usually sufficiently repetitive to produce dynamic amplification of oscillation of the structure (resonance).

On the other hand, dynamic phenomena due to aerodynamic or aeroelastic structural behaviour in the case of steady wind are possible.

In certain cases, in fact, the structure can oscillate perpendicularly to the wind direction as a result of so called von Karman vortices, which form alternately at separation points of the incident flow (Fig. 3.31a). Fig. 3.31b shows their effect on the overall behaviour of the structure; Fig. 3.31c shows a possible shape variation which may also be caused by these vortices.

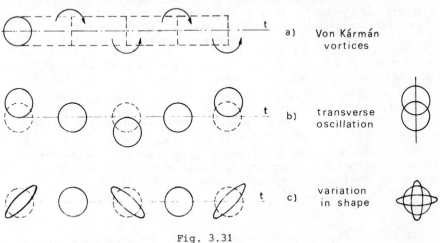

a) Von Kármán vortices
b) transverse oscillation
c) variation in shape

Fig. 3.31

In other cases if the structure oscillates transversely, the relative wind speed and its angle of incidence on the construction vary. Incidence angle variation might cause an increase in wind forces action and therefore a flutter. This phenomenon is typical of highly flexible construction, such as cable-stayed structures and suspension bridges.

3.4.2.2 Design speed

The first step towards determining wind effects on a building is to define design speed: $v_d = \alpha_1 \, \alpha_2 \, \alpha_3 v_{ref}$. It depends on:

Mean reference wind speed in the geographical area in which the building is located
Return period Δt_o chosen for structural control (coefficient α_3)
Influence of site topography (coefficient α_1)
Site roughness, building height and influence of gusts on the building (coefficient α_2).

Reference speed in the geographical area in which the building is located is determined from codes the values in which have been deduced from available statistical data. Generally, the notation reference speed value v_{ref} indicates the value as given in $|12|$: i.e. the value averaged over 10 minutes, at 10 m above the ground, having a return period $\Delta t_o = 50$ years and referred to an open flat terrain without any obstacles.

If the measurement is carried out at a site that is not free from obstacles, the surveyed data must be modified according to the relation:

$$v_{ref} = \frac{v}{k_1 \, \log(z/z_o)}$$

in which z (in m) is the height above ground (10 m) at which V is measured and k_1 and z_o, (Fig. 3.34) depend on the roughness classification of the site at which measurement is carried out.

Coefficient α_3 (Fig. 3.29) depends on the return period Δt_o assumed in controls. It is a function of the design life Δt_s or structural element and of the consequences that would ensue a possible attainment of limit state. One can assume that:

The structure must resist elastically winds having a return period of the order of its design life Δt_s
The structure must survive winds having a return period of the order of twenty times its design life Δt_s

The latter statement is justified by the fact that probability p of an event having greater intensity than that defined by return period Δt_o taking place during a number of years Δt_s is:

$$p = 1 - (1 - \frac{1}{\Delta t_o})^{\Delta t_s} \simeq \frac{\Delta t_s}{\Delta t_o}$$

If $\Delta t_o = 20\Delta t_s$, one obtaines $p \simeq 0.05$ for any value of Δt_s: wind speed having a return period $\Delta t_o = 20\Delta t_s$ thus acquires a similar meaning to that of the characteristic value for permanent loads for the 95% fractile.

96

The consequences of reaching a particular limit state must be taken into account when defining return time. It is obvious that the attainment of limit state for a roof purlin or a façade supporting structure has less important consequences than those due to exceeding the limit state for a beam bearing a slab, a column or a bracing. One can therefore define as secondary those elements for which attainment of limit state does not entail damage to other elements of the structure.

On the other hand, from an operative point of view, difficulties may arise when dynamic wind effects on the building are examined. In fact, it is possible to assess such effects with any degree of reliability only in the elastic phase for serviceability limit states; it is extremely difficult to calculate the collapse situation (ultimate limit state).

Given our present knowledge of the matter, at least as far as design is concerned, the same approach can be applied as for earthquakes: the dynamic behaviour of structure in the elastic field is studied for a reduced wind value (i.e. having a return period of 50 years). Ductility of the material and of its connections is assumed to allow an energy dissipation during elastic-plastic oscillations so that collapse loads are 4-9 times higher than those corresponding to the elastic strength of the structure: thus the structure is able to bear, at collapse, wind speeds of the order of 2-3 times those assumed.

Briefly, as more precise methods or rules are lacking, it appears reasonable to adopt the following values of return periods Δt_o as functions of structure design life Δt_s:

(a) Static controls in the elastic field with reference to serviceability limit states

$$\text{for all elements} \quad \Delta t_o = \Delta t_s/5$$

(b) Static controls with reference to collapse or to conventional elastic ultimate limit states,

$$\text{for secondary elements} \quad \Delta t_o = \Delta t_s$$
$$\text{for non-secondary elements} \quad \Delta t_o = 20\Delta t_s$$

(c) Dynamic controls with reference to ultimate limit states, performed in the elastic field on ductile structure

$$\Delta t_o = \Delta t_s$$

Coefficient α_1 is a topography factor and takes into account the effects of the surrounding terrain. It may as in |14| be assumed equal to:

1.1 for very exposed hill slopes and crests or when valley configurations produce tunnelling of the wind

0.9 for steep sided enclosed valleys sheltered from all winds, provided distance between the highest part of the building and the imaginary line across the top of the valley is greater than 1/50 of the valley's smallest width at the top (Fig. 3.32)

1.0 in all other cases

Coefficient α_2 takes into account the variation in wind speed over open flat terrain due to height above ground, ground roughness and mean duration of a defined gust. Ground roughness can be divided into categories as follows |12,14|:

97

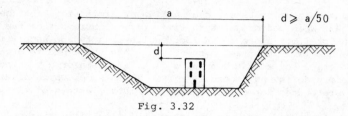

Fig. 3.32

1. Sea or lake shore with at least 5 km fetch over the water
2. Flat terrain with some isolated obstacles (trees, buildings), e.g. flat coastal fringes, fens, airfield and grassland, moorland or farmland without hedges or walls around the fields
3. Rural area with buildings, trees, hedges, e.g. most farms and country estates with the exception of those parts that are well wooded
4. Urban, industrial or forest area, e.g. well wooded park-land and forest areas, towns and their suburbs and outskirts of large cities. The general level of roof tops and obstructions is assumed at about 10 m but the category will include building areas generally apart from those that qualify for category 5
5. Centres of large cities with general roof height of about 25 m or more and streets relatively narrow

It is obvious that the effects of ground roughness on wind speed continuously vary and there can be doubts as to the category in which a building should be included if it is in an area at the changing point between two categories. The following approach may be adopted (Fig. 3.33). If the distance of the building from the boundary between the two categories is equal to a, then for:

$a \leq 500$ m   speed is calculated on the basis of the lower category (i.e. that having higher speeds)

$a > 5000$ m   speed is calculated on the basis of the category of the area in which the building is located

$500 < a < 5000$ m   the building is assumed to be in the higher category area (i.e. with lower speeds) if its height $h < 0.08\ a$ and in the adjacent one if $h \geq 0.08\ a$.

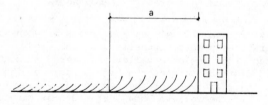

Fig. 3.33

The coefficient α_2 takes into account the time interval which is considered when defining mean speed of the gusts affecting the building and the height z from ground level:

$$\alpha_2 = \frac{v(z)}{v_{ref}} \sqrt{1 + k_3 I(z)} = k_1 \log \frac{z}{z_o} \sqrt{1 + \frac{k_2 k_3}{k_1 \log(z/z_o)}}$$

where:

v_{ref} reference wind speed

$v(z)$ mean wind speed at height z (in m)

$I(z)$ turbulence intensity at height z

k_3 a coefficient depending on the time interval in which the defined gust is averaged

k_1, k_2, z_o coefficients depending on ground roughness category, as given in Fig. 3.34

Category	1	2	3	4	5
z_o	0.005	0.05	0.30	1.0	2.50
k_1	0.16	0.19	0.23	0.26	0.29
k_2	0.16	0.19	0.22	0.23	0.23

Fig. 3.34

The time interval over which the gust should be averaged essentially depends on structure response: it should therefore be defined, on the basis of structure dynamic analysis, in the elastic field if serviceability limit states are concerned and in the elastic plastic field if an ultimate limit state is considered. As has already been pointed out, at our present state of knowledge only a conventional definition of coefficient k_3 is possible and must be accepted |14,15|.

To determine wind effects with reference to serviceability limit states, one must not assume any energy dissipation by the structure. This implies taking into account speed values averaged over time intervals of the order of seconds ($5 \leq k_3 \leq 7$).

To determine wind effects with reference to ultimate limit states, one can count on an energy dissipation by the structure. This implies that as a first approximation, speed can be averaged over time intervals of the order of minutes ($0 \leq k_3 \leq 1$).

For the sake of simplicity, the same value of coefficient k_3 may be assumed both for serviceability and ultimate limit states. In this case |11,13| $k_3 = 7$ is usually adopted, although high values of design speed are against common sense and engineering practice. Therefore codes |11,14| adopt high values for k_3 but take a low return period (50 years) for ultimate limit states also in order to give reasonable values for design speed. Furthermore this choice is contrary to the results of a correct reliability analysis because assuming a return period of the magnitude of the lifetime of the structure one may find that a building on average collapses once during its expected life!

Finally to determine the coefficient $\alpha_2(z)$, reference must be made to a height z measured form the ground surrounding the building. When the building is on or near an escarpment height z to be applied in calculating α_2 can be assessed starting from a fictitious ground level, such as that defined for example in Fig. 3.35 |14|.

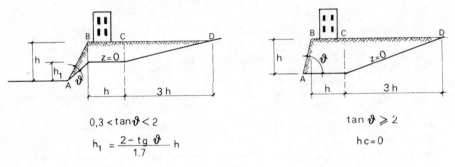

$$0,3 < \tan \vartheta < 2$$

$$h_1 = \frac{2 - tg\,\vartheta}{1.7}\,h$$

$$\tan \vartheta \geqslant 2$$

$$hc = 0$$

Fig. 3.35

3.4.2.3 Dynamic pressure

Once wind design speed

$$v_d = \alpha_1 \alpha_2 \alpha_3 v_{ref}$$

is defined, it is possible to determine the value of dynamic wind pressure p as that corresponding to a transfer of all the kinetic energy into pressure energy. It follows that:

$$p = \rho\, v_d^2 / 2$$

where ρ is air density, to be assumed equal to 1.225 kg m^{-3}. If v is expressed in ms^{-1} and p in Nm^{-2}, one has:

$$p = v_d^2 / 16$$

3.4.2.4 Static effects

Pressure on a surface exposed to wind and therefore the load w to be considered in structural design varies from point to point and also depends on the angle of incidence of the wind. In general the wind load can be expressed by the formula

$$w = c_p\, p$$

where p is the dynamic wind pressure defined above and c_p is a coefficient indicating either a pressure (if it is positive) or a suction (if it is negative).

Coefficients c_p are determined experimentally by analysis on a model in a wind tunnel. They are dealt with in many different ways in codes or specialist books |11,12,14,16| and are distinguished, according to the effects they characterize, into:

Local pressure coefficients $c_{p,1}$, applicable in the calculation of maximum pressure or suction peaks on cladding and their supporting elements

Internal pressure coefficients $c_{p,i}$, applicable when defining wind effects on buildings, the openings and windows of which might cause

internal pressure alterations
External pressure coefficients $c_{p,e}$, applicable when defining
mean pressures on façade or roofing surfaces of given area
(indicatively comprised between 10 and 100 m^2)
Overall pressure coefficients $c_{p,f}$, applicable to the calculation
of overall wind effects on entire buildings or on considerable
parts thereof

Local pressure coefficients $c_{p,1}$ are relatively high, inasmuch as
they express turbulence effects on boundary areas of buildings, such
as corners, projections, balustrades, and areas near discontinuities.
They are essential to determine pressure peaks on such surfaces.
Internal pressure coefficients $c_{p,i}$ essentially depend on building
permeability, i.e. the percentage of openings while the wind is
acting. These, together with the $c_{p,e}$ coefficients are required to
determine wind effects on windward walls and structural elements
(e.g. columns).
External pressure coefficients $c_{p,e}$ depend on the aerodynamic
behaviour of the building and are therefore vastly influenced by its
shape and dimensions, as well as by the degree of wall rugosity.
They are most often specified in the various codes for buildings
having parallelepiped shape, roofs variously inclined to the horizontal,
barrel vaults or domes, and cylinders at or above ground level.
To calculate the overall effects of wind on entire buildings, one
of the following approaches can be chosen:

Vectorially sum contributions of the individual surfaces, evaluated
applying coefficients $c_{p,e}$ and $c_{p,i}$
Directly calculate resulting action by multiplying dynamic pressure
$p = v_d^2/16$ by building surface area in a perpendicular plane to
wind direction and by coefficient $c_{p,f}$ depending on aerodynamic
conditions of the site and of the building

It is obvious that the first method errs on the side of safety.
Coefficients $c_{p,e}$ are in fact mean pressure coefficients assessed for
relatively small areas. They also cover the effects of vortices
having smaller dimensions than those of the building and which do not
contribute to overall action on the building itself.
The second method is more accurate but it cannot always be applied
due to lack of data on coefficients $c_{p,f}$ which must be determined on
the basis of wind tunnel tests. Codes generally indicate coefficients
for parallelepiped-shaped buildings, variously inclined roofs, sheds
and canopy roofs, lattice towers having triangular and square bases,
series of plane lattice frames, silos and chimneys with ribbed and
smooth walls.

3.4.2.5 Dynamic effects
It is not possible to indicate general methods for taking account of
the dynamic effects of wind on particularly flexible structures such
as suspension or cable stayed bridges. In such cases, it is often
useful to carry out experimental researches. In the case of other
particularly deformable, but more traditional structures, such as tall
buildings (> 60 m), antennae and masts, dynamic controls can be
achieved by more approximate methods which provide sufficiently

accurate results $|16|$. As a first approximation, in order to assess increase in stress and strain due to building oscillations in the direction of the wind, the load per unit surface area $w = c_{p,f}$ can be multiplied $|16|$ by a coefficient:

$$\beta = 1 + \xi\tau$$

where:

    τ is a dynamic coefficient and a function of the natural period of vibration T of the structure

    ξ is a pulsation coefficient, and a function of height z as defined in 3.4.2.2.

Loads normal to the wind direction might occur on a structure because of von Karman vortices (Fig. 3.31). Such a stream can take place only if wind speed is relatively low and constant in time and the building surface is smooth.

The danger therefore does not exist if the building has a rough or discontinuous surface, such as those of lattice girders.

REFERENCES

1. Ferry Borges, J. and Castanheta, M. (1971) *Structural Safety*, Laboratorio National de Engenharia Civil, Lisbon
2. CEB-CECM-CIB-FIP-IABSE-RILEM (1972) *Basic Notes on Actions*, CEB Bulletin n. 112, Paris
3. ATC-Applied Technology Council (1978) *Tentative Provisions for the Development of Seismic Regulations for Building*, National Bureau of Standards (US) Special Publication 510, Washington
4. Clough, R. W. and Penzien, J. (1975) *Dynamic of Structures*, McGraw Hill, New York
5. IAEE-International Association for Earthquake Engineering (1973) *Earthquake Resistant Regulations - A World List*, Gakujutsu Bunken Fukyu-kai, Tokyo
6. Green, W. E. (1972) *Stochastic Models and Live Load Surveys*, Proceedings of International Conference on Planning and Design of Tall Buildings, Vol. 1b, 35-58, Lehigh University, Bethlehem, Pennsylvania
7. Council on Tall Buildings, Group CL, (1979) *Tall Building Criteria and Loading*, Volume CL of Monograph on Planning and Design of Tall Buildings, ASCE, New York
8. Apeland, K. (1972) *Reduction of Live Loads and Combinations of Loads*, Proceedings of International Conference on Planning and Design of Tall Buildings, Vol. 1b, 85-96, Lehigh University, Bethlehem, Pennsylvania
9. Kogan, J. (1976) *Crane Design. Theory and Calculation of Reliability*, Wiley, New York
10. Davenport, A. G. (1972) *Theme Report: Wind Loading and Wind Effects*, Proceedings of International Conference of Planning Design of Tall Buildings, Vol. 1b, 335-364, Lehigh University, Bethlehem, Pennsylvania
11. SACHS, P. (1978) *Wind Forces in Engineering*, Pergamon Press, London
12. CECM-ECCS (1979) Calcul des effects du vent sur les constructions, *Construction Métallique*, 3, 27-96

13. Building Research Station (1970) *The Assessment of Wind Loads*, Digest 119, Her Majesty's Stationery Office, London
14. BSI (1972) *Code of Basic Data for the Design of Buildings.* Chapter V, *Loading*, Part 2, *Wind Loads*, British Standard Institution, London
15. CNR (1880) *Azioni sulle costruzioni*, Consiglio Nazionale delle Ricerche, Roma
16. DTU (1976) *Règles Définissant les Effects de la Neige et du Vent sur les Constructions et Annexes*, Eyrolles, Paris

C H A P T E R   F O U R

The Material

4.1 SHAPES

The steel industry provides the designer with a large range of products for steel constructions.

A building is the result of a combination of such products and the design problems consists in making the most rational choice of the structural components and of their assembly.

Except for certain special elements (e.g. bearings or particular types of constraint) which are obtained by forging or casting methods, all steel products come from the rolling process. On the one hand steel plates provide the basic elements which make possible any built up shape. On the other hand steel profiles, whose types and dimensions have been pre-designed, provide prefabricated structural elements for satisfying any constructional requirement.

Static and aesthetic considerations have always required a variety of sections. In the early days of steel construction they were obtained by combining elementary shapes by means of rivets (Fig. 4.1). Now steel building design is based on:

Standard profiles (Fig. 4.2), including those of large dimensions (jumbo profiles)
Sections composed by welding rolled plates (Fig. 4.3)
Thin walled cold formed shapes obtained by bending cold rolled thin plates or sheets (Fig. 4.4)

In Figs 4.5, 4.6 the most common Standard European, British and US I hot rolled sections are compared with reference to their weight g and depth h |1,2|.

Welding and cold forming allow a much greater flexibility than hot rolling which gives only a limited range of shapes. The most varied and rational sections can be obtained in such a way by providing maximum material exploitation and therefore maximum lightness of the steel construction.

Thin walls can, however, entail dangers related both to corrosion and local buckling phenomena (see 9.6).

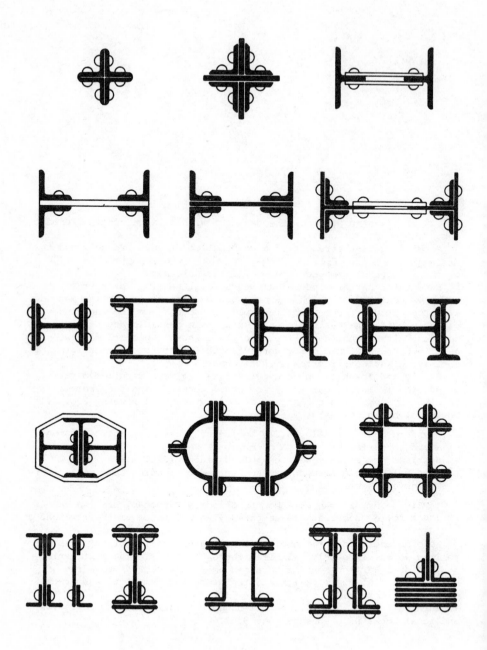

Fig. 4.1

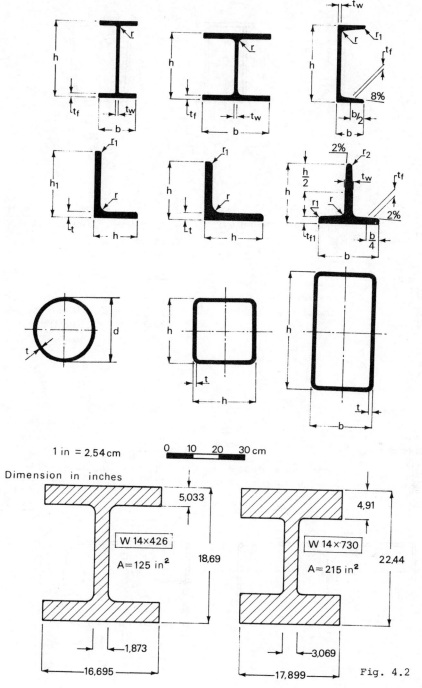

1 in = 2,54 cm

Dimension in inches

W 14×426
A = 125 in²

5,033

18,69

1,873

16,695

W 14×730
A = 215 in²

4,91

22,44

3,069

17,899

0   10   20   30 cm

Fig. 4.2

107

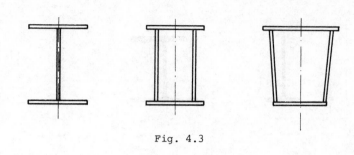

Fig. 4.3

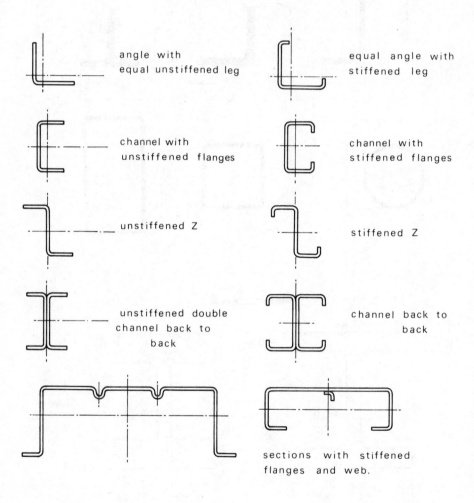

angle with
equal unstiffened leg

equal angle with
stiffened leg

channel with
unstiffened flanges

channel with
stiffened flanges

unstiffened Z

stiffened Z

unstiffened double
channel back to
back

channel back to
back

sections with stiffened
flanges and web.

Fig. 4.4

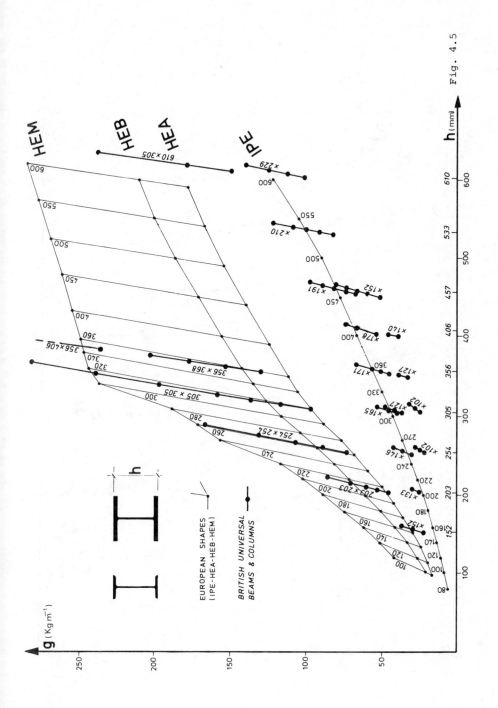

Fig. 4.5

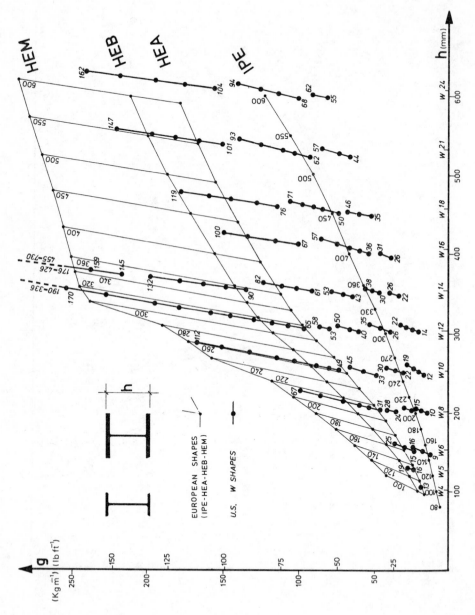

Fig. 4.6

4.2 STRUCTURAL IMPERFECTIONS

Structural or mechanical imperfections of steel profiles substantially consist of |3-23|:

Presence of residual stresses
Non-homogeneous distribution of mechanical characteristics over the cross section

Residual or locked in stresses represent the balanced internal stress state in steel bars as a consequence of the industrial manufacturing procedures. They occur in bodies undergoing non-uniform plastic deformations. If no external forces oppose them, residual stresses are always elastic. The non-homogeneous deformation condition, which creates residual stresses in steel sections, is due to thermal (cooling, welding, flame cutting) and mechanical (cold rolling, straightening) industrial procedures.

The non-homogeneous distribution of mechanical characteristic over the cross sections of steel bars is also due to their production technique. Among the various mechanical characteristics, the structural behaviour of a steel bar is most deeply influenced by the variation in yield stress.

All the most recent trends in the evaluation of the load bearing capacity of steel members agree in taking such imperfections into account. The assumption of perfectly straight ideal bars made of isotropic homogeneous materials free from internal stresses has been abandoned, because such bars do not exist in practice. In a more realistic interpretation of their physical nature, they have been replaced by industrial bars with their inevitable random imperfections (mechanical and geometrical) due to the manufacturing process.

Structural imperfections of the three main types of structural steel elements, hot rolled, cold formed and welded sections are discussed in 4.2.1 to 4.2.3.

Residual stresses which have to be considered are mostly longitudinal stresses, as they act in the same direction as the stresses due to external loads. Such superimposition almost always plays a lowering rôle by reducing the load bearing capacity of bars subjected to buckling phenomena.

4.2.1 *Hot rolled sections*

In these sections residual stresses are, due to the cooling process which follows the hot rolling, of a thermal nature and their distribution can be afterwards modified by straightening, a mechanical process.

Consider the cooling process of a hot rolled I profile (Fig. 4.7a).

Starting from the end rolling temperature T_o, equal to approx. 600°C, differences in temperature of the various parts of the cross section begin to take place. The most exposed parts (flange edges, mid-web) cool faster than the other parts (flange to web joints), which are better protected thermally. At an intermediate phase of the cooling (T_1), as the tendency of the colder areas to shorten is hindered by the warmer areas, a longitudinal residual stress distribution arises as shown in Fig. 4.7b.

111

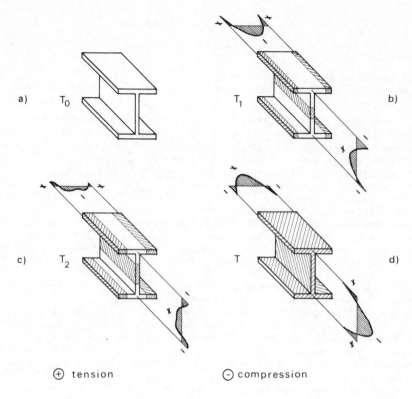

a) T_0

b) T_1

c) T_2

d) T

⊕ tension   ⊖ compression

Fig. 4.7

At this point, the warmer areas due to the compressive stress imposed on them by the contraction of the faster cooling areas, undergo plastic strain. This reduces the residual stress already induced, as shown in Fig. 4.7c for an intermediate temperature (T_2).

The complete cooling of the most exposed areas prevents contraction of the still warm fibres, which therefore undergo plastic deformation. Consequently, once the cooling is complete, the first cooled areas are in compression, while the last are in tension, as shown in Fig. 4.7d.

A similar process would take place in any metal element which cools off, even a compact section (Fig. 4.8).

The intensity of such residual stresses depends on the material stress–strain relationship at the different temperatures between the initial value (T_o) and the final one (T), as well as on the non-uniform degree of deformation which takes place during the cooling process according to the temperature distribution over the cross section.

Such a degree of non-uniformity increases as thermal conductivity k decreases and as the specific heat c, of the metal, its coefficient of thermal expansion α and its specific weight γ increase. These parameters can be combined in the ratio $k/\gamma c$, upon which the temperature gradient over the cross section of a steel bar during cooling depends.

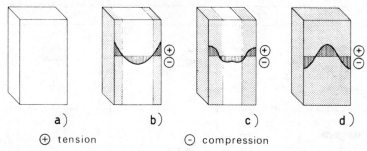

a)   b)   c)   d)

⊕ tension   ⊖ compression

Fig. 4.8

As such a ratio can be considered practically the same for any quality of steel, independent of strength and other mechanical characteristics, distribution of thermal residual stresses in hot rolled steel bars mainly depends on the shape of the cross section.

In the case of I sections, in fact, experience has so far shown that longitudinal residual stresses due to cooling of the flanges and of the web are closely related to the following geometrical ratios (Fig. 4.9):

$$h/b \quad ; \quad t_f/h \quad ; \quad t_f/b \quad ; \quad t_w/h \quad ; \quad t_w/b$$

where h is the depth of cross section, b is the flange width, t_w is the web thickness and t_f is the flange thickness.

In fact the compactness of each cross section, which can be defined by the above parameters, obviously influences the thermal radiation from the various surfaces, and consequently the temperature gradient across the section during cooling and the final residual stress distribution. Hence, thick sections having $h/b < 1.2$ are affected by tensile residual stresses in the middle of the flanges and compressive ones at their edges, while residual stresses at mid-web can be either tensile or compressive, depending on ratios of thicknesses and overall dimensions of the individual parts. On the contrary, in slender profiles having $h/b > 1.5$, distribution takes place with predominant tensile residual stresses in the flanges and compressive ones in the web.

Typical distribution of residual stress for various types of I section are shown in Fig. 4.9: they synthesize the results of many US and European tests |11|. Compressive residual stresses (especially at flange edges) are an unfavourable circumstance with reference to buckling phenomena (see Chapter 9).

If the out of straightness of a hot rolled bar after cooling exceeds the permitted tolerance, cold straightening is applied and this causes residual stresses of a mechanical origin. Usually a bending moment is applied in the plane of the flange, which plastically deforms the bar. Once the bar is unloaded, a residual stress distribution arises there, generally having an asymmetrical pattern.

During straightening, thermal residual stresses already present in the bar influence the yielding of the most stressed points, where superposition of both effects exceeds the elastic limit.

113

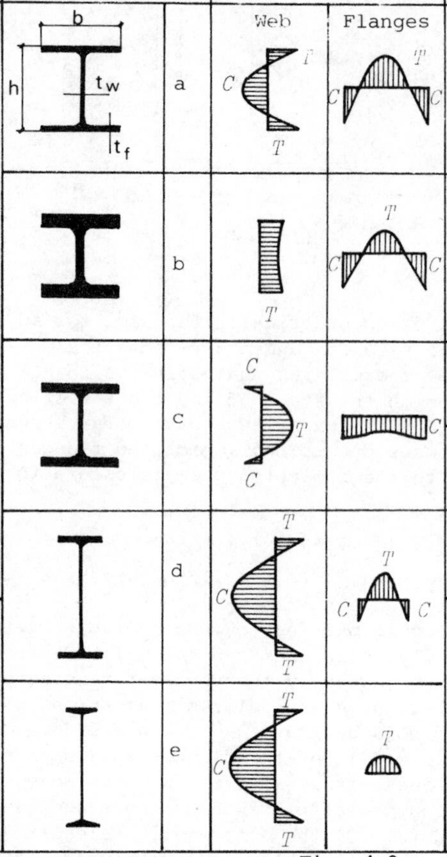

h/b	Profile	Residual stresses Web	Residual stresses Flanges	t_w/h	t_w/b	t_f/h	t_f/b
<1.2	a			0.032 to 0.040	0.032 to 0.040	0.045 to 0.061	0.045 to 0.080
<1.2	b			0.075 to 0.100	0.078 to 0.112	0.091 to 0.162	0.093 to 0.182
>1.2	c			0.062 to 0.068	0.068 to 0.073	0.104 to 0.114	0.113 to 0.121
>1.2	c			0.031 to 0.032	0.042 to 0.043	0.048 to 0.051	0.062 to 0.080
<1.7	d			0.030	0.046	0.051	0.077
>1.7	e			0.018 to 0.028	0.039 to 0.056	0.025 to 0.043	0.063 to 0.085

Fig. 4.9

The straightening procedure most commonly used in the steel industry consists of introducing the cold bars into a train of rolls which submits them to alternate bending (Fig. 4.10), a process useful in relieving residual stresses the values of which are considerably lower after straightening than before (Fig. 4.11).

An interesting comparison of the results of the main tests on residual stresses in hot rolled I bars can be done by means of the classification criterion proposed by Ketter and reported by Young |16| using the dimensionless parameter bt_w/ht_f, b and h being cross sectional width and depth, t_w and t_f web and flange thickness respectively. Such a classification has the advantage of separating I sections into two classes: beam sections ($bt_w/ht_f < 0.5$) and column sections ($bt_w/ht_f > 0.5$).

Fig. 4.12 shows the maximum values of residual stresses at flange edges (Fig. 4.12a) and at mid-web (Fig. 4.12b) derived from results of tests made in the UK (open circles) and Italy (asterisks),

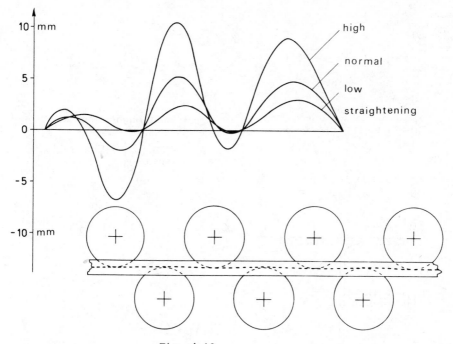

Fig. 4.10

together with others from the US, Australia, Japan and Belgium (full circles).

From this comparision it appears, synthetically, that residual stress intensity in hot rolled sections depends not only on the cross sectional shape but also on both cooling conditions and cold straightening procedures, which generally vary from country to country.

The existence of tensile residual stresses at mid-web (Fig. 4.12b) must be attributed to the US practice of cooling I sections as closely packed, while the same tensile residual stresses at flange ends (Fig. 4.12a) indicate the effect of straightening by rolling.

The problem of defining models of residual stresses has been studied by many authors in various countries |4-12| with different results including those shown in Fig. 4.13, namely the US model with a bi-triangular distribution in the flanges and a constant tension in the web, the British model with parabolic distribution both in the web and in the flanges and the Australian model with constant tension in the flanges, but triangularly variable in their thickness.

A model similar to the British one, adopted by CECM-ECCS |30|, is characterized by linear distribution at the ends and a parabolic one in the middle both of flanges and web (Fig. 4.14). Such a distribution is defined by the following values:

$\sigma_{rc,1}$ = compressive residual stress at flange edges
$\sigma_{rc,2}$ = compressive residual stress at mid-web
$\sigma_{rt,1}$ = tensile residual stress at flange/web joint

115

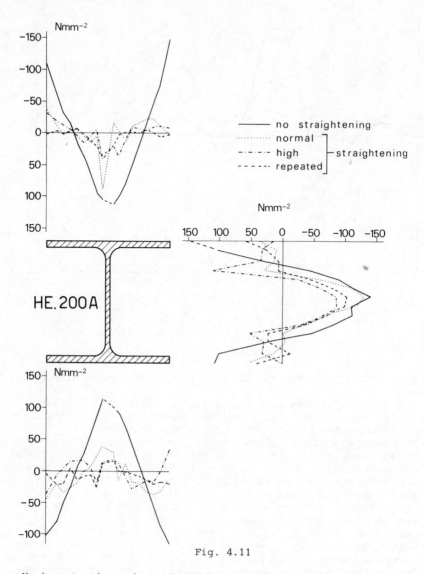

Fig. 4.11

Maximum testing values should be assumed for $\sigma_{rc,1}$ and $\sigma_{rc,2}$. Values for $\sigma_{rt,1}$ can be determined because the distribution overall is in equilibrium, as guaranteed by the following equation:

$$(3\chi + 8\beta)\sigma_{rt,1}^3 + (16\beta\sigma_{rc,2} + 1.5\chi\sigma_{rc,1})\sigma_{rt,1}^2$$

$$- (2\beta\sigma_{rc,1}^2 + 8\chi\sigma_{rc,2}^2)\ \sigma_{rt,1} = 6\beta\sigma_{rc,1}^2\sigma_{rc,2} + 4\chi\sigma_{rc,1}\sigma_{rc,2}^2$$

where:
$$\chi = t_w/t_f \quad \text{and} \quad \beta = \frac{b}{h - t_f}$$

Maximum values (in N mm^{-2}) characterizing the distribution proposed by Young (Fig. 4.13) are:

$$\sigma_{rc,1} = 165(1 - \frac{\chi}{2.4\beta}) \; ; \; \sigma_{rc,2} = 100(1.5 + \frac{\chi}{2.4\beta}) \; ; \; \sigma_{rt,1} = 100(0.7 + \frac{\chi}{2\beta})$$

The above values satisfy equilibrium for the hypothesis of a parabolic distribution in both flange and web, provided that yield stress is not exceeded.

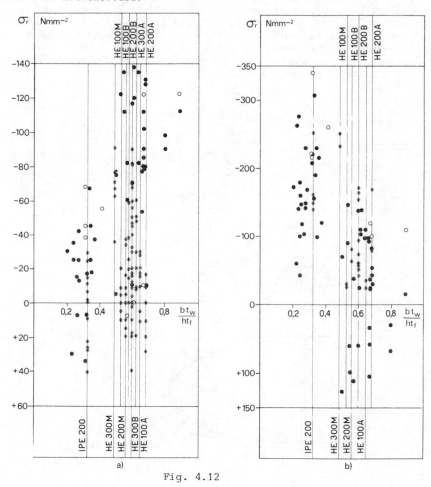

Fig. 4.12

Recent experimental researches to determine residual stresses in jumbo rolled sections have shown a great scatter across the flange thickness (Fig. 4.15). The corresponding maximum compressive stress $\sigma_{rc,1}$, given by Alpsten [21], can be interpreted by the following empirical formulae (values expressed in Nmm^{-2}):

Light shapes $\quad \sigma_{rc,1} = 180 \dfrac{bt_w}{ht_f} - 53$

Medium shapes $\quad \sigma_{rc,1} = 290 \dfrac{bt_w}{ht_f} - 58$

Thick shapes $\quad \sigma_{rc,1} = 450 \dfrac{bt_w}{ht_f} - 23$

In the case of hot rolled plates residual stress distribution due to cooling can still be assumed to be parabolic, with compression at the ends and tension at the centre (Fig. 4.16). As the distribution must obviously be in equilibrium, the parabolic assumption requires that compressive stresses (σ_{rc}) at the plate edges are twice the tensile stresses at mid-width (σ_{rt}). The magnitude of such stresses is mainly influenced by two parameters:

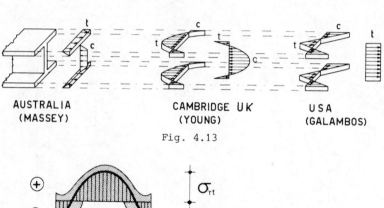

AUSTRALIA
(MASSEY)

CAMBRIDGE UK
(YOUNG)

USA
(GALAMBOS)

Fig. 4.13

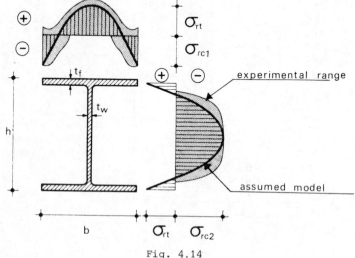

Fig. 4.14

Width to thickness ratio: b/t

Cross sectional area to perimeter ratio $r_{eff} = \dfrac{2A}{u} = 2bt/2(b + t)$

The latter determines the variable temperature distribution during cooling.

Alpsten's researches on the relationship between σ_{rc} and parameters r_{eff} (expressed in mm) and b/t might suggest the following empirical formula (valid for $r_{eff} \geq 6.5$ mm)

$$\sigma_{rc} = 0.18(\frac{r_{eff}}{2})^{3/2}(\frac{b}{t}) \quad N \ mm^{-2}$$

Another less accurate, but still conservative formula for each value of r_{eff} and valid for $b < 28t$ is:

$$\sigma_{rc} = 220(1 - \frac{b}{37t}) \quad N \ mm^{-2}$$

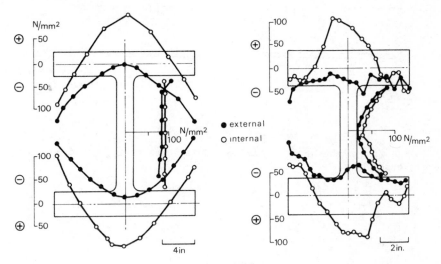

● external
○ internal

Fig. 4.15

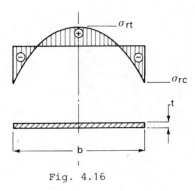

Fig. 4.16

119

Whenever plates are obtained by flame cutting, the concentrated thermal effect at the cut end sensibly modifies the previous distibution of residual stresses due to cooling.

In fact, tensile residual stresses are produced in the fibres close to the heated area, where they usually reach yield stress f_y. In the remaining parts, equilibrium imposes compressive residual stresses, distributed as shown in the models of Fig. 4.17 for one and two cuts respectively.

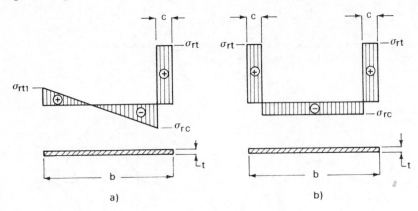

Fig. 4.17

In the first case (Fig. 4.17a) residual stresses $\sigma_{rt,1}$ and σ_{rc} can be expressed by:

$$\sigma_{rt,1} = \sigma_{rt} \frac{c(2b + c)}{(b - c)^2}$$

$$\sigma_{rc} = \sigma_{rt} \frac{c(4b - c)}{(b - c)^2}$$

whereas in the second case (Fig. 4.17b) the uniform compressive stress is expressed by:

$$\sigma_{rc} = \sigma_{rt} \frac{2c}{b - 2c}$$

where σ_{rt} can be considered as coincident with yield stress f_y and constant in the part having width c.

For the values of c (in mm) Young and Dwight $|10|$ suggest the following empirical expression, which is function of thickness t (in mm) and of yield stress f_y (in N mm^{-2}):

$$c = \frac{235}{f_y} 4.6 \sqrt{t}$$

Now consider the second aspect of structural imperfections, the variation of mechanical characteristics over the cross section. The

120

1 mean values in the flanges
2 mean values in the web
3 web/flanges ratio

	HE 100 A			HE 100 B			HE 100 M		
	1	2	3	1	2	3	1	2	3
f_y	300.2	339.4	0.88	318.4	353.8	0.89	253.5	283.3	0.89
f_t	446.6	459.3	0.97	459.0	488.0	0.94	433.5	470.7	0.92
ε_t	34.1	30.8	1.10	35.7	29.3	1.21	43.7	34.2	1.28
Vickers	140	124	1.12	149	159	0.93	125	122	1.02
Charpy	14.8	16.7	0.88	18.7	15.6	1.20	14.3	19.4	0.73

	HE 200 A			HE 200 B			HE 200 M		
	1	2	3	1	2	3	1	2	3
f_y	267.8	304.3	0.88	283.6	343.0	0.82	244.9	329.6	0.74
f_t	444.7	454.1	0.97	472.3	475.3	0.89	443.3	481.0	0.92
ε_t	39.9	34.7	1.15	41.0	33.5	1.22	43.4	37.4	1.16
Vickers	124	129	0.96	142	149	0.95	127	149	0.84
Charpy	15.2	15.9	0.95	14.6	11.4	1.27	12.0	8.5	1.41

	HE 300 A			HE 300 B			HE 300 M		
	1	2	3	1	2	3	1	2	3
f_y	270.6	294.5	0.91	246.9	265.6	0.92	230.8	246.9	0.93
f_t	447.1	447.1	1.00	446.8	434.5	1.02	466.7	473.0	0.99
ε_t	41.2	33.3	1.23	44.7	41.6	1.07	44.4	45.9	0.96
Vickers	132	134	0.98	131	138	0.95	133	132	1.01
Charpy	11.7	9.2	1.27	16.2	17.2	0.94	10.5	10.2	1.03

Fig. 4.18

main properties are:

Yield stress f_y
Tensile strength f_t
Elongation at rupture ε_t
Hardness (Vickers)
Toughness (Charpy)

These all have a non homogeneous distribution, a behaviour which has been emphasized by several tests. Some results are illustrated in Fig. 4.18 for wide flanged I sections of the European series HE |12|. The scatter among mean values of the various characteristics in flanges and in web is significant. A logical interconnection between the different magnitudes is generally noticeable;toughness and ultimate elongation are lower whenever strength,and therefore hardness, are greater. This is confirmed by micrographic examination,which shows

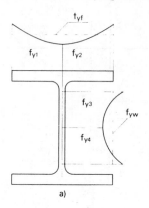

a)

f	HEA 100	HEB 100	HEM 100	HEA 200	HEB 200	HEM 200	HEA 300	HEB 300	HEM 300	IPE 200
1) f_{y1}/f_{yf}	1,06	1,07	1,07	1,04	1,09	1,06	1,07	1,15	1,80	1,02
2) f_{y2}/f_{yf}	0,90	0,95	0,95	0,99	0,95	0,94	1,10	0,97	0,99	0,97
3) f_{y3}/f_{yw}	0,99	1,02	1,02	1,23	1,31	1,36	1,08	1,08	1,00	0,95
4) f_{y4}/f_{yw}	0,96	0,99	0,99	0,97	0,85	0,82	0,98	1,00	1,05	1,03
5) f_{yf}	30,02	31,84	25,35	26,78	28,36	24,49	27,06	24,69	23;08	30,20
6) f_{yw}	33,94	35,38	28,33	30,42	34,30	32,96	29,45	26,56	24,69	36,80
7) f_{yf}/f_{yw}	0,88	0,90	0,89	0,88	0,82	0,74	0,92	0,93	0,93	0,82
8) f_{ym}	31,98	33,61	26,84	28,60	31,33	28,72	28,25	25,62	23,88	33,50
9) f_{yc}	30,98	32,65	25,95	27,49	29,73	26,30	27,66	25,13	23,40	32,87

b)

Fig. 4.19

differences in the size of metal grains related to the thickness of
the individual parts of the section. Grains are larger in the thicker
parts, i.e. the flanges, than in the web and consequently yield stress
is generally lower in the flanges than in the web.

Similar differences were found among I sections having the same
depth but different flange and web thicknesses. The more slender
sections (HEA) have a finer crystalline structure, but, due to a higher
reduction ratio in rolling, have more marked distribution bands. On
the other hand, slower cooling of the thick sections (HEM) entails
grains being bigger. The best compromise is found in sections in the
normal series (HEB), having medium size grains and almost non-existant
rolling bands. This corresponds also to the toughness values which are
usually higher for the HEB sections.

Variable yield stress distribution over the cross section can be
represented by the qualitative pattern of Fig. 4.19a. Table (Fig.
4.19b) contains some test results of the main parameters of yield
stress distribution in European I sections (HE and IPE).

4.2.2 *Cold formed sections*

Thin walled profiles can be obtained by cold forming of cold rolled
sheets or of hot rolled plates.

Cold rolling produces mechanical residual stresses which vary
across the sheet thickness (Fig. 4.20). In fact, the outer fibres tend
to elongate, while the centre tends to remain undeformed but some
deformation between surface and centre must take place through the
thickness. The internal fibres resist the elongation of the external
ones, which in turn try to stretch the internal ones. The result is a
residual stress distribution with compression on the surface and
tension within the thickness.

In hot rolled plates, residual stresses are produced by cooling
and their distribution is similar to that of cold rolled sheets. In
fact, the areas which cool off first (surface) undergo compressive
residual stresses, whereas the last cooled areas (centre) are in
tension.

Thin walled sections, whether formed from cold or hot rolled sheets,
have residual stresses distributed through the thickness of the
individual parts of which they are composed. Within the thickness,

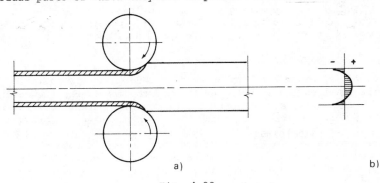

a)   b)

Fig. 4.20

the resultant of such a distribution is nil and its effect
on the overall behaviour of a thin walled bar can therefore be
neglected in most cases. It must be considered just as a local effect
which can increase the danger of local buckling in the thin wall.

In conclusion, from the point of view of residual stresses, thin
walled cold formed sections are more accommodating compared with
similar hot rolled sections.

The second structural imperfection, the variation of mechanical
characteristics over the cross section, seems to be particularly
important for these profiles, due to the strain hardening effect
which the material undergoes during cold forming. This procedure, in
fact, produces an increase in the elastic limit of the material
compared to its original value, and its increment is proportional to
the severity of folding, expressed as the ratio between fillet radius
r and sheet thickness t (Fig. 4.21).

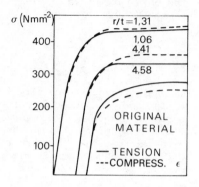
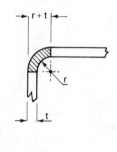

Fig. 4.21

Increase in strength due to strain hardening produces, as an
immediate negative consequence, a lowering of toughness. Tests
made in the US have shown up to 50% increase in material
elastic limit in the folded area.

In the case of profiles obtained by more than one folding operation
(Fig. 4.22), a highly variable distribution of yield stress f_y (full
lines) and of tensile strength f_t (dashed lines) arises over the cross
section. It is characterized by peaks corresponding to the folds and
minimum values at the centre of the individual parts of the profile.

4.2.3 *Welded sections*

Structural elements built up by welding undergo a very variable
thermal treatment, due to the heating of the zones near the weld where
the fibres tend to elongate but are hindered by parent material remote
from the weld and therefore colder. As a consequence of this constraint,
residual stresses arise. They reach the yield strength of the material
at the temperature in the melted zone and in its immediate
surroundings, causing plastic contraction. Once cooling has taken
place, due to the permanent contraction, tensile residual stresses are
produced in the weld and in the fibres near it, up to the maximum

124

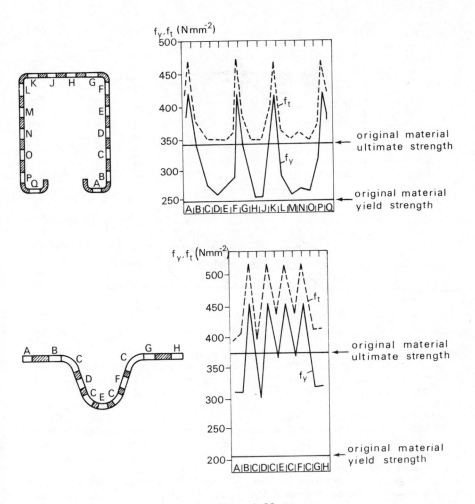

Fig. 4.22

value of material yield strength. To restore equilibrium, compressive stresses are produced in the farthest points of the parent material.

Consider the residual stress distribution in the case of a butt weld joining two plates completely free to more in their own plane.

Tranverse stresses $\sigma_{r\perp}$ will be induced perpendicular to the weld, together with longitudinal stresses $\sigma_{r\parallel}$ parallel to the joint axis (Figs 4.23a,b).

While the transverse stresses generally do not have very high values (of the order of 100 Nmm^{-2}), the longitudinal stresses, due to the severe constraint conditions operating in the longitudinal direction, reach very high values, and may attain material yielding. The longitudinal stresses are more important with regard to the structural behaviour of welded members because they usually act in the

125

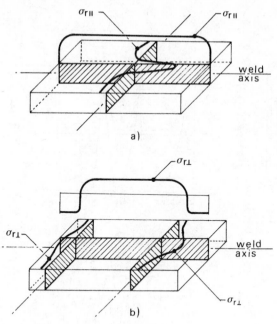

a)

b)

Fig. 4.23

same direction as the stresses caused by applied loads.

The amount and distribution of residual stresses in welded joints depend on many factors, among the most important of which are welding procedure, number of welding passes, thickness and geometry of welded pieces.

Although it also has a thermal origin, residual stress formation in welded profiles follows a phenomenon which is quite different from that of hot rolled profiles. In the case of welded shapes, residual stresses and plastic deformations are the consequence of locally concentrated heating, whereas with hot rolled sections, the thermal state is initially uniform throughout immediately after rolling and only becames variable during cooling, when residual stresses are induced.

The relationship between cross sectional shape and residual stress distribution, which generally exists in the case of rolled sections, is not therefore so well defined for welded members since it is also influenced by the welding process and the succession of passes in relation to the thickness of the joined pieces.

The influence of the welding process depends most of all on the amount of heat required and the depth of penetration. The greater both are, the larger are the zones of the cross section where tensile stresses reach yield and consequently, for equilibrium, also compressive stresses are higher.

It must also be pointed out that in all steel plates to be joined there are, before welding, residual stresses due to rolling and/or flame cutting.

A characteristic residual stress distribution for welded I sections is shown in Fig. 4.24. Tensile stresses are maximum inside the weld, where they reach the yield limit of the weld material. Such a distribution can be quantified by means of the values in the table. They are the result of many researches carried out in the US by the Column Research Council (CRC) |5| and in Europe by the European Convention for Constructional Steelwork (ECCS), on welded profiles made of steel grade Fe 360 |30|.

A quantitative analysis of the effect of the various factors influencing residual stress distribution due to welding can be carried out on a joint consisting of two plates welded at their middle (Fig. 4.25). Although it is a very simple example, it is certainly significant, as welded sections generally consist of various plates.

The distribution model can be assumed as composed of two blocks having constant value. In the middle across the weld gap 2 c a tensile residual stress σ_{rt} equal to yield stress is reached, whereas in the lateral zones compensating compressive residual stresses σ_{rc} arise. The half width c of the tension area for a single pass of continuous welding has been determined by Young |20| according to the following formula:

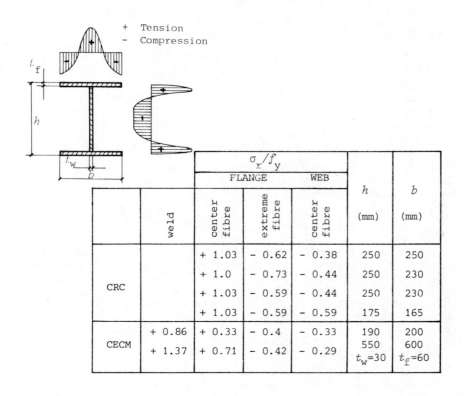

		weld	FLANGE center fibre	FLANGE extreme fibre	WEB center fibre	h (mm)	b (mm)
CRC			+ 1.03	− 0.62	− 0.38	250	250
			+ 1.0	− 0.73	− 0.44	250	230
			+ 1.03	− 0.59	− 0.44	250	230
			+ 1.03	− 0.59	− 0.59	175	165
CECM		+ 0.86	+ 0.33	− 0.4	− 0.33	190	200
		+ 1.37	+ 0.71	− 0.42	− 0.29	550 t_w=30	600 t_f=60

Fig. 4.24

127

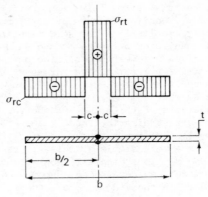

Fig. 4.25

$$c = 50 \, \alpha \, \frac{A}{\Sigma t} \frac{235}{f_y} \quad ; \quad \sigma_{rc} = \sigma_{rt} \frac{2c}{b - 2c}$$

where:

α is the process efficiency factor
A is cross sectional area of added metal
Σt is the sum of the plate thicknesses meeting at the weld.

Process efficiency factor values depend on the adopted welding process. Some suggested values are:

Submerged arc $\alpha = 0.90$
Cored wire CO_2 $\alpha = 0.85$
Manual $\alpha = 0.80$
Fusiarc $\alpha = 0.75$
MIG: spray $\alpha = 0.62$
    dip $\alpha = 0.42$

In the case of discontinuous welding, Kamtakar [21] suggested a modification to the above expression for c, in terms of the length of the welding stretch L and the interval a between adjacent stretches:

$$c' = cL/(L + a)$$

In the case of a weld deposited between two previously flame cut plates, the corresponding tensile zones are not additive, as welding heating tends to relieve the stresses due to flame cutting. The following empirical formula by Young and Dwight [10] relates the two effects and provides a reliable assessment of overall width of the area in tension:

$$c = \left\{ \frac{235}{f_y} \, (4.6 \, \sqrt{t})^4 + (50 \, \frac{A}{\Sigma t})^4 \right\}^{1/4}$$

A relation of the same type may be applied in order to take account of the superposition of many welding passes of the same size. If each pass produces a width equal to c, n equal passes produce a width equal

128

to:

$$c_n = c\, n^{1/4}$$

The welded flange to web connection in an I profile is an example of two fillet welds mutually separated by the web thickness. If they are laid simultaneously, the general expression for c is still valid, A being equal to the sum of the cross sectional areas of both welds. If, on the contrary, the welds are laid successively, the expression for c is:

$$c_2 = 2^{1/4}\, c \;\simeq\; 1.2c$$

In the case of two parallel welds set at a distance a, if $a > 2c$, one can still consider two separate tension blocks each having a half width c. If $a \leq 2c$, there is only tension block having half width equal to:

$$c_2 = c + 0.5a$$

The above results are based on tests on rather slender shapes.

In the case of thick elements, such as welded jumbo profiles, Alpsten and Tall |7,9| showed that welding residual stress intensity tends to increase as the piece dimensions increase and that furthermore it is no longer possible to neglect residual stress variation across the thickness (Fig. 4.26).

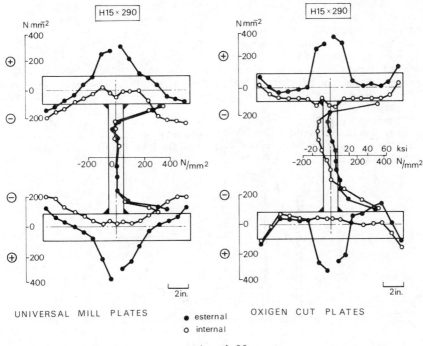

UNIVERSAL MILL PLATES   ● esternal   OXIGEN CUT PLATES
                         ○ internal

Fig. 4.26

In welded profiles, the variation of mechanical properties over the cross section is not a problem requiring any especial attention. Welding, in fact, produces a local effect due to metallurgical modification caused by the thermal treatment in the weld affected zones. This fact must be given particular attention with reference to brittle fracture phenomena (see 5.1.3). On the other hand, overall section behaviour depends on the mechanical properties of the various types of steel plate used to compose the section itself. Mechanical properties, mainly yield stress and tensile strength, vary with thickness; therefore their distribution within the cross section depends on the thicknesses of the steel plates of which the section is made.

4.3 GEOMETRICAL IMPERFECTIONS

The notion of geometrical imperfections |23-26| includes all the shape variations in structural members with reference to their ideal geometry. They may be transverse when related to the cross section, or longitudinal whenever they concern the bar axis.

Progressive deterioration of shaping rolls in the mill may cause thickness variations in all the parts of which a rolled shape consists.

In the case of welded profiles, dimensional deviation can depend either on thickness variations of the steel plate components or on overall dimensional (depth, width) variations due to the manufacturing procedure.

Further geometrical imperfections affecting the cross section consist of out of squareness of sections and out of flatness of sectional elements.

Dimensional variations of about 5000 I sections (HE) |23| rolled at many European steel mills are shown in Fig. 4.27. The sections examined were rolled in accordance with the appropriate Euronorm and were predominantly of light (HEA) and medium (HEB) size; some specimens of the heavier (HEM) range with flange thickness in excess of 25 mm were also included.

It can be seen that variations in flange width and in section depth are very small compared to thickness variations. It is also apparent that flanges tend to be thinner and webs tend to be thicker than their nominal values.

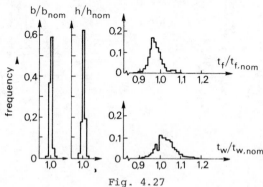

Fig. 4.27

Histograms of cross sectional properties (area A, moment of inertia I, section modulus W and plastic modulus Z) for the same 5000 sections are shown in Fig. 4.28. Their patterns are all very similar; properties about the strong axis are closer to nominal values than the those about the weak axis.

The above shows that the most important geometrical imperfection in the transverse direction is the variation in flange thickness.

Dimensional variations of plates were emphasized by a series of measurements on steel plates having thicknesses varying between 1/4 and 2 in.|26|.

Fig. 4.29 shows the coefficients of variation and standard deviations as a function of nominal thickness t_{nom}: s is practically constant up to 25 mm thickness.

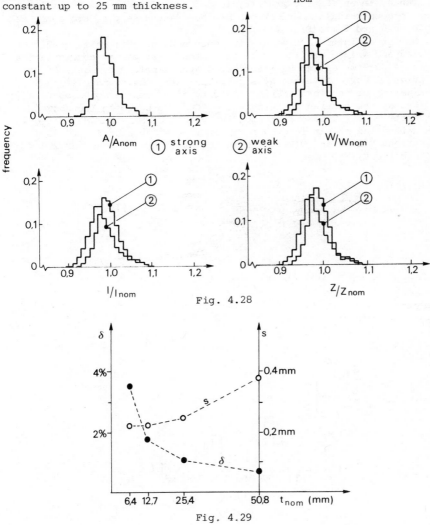

Fig. 4.28

Fig. 4.29

131

Longitudinal geometrical imperfections essentially consist of deviation of the axis from its ideal perfectly straight position, caused by the manufacturing process, and added to due to lack of precision in assembly.

The effects of such imperfections on the load bearing capacity of bars can be defined by two parameters (Fig. 4.30):

Load eccentricity e at bar ends

Mid-span deflection v_o characterizing transverse displacement of a bar affected by initial curvature

Load eccentricity can be due to variations in the shape of the cross section causing displacement of the centre of gravity. For instance, in an IPE 160 profile, loaded at mid-web, eccentricities due to geometrical imperfections of the transverse type are comprised between 0 and 2 mm (Fig. 4.31).

As far as initial curvature is concerned, several measurements on rolled and welded profiles have indicated a variety of configurations (Fig. 4.32), which can be interpreted approximately by means of a sinusoidal curve without any significant error being made in the evaluation of the load bearing capacity of a compression member |24|.

Researches carried out within the ECCS are based on the above assumption |25|. Some corresponding values of mid-span deflections as a function of bar length are given in the Table of Fig. 4.33 (sections DIE 20 and DIR 20 are analogous to HE 200 A and HE 200 M respectively).

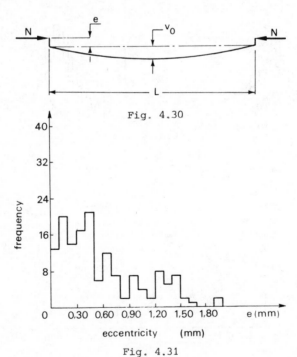

Fig. 4.30

Fig. 4.31

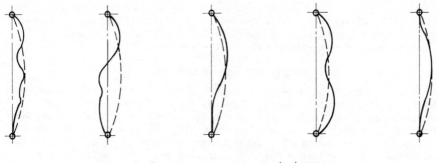

- - - assumed shape   ——— real shape

Fig. 4.32

Profile	L/v_o		
	Minimum	Average	Maximum
IPE 160	2700	4400	7200
IPE 200	1390	3800	10000
DIE 20	1690	3700	7600
DIR 20	2260	5800	9500
Hollow d/t = 18	3050	7000	33000
Hollow d/t = 11	550	2000	3700

Fig. 4.33

Statistical distribution of ratio v_o/L for IPE 160 is shown in Fig. 4.34. The mean value v_{om} is 0.0085 L and the standard

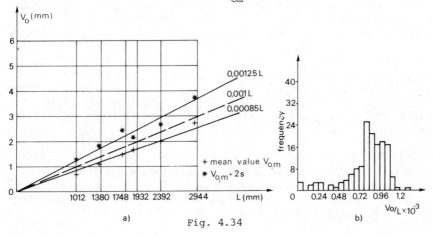

a)   Fig. 4.34   b)

133

deviation s is $0.00020\,L$. Therefore, the assumption of a conventional initial displacement of $L/1000$ seems to be justified and it has now been universally adopted to characterize longitudinal geometrical imperfection of all structural steel elements.

4.4 QUALITY AND LABORATORY CONTROLS

Chemical, physical and mechanical properties of steel are determined by means of laboratory tests. Some of them are conventional and carried out in accordance with the requirements of standards and codes dealing with the acceptance of structural steel (e.g. tensile proof test, toughness, bending). Other tests are made as necessary to determine in detail some special characteristic with which some more particular problem is concerned (e.g. measurement of residual stresses).

Depending on load application, laboratory tests can be classified into two groups |27|:

Static tests, whenever the load application gradient versus time is so small as to be negligible

Dynamic tests, whenever the load is applied with a high gradient (impact load) or according to a variable law which is periodically repeated (cyclic load for fatigue tests).

Some of the main laboratory tests normally carried out to check structural steel quality are briefly discussed in the following sections.

4.4.1 *Chemical analysis*

The main purpose of this is to limit carbon percentage and to reduce to within very restricted limits sulphur and phosphorus impurities. This control is particularly important with reference to material welding properties.

4.4.2 *Macrographic examination*

This is performed by means of the so called Bauman print and provides an index of the degree of de-oxidation, to avoid the use of effervescent steel in welded constructions.

4.4.3 *Micrographic examination*

Crystalline structure and grain dimensions are examined through a microscope for metals. Material micrographic composition is strictly related to its mechanical properties (strength, toughness, ductility, etc.), and to their modification due to thermal treatment (quenching, welding, etc.).

4.4.4 *Complete tensile test*

This is certainly the best known and most commonly used test. A mono-axial loading condition is applied to a standard specimen, as shown in Fig. 4.35. Load-elongation ($F - \Delta L$) diagrams are obtained from which, knowing the testing equipment data and geometry of the specimen, it is possible to deduce stress-strain diagrams, by dividing force F by nominal area A_o of specimen and elongation ΔL by reference length L_o.

The stress-strain diagram for structural steel has the typical form shown in Fig. 4.36. For the first part OP, where $0 < \sigma < f_o$ where the law is linear, is defined by the normal elastic modulus (Young's modulus): $E = \sigma/\varepsilon$.

This perfectly elastic behaviour has its upper limit at P (stress f_o) called the limit of proportionality.

In the second part PE where $f_o < \sigma < f_e$ the behaviour is still elastic, but not linear, and is defined by the instantaneous modulus:

$$E_t = d\sigma/d\varepsilon$$

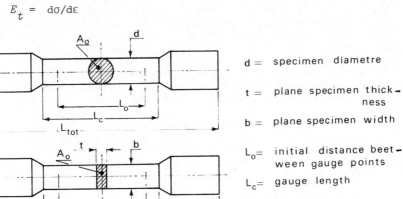

$d =$	specimen diametre
$t =$	plane specimen thickness
$b =$	plane specimen width
$L_o=$	initial distance beetween gauge points
$L_c=$	gauge length
$L_{tot}=$	total length
$A_o=$	initial area
$A_u=$	minimum area after failure
$L_u=$	final distance beetween gauge points

$$L_o = 5,65\sqrt{A_o}$$

Fig. 4.35

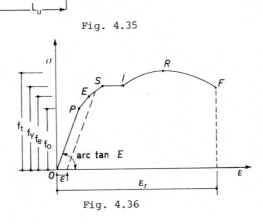

Fig. 4.36

135

called the tangent modulus. The upper limit f_e of this interval is called elastic limit stress.

Continuing with the test load to the interval ES where $f_e < \sigma < f_y$ the behaviour is no longer elastic and if the specimen is unloaded, a permanent deformation ε_r remains. The unloading diagram is still straight and parallel to OP . In practice it is difficult to find the two limits f_o and f_e and generally they can be considered as coincident.

From point S (stress f_y), mild steel undergoes yielding which is evident through a spontaneous and large elongation without increase in stress, represented by the horizontal line SI.

Once yielding is complete (point I in Fig. 4.36), an increase in strength takes place in the material, due to strain hardening, corresponding to the interval IR; at R the stress f_t is known as the ultimate or tensile strength of the steel. After this point the diagram falls away and stops at elongation value ε_t, called the strain at rupture, corresponding to the specimen breaking.

In the case of high strength steels the σ-ε diagram (Fig. 4.37) does not contain the horizontal plateau corresponding to yielding; instead of the yield stress an elastic limit is conventionally assumed at that stress which corresponds, after unloading, to a residual deformation ε_r equal to 0.1% $(f_{0.1})$ or to 0.2% $(f_{0.2})$.

For all kinds of steel Young's modulus always lies between 200 and 210 N mm^{-2} and it can, therefore, be considered constant in practice.

For ordinary mild steel the yield stress varies from 240 to 360 N mm^{-2}.

For high strength steels ultimate or tensile strengths can be from 500 to 950 N mm^{-2}, with corresponding yield stresses from 400 to 700 N mm^{-2}. The amount of elongation at rupture is an important index of material ductility. It varies from 21 to 26% in the case of mild steel and from 14 to 22% for high strength steels. Generally speaking as steel strength grows ductility decreases.

The complete tensile test can be performed at different temperatures, in order to assess the variability of mechanical properties with temperature |29|. These data are important in respect of the behaviour of steel structures in situations of extreme temperatures and the problem of their fire resistance.

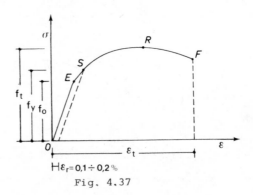

Fig. 4.37

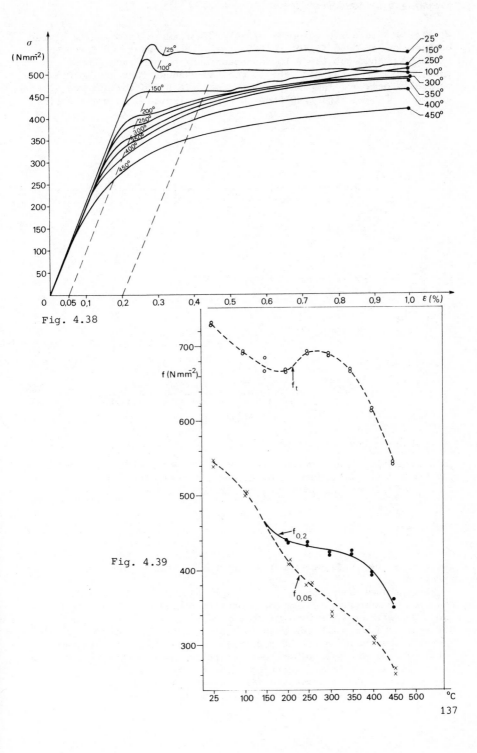

Fig. 4.38

Fig. 4.39

Fig. 4.38 shows aι. example of σ-ε diagrams as temperature varies between 25° and 450°C, for Fe 600 steel. From about 200°C the yielding phenomenon tends to disappear and the σ-ε law becomes continuous. For the same steel, Fig. 4.39 shows that strength generally decay   temperature increases. A strength increase appears around 250°C and Fig. 4.40 shows a similar increase in the elastic modulus.

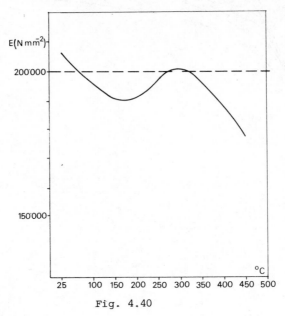

Fig. 4.40

4.4.5 *Stub column test*

This test, which was imported from US |28|, has a great interest for the prediction of structural behaviour of steel elements. The stub must have a length L, according to these formulae:

$$L \leq \begin{cases} 20 \ i_{min} \\ 5h \end{cases} \qquad L \geq \begin{cases} 2h + 250 \ mm \\ 3h \end{cases}$$

where h is the cross sectional depth and i_{min} the minimum radius of gyration.

The specimen undergoes overall compression (Fig. 4.41) and, by means of an adequate system for measuring shortening, the test provides the average relationship between mean applied stress and deformation from which the mean tangent modulus E_{tm} referred to the overall cross section can be deduced. The stress-strain diagram obtained from a stub column test generally has a lower limit of proportionality than that corresponding to the tension test on specimens from web and flanges (Fig. 4.42). This is due to the presence of structural imperfections (residual stresses and variable

138

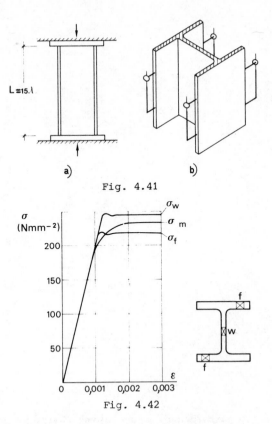

Fig. 4.41

Fig. 4.42

distribution of yield stress over the cross section), which lower the compressive strength (as pointed out in 4.2).

4.4.6 *Hardness test*

This test is performed with special equipment which mainly differs in the penetration system (Brinell, Vickers, Rockwell). It is based on measuring the penetration print diameter of a steel sphere after a load F is applied to the specimen for a given time. Brinell's hardness is evaluated through the formula:

$$H_B = \frac{2F}{nd\{d - \sqrt{(d^2 - d_o^2)}\}} \qquad |N \ mm^{-2}|$$

where d is the diameter of the sphere and d_o is the print diameter.

Hardness values are an index of superficial resistance to penetration and they are roughly proportional to mechanical strength.

4.4.7 *Toughness test*

This control gives a measure of the resistance of the material against brittle fracture.

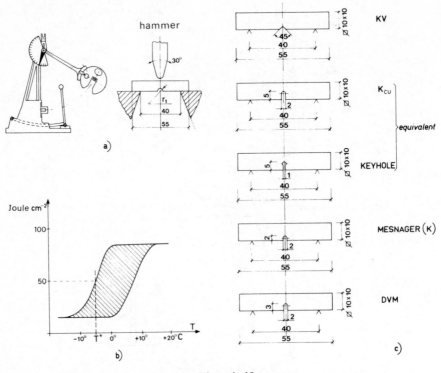

Fig. 4.43

It is generally performed by means of the Charpy pendulum on a specimen of standard shape (Fig.4.43).

A special hammer dropped in an arc from height h_o produces the rupture of the specimen due to the impact load and then rises again to height h. The difference $(h_o - h)$ is proportional to the energy expended in rupturing the specimen which, referred to the rupture area, gives a conventional value of toughness (expressed in Nm cm^{-2} or in joules cm^{-2}).

Toughness usually decreases as mechanical strength increases and it is strongly influenced by temperature.

The diagram of toughness against temperature (Fig. 4.43b) shows that there is a so-called transition temperature T below which toughness decays to such low values as are considered unacceptable for structural use. As low toughness means tendency to brittleness, it is obviously important to use steel having a relatively low transition temperature, guaranteeing that under service conditions toughness values should be in the upper part of the diagram.

It is important to point out that toughness values depend on specimen shape (Fig. 4.43c) which varies according to the different standard methods of test (Charpy V, Mesnager, D V M, etc.), as brittle rupture is influenced by them.

140

Therefore, neither toughness nor transition temperature values are objective experimental data; they represent only a term of comparison and conventional reference.

Transition temperature strictly depends on the chemical composition of the steel. By operating on carbon and manganese constraints, transition temperatures as low as -35 °C can be obtained.

4.4.8 *Bend test*

This test is performed by submitting the specimen to plastic deformation by bending through 180°, to evaluate the capability of the material to undergo large deformations at room temperature.

4.4.9 *Fatigue test*

It has been known since prehistoric times that a piece of wood or metal could be broken if it were bent repeatedly in opposite directions, provided that the materials could be bent sufficiently.

However, the fatigue phenomenon was recognized with a certain degree of surprise and it was first studied just about a century ago. It then became very important, mostly in the field of mechanical and transportation engineering. Only recently has interest in fatigue phenomena entered the field of civil engineering.

Fatigue tests emphasize the lowering of the mechanical strength of a material after it has undergone a cycle of stresses having an oscillating intensity in time.

These tests can be performed in various ways, e.g. by rotating a specimen around its own axis at constant speed, while any load remains vertical (Fig. 4.44a). The bending loads at each point varies

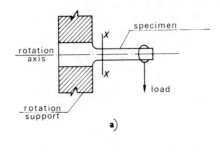

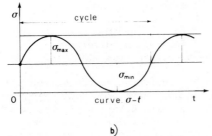

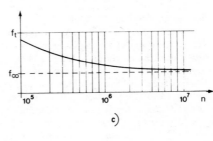

Fig. 4.44

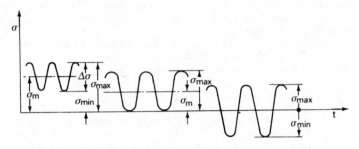

Fig. 4.45

sinusoidally with time t so that the extreme fibres of the section are subjected to stresses varying from a maximum tension σ_{max} and a maximum compression σ_{min} (Fig. 4.44b).

If the corresponding values of $f_t = \sigma_{max}$ and of n are plotted on a semi-logarithmic diagram (n being the number of cycles producing rupture), the curve of Fig.4.44c is obtained. It shows that for a given value of f_∞ it is possible for the specimen to undergo $n = \infty$ cycles without rupture taking place.

The number of cycles for which the curve of f_t against n is asymptotic can vary from a few million to many million, depending only on material quality.

The first systematic studies on fatigue were carried out by Wöhler. The results he obtained, which are still fully valid, can be outlined by the following laws (Fig. 4.45):

(a) Whenever a material undergoes repeated loading, rupture can take place at stresses below the static strength of the material

(b) For given maximum stress the number of cycles necessary to attain rupture is greater as the range of stress $\Delta\sigma$ is smaller

(c) Stress can oscillate indefinitely without producing rupture between zero and a given maximum value called initial strength

(d) Stress can alternate indefinitely without producing rupture beween two given equal and opposite limits, having an overall value below the initial strength; this is called alternating force strength.

4.4.10 Residual stresses tests

The experimental methods for measuring residual stresses in structural steel elements can be divided into sectioning methods and non-destructive methods.

Sectioning methods are based on dividing the specimen into pieces, for which strains (mechanical methods) (Fig.4.46) or displacements (deflection methods) (Fig.4.47) are measured, and from which the internal stress state can be deduced.

Kalatotsky's experiments (1888) can be considered as the starting point of such methods. Since then many authors have proposed other methods based on the sectioning technique, which consists of cutting the specimen into strips and releasing residual stresses. The new equilibrium condition corresponds to a shape modification and, as such

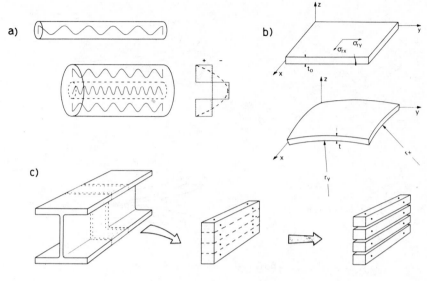

Fig. 4.46

unloading processes are essentially elastic, it is therefore possible using Hooke's law, to use a deformation measurement in order to deduce the stress state which caused them.

In 1911 Bauer and Heyn developed a procedure for measuring longitudinal residual stresses in a cylinder by considering it as made of a central spring in compression and a perimetral one in tension (Fig.4.46a). Once the central spring of the cylinder is removed, it will stretch by an amount which is directly proportional to the force which the perimetral spring imposed on it. Applying successive material removals starting from the centre, the authors proposed the following formula:

$$\varepsilon_r = E \ (A \ \frac{d\varepsilon}{dA} - \varepsilon)$$

which expresses longitudinal stresses as a function of axial strains (ε), as well as of the removed areas (A) and those remaining (dA). This method, which neglected transverse and radial residual stresses, can entail errors of about 30%.

In order to take account of the complete stress state, Mesnager proposed a similar method (1919), later (1927) modified by Sachs. Such a method, which is still limited to cylindrical bodies, assumes stress distribution to be radial-symmetrical. By means of experimental operations, which are perfectly similar to the previous one, and by measuring longitudinal deformations ε_1 and tangential ones ε_2 after each removal of internal parts, one attains the complete stress state consisting of longitudinal ($\sigma_{r,1}$), tangential ($\sigma_{r,2}$) and radial ($\sigma_{r,3}$) residual stresses, as functions of the initial area (A_o) of the cylinder and the area (A) of the removed cylindrical portion:

143

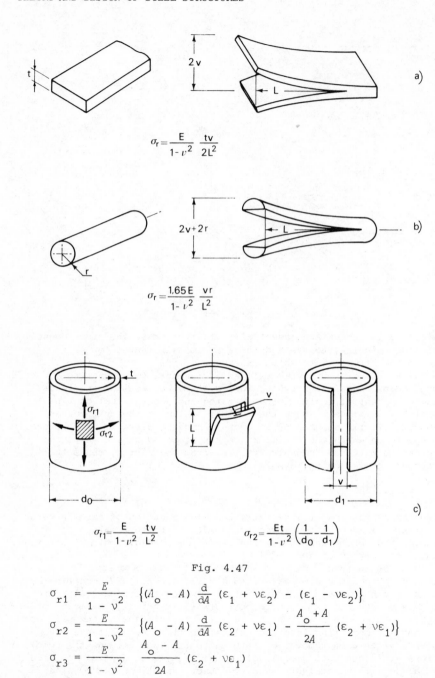

$$\sigma_r = \frac{E}{1-\nu^2} \frac{tv}{2L^2}$$

$$\sigma_r = \frac{1.65\,E}{1-\nu^2} \frac{vr}{L^2}$$

$$\sigma_{r1} = \frac{E}{1-\nu^2} \frac{tv}{L^2} \qquad \sigma_{r2} = \frac{Et}{1-\nu^2}\left(\frac{1}{d_0} - \frac{1}{d_1}\right)$$

Fig. 4.47

$$\sigma_{r1} = \frac{E}{1-\nu^2}\left\{(A_o - A)\frac{d}{dA}(\varepsilon_1 + \nu\varepsilon_2) - (\varepsilon_1 - \nu\varepsilon_2)\right\}$$

$$\sigma_{r2} = \frac{E}{1-\nu^2}\left\{(A_o - A)\frac{d}{dA}(\varepsilon_2 + \nu\varepsilon_1) - \frac{A_o + A}{2A}(\varepsilon_2 + \nu\varepsilon_1)\right\}$$

$$\sigma_{r3} = \frac{E}{1-\nu^2}\,\frac{A_o - A}{2A}(\varepsilon_2 + \nu\varepsilon_1)$$

144

where ν is Poisson's ratio and E is Young's modulus.

Another method, used to determine bi-axial residual stresses in steel sheets, was developed by Trenting and Read (1951). The experimental operation consists in reducing the thickness of a specimen glued to a rigid plane (Fig.4.46b), by successively setting thin layers free, measuring the longitudinal (r_x) and transverse (r_y) curvatures and introducing them into the following parameters:

$$k_x = \frac{1}{r_x} + \frac{\nu}{r_y} \qquad ; \qquad k_y = \frac{1}{r_y} + \frac{\nu}{r_x}$$

The stress state is then expressed through the following relation-ship:

$$\sigma_{rx} = -\frac{E}{6(1-\nu^2)}\left\{ (t_o + t)^2 \frac{dk_x}{dt} + 4(t_o + t)k_x + 2\int_{t_o}^{t} k_x dt \right\}$$

$$\sigma_{ry} = -\frac{E}{6(1-\nu^2)}\left\{ (t_o + t)^2 \frac{dk_y}{dt} + 4(t_o + t)k_y + 2\int_{t_o}^{t} k_y dt \right\}$$

where t_o is the initial thickness and t the final one.

In the case of steel profiles, the universally adopted mechanical method is the sectioning test that was first applied in the US by the Column Research Council (CRC) around 1950 and later introduced into Europe. It consists in cutting the various parts of the profile into parallel longitudinal strips and in measuring deformations by means of mechanical extensometers (Fig. 4.46c). If L_o is the basic length actual length of the specimen must be $L = L_o + 3b$ (b being its maximum transverse dimension), as a precaution against edge effects.

With this method only the longitudinal components of the residual stress state are measured. They are, however, the most important ones as far as structural behaviour is concerned, because they are superimposed upon the stresses caused by external loads and therefore influence the load bearing capacity of the member. In the case of sections consisting of very thick elements, especially if welded, it is also important to know the longitudinal stress variation in the thickness. These measurements have recently been performed in the US (1970) on jumbo profiles |9|, applying a procedure which is substantially similar to the sectioning test, but makes cuts through both thickness and width of the single plates comprising the section.

All the foregoing procedures are mechanical methods and are generally quite laborious, as they require long cutting and measuring operations.

In many cases it is preferable to adopt deflection methods; while not as accurate, they are certainly faster because they require only one cut in the specimen and later the measurement of a conveniently chosen displacement from which it is possible to deduce the stress state which caused it, according to the formulae given in Fig. 4.47.

Among the non-destructive methods for determining residual stresses, it is worth mentioning the X-ray method which uses the interatomic spaces of the various planes as a basis from which to measure the deformations, relating their variations to the residual stress state.

145

Due to the low penetrating power of X-rays into metal, this method is resticted to the analysis of mono- or bi-axial superficial states.

The so-called hole drilling method is considered a semi-destructive one. It consists essentially of releasing the residual stresses in the specimen by drilling small holes (8 mm maximum diameter) at an appropriate number of points and measuring the resulting deformations. This method is particularly useful for thin hollow sections where residual stresses can be considered constant throughout the thickness.

This type of control was first adopted by J. Mathar (1932) and later made applicable by W. Birkenfeld (1967), thanks to the introduction of special electric extensometers (Fig. 4.48) formed of three radial spirals enclosing a space where a hole can be drilled |22|. Drilling the hole releases residual stresses and the resulting deformations are measured by each extensometer and expressed as follows:

$$\varepsilon_\alpha = k_1(\sigma_{r1} + \sigma_{r2}) + k_2(\sigma_{r1} + \sigma_{r2}) \cos 2\alpha$$

$$\varepsilon_{+\beta} = k_1(\sigma_{r1} + \sigma_{r2}) + k_2(\sigma_{r1} - \sigma_{r2}) \cos 2(\alpha + \beta)$$

$$\varepsilon_{-\beta} = k_1(\sigma_{r1} + \sigma_{r2}) + k_2(\sigma_{r1} - \sigma_{r2}) \cos 2(\alpha - \beta)$$

$$k_1 = -\frac{1+\nu}{2E} \frac{r_o^2}{r_1 r_2}$$

$$k_2 = \frac{2}{E} \frac{r_o^2}{r_1 r_2} \left\{ -1 + \frac{1+\nu}{4} r_o^2 \frac{r_1^2 + r_1 r_2 + r_2^2}{r_1^2 r_2^2} \right\}$$

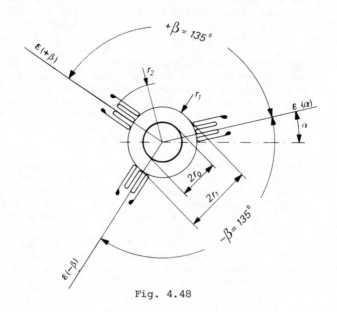

Fig. 4.48

where ν is Poisson's ratio and E is Young's modulus.

The other symbols are self-explanatory in Fig. 4.48.

The principal components σ_{r1} and σ_{r2} of the residual str their inclination α are the unknowns of the problem, but th obtained from the three equations above as functions of the deformation components ε_α, $\varepsilon_{+\beta}$ and $\varepsilon_{-\beta}$. They are given by:

$$\left.\begin{array}{c}\sigma_{r1}\\\sigma_{r2}\end{array}\right\} = \frac{E}{8}\ \frac{\varepsilon_{+\beta} + \varepsilon_{-\beta} - 2\varepsilon_\alpha\ \cos 2\beta}{k_1\ \sin^2\beta}$$

$$\pm\ \frac{\sqrt{(2\varepsilon_\alpha - \varepsilon_{+\beta} - \varepsilon_{-\beta})^2 + \beta(\varepsilon_{-\beta} + \varepsilon_{+\beta})^2\tan^2\beta}}{k_2\ \sin^2\beta}$$

where:

$$\tan 2\alpha = \frac{\varepsilon_{-\beta} - \varepsilon_{+\beta}}{2\varepsilon_\alpha - \varepsilon_{+\beta} - \varepsilon_{-\beta}}$$

4.5 STRUCTURAL STEELS

Carbon steel qualities to be used in structural applications are unified in Europe. Their yield strengths vary from 235 to 355 N mm^{-2}.

In addition to the above qualities there are steel grades of higher strength. Generally higher strength steels give lower values for the ultimate elongation ε_t and for the ratio f_t/f_y; they must therefore be used cautiously, especially when plastic redistribution is required in the structure.

From a statistical point of view, it is interesting to note the values of the variation coefficient δ. For yield stress, Fig. 4.49 shows variation coefficients for structural steel types used in various countries. The figure represents the results of 100 series of tests, each series having a specimen population of between 20 and 2000.

If the values of δ in Fig. 4.49 are divided into two groups with reference to steel strength, having nominal yield stress above and below 300 N mm^{-2}, one obtains the histogram in Fig. 4.50.

For $f_{y,nom} < 300$ N mm^{-2}, δ lies between 2% and 11% with a mean value near 7.5%, which is the value indicated by ECCS. For $f_{y,nom} > 300$ N mm^{-2}, δ is more restricted and the corresponding mean value is about 6%. These results suggest that lower variation coefficients may be expected as steel strength increases.

147

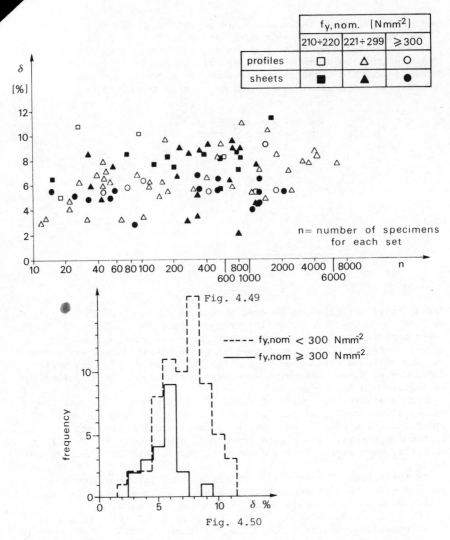

	$f_{y,nom.}$ [Nmm^2]		
	210÷220	221÷299	⩾300
profiles	□	△	○
sheets	■	▲	●

Fig. 4.49

Fig. 4.50

Whenever a serviceability or an ultimate limit state is reached in a structure, it is necessary to check whether the stress state at each point P given by the components $p_{ij}(P)$ of the stress tensor is compatible with the material elasticity. In other words, it is always necessary to know the value of the factor by which to multiply the components $p_{ij}(P)$ in order to achieve material yielding at the point P.

Such a comparison (strength check) can be performed only on the basis of a given yielding criterion.

Strength criteria applied in practice are based on simple, fast and inexpensive laboratory tests. In the case of steel, the experimental parameter to be introduced into the theoretical yield criterion, in order to compare calculated with permissible stresses is given by the simple mono-axial tensile test.

Of all the different criteria, the structural codes unanimously adopt that of Huber-Hencky-Von Mises for steel. If the overall elastic potential energy is expressed as the sum of energy ϕ_D due to shape variation and of energy ϕ_V due to volume variation according to the above criterion, material rupture is due to the energy ϕ_D only. Therefore, the overall stress tensor:

$$
P_{ij} = \begin{vmatrix} \sigma_1 & 0 & 0 \\ 0 & \sigma_2 & 0 \\ 0 & 0 & \sigma_3 \end{vmatrix}
$$

expressed in terms of principal components σ_i can be broken down into the sum of tensors:

$$
P_{ij} = \begin{vmatrix} \sigma_1 - \sigma_m & 0 & 0 \\ 0 & \sigma_2 - \sigma_m & 0 \\ 0 & 0 & \sigma_3 - \sigma_m \end{vmatrix} + \begin{vmatrix} \sigma_m & 0 & 0 \\ 0 & \sigma_m & 0 \\ 0 & 0 & \sigma_m \end{vmatrix}
$$

The first term is defined as the stress deviation and corresponds to the distorting energy ϕ_D. The second term, defined as the hydrostatic tensor, corresponds to energy due to volume modification ϕ_V in which it is assumed that

$$
\sigma_m = (\sigma_1 + \sigma_2 + \sigma_3)/3
$$

Distorting energy for tri-axial stress state is given by:

$$
\phi_D = \left| (\sigma_1 - \sigma_2)^2 + (\sigma_2 - \sigma_3)^2 + (\sigma_3 - \sigma_1)^2 \right| /12G
$$

In a mono-axial stress state it can be simplified into:

$$
\phi_D = \sigma^2 /6G
$$

By applying the same factor of safety against failure to both the mono-axial and the tri-axial stress state and assuming $\sigma = \sigma_{id}$, one obtains the equality:

$$
\sigma_{id} = \sqrt{\{(\sigma_1 - \sigma_2)^2 + (\sigma_2 - \sigma_3)^2 + (\sigma_3 - \sigma_1)^2\}/2}
$$

$$
= \sqrt{\sigma_1^2 + \sigma_2^2 + \sigma_3^2 - \sigma_1\sigma_2 - \sigma_2\sigma_3 - \sigma_1\sigma_3}
$$

which expresses the equivalent stress σ_{id} (ideal stress) according to the Huber-Henckey-Von Mises criterion.

In practice one finds, almost exclusively, a bi-axial stress state. The ideal stress expression then becomes:

149

$$\sigma_{id} = \sqrt{\sigma_1^2 + \sigma_2^2 - \sigma_1\sigma_2}$$

or

$$\sigma_{id} = \sqrt{\sigma_x^2 + \sigma_y^2 - \sigma_x\sigma_y + 3\tau_{xy}^2}$$

In the most frequent case of bending accompanied by shear, as $\sigma_y = 0$

$$\sigma_{id} = \sqrt{\sigma_x^2 + 3\tau_{xy}^2}$$

In the case of pure shear, as $\sigma_x = \sigma_y = 0$:

$$\sigma_{id} = \sqrt{3}\,\tau$$

REFERENCES

1. BCSA (1980) *British Sections A Guide to Replacement by Continental Sections*, Publication n. 6/80, London
2. AISC (1980) *Manual of Steel Construction*, 8th Edition, AISC, Chicago
3. Dieter, G. E. (1961) *Mechanical Metallurgy*, McGraw Hill, New York, 393-418
4. Mas, B. and Massonnet, Ch. (1966) The belgian contribution to the experimental work of ECCS, *Acier Stahl Steel*, 9, 385-392
5. Tall, L. (1966) *Welded Built Up Columns*, Fritz Engineering Laboratory, Report 249.29, Lehigh University, Bethlehem, Pennsylvania
6. Alpsten, G. A. (1967) *Residual Stresses in Hot Rolled Steel Profiles*, Inst. of Struc. Eng. and Bridge Building, Roy. Inst. of Techn., Stockholm
7. Tall, L. and Alpsten, G. A. (1969) *On the Scatter in Yield Strength and Residual Stresses in Steel Members*, IABSE Symposium on Concepts of Safety of Structures and Methods of Design, London 151-164
8. Alpsten, G. A. (1970) Residual stresses and mechanical properties of cold straightened H shapes, *Jernkontorets Annales*, 154, 255-283
9. Alpsten, G. A. (1970) *Residual Stresses in a Heavy Welded Shape 23H681*, Fritz Engineering Laboratory Report 337.9, Lehigh University, Bethlehem, Pennsylvania
10. Young, B. W. and Dwight, J. B. (1971) *Residual Stresses and their Effect on the Moment Curvature Properties of Structural Steel Sections*, CIRIA Tech. Note 32, London
11. Mazzolani, F. M. (1972) *Analisi sperimentale delle tensioni residue nei profilati metallici*, Atti del 1° Convegno AIAS, Palermo, 297-309
12. Daddi, I. and Mazzolani, F. M. (1972) Determinazione sperimentale delle imperfezioni strutturali nei profilati di acciaio, *Costruzioni Metalliche*, 5, 374-394
13. Alpsten, G. (1972) *Prediction of Thermal Residual Stresses in Hot Rolled Plates and Shapes of Structural Steel*, Swedish Inst. of Steel Const. Report 16.4, Stockholm

14. Alpsten, C.A.(1972) *Residual Stresses, Yield Stress and Column Strength of Hot Rolled and Roller Straightened Steel Shapes*, Proceedings of International Colloquium on Column Strength, Paris 39-59

15. Alpsten, G. A.(1972) *Variations in Mechanical and Cross Sectional Properties of Steel*, Proceedings International Conference on Planning and Design of Tall Buildings, Vol. 1b, Lehigh University Bethlehem, Pennsylvania, 755-807

16. Young, B. W. (1972) *Residual Stresses in Hot Rolled Members*, Proceedings of International Colloquium on Column Strength, Paris, 25-38

17. Daddi, I. and Mazzolani, F. M. (1973) Il livellamento delle tensioni residue sotto carichi oscillanti, *Costruzioni Metalliche*, 2, 115-127

18. Tebedge, N. and Tall, L. (1973) *Residual Stresses in Structural Steel Shapes: Summary of Measured Values*, Fritz Engineering Laboratory Report 337.34, Lehigh University, Bethlehem, Pennsylvania

19. Young, B. W., Elliot, R. and Bowers, G. (1973) *Residual Stresses and Measurement of Tolerances*, Proceedings International Cohference on Steel Box Girder Bridges, London

20. Young, B. W. (1974) The effect of process efficiency on the calculation of weld shrinkage forces, *Proceedings Institution Civil Engineers*, 57, Dec., 685-692

21. Kamtakar, A. G. (1974) *An Experimental Study of Welding Residual Stresses*, Cambridge University Engineering Department, Tech. Report 39

22. Mazzolani, F. M. and Corsaro, I. (1974) *Indagine sperimentale sulle autotensioni in tubi a gomito di caldaie di recupero*, Bollettino Tecnico Finsider n. 330, Genova

23. Alpsten, G. A. (1970) *Variations in Strength and Cross-Sectional Dimensions of Structural Shapes*, Proceedings Stalbyggnad Nordiska Forskningsdagar, Swedish Institute of Steel Construction, 2.1-2.17

24. Ersvik, O. and Alpsten, G. A. (1970) *Experimental Investigation of the Column Strength of Wide-Flange Shapes ME 200 A, Roller Straightened in Different Ways*, Report 19.3, Swedish Institute of Steel Construction, Stockholm,

25. Sfintesco, D. (1970) Experimental basis of the European column curves, *Costruzioni Metalliche*, 3, 5-12

26. Baker, M. J. (1972) *Variability in the Strength of Structural Steel. A Study in Structural Safety. Part 1, Material Variability*, CIRIA, Techn. Note 44, London

27. Giangreco, E. and Mazzolani, F. M. (1974) *Costruzioni Metalliche*, Enciclopedia dell'Ingegneria, ISEDI, Milan

28. Ching, K. Yu and Tall, L. (1971) Significance and application of stub column test results, *Journal of the Structural Division*, ASCE, 97, ST 7, July, 1841-1861

29. Di Bartolo, S. (1972) *Lo snervamento a caldo degli acciai*, Bollettino Tecnico Finsider, 301

30. CECM ECCS (1977) *Introductory Report*, Second International Colloquium on Stability, Liège

CHAPTER FIVE

Welds

5.1 TECHNOLOGY OF WELDS

The practice of connecting steel plates by welding started at the beginning of this century when it was recognized that inconveniences associated with the use of rivets could be avoided if the parts were monolithically joined after local melting.

The most obvious difficulty lay in the fact that heating had to be limited to prevent modification of the micrographic structure of the plates, thus making them unfit for use.

The electric arc was therefore adopted, providing a concentrated source of a remarkable quantity of heat. At first two carbon electrodes were used set very near to the pieces to be joined (Zrener). Next, the arc was formed between one carbon electrode and the piece to be welded (Bernados), and finally the carbon electrode was replaced by a metal wire coated with insulating material (Kielberg, 1908). This new connecting technique rapidly spread throughout the USA and then also in Europe. Laboratory tests have always confirmed its complete reliability. Subsequently, technology has led to the development of many other welding procedures related to the nature of the pieces to be joined and to the use for which they are intended.

Currently, about forty different welding systems are available to melt different constructional requirements |1,2|.

5.1.1 *Flame cutting*

Steel plates before welding are normally cut by means of an oxy-acetylene or oxy-propane torch, which provides the heat source needed to produce a concentrated high temperature and a jet of oxygen, which generates with the metal a strong isothermic reaction capable of melting the resultant oxide.

If the torch advances automatically at a regular speed and if oxygen purity is sufficient, the oxy-cut surface is regular enough to enable welding to be carried out without further preparation. Using special shapes to guide the torch, curved cuts of any desired form can also be obtained. The torch can be inclined when necessary to obtain oblique cuts to form the grooves required between the edges

153

to be welded.

Cutting with ionized gases (plasma cutting), due to the extremely high temperatures it can reach, obtains good results also in cases where oxy-cutting is inadequate.

5.1.2 *Welding processes*

Autogenous fusion welding processes vary substantially according to the applied heat source and to how the molten pool is protected against the atmosphere.

(a) *Oxy-acetylene welding*. The heat source consists of the oxy-acetylene flame produced by the combustion of acetylene (C_2H_2) with oxygen. These two gases are fed to the torch separately, mix together inside it and emerge at its tip, where the combustion takes place. The flame forming at high temperature (about 3100°C) also produces reducing gases (CO and H_2) which protect the pool. This process - the first to be adopted industrially - is now much less common than in the past.

(b) *Shielded metal arc process*. (Fig. 5.1a). The heat source is the electric arc which struck between the electrode and the parent metal and the heat developped causes the quick melting of both. The melting of the coating of the cylindrical electrode (length: 350 to 450 mm.) produces some gases which protect the area surrounding the arc and the weld pool.

Depending on the composition of the coating, the most common types of electrode can be basic (excellent mechanical and metallurgical characteristics), acid (good mechanical characteristics), or cellulose (special uses, such as fixings). This process is applicable to a wide range of structural applications.

(c) *Submerged arc process*. (Fig. 5.1b). The heat source is again the arc struck between the electrode and the parent metal. The electrode consists of a continuous wire wound on a drum or spool. An automatic feeding device makes it advance at the same speed at which it is melted.

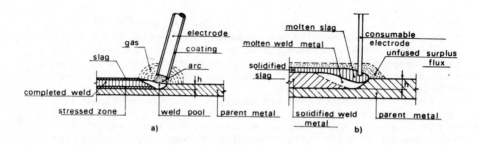

Fig. 5.1

Arc area protection is provided by a granulated flux distributed on the joint to form an accumulation inside which the arc strikes, thus being submerged and invisible. This process is highly productive.

(d) *Gas shielded metal arc process with consumable electrode* (MIG and MAG). This too is a continuous wire process in which arc protection is provided by an inert gas (e.g. argon: MIG), or by a chemically active gas (e.g. carbon dioxide or a mix of the latter with argon: MAG).

Pool dimensions are smaller than in the previous case and productivity is lower. The MAG process is used to weld mild steel and low-alloy steel.

(e) *Gas shielded metal arc process with non-consumable electrode* (TIG). The heat source still consists of the arc. It strikes between a tungsten element and the parent metal. Arc area protection is provided using argon. The only purpose of the electrode is to allow arc formation. The weld metal must be introduced separately in the form of rods, as in the oxy-acetylene process. This process is used to weld materials where the weld pool is easily oxidizable, such as stainless steel and aluminium alloys.

(f) *Electroslag welding*. The parts to be welded are mounted vertically, with a given distance between them. Copper sliding blocks are set across the gap so that a kind of vertical 'melting-pot' (having a rectangular section) is thus obtained. In its lower part a molten flux is provided with particular electrical characteristics, into which the continuous wire electrode sinks. Both the electrode and the parent metal are connected to a current generator. According to the Joule effect, the current passing through the molten flux keeps it at a sufficiently high temperature to melt the edges of the parent material and the wire. The copper sliding blocks are prevented from melting by an internal water circulation system. This process offers very high productivity and is used for vertical welding of very thick steel parts, with one only weld pass.

5.1.3 *Consequences of metallurgic phenomena*

There are, essentially, two metallurgical phenomena |1,2|: solidification of the material that has melted during the various weld passes and heat treatment of the parent metal around the weld. Welding is characterized by small masses of metal, rapidly melted and rapidly cooled due to heat absorption by the surrounding metal. It therefore consists of thermal cycles with a high rate of cooling which produces zones of high hardness, expecially in the material close to the weld. These hard zones in the parent metal near the weld can lead to so-called cold cracking (Fig. 5.28).

The origin of such cracks is acribed to the hydrogen absorbed by weld material in the molten state and by the adjacent area of parent material brought to high temperature. This hydrogen generally comes from the arc atmosphere of the coated electrodes, except in the case of electrodes with a basic coating |3|.

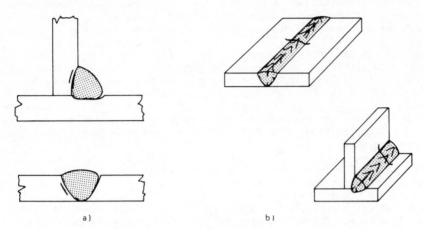

Fig. 5.2

To prevent this fault, the thermal cycle should be moderated by appropriately pre-heating the part to be welded and using basic electrodes. These precautions are strongly recommended whenever steel having a high elastic limit is being welded, as very hard areas could be easily formed.

Another dangerous phenomenon is that of hot cracking which can occur in the molten area whenever the content of parent material is high (Fig. 5.2b). Thrse cracks form during solidification because of impurity segregations thickening in preferential zones of the molten area and solidifying at a lower temperature than steel, entailing material dissociation due to shrinkage stresses.

The metallurgical structure of molten areas is strictly related to its mechanical characteristics, particularly toughness.

When welding is performed by means of a few large weld passes, recrystallization is coarse and toughness low. It can possibly be corrected using special reactive powders. When, however, welding is performed in many weld passes of limited section, a high toughness results, partly due also to the useful normalizing action which each pass produces on the proceding one.

5.1.4 *Consequences of thermal phenomena*

Welding processes are always accompanied by a development of a great quantity of heat, which causes considerable thermo-plastic deformations in the connected steel elements |1,4|, due to the lowering at high temperatures of yield limit and of elastic modulus (see 4.4.4).

The physical nature of such phenomena can he demonstrated by reference to the model of a bar of length L having in the middle a welded section L_o (Fig. 5.3). Consider the limit of deformation due to transverse contraction and the internal stresses induced by totally preventing contraction.

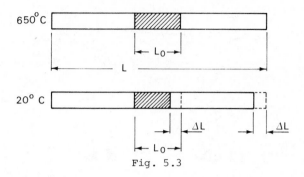

Fig. 5.3

As a consequence of cooling from the temperature after welding (650 °C) to ambient temperature (20 °C), the weld metal undergoes a transverse contraction ΔL. According to some experiments on steel plates 14 mm thick transverse contraction (or shrinkage) is of the order of about 20% of the initial length, i.e. $\Delta L = 0.18\ L$. If from the beginning the bar is prevented from shortening, an axial force N arises. It can be evaluated according to the following compatibility condition:

$$\Delta L = \frac{NL}{E_m A} = \frac{\sigma L}{E_m}$$

where E_m is a mean value of the elastic modulus for the considered temperature range (650 to 20 °C). If one assumes $E = 2.1 \times 10^5$ N mm^{-2} and $E_m = 0.75\ E$ then:

$$\sigma = \frac{L_o}{L} \times 0.18 \times 0.75 \times 2.1 \times 10^5 = 280 \times 10^2\ \frac{L_o}{L}\ \text{(N mm}^{-2})$$

Such a stress could reach the yield limit when L_o/L is of the order of magnitude 1/100.

Summing up, this schematic example shows that, as a thermal consequence of welding, either relatively important deformations take place or internal (residual) stress of considerable intensity is induced.

Both these deformations and stresses are perpendicular to the seam and, if the deformations are not prevented, transverse shrinkage phenomena take place, producing angular deformation of the welded parts (Fig. 5.4b).

Deformation also takes place along the beam but this longitudinal shrinkage cannot take place entirely freely, because it is prevented by the elasticity of the surrounding cold parts and therefore longitudinal stresses are always present (Fig. 5.4a).
The reader is referred to 4.2.3 for a study of longitudinal residual stress in welded elements, here only mention is made of most common systems applied in building practice to keep deformations and residual stresses due to welding within reasonable limits. In the preventive phase, they are: initial pre-bending, clamping the parts in the seam

157

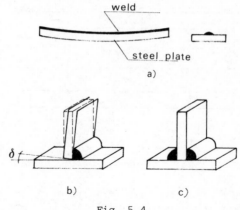

a)

b)   c)

Fig. 5.4

deposition phase, pre-heating, and a rational study of welding
sequences.

Once the welding has been carried out, concentrated heating can be
applied to the distorted parts, or the whole fabrication can be stress
relieved by subjecting it to slow heating in a furnace to a uniform
temperature overall followed by very slow cooling.

5.1.5 *Welding defects and controls* |5,6,7,8|

The worst defect which can affect a weld is the presence of cracks
(see 5.1.3).

Other defects include blow holes which are metallurgical defects
due to unforeseen reactions which arise in the molten pool because of
operational imperfection and slag inclusions, which are generally
caused by improper shape or sequence of weld pass.

A further possible defect is lack of penetration due to the failure
of the weld metal to penetrate and fill the root of the weld at the
vertex of the grooves (Fig. 5.5a) or at the centre (Fig. 5.5b). The
vertex of a V-shaped or X-shaped or any other setting is always that
part at which the edges to be welded are closest and it is not so easy
to obtain a regular and complete melting of the shaped edge by the
first weld pass and faulty edge preparation can only make matters
worse. The defect can be eliminated if it is possible to back gouge
and apply a sealing pass.

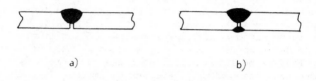

a)   b)

Fig. 5.5

Another defect is sticking, i.e. molten filler material being deposited on the parent metal before it has melted, so that an oxide coating is interposed between the edge and the melted area. This takes place in welding processes having low or poorly concentrated thermal input.

The most common way of detecting internal welding defects is by X-ray examination, using the X-rays produced by special radiographic equipment, or by γ-rays spontaneously released by some radioactive source.

Internal defects, such as porosity, blow holes, slag inclusion, lack of penetration, cracks, appear as darker stains on the film. They are interpreted by comparing them to sample defects, corresponding to radiographic standards set out in codes or previously agreed upon as a condition for welding approval.

Controls can be performed also by means of a thin ray of ultrasounds produced by a probe containing a quartz crystal, which is vibrated at its fundamental frequency by means of a variable high frequency electic field. The probe is laid on the surface of the part to be examined, in order to transmit a series of ultrasonic impulses to it. These impulses can be reflected from obstacles consisting of other surfaces of the specimen or defects. By moving the probe along the specimen, if there is no defect, clearly distinct luminous vertical lines appear on the screen of a cathode ray oscillograph. These lines correspond to the pulse at the top surface and the echo from the bottom. If instead the beam meets a defect, intermediate lines appear on the screen, from which the location and size of the defect can be assessed.

Mention should also be made of some simpler methods, such as magnetoscopic (magnaflux) examination and examination by penetrants.

The former consists of two wire-bearing poles which when put in contact with the weldment create a magnetic field which, by means of magnetic powders, allows detection of cracks near the surface.

Liquid penetrants, because of their low surface tension can penetrate even into very narrow superficial cracks, which are practically invisible to the naked eye. Their presence is detected by means of a thin coat of a revealing liquid applied over the area after the surplus penetrating liquid has been removed from the surface.

5.2 STRENGTH OF WELDED JOINTS

Strength verification mainly refers to two different kinds of joint (Fig. 5.6): butt welds and fillet welds.

5.2.1 *Butt welds*

In a butt weld which is free from internal defects and axially loaded perpendicularly to its own axis, the stress state can be considered equal to that of a homogeneous piece (Fig. 5.7).

Consequently, the criterion generally adopted by modern codes is to assume its resistant cross section as having an area given by the weld length times the connected thickness. The corresponding strength can be considered equal to or lower than that of the parent material,

159

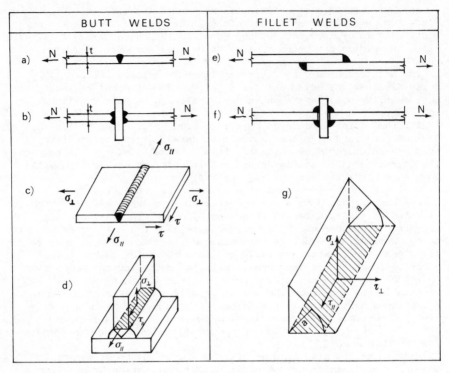

BUTT WELDS	FILLET WELDS

Fig. 5.6

according to the value assumed for the efficiency coefficient, defined as the ratio between the design strength of the weld $f_{d,w}$ and that of the parent metal f_d. One therefore assumes:

$$f_{d,w} = \eta_w \, f_d$$

Such a criterion derives from the allowable stress method, as it was first defined as the ratio between the allowable stress of the weld $\sigma_{w,adm}$ and that of the parent material σ_{adm}.
The various codes assume a value varying from 0.6 to 1 for

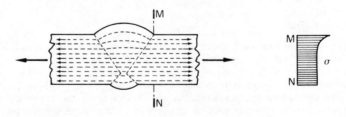

Fig. 5.7

coefficient η_w. This is related to the technological parameters characterizing weld joint performance, including the degree of control, welding execution (in the shop or on site), the type of joint (with or without back gouging).

Briefly, the efficiency coefficient η_w reduces allowable stresses in welds in order to compensate for lack of control and as an indirect precaution against presumed but unknown defects. It is therefore obvious that statistical meaning can be given to it. It can be considered only as an empirical coefficient justified by experience. For this reason the IIW (International Institute of Welding), and later the ECCS (European Convention for Constructional Steelwork), have recently suggested the adoption of an efficiency coefficient equal to one, at the same time recommending suitable tests for the qualification of welders and of the welding process before the actual work is begun, together with the adoption of adequate inspection and control methods, in order to have a reasonable degree of assurance that welded joint quality be sufficient to restore at least original strength of the parent metal.

If an efficiency coefficient of 1. is assumed, it is obvious that strength verificacion for a butt weld is unnecessary as it is superseded by parent metal control.

In general, the calculation method for a butt welded joint can be based on the following assumptions.

The resistant cross section of a weld is the longitudinal one (throat section) having as its length the entire welding length and as its height the smaller of the two connected thicknesses, measured in proximity to the welding in the case of butt welds (Fig. 5.6a), or else the thickness of the completely penetrated element in the case of T-shaped or cross-shaped joints (Fig. 5.6b).

The following stresses are considered to act on the throat section (Fig 5.6c,d):

$\sigma_\perp$ is the tensile or compressive stress acting perpendicularly to the throat section

τ is the shear stress acting parallel to the throat section axis

$\sigma_\parallel$ is the tensile or compressive stress acting parallel to the seam axis, referred to the cross section of the weld joint (i.e. parent metal and weld) perpendicular to such axis

On these bases, strength verifications are carried out adopting the same yield criterion used for the parent material, i.e. the Hencky-Von Mises criterion:

$$\sigma_{id} = \sqrt{\sigma_\perp^2 + \sigma_\parallel^2 - \sigma_\perp \sigma_\parallel + 3\tau^2} \leq \eta_w f_d$$

5.2.2 *Fillet welds*

Many studies have examined the problem of checking the strength of a fillet weld once the stress state created in it by external loads is known.

The various calculation methods are all based on the simplifying hypothesis that stresses are uniformly distributed within the seam throat section. Most of them consider the throat section overturned on one side of the seam.

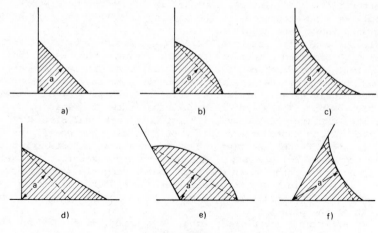

Fig. 5.8

In 1961 ISO approved a calculation method proposed by IIW which assumed the throat section in its actual position as the resisting section.

When this method first appeared, it was considered appropriate progressively to replace all other methods as the ISO recommendation would be adopted by the various constituent Countries. However, it seemed to be too complicated for practical use and some important codes, such as that of the AISC (USA, 1969) and DIN 4100 (Germany, 1968) followed simpler calculation methods considering the throat section as ouverturned on one of the two sides of the seam.

5.2.2.1 Definition

Assume the resistant cross section of a fillet weld to be the throat section, given by the production of the throat depth times the seam effective length.

Throat depth a is the smallest length of the triangle inscribed in the seam cross section (Fig. 5.8).

The effective length L coincides with the overall length of the seam, provided that, obviously, faulty ends due to tailing off are eliminated.

Furthermore, the following stress components are taken into account:

(a) Stresses referred to the throat section, overturned on the plane of either side of the seam (Fig. 5.9a):

$n_\perp$ is the tensile or compressive stress acting perpendicularly to the plane containing one side of the seam, referred to the area of the throat section, overturned on the same plane

$t_\perp$ is the shear stress acting perpendicularly to seam axis, lying on a plane containing one side of the seam, referred to the area of the throat section, overturned on the seam plane

$t_{||}$ is the shear stress acting parallel to the seam axis, lying on a

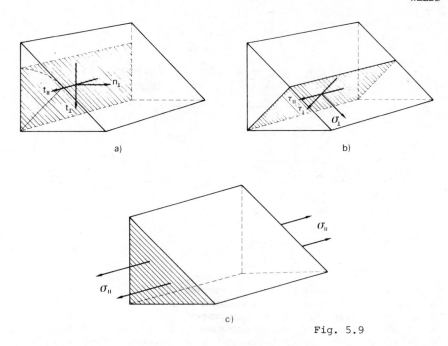

Fig. 5.9

plane containing one side of the seam and referred to the area of the throat section, overturned on the same plane

(b) Stresses referred to the throat section in its actual position (Fig. 5.9b):
$\sigma_\perp$ is the tensile or compressive stress acting perpendicularly to the throat section, considered in its actual position
$\tau_\perp$ is the shear stress acting perpendicularly to the seam axis, lying in the throat section, considered in its actual position
$\tau_\parallel$ is the shear stress acting parallel to the seam axis, lying in the throat section, considered in its actual position

(c) Stress referred to the cross section of the seam (Fig. 5.9c):
$\sigma_\parallel$ is the tensile or compressive stress acting parallel to the seam axis on its cross section

5.2.2.2 Stress state in a fillet weld
Whereas stresses in a butt weld are very close to those in a continuous piece, in a fillet weld the actual stress distribution in the plane of the cross section of the weld is considerably more complicated. The stress state changes from one point to another and - especially at seam vertex and edge - considerable stress peaks are present, due to the sensible deviation of the stress lines passing from one piece to the other (Fig. 5.10). However, as for butt welds, stresses in fillet welds are usually considered uniformly distributed in the throat section. Substantially, this assumption is based on satisfactory ductility and toughness of material, which are checked

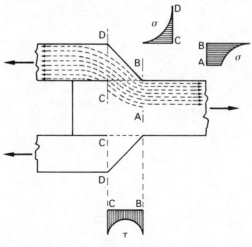

Fig. 5.10

by means of appropriately related control of welding material and qualification tests of the welding process.

The authors consider that stress $\sigma_{||}$ has no influence on the behaviour of welded joints because it is not present in the weld - the weld itself just connects the various parts together - but it is due to the external action undergone both by the weld and by the connected elements. In support of this, consider a fillet welded I-beam. The seams are loaded by shear stresses ($\tau_{||}$) which act against the sliding between flanges and web and therefore represent the stresses due to the connecting function. Stresses due to bending under external loading act on the overall I-beam cross section and in the weld, they are stresses of the type $\sigma_{||}$.

Opinions have been divided as to whether it is or is not necessary to take the stresses $\sigma_{||}$ into account in the strength calculation of a fillet welded joint and the problem was strongly discussed some time ago in both IIW and ISO Committees, some countries were in favour of taking $\sigma_{||}$ into account because it was on the safe side and because it led to a calculation method which was uniform with that for the parent metal; other countries were against it. The latter prevailed because experimental data justified their view and $\sigma_{||}$ was ignored.

Further tests have shown that in fact even high values of $\sigma_{||}$ do not reduce the load bearing capacity of welded joints in the case of static loads and of ductile and tough materials.

The most important codes, e.g. DIN 4100 (Germany, 1968), AISC (USA, 1969) and BS 153 (Great Britain, 1966), as well as French, Italian, Dutch, Hungarian and Russian recommendations ignored $\sigma_{||}$ in the calculation of the strength of fillet welds.

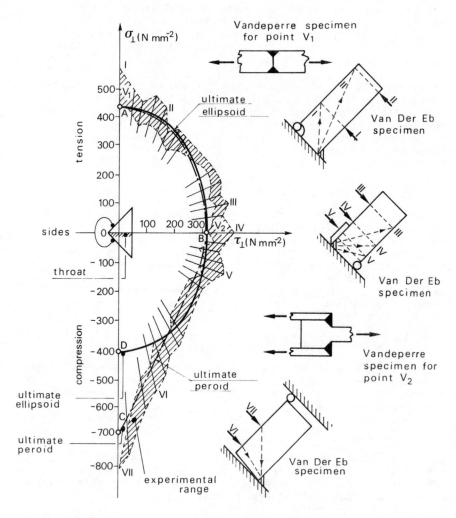

Fig. 5.11

5.2.2.3 Recommendations and calculation methods |1,9,10,11|
Experimental bases. Many tests were carried out to anlyse fillet weld
strength as stresses varied, in order to trace their ultimate surface
in the space of co-ordinates $\sigma_\perp$, $\tau_\perp$, $\tau_{||}$, referred to the throat
section in its actual position.

Among the earliest tests, mention should be made of those on fillet
welds only perpendicularly and tangentially loaded (Vandeperre,
Belgium) and those on fillet welds undergoing internal actions in any
direction in the plane perpendicular to the seam axis (Van der Eb,
Holland, 1952-53). The ultimate surface corresponding to Van der Eb's
tests was called 'peroid' (Fig. 5.11).

A few years later (1968), an international series of tests on fillet welds was prompted by the 15th Commission of IIW, with the participation of France, Belgium, Holland, Italy, Sweden, Canada, Japan and the USA. A study of the corresponding results influenced the choice of the various calculation methods.

ISO calculation method. This method is based on the principle of considering an ultimate domain for the fillet welds which is as close as possible to the rupture 'peroid',but able to be expressed in an equation.

The proposed equation ($f_{u,w}$ being rupture stress referred to the seam throat section) is:

$$1 = \frac{\sigma_\perp^2}{f_{u,w}^2} + \frac{\tau_\perp^2}{(0.75f_{u,w})}$$

representing an ellipse having its major semi-axis coincident with the upper semi-axis of the 'peroid' section in the plane $\sigma_\perp$, $\tau_\perp$, and the 'peroid' semi-axis as its minor semi-axis.

By rotating such an ellipse around its major axis ($\sigma_\perp$), an ellipsoidal surface is obtained, as the ultimate domain, which conveniently replaces the 'peroid', if the small part in compression is neglected, where the ellipsoid is more conservative.

The equation of the ellipsoid of revolution is therefore:

$$1 = \frac{\sigma_\perp^2}{f_{u,w}^2} + \frac{\tau_\perp^2}{(0.75f_{u,w})^2} + \frac{\tau_{||}^2}{(0.75f_{u,w})^2} \tag{5.1}$$

This equation enables one to determine whether a plury-axial stress state characterized by generic values of $\sigma_\perp$, $\tau_\perp$ and $\tau_{||}$ is acceptable or not. Now $1/0,75^2 \cong 1.8$ and if this is adopted, the equation (1) gives the verifiction condition:

$$\sqrt{\sigma^2 + 1.8(\tau_\perp^2 + \tau_{||}^2)} \le f_{u,w} \tag{5.2}$$

From a formal point of view, it is convenient to express the verification condition in an analogous manner to that adopted for the parent metal. For this purpose, the first term of the above relation can be used as equivalent 'ideal' stress:

$$\sigma_{id} = \sqrt{\sigma^2 + 1.8(\tau_\perp^2 + \tau_{||}^2)} \tag{5.3}$$

The second term of (2) indicate the ultimate tensile strength of the fillet weld. This depends on electrode quality, which must be chosen in relation to the type of steel of which the elements to be welded are made. In general, therefore, the ultimate tensile strength of the weld $f_{u,w}$ (conventionally assuming $\gamma_w = 1$) can be expressed as a function of the design strength f_d of the parent metal:

$$f_{u,w} = \beta_w f_d \tag{5.4}$$

wherein β_w is the efficiently coefficient of the fillet weld. The general verification, therefore, becomes:

166

$$\sigma_{id} = \sqrt{\sigma_\perp^2 + 1.8(\tau_\perp^2 + \tau_\parallel^2)} \le \beta_w f_d \qquad (5.5)$$

ISO has recommended this formula in the more general form:

$$\sigma_{id} = \sqrt{\sigma_\perp^2 + k_w(\tau_\perp^2 + \tau_\parallel^2)} \qquad (5.6)$$

where parameter k_w has not been fixed, so that the codes of the various countries are free to adopt different values of k_w, according to the chosen safety factor in shear. Some codes adopt $k_w = 3$, in order to maintain, also for fillet welds, a combined stress formula similar to that of the Huber-Henky-von Mises criterion already used both for the parent metal and for butt welds. In this case, the formula becomes:

$$\sigma_{id} = \sqrt{\sigma_\perp^2 + 3(\tau_\perp^2 + \tau_\parallel^2)} \le \beta_w f_d^< \qquad (5.7)$$

representing an ellipsoid having semi-axis $\tau_\parallel = \tau_\perp = 0.58\,\sigma$. Further tests have shown that the ellipsoid of ultimate strength cannot be considered as a solid of revolution, semi-axis $\tau_\parallel$ being longer than $\tau_\perp$. A generalization of the ISO formula (5.6) was therefore proposed as

$$\sigma_{id} = \sqrt{\sigma_\perp^2 + k_{w,1}\tau_\perp^2 + k_{w,2}\tau_\parallel^2} \le \beta_w f_d \qquad (5.8)$$

where $k_{w,1}/k_{w,2}$ is about 3/2. Formula (7) therefore becomes:

$$\sigma_{id} = \sqrt{\sigma_\perp^2 + 3\tau_\perp^2 + 2\tau_\parallel^2} \le \beta_w f_d \qquad (5.9)$$

representing an ellipsoid having semi-axes $\tau_\parallel = 0.76\sigma_\perp$ and $\tau_\perp = 0.58\sigma_\perp$.

This formula is closer than the ISO one to experimental results, but it still requires considerable calculation time to determine stresses $\sigma_\perp$, $\tau_\perp$, $\tau_\parallel$, which refer to the throat section in its actual position (Fig. 5.9b).

Formulae referred to the overturned throat section. Combination formulae are less complicated to apply when they are in terms of stresses $n_\perp$, $t_\perp$ and $t_\parallel$ which refer to the throat section overturned on either of the two sides of the seam (Fig. 5.9a).

It is obviously possible to express design formulae (5.7) and (5.9) also in terms of $n_\perp$, $t_\perp$ and $t_\parallel$ by rotating the reference axes, but this would entail considerable formal complications. The relations between the components $\sigma_\perp$, $\tau_\perp$, $\tau_\parallel$ and the components $n_\perp$, $t_\perp$, $t_\parallel$, in the assumption that the seam has an isosceles triangle cross section, are the following:

$$\sigma_\perp = (n_\perp + t_\perp)/\sqrt{2} \qquad ; \qquad n_\perp = (\sigma_\perp - \tau_\perp)/\sqrt{2}$$
$$\tau_\perp = (t_\perp - n_\perp)/\sqrt{2} \qquad ; \qquad t_\perp = (\sigma_\perp + \tau_\perp)/\sqrt{2}$$
$$\tau_\parallel = t_\parallel \qquad ; \qquad t_\parallel = \tau_\parallel$$

Formulae (5.7),(5.9) therefore become:

$$\sigma_{id} = \sqrt{2(n_\perp^2 + t_\perp^2) + 3t_\parallel^2 - 2n_\perp t_\perp} \le \beta_w f_d \qquad (5.7')$$

$$\sigma_{id} = \sqrt{2(n_\perp^2 + t_\perp^2) + 2t_\parallel^2 - 2n_\perp t_\perp} \le \beta_w f_d \qquad (5.9')$$

The fastest way to simplify calculations consists in interpreting the ultimate limit surface by means of a solid the equation of which does not vary if the reference axes rotate through 45°, which is equivalent to overturning the throat section on one side. The solid having this property is the sphere and its equation can be written in the following form in terms of $\sigma_\perp$, $\tau_\perp$, $\tau_\parallel$:

$$\frac{\sigma_\perp^2}{(\chi f_{u,w})^2} + \frac{\tau_\perp^2}{(\chi f_{u,w})^2} + \frac{\tau_\parallel^2}{(\chi f_{u,w})^2} = 1 \qquad (5.10)$$

or else, using $n_\perp$, $t_\perp$ and $t_\parallel$ components:

$$\frac{n_\perp^2}{(\chi f_{u,w})^2} + \frac{t_\perp^2}{(\chi f_{u,w})^2} + \frac{t_\parallel^2}{(\chi f_{u,w})^2} = 1 \qquad (5.10')$$

Value $\chi f_{u,w}$ indicates the sphere radius and the value of ultimate strength in the most favourable direction for the material i.e. for $\tau_\perp$. These formulae obviously do not take into account the different behaviour of the weld in respect of axes $\tau_\perp$ and $\tau_\parallel$.

Equations (5.10), (5.10'), according to (5.4), give the following verification conditions:

$$\sigma_{id} = \frac{1}{\chi}\sqrt{\sigma_\perp^2 + \tau_\perp^2 + \tau_\parallel^2} \le \beta_w f_d \qquad (5.11)$$

$$\sigma_{id} = \frac{1}{\chi}\sqrt{n_\perp^2 + t_\perp^2 + t_\parallel^2} \le \beta_w f_d \qquad (5.11')$$

Although with different radii spherical strength domains have been adopted by the following codes:
British (BS-153, 1966), assuming $\chi = 0.58$
American (AISC, 1969), assuming $\chi = 0.61$
German (DIN 4100, 1968), assuming $\chi = 0.70$

Cut sphere. When the CNR-UNI 10011-67 Recommendations were revised |11|, the Istituto Italiano della Saldatura rigorously studied the problem of simplifying fillet weld calculation, starting from a strength domain of a spherical type, in order to determine stresses acting on the throat section more quickly. One, however, sought to maintain the interpretation of the different behaviour of shear stresses $\tau_\perp$ and $\tau_\parallel$, but remaining on the safe side. This was done by using a 'cut sphere' (Fig.5.12a), i.e. a sphere having the same radius as the German sphere (DIN 4100, 1968), but cut by two pairs of planes perpendicular to axes $\sigma_\perp$ and $\tau_\perp$ and passing by points $\tau_\perp = 0.58 f_{u,w}$ and $\sigma_\perp = 0.58 f_{u,w}$. The 'cut sphere' section with plane $\sigma_\perp$, $\tau_\perp$ is shown in Fig.5.12b. For the equivalent stress σ_{id} the lower of the following values must be assumed:

168

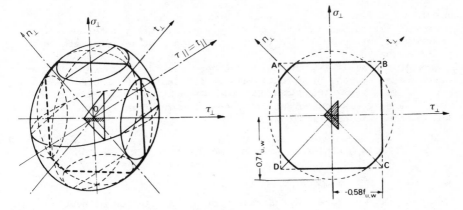

Fig. 5.12

$$\sigma_{id} = \frac{1}{0.70} \sqrt{t_\perp^2 + n_\perp^2 + t_\parallel^2} \leq \beta_w f_d \tag{5.12}$$

$$\sigma_{id} = \frac{1}{0.58\sqrt{2}} (|t_\perp| + |n_\perp|) \leq \beta_w f_d \tag{5.13}$$

Comparison among the various formulae. Fig. 5.13 shows the sections in the two planes $(n_\perp - t_\perp)$ and $(n_\perp - t_\parallel)$ of the above strength domains and, in particular, those defined by ISO formula (5.7'), the generalised elipsoid (5.9'), the spherical formula (5.11') with $\chi = 0.58$ (GB), $\chi = 0.61$ (USA), $\chi = 0.70$ (DBR), and the cut sphere (5.12), (5.13).

The British sphere is internal to the ISO ellipsoid (5.7') and (5.9') and it is everywhere safe. The German sphere is tangential to the generalised ellipsoid (5.9') in the plane $n_\perp$, $t_\perp$ but intersects it in the plane $\sigma_\perp$, $\tau_\perp$ at the bisector lines of the four squares.

Consequently, the German sphere is less penalising than the British

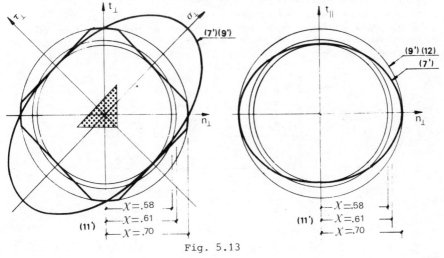

Fig. 5.13

one with reference to joints loaded according to axis $\sigma_\perp$ (i.e., perpendicular to the throat section) and it correctly interprets the strength of joints loaded in the $\tau_\parallel$ direction, but is is unsafe in the $\tau_\perp$ direction.

The 'cut sphere' is tangent to the generalised ellipsoid (5.9') in the plane $n_\perp$, $t_\parallel$ and practically always internal to it in the plane $\sigma_\perp$, $\tau_\perp$. Thus, while less penalising than the British sphere, it is always safe; it is however stricter than the ISO formulae and the German sphere along the $\sigma_\perp$ axis.

5.3 EFFECTS OF INTERNAL ACTIONS AND STRENGTH VERIFICATIONS

5.3.1 *Calculation methods*

The foregoing section shows that the various calculation methods for welded joints can be classified, substantially, in two groups |10|: those using an ultimate strength domain having the shape of an ellipsoid and those using a spherical one. The 'cut sphere' method |11| can be considered as a variant of the latter class.

ECCS Recommendations suggest the adoption of the β_w values given in the table of Fig. 5.14 as a function of yield stress f_y of the material of which the welded elements are made.

The same Recommendations allow the following strength domains to be adopted:

Ellipsoid ISO with $k_w = 3$ and thus formulae (5.7) or (5.7')

Sphere having its radius equal to $f_{w,u}/\sqrt{3} = 0.58 f_{u,w}$ and thus formulae (5.11) or (5.11') with $\chi = 0.58$

5.3.2 *Joints under tension*

In the case of side welds (Fig. 5.15), the throat section of each seam can be considered overturned on the seam side lying in the contact plane of the two welded parts.

The two throat sections of each pair of seams therefore lie on the same plane and undergo a shear stress which can be considered as uniformly distributed in the resistant section, according to the basic assumption. Therefore, considering the overturned throat sections, one

β_w	f_y (N mm^{-2})
1.43	$\leq$ 265
1.25	> 265 $\leq$ 295
1.18	> 295 $\leq$ 355
1.00	> 355 $\leq$ 390

Fig. 5.14

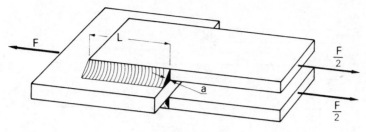

Fig. 5.15

has:

$$t_{||} = \frac{F}{\Sigma La}$$

in this case: $\Sigma La = 4La$.

In the case of front welds (Fig. 5.16a), the throat section can be considered overturned on the plane of either side of the seams. Depending on whether the throat section is overturned on x or y axis (Fig. 5.16b), one has:

(y axis) $n_{\perp y} = \dfrac{F}{2La}$; $t_{\perp y} = 0$

(x axis) $n_{\perp x} = 0$; $t_{\perp x} = \dfrac{F}{2La}$

For the calculation of joints with inclined welds (Fig. 5.17), a good approximation is obtained if only bi-axial stresses are taken into account in the throat section.

The throat section is overturned on the seam side which is set transversely to the direction of loading and the load F is resolved into components $N = F \sin \theta$ and $V = F \cos \theta$. Then

$$n_{\perp} = \frac{N}{2La} = \frac{F \sin \theta}{2La} \quad ; \quad t_{||} = \frac{V}{2La} = \frac{V \cos \theta}{2La}$$

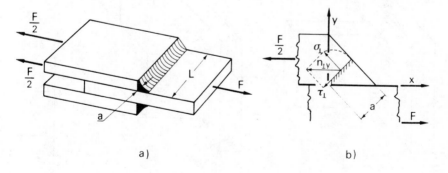

a)   b)

Fig. 5.16

171

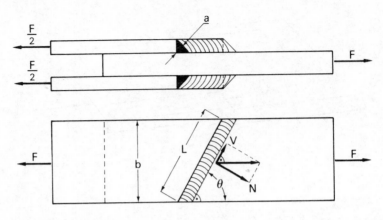

Fig. 5.17

5.3.3 *Joint under combined bending and shear*

The joint of Fig. 5.18 is undergoing a shear force and a bending moment

$$V = F \quad ; \quad M = FL$$

The resistant section of weld lies in the vertical plane and consists of two rectangular sections corresponding to the overturned throat sections having dimensions a x h, and therefore it has an area $A = 2ah$ and section modulus $W = 2ah^2/6$.

Maximum stress due to the bending moment is:

$$n_{\perp max} = \frac{M}{W} = FL \frac{3}{ah^2}$$

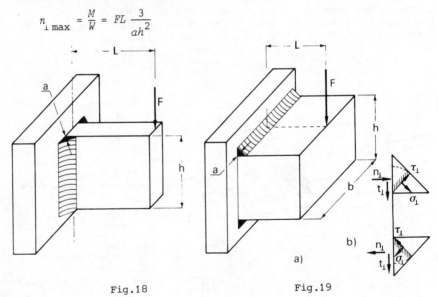

Fig.18   Fig.19

This result is correct on the basic assumption that the neutral axis of the joint bisects the seam length.

The shear force is generally considered uniformly distributed on the throat section. This assumption gives:

$$t_{\parallel} = \frac{F}{2ah}$$

If there are horizontal welds (Fig.5.19), first consider the effect of bending moment $M = FL$ alone. Bending stresses act on the resistant section obtained by overturning the throat sections of the two seams on the side lying in the plane of the joint. In this case, if ah is the section modulus of the overturned throat sections and this can be considered equal to that of two concentrated masses, $W = bah$. Bending stress is therefore:

$$n_{\perp} = \frac{M}{W} = \frac{FL}{bah}$$

The joint is also subject to a shear force $V = F$ producing tangential stresses that can be considered uniformly distributed over the throat section. Hence:

$$t_{\perp} = \frac{F}{2ba}$$

A tipical example of a combination of vertical and horizontal welds is the beam-to-column connection shown in Fig. 5.20a. Calculation can be carried out as for the preceding types of joint, provided that the various parts of the joint are of more or less the same stiffness and fillet weld dimensions are adequate for flange and web thicknesses of the I-beam to be connected.

The resistant section is comprised of the various throat sections of all the seams, overturned on the joint plane, and it must bear a shear force $V = F$ and a bending moment: $M = FL$.

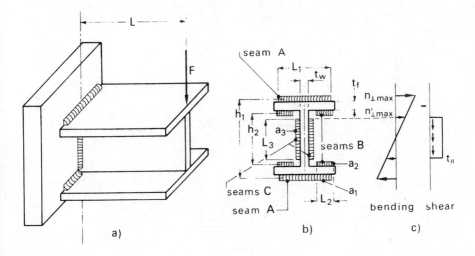

Fig. 5.20

173

Conventionally it supposes that the shear force is carried by the web seams alone, the tangential stresses being uniformly distributed over their throat sections and having a value (Fig. 5.20c):

$$t_{\parallel} = F/2a_3L_3$$

Bending stresses can be evaluated by considering the overall area of the throat section corresponding to seam 'A' 'B' and 'C' so that the maximum stress value is:

$$n_{\perp max} = M/W$$

Strength checks must be made for the most highly stressed parts of the joint. They are the external flange seams 'A' and the ends of the web seams 'C'.

Another simplified calculation method separates the effects of bending and shear, by assuming that the web seams alone resist the shear forces and the flange seams resist only the bending moments.

5.3.4 *Joints under combined torsion, bending and shear*

The various methods for calculating torsionally loaded welded joints are generally based on simplifying assumptions. The two most commonly applied ones are the 'polar moment' method and the 'two forces' method.

The 'polar moment' method extends to the case of fillet welds the principles of material strength theory, which are strictly valid only for circular sections.

All the throat sections are overturned on the side lying in the plane of the connection. The centroid of the overturned sections G is assumed as the centre of rotation and the polar moment of inertia I_o of the overturned throat sections is evaluated with reference to that point (Fig. 5.21).

The maximum load S_{max} is thus supposed to act at the farthest seam point from G, in a direction perpendicular to radius r_{max} joining Q to G.

The value of S_{max}, therefore, is:

$$S_{max} = \frac{Tr_{max}}{I_o}$$

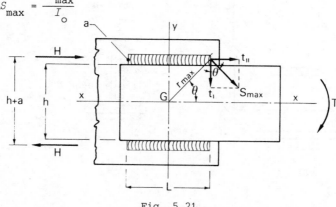

Fig. 5.21

T being the torsional moment.

By resolving S_{max} in two directions, one parallel to and the other normal to the direction of welding values for $t_{||}$ and $t_{\perp}$, can be obtained and used in strength checks. On the overturned throat section one has:

$$t_{\perp} = S_{max} \cos \theta \quad ; \quad t_{||} = S_{max} \sin \theta$$

The 'two forces' method considers the torsional moment T as replaced by a couple given by two opposing forces H acting in the parallel seams, so that, for rotational equilibrium:

$$H(h + a) = T$$

In the seams these forces produce only a tangential stress:

$$t_{||} = \tau_{||} = H/aL$$

Because of its simplicity , this method is used to illustrate the following few significant examples.

In the case of side welds (Fig.5.22), torsional moment $T = Fa$ is replaced by a couple of force H acting at the middle of the seam sections. Since -as is usual in practice - $a \ll h$, $H = Fe/h$.

The two forces H produce tangential stresses in the seams expressed by:

$$\tau_{||} = \frac{H}{La} = \frac{Fe}{hLa}$$

The value of the stresses due to shear $V = F$, also considered uniformly distributed in the seam, is:

$$\tau_{\perp} = \frac{F}{2La}$$

In the case of front welds (Fig. 5.23), the torsional moment Fe is replaced by the couple formed by the two opposite forces V acting on the seams, so that $V = Fe/z$.

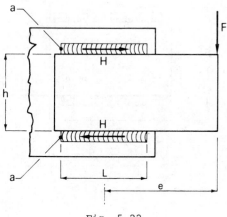

Fig. 5.22

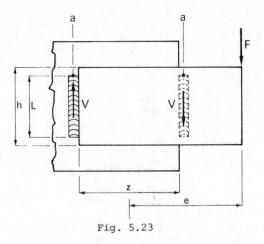

Fig. 5.23

These two forces produce tangential stresses in the seams, equal to:

$$t'_{\|} = \frac{V}{La} = \frac{Fe}{zLa}$$

The tangential stresses due to shear force F can also be considered uniformly distributed in the welds, their value being:

$$t''_{\|} = \frac{F}{2La}$$

In the case of the combination of two side welds and one front weld (Fig. 5.24), the torsional moment Fe produces tangential stresses in the two side seams of:

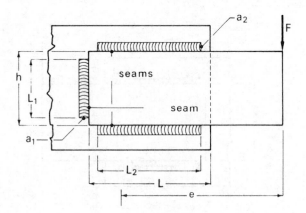

Fig. 5.24

$$t_{||} = \frac{Fe}{ha_2L_2}$$

As side welds are less fit to bear shear forces, the direct shear action from F can be entrusted only to the front weld. Its value is:

$$t_{||} = \frac{F}{a_1L_1}$$

By using these stresses, strength checks are separately given for side and front welds.

If the seam form a closed section undergoing torsion (Fig. 5.25), the tangential stresses in the seams can be evaluated by means of Bredt's classical formula for thin closed sections:

$$t_{||} = \frac{T}{2Aa}$$

where:
A is the area defined by the segments axially bisecting the seam throat sections, overturned on the connection plane
a is seam throat depth.

If the section is not completely welded on its perimeter, the connection can be considered as a pair of vertical welds and a pair of horizontal welds. The torsional moment can than be distributed proportionally to maximum moments T_1 and T_2 which each pair of welds can bear:

$$T_1 = t_{||}a_1L_1L \qquad ; \qquad T_2 = t_{||}a_2L_2h$$

As $T = T_1 + T_2$, one can assume

$$t_{||} = \frac{T}{LL_1a_1 + hL_2a_2}$$

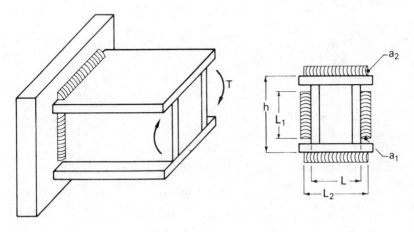

Fig. 5.25

177

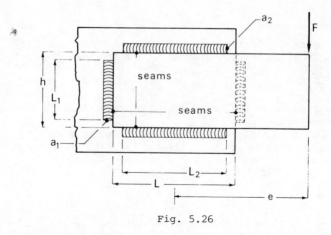

Fig. 5.26

in all the seams. (Suffixes 1 and 2 refer to vertical and horizontal welds respectively).

If the above is extended to the case of Fig.5.26, a good approximation of the effect of shear $V = F$ is obtained by considering stresses $t_\perp$ due to F in the horizontal welds as nil, the shear being entirely absorbed by vertical seams. Hence, for the vertical seams:

$$t_\perp = \frac{Fe}{a_1 L_1 L + a_2 L_2 h} + \frac{F/2}{a_1 L_1}$$

and for the horizontal seams:

$$t_{||} = \frac{Fe}{a_1 L_1 L + a_2 L_2 h}$$

It is thus possible to balance the external forces by means of $t_{||}$ stresses alone.

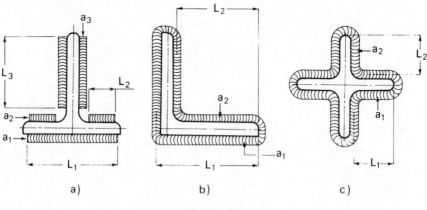

a)   b)   c)

Fig. 5.27

178

Cross sections such as angles, tees and cruciform shapes (Fig. 5.27), in which the shear centre is at the intersection of the centre lines of the elements forming them, have a low degree of torsional stiffness.

Moment of inertia for uniform torsion (see 8.2.5) for the welding resistant section, as in the case of open profile section formed by thin-walled elements, is expressed by:

$$I_T = \frac{1}{3} \Sigma_i L_i a_i^3$$

where the seam is extended to the various parts of the overturned throat section having dimensions a_i x L_i.

Tangential stresses due to uniform torsion vary linearly with the thickness a of each seam of the resistant section and they have a direction parallel to the seam axis. Their maximum value is:

$$t_{\parallel max} = \frac{T}{I_T} a_{max}$$

and it occurs in the seam having the thickest throat section.

REFERENCES

1. Costa, G., Daddi, I. and Mazzolani, F. M. (1977) *Collegamenti saldati*, CISIA, Milan
2. Pratt, J. L. (1979) *Introduction to the Welding of Structural Steelwork*, Constrado, Croydon
3. Granjon, H. (1972) La formazione di cricche a freddo nella saldatura degli acciai, *Rivista Italiana della Saldatura*, 3, 159-172
4. Daddi, I. and Mazzolani, F. M. (1972) Determinazione sperimentale delle imperfezioni strutturali nei profilati di acciaio, *Costruzioni Metalliche*, 5, 374-394
5. Costa, G. (1976) Principali difetti dei giunti saldati, *La Metallurgia Italiana*, Atti notizie, 101-107
6. McMaster, R. C. (1959) *Non-Destructive Testing Handbook*, The Ronald Press Company, New York
7. Welding Inspection (1968) American Welding Society, New York
8. Xarren McGonnagle, J. (1969) *Non-Destructive Testing*, Gordon and Breach Science Publishers, New York
9. Blodgett, O. W. (1966) *Design of Welded Structures*, The James F. Lincoln Arc Welding Foundation, Cleveland, Ohio
10. IIW - Commission XV (1976) Design rules for arc-welded connections in steel submitted to static load, *Welding in the World*, 14, n. 5/6
11. Guerrera, U. (1971) *Esperimenti alla base delle modifiche alle Norme CNR-UNI 10011-67*, Allegato III al Doc. CUS-UNIMET (GL1), 6

C H A P T E R   S I X

Bolts

6.1 TECHNOLOGY OF BOLTS

6.1.1 *Classification of bolts*

Bolts are connecting elements consisting of:

A metal pin having a head (usually hexagonal) and a partially threaded shank (Fig. 6.1a)

A nut (usually hexagonal, Fig. 6.1b)

Washers (usually round, Fig. 6.1c)

Where vibrations occur and the nut might loosen, lock nuts (Fig.

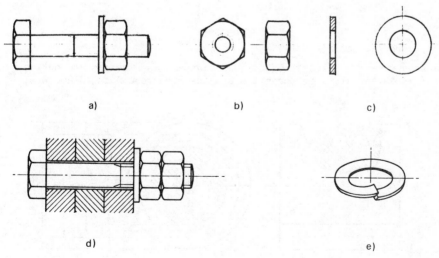

a)   b)   c)

d)   e)

Fig. 6.1

6.1d) or spring washers (Fig. 6.1e) should be used. Bolts are divided
into various grades according to the material of which they are made.
Mechanical properties of European bolts shown in Fig. 6.2 are as
follows:

Minimum tensile strength f_t (in N mm^{-2}) measured on a machined test
piece or on an entire bolt

Minimum yield strength f_y (or the conventional 0.2% proof strength
corresponding to a permanent set of 0.2%, measured on a specimen

Minimum yield strength $f_{y,N}$ (or 0.2% proof strength) measured on
the entire bolt

Ultimate strain ε_t.

The lower value between $f_{y,N}$ and 0.70 f_t is conventionally assumed
as the characteristic value $f_{k,N}$ of tensile strength. High strength
bolts (grade 8.8, 10.9, 12.9) usually have the shank completely
threaded.

Grade	f_t	f_y	$f_{y,N}$	ε_t	$f_{k,N}$
4.6	390	205	–	25	235
5.6	490	295	275	20	275
6.6	590	350	335	16	335
8.8	785	630	570	12	550
10.9	980	880	775	8	685
12.9	1175	1060	930	8	820

Fig. 6.2

6.1.2 *Bolt geometry*

To evaluate the strength of bolted joints it is necessary to define
the resistant areas of the bolts and their corresponding diameters.

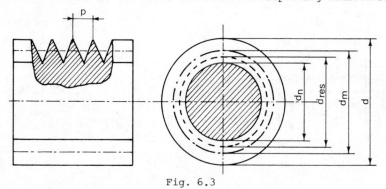

Fig. 6.3

This can be done by reference to Fig. 6.3 |1|, where:

p = thread pitch
d = nominal diameter of the bolt
d_n = diameter of the core of the shank
d_m = average diameter
d_{res} = resistant diameter = $(d_n + d_m)/2$

From the above the following areas are derived:

METRIC BOLTS

d mm	P mm	A_{res} mm^2	A mm^2	A_{res}/A	d mm	P mm	A_{res} mm^2	A mm^2	A_{res}/A
8	1.25	38.6	50.3	0.77	33	3.5	694	855	0.81
10	1.50	58.0	78.5	0.74	36	4.0	817	1018	0.80
12	1.75	84.3	113	0.75	39	4.0	976	1195	0.82
14	2.00	115	154	0.75	42	4.5	1120	1385	0.81
16	2.00	157	201	0.78	45	4.5	1310	1590	0.82
18	2.50	192	254	0.75	48	5.0	1470	1810	0.81
20	2.50	245	314	0.78	52	5.0	1760	2124	0.83
22	2.50	303	380	0.80	56	5.5	2030	2463	0.82
24	3.00	353	452	0.78	60	5.5	2360	2827	0.83
27	3.00	459	573	0.80	64	6.0	2680	3217	0.83
30	3.50	581	707	0.82	68	6.0	3060	3632	0.84

US BOLTS

d inch	P mm	A_{res} mm^2	A mm^2	A_{res}/A	d inch	P mm	A_{res} mm^2	A mm^2	A_{res}/A
1/4	1.27	20.6	31.6	0.65	1 1/4	3.6	625	792	0.78
3/8	1.6	50.3	71.0	0.71	1 3/8	4.2	748	958	0.78
1/2	2.0	91.6	126	0.72	1 1/2	4.2	910	1140	0.80
5/8	2.3	146	198	0.74	1 3/4	5.1	1226	1551	0.79
3/4	2.5	215	285	0.76	2	5.6	1613	2027	0.80
7/8	2.8	298	388	0.77	2 1/4	5.6	2097	2565	0.82
1	3.2	391	506	0.77	2 1/2	6.4	2581	3167	0.81
1 1/8	3.6	492	641	0.77	2 3/4	6.4	3181	3832	0.83

Fig. 6.4

THEORY AND DESIGN OF STEEL STRUCTURES

$A = \pi d^2/4$ - area of the non-threaded part of the shank
$A_{res} = \pi d^2_{res}/4$ - resistant area

For standard metric threads having a triangular fillet:

$d_m = d - 0.64952\ p$ and $d_n = d - 1.22687\ p$

The table of Fig. 6.4 shows geometrical characteristics of metric and U.S. standard bolts normally used in steel construction.

6.1.3 *Bolt tolerances*

Tolerances are necessary both for hole/bolt clearance and for the length of the non-threaded part of the shank.

As a general rule hole/bolt clearance should be as small as possible to minimize joint slip before getting contact between plates and bolt shanks.

If hole diameter is ϕ, the usual permitted hole/bolt clearance is:

$\phi - d \leq 2$ mm   for $d \leq 24$ mm

$\phi - d \leq 3$ mm   for $d > 24$ mm

Joint slip can be practically avoided using 'fitted' holes allowing a clearance of 0.1 - 0.2 mm, but difficulties can arise during assembly.

Optimum length of the non-threaded part of the shank is equal to the thickness of the plates to be joined, so that all contact between plates and bolt takes place along the non-threaded part. It is therefore good practice to have threading start inside the washer (Fig. 6.5a). Shear strength may then be evaluated on the basis of shank area A, while tensile strength has to be computed on the basis of resistant area A_{res}. If threading begins inside the thickness of the connected plates (Fig. 6.5b), or if the shank is completely threaded both shear and tensile strength have to be evaluated with reference to A_{res} |1|.

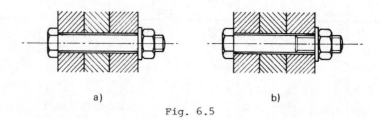

a)   b)

Fig. 6.5

6.1.4 *Hole spacing*

The ultimate strength of bolted connections may be evaluated assuming some hypothesis on the redistribution of internal forces as suggested by experimental evidence.

Therefore the following dimensional limitations on geometry of
184

connection have to be fulfilled in order to keep within the bounds derived from that experimental evidence (Fig. 6.6):

Hole spacing in the direction of the applied force:

Elements in compression $15t_1 \geq p \geq 2.5d$
Elements in tension $25t_1 \geq p \geq 2.5d$

Hole distance from the free edge in the direction of the applied force:

$$3d > a > 1.2d$$

Unstiffened edge $a \leq 6t_1$
Stiffened edge $a \leq 9t_1$

Hole distance from the free edge normal to the applied force:

Elements in compression $a_1 \geq 2.0d$
Elements in tension $a_1 \geq 1.5d$
Unstiffened edge $a_1 \leq 6t_1$
Stiffened edge $a_1 \leq 9t_1$

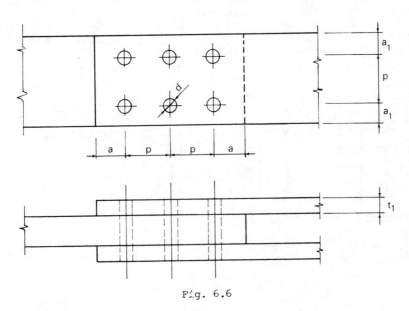

Fig. 6.6

Only by testing can the adequacy be checked for connections not complying with the above limitations.

6.1.5 *Tightening*

Tightening involves the application of a twisting moment to the joint before external forces are applied. This produces nut shortening and shank elongation. Only a portion of the twisting moment is absorbed by friction between plate and bolt on one hand, and plate and nut on the other. The remaining part is carried by the shank. Thus, once

185

the bolt is tightened, the joint is loaded by self-stresses consisting of:

A bolt tension, balanced by compression in the plates

A torsion in the bolt, balanced by the plate/bolt friction

Tightening increases joint performance with reference to serviceability limit states:

In shear joints it prevents plates slipping and, therefore, inelastic settlements in the structure

In tension joints it prevents plate separation (reducing corrosion dangers) and it improves fatigue resistance (see 7.3.6)

However tightening must not exceed a certain limit, to avoid attaining joint ultimate capacity. This is illustrated by the following three experiments.

Experiment No. 1. Tighten a bolt connecting two practically rigid plates until contact is obtained. Then apply (Fig. 6.7a) a tensile load N to the plates and measure bolt elongation $\Delta L = \Delta L(N)$. From the results (Fig. 6.7c, curve 1), an elastic limit load $N_{e,1}$, an ultimate load $N_{u,1}$, an elongation ΔL_1, corresponding to that load, and an ultimate elongation $\Delta L_{t,1}$ are derived.

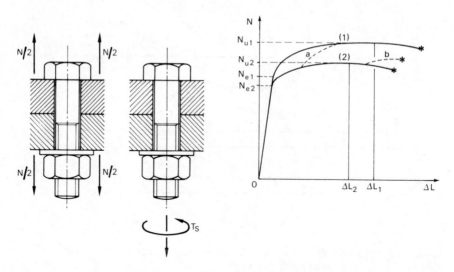

Fig. 6.7

Experiment No. 2. Tighten the bolt until the two plates are in contact and then apply a twisting moment T_s to the nut (Fig. 6.7b). Measure shank tension N and its elongation ΔL: curve 2 of Fig. 6.7c will be obtained, which gives an elastic load $N_{e,2}$, an ultimate load $N_{u,2}$ an elongation ΔL_2 corresponding to that load and an ultimate elongation $\Delta L_{t,2}$.

186

Experiment No. 3. Repeat experiment No. 2 until shank elongation is $\Delta L < \Delta L_2$. Then stop turning the nut and apply a force N to the plates, following the same procedure as in experiment No. 1. Measuring axial force in the bolt and its elongation, dashed line (a) of Fig. 6.7c will be obtained.

If experiment No. 2 is repeated until shank elongation is $\Delta L \geq \Delta L_2$ and then one stops turning the nut and applies force N to the plates, dashed line (b) of Fig. 6.7c is obtained.

Qualitative conclusions which can immediately be drawn from the three experiments are: Curve (1) shows the behaviour of a simply tensioned bolt. Experiment No. 2 shows that, as long as strains are small, the twisting moment induced by tightening is practically entirely absorbed by the friction between nut and plates and does not load the shank. As strains grow, the part of the twisting moment loading the shank increases and the ultimate load of the joint is reduced by about 20%, and ultimate elongation by about 30%. If, however, an external tension ΔN is applied to a tightened bolt at a lower value than $N_{u,2}$ corresponding to the ultimate value for tension and twisting, the part of the twisting moment that would normally have loaded the shank tends to decrease due to plate separation and thus becomes uninfluential for joint ultimate performance. It does however becomes influential if the tightening caused an elongation ΔL in the shank greater than ΔL_2 corresponding to the ultimate load for tension and twisting.

Therefore, tightening does not influence joint ultimate performance for statical loads provided it is duly limited and thus controlled.

Analysing experiment No. 2 from a quantitative point of view, let $\sigma = N/A_{res}$ be the mean axial stress and $\tau = T_{s,1} \, d/2 \, I_o$ be the maximum tangential stress corresponding to the portion $T_{s,1}$ of the applied twisting moment loading the shank. The strength limit can be defined by the following formula $|1|$:

$$\sigma_{id} = \sqrt{\sigma^2 + 3\tau^2} = \eta\sigma$$

The coefficient η depends on thread pitch, on friction between shank and nut and between plate and bolt head or nut. Experiments carried out by many researchers on metric, Whitworth and UNC threads have shown that η lies between 1.15 and 1.25.

For safety, tightening can be limited to a value that does not allow the shank tensile force N to exceed the elastic limit $N_{e,2}$ of Fig. 6.7c; that is: $\eta\sigma \leq f_e$, f_e being the bolt elastic limit. Therefore, if $\sigma = f_e/\eta = 0.8 \, f_e$, tightening certainly does not influence joint ultimate load: in fact for $N < N_{e,2}$ bolt elongation is certainly lower than ΔL_2.

For tightening bolts, one of the two following methods may be used $|2-4|$:

1st method. A dynamometric wrench is used. The preload N_s induced in the bolt shank due to the twisting moment T_s may be expressed by the following relation:

$$T_s = \chi N_s d$$

where χ is a coefficient depending on material and on surface conditions. Generally the value $\chi = 0.20$ is assumed. The torque to be applied is therefore:

$$T_s = 0.20 \, N_s d$$

As $f_e = f_{k,N}$ is conventionally assumed: $N_s = 0.80 \, f_{k,N} \, A_{res}$ and $T_s = 0.16 \, f_{k,N} \, A_{res} \, d.$

2nd method. Nut rotation is checked. For this purpose the nut is tightened manually or with an impact wrench until the plates between the head and the nut are in contact. Then the nut is given a part turn of 90° to 120° with tolerances of 60° upwards (i.e. 0.25 to 0.50 turn).

The second method is empirical, but it offers two substantial advantages over the first:
(a) It is easier to apply
(b) It is more reliable, because it gives a lower scatter in results.

In order to justify the latter statement, consider Fig. 6.8 illustrating the tension induced in the shank by tightening, against the number of turns of the nut after plate contact has been established.

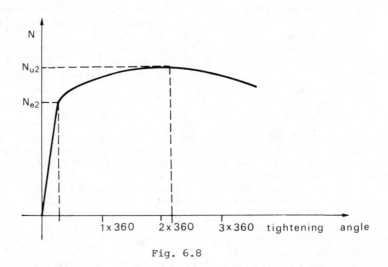

Fig. 6.8

As shown earlier, the value of $N_s = 0.8 \, A_{res} \, f_e$ practically corresponds to the attainment of bolt elastic limit, taking into account the portion of twisting moment acting on the shank. Therefore $N_e = N_{e,2}.$

Applying 1st method, the safety coefficient against bolt failure under tension and twisting is:

188

$$N_{u,2}/N_{e,2} = 1.2 \text{ to } 1.4$$

Conversely, for an elongation ΔL_2 corresponding to load $N_{u,2}$, 2 to 2.5 turns of the nut are necessary, against 0.25 to 0.50 turn imposed when tightening. The safety coefficient against bolt failure due to an excessive turning of the nut is therefore greater than $2/2.50 = 4$.

Sometimes it is necessary to check that the preload in the bolt corresponds to the applied twisting moment T_s. In this case, using a dynamometric wrench, one of the two following methods may be used:
(a) The bolt is turned through a further 10°: the measured value of the torque applied to obtain such a rotation must not be less than T_s.
(b) Nut position with reference to the plates is marked. The nut is untightened by about 1/6 turn. A torque not less than T_s must be required to bring the nut to its initial position.

6.2 BOLTED JOINT STRENGTH

The strength of bolted joints can conventionally be evaluated by means of formulae that interpret their true behaviour as shown in tests |5,6|. For this purpose joints can be classed as those with:
(a) Bolts in shear only
(b) Bolts in tension only
(c) Bolts simultaneously in tension and shear.

For each type of joint, strength must be distinguished with reference to:

Serviceability limit states
Ultimate limit states.

The latter corresponds to joint failure. The former defines possible limitations to joint deformability such as slipping with ensuing settlement due to the hole/bolt clearance for shear bolts and loss of pressure between the plates leading to plate separation for tension bolts.

6.2.1 *Joints with bolts in shear only*

Shear bolts are typical of steel constructions and are found for example when cover plates are used at a splice in an element to restore continuity (Fig. 6.9). The simplest joint of this kind is

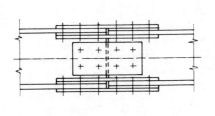

Fig. 6.9

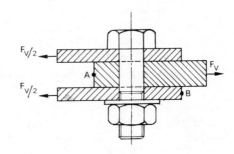

Fig. 6.10

189

shown in Fig. 6.10: an external force F applied to one plate is transmitted to two plates through only one bolt connecting all three plates.

Fig. 6.11a shows the typical behaviour of the joint plotting the relative displacement ΔL between points A and B in Fig. 6.10 as a function of applied load F_V. Four distinct phases can be observed.

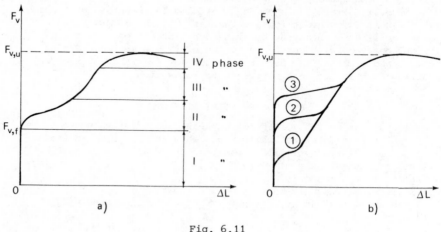

Fig. 6.11

1st phase: Load increases but no relative displacement is observed; force transmission therefore takes place by friction between the plates. The phase ends when the friction limit of the joint is reached.

2nd phase: For $F_V = F_{V,f}$ slipping of the joint suddenly begins due to the clearence between bolt and hole. It stops when the shank comes into contact with the plates. During this phase the applied load is practically constant.

3rd phase: Elastic behaviour is evident: ΔL is practically proportional to F_V. The elastic phase ends when the elastic limit is reached either in the connected plates or in the bolt.

4th phase: Plastic range is reached; considerable deformation takes place for slight load increases, and the joint fails at a load value $F_{V,u}$. At ultimate load the following mechanisms are possible:

Collapse due to bolt failure (Fig. 6.12a)
Collapse due to hole failure (Fig. 6.12b,c)
Collapse for tension failure of plates (Fig. 6.12d).

If the experiment is repeated varying the bolt tightening or the superficial treatment of the plates, the joint behaviour is qualitatively analogous to that of Fig. 6.11b. Theoretically only the load $F_{V,f}$ at which slipping takes place and the elastic phase begins, should change. Nevertheless, for the failure mechanisms shown in Figs 6.12b,c,d tightening may give a small increase of ultimate load and even a great increase of ultimate elongation ΔL_t (see 6.2.1.5). In most cases it is possible to define:

190

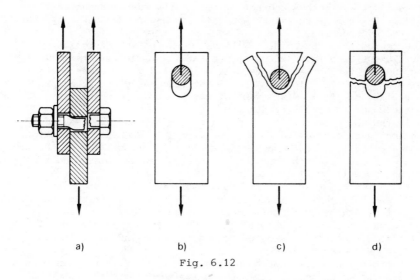

a)   b)   c)   d)

Fig. 6.12

Serviceability limit force $F_{V,f}$ as that which characterizes the onset of joint slipping and which essentially depends on friction between the plates

Ultimate limit force $F_{V,u}$ as the load at joint failure

6.2.1.1 Serviceability limit state: joint slipping

If inelastic settlements of connections must be avoided in order to reduce the deformability of the structure or to fulfil some functional requirement, the slipping resistance $V_f > 0$ has to be obtained. Its value depends upon bolt tightening and surface treatment. At present only a few statistical studies |4,6| have been carried out on the elementary joint of Fig. 6.10: actual codes therefore quote an average value $V_{f,m}$ and correct it by applying a safety coefficient γ_f of between 1.2 and 1.4.

For an external friction load F_f constant in time high strength bolts (grades 8.8 and 10.9) are used in order to ensure that relaxation phenomena do not significantly diminish bolt preloading

Design load $V_{f,o}$ corresponding to joint slipping can be expressed by:

$$V_{f,o} = \mu \, N_s \, n_f / \gamma_f$$

where:

n_f is the number of surfaces in contact
μ is surface friction coefficient
N_s is the preload of bolt due to tightening

The friction coefficient μ essentially depends on the treatment of the joint surfaces and on their degree of cleanliness when the joint is assembled.

Surfaces must be properly prepared and in particular must be free from paint, oil or rust. Possible oil smears must be removed by means of chemical detergents and not by flame; the rust coating that might form in the time interval between blast cleaning and assembly must be removed by brushing.

Numerous tests performed |7| have shown that the friction coefficient associated with the preload values imposed by either of the two tightening methods described in 6.1.5 does not exceed the value μ_{max} = 0.60. Thus, μ = 0.45 can reasonably be assumed for joints assembled immediately after surface treatment and μ = 0.30 for surfaces not carefully treated.

If the above values of the friction coefficient are assumed, tightening is controlled, and γ_f = 1.25 ECCS-CECM Recommendations estimate that $V_{f,o}$ corresponds to a slipping probability of the order of 10^{-3} to 10^{-4}. Such a probability level can be considered acceptable for serviceability limit states.

If one wishes to use a friction coefficient μ > 0.45 or to use different surface treatments (galvanized or special painted surfaces) adequate tests of joint efficiency must be performed. For this purpose and in accordance with the ECCM-CECM Recommendations |1| at least five specimens analogous to one of those indicated in Fig. 6.13 have to be tested.

The relative displacements of plates a and c with reference to plate b are measured. In this way slipping of both the upper and lower parts of the specimen is determined so that two significant results are obtained from each test.

Slipping value V_f is defined as the value corresponding to a relative slipping of 0.15 mm and the friction coefficient is defined by the following relation:

$$\mu = V_{f,o}/4N_s$$

Four of the five specimens must be tested using normal load increments of the order of 10 to 20 kN per minute; the fifth specimen must satisfy a long term requirement. It must carry a load equal to 90% of the average of the slipping loads for the first four specimens and must remain loaded for three hours. If a relative displacement $\Delta L \leq 2 \times 10^{-3}$ mm takes place it is unloaded and then reloaded to determine its slipping load as with the four previous specimens.

If the coefficient of variation of the ten values V_f thus determined does not exceed 0.08 the minimum value may be assumed as the design value $V_{f,o}$. Otherwise, tests must be performed on n further specimens $n > (100 \delta)^2/(2 \times 2.5^2)$.

If, however, the long term test fails, at least three more specimens must be tested to prove that the load associated with the friction coefficient μ chosen for the joint design cannot cause a relative displacement ΔL = 0.3 mm throughout the life of the structure. For this purpose, a displacement-time logarithmic curve can be extrapolated by a tangent, as shown in Fig. 6.14.

Once the design life of the structure Δt_s is defined, it is possible to interrupt tests at time t_i for which the tangent to the experimental curve passes through the point defined by an abscissa $t = \log \Delta t_s$ and an ordinate ΔL = 0.3 mm. In the example shown in the figure, the experimental behaviour of specimens 1 and 2 is satisfactory, but that of specimen 3 is not acceptable.

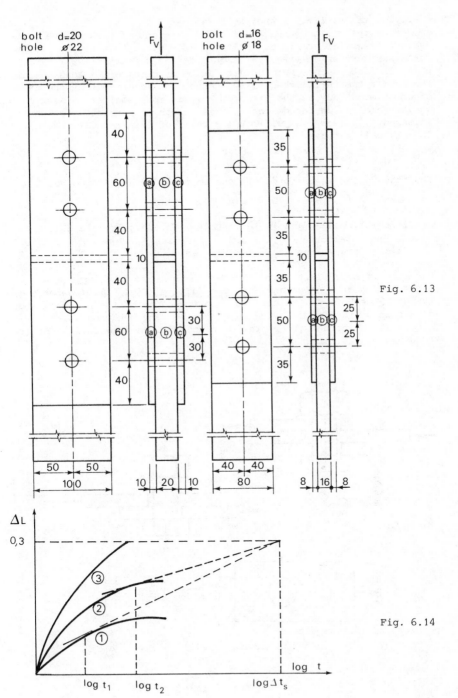

Fig. 6.13

Fig. 6.14

6.2.1.2 Ultimate limit state: bolt failure

To define the design strength of each resistant section of the bolt against shear failure (Fig. 6.12a) it is useless to apply the Huber von Mises' criterion (see 4.6) and to define a coefficient $\gamma_m \geq 1$. A bolt, in fact, cannot be considered a beam, because its length - i.e., the sum of the thicknesses of the connected plates - is of the same order of magnitude as the diameter. It is more logical to assume a conventional value for design shear strength $f_{d,v}$ related to tensile strength $f_{k,N}$ (see Fig. 6.2) on the basis of experimental results. Codes usually assume:

$$f_{d,v} = 0.70 \; f_{k,N}/\gamma_v \qquad \text{with} \qquad \gamma_v = 1$$

Thus the design shear force for one bolt can be assumed to be:

$$V_{d,o} = f_{d,v} \; A = 0.70 \; f_{k,N} \; A$$

or, if the threaded part is in contact with the joint plates:

$$V_{d,o} = f_{d,v} \; A_{res} = 0.70 \; f_{k,N} \; A_{res}$$

6.2.1.3 Ultimate limit state: hole failure

The strength of a joint against hole failure (Fig. 6.12b) depends on the distance between the bolt and the free edge of the plate in the direction of the force acting on the bolt. The behaviour also varies according to whether the plate is in compression (Fig. 6.15a) or in tension (Fig. 6.15b).

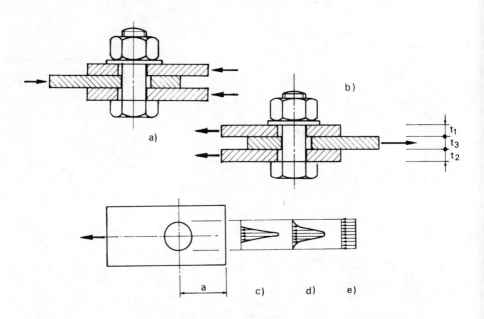

Fig. 6.15

In any case, a conventional value is assumed for the bearing pressure between bolt and plate. The actual bearing pressure distribution is as indicated in Fig. 6.15c in the elastic field and in Fig. 6.15d in the elastic-plastic field. Calculations, however, must be referred to a mean value (Fig. 6.15e), as this is the value to which experimental results are referred.

If t_{min} is the overall thickness of the plates in which contact pressure has the same direction (t_{min} is the lower of t_3 and $t_1 + t_2$, see Fig. 6.15) design strength $V_{d,b}$ against hole failure can be assessed as follows:

$$V_{d,b} = f_b \, d \, t_{min} = \alpha \, f_d \, d \, t_{min}$$

The conventional design bearing strength f_b is thus related to the design strength f_d of the plate material by a coefficient $\alpha \geq 1$, which is a function of hole distance a from the free edge in the direction of the force acting on the bolt. Its value $|1|$ can be assumed to be 3.0 if the plates are in compression (Fig. 6.15a).

If the plates are in tension (Fig. 6.15b) α has the value α_1, or, if bending effects in the joint can be excluded and at the same time ovalization of the hole (even a significant one) is allowed, α_2.

According to the magnitude of a, α_1 and α_2 may be assumed as shown in the table of Fig. 6.16.

	α_1	α_2
$a \geq 3d$	2.5	2.5
$3d > a \geq 2d$	2.0	2.5
$2d > a \geq 1.5d$	1.4	1.75
$1.5d > a \geq 1.2d$	1.0	1.25

Fig. 6.16

6.2.1.4 Ultimate limit state: plate failure
Design strength against tensile failure of the plate (Fig. 6.12d) is also treated conventionally in practice.

In fact, stress distribution in the neighbourhood of a hole is as

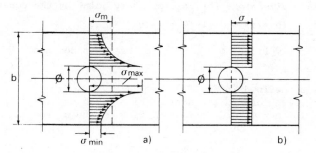

Fig. 6.17

195

shown in Fig. 6.17a in the elastic field. Plastic redistribution at failure (Fig. 6.17b) justifies the use of a mean value of ultimate force equal to:

$$V_{d,t} = f_d A_{res}$$

where:

f_d is the design tensile strength of the plate

$A_{res} = t_{min} (b - \phi)$ is the stress resistant area.

If there are more than one bolt, the resistant area computation might become complex: it must be made on the basis of ultimate load for tension and shear as a function of the possible failure path (Fig. 6.18a).

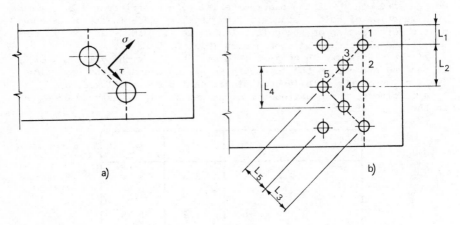

Fig. 6.18

Following an empirical rule, always on the side of safety, the resistant area is that corresponding to the shortest path passing through one or more holes. Thus it is the minimum of those shown in Fig. 6.18b, e.g., those of lengths $2L_1 + 2L_2$; $2L_1 + 2L_3 + L_4$; $2L_1 + 2L_3 + 2L_5$, obviously deducting 3, 4 and 5 holes respectively. Other methods for working out the resistant area are referred to in 8.2.1.

On the basis of the above considerations it is possible to define optimum plate dimensions for an elementary joint by the equation:

$$f_d t_{min} (b - \phi) = \alpha(a) f_d t_{min} d$$

Assuming for simplicity $\phi = d$ (Fig. 6.19a) this becomes:

$$b/d = \alpha + 1$$

where α is as defined in 6.2.1.3.

Optimum values of b/d for a range of a/d values are given in Fig. 6.19b both for $\alpha = \alpha_1$ and for $\alpha = \alpha_2$ (for α_1 and α_2 see Fig. 6.16).

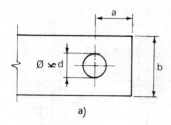

a/d		3	2	1,5	1,2
b/d	$\alpha=\alpha_1$	3,5	3,0	2,4	2,0
	$\alpha=\alpha_2$	3,5	3,5	2,75	2,0

a)   b)

Fig. 6.19

6.2.1.5 Ultimate limit state: effects of friction

In some cases bolt shear strength is considerably higher than the strength against hole or plate failure. If one takes advantage of friction transmission, the hole/bolt bearing stress can be limited and consequently the force acting on the net section will also be limited.

In such a case, however, joint slipping should be considered as an ultimate limit state and not as in 6.2.1.1. In fact, its attainment could perhaps cause the hole and/or net section to fail suddenly. It is therefore advisable to be more cautious when assessing the force that can be transmitted by friction (see 6.2.1.1) by adopting a coefficient $\gamma_f > 1.25$. On the other hand one may assume that only a fraction βF_v of the external force F_v acts on the hole and therefore on the net section. Consequently, the ultimate load V_u has to comply with the following limitations:

Slipping $F_{u,V} \leq V_{f,o} = \mu\, N_s\, n_f/\gamma_f$; $F_{u,V} \leq V_{d,o}$

Bearing $\beta\, F_{u,V} \leq V_{d,b} = \alpha\, f_d\, d\, t_{min}$

Plate failure $\beta\, F_{u,V} \leq V_{d,t} = f_d\, A_{res}$

Actual codes suggest $\gamma_f = 1.5$ and $\beta = 0.5$ to 0.7.

6.2.2 *Joints with bolts in tension only*

Typical joints with bolts in tension are found, among others, whenever the continuity of a structural member is to be restored by means of flanged connections (Fig. 6.20).

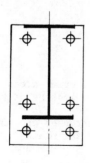 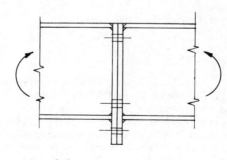

Fig. 6.20

197

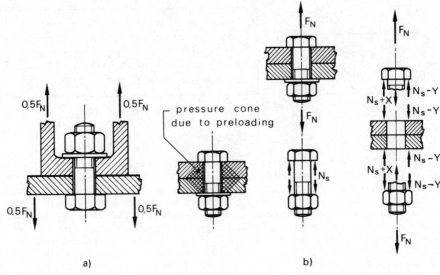

a)   b)

Fig. 6.21

To understand their behaviour, consider the joint formed by two elements connected by only one bolt and loaded by an external force F_N (Fig. 6.21).

Before the external load F_N is applied, each bolt transmits to the joint flanges a total compressive force equal to the preload N_s present in the bolt shank due to tightening: such a force corresponds to an initial elongation of the bolt shank. As external load F_N begins to act, the bolt shank force increases by a fraction X causing a slight elongation of the bolt and a reduction Y in the compression on the flanges. If Y is smaller than N_s, the parts remain in contact and bolt elongation ΔL_1 coincides with plate decompression ΔL_2. They are:

$$\Delta L_1 = X/k_1 \qquad ; \qquad \Delta L_2 = Y/k_2$$

where k_1 and k_2 are, respectively, bolt and plate extensional stiffnesses.

Bolt stiffness is given by:

$$\frac{1}{k_1} = \frac{L_1}{EA} + \frac{L_2}{EA_{res}}$$

A and A_{res} being the shank section and resistant section areas respectively, while L_1 and L_2 are respectively the lengths of the non-threaded and threaded parts of the shank.

It is not easy to determine plate stiffness: stresses are three-dimensional and depend on contact area extension; for plate thicknesses t greater or equal to bolt diameter d experimental evidence has shown that $k_2 \geq 10\ k_1$.

198

Due to bolt equilibrium, $X + Y = F_N$ and due to compatibility, $\Delta L_1 = \Delta L_2 = X/k_1 = Y/k_2$, therefore:

$$X = \frac{F_N}{1 + k_2/k_1} \le \frac{F_N}{11} \quad ; \qquad Y = (1 - \frac{1}{1 + k_2/k_1})\ F_N \ge \frac{10}{11} F_N$$

The increase in tension in the shank is thus less than 10% of the applied external tension F_N. When the preload in the bolt N_s is greater than the elastic limit, bolt stiffness k_1 tends to zero: increase X is thus absolutely negligible.

The above relations are valid as long as the plates remain in contact: i.e., for $Y < N_s$. If $Y > N_s$, the plates separate and the bolt carries the entire external force F_N. Separation therefore begins at $F_N \ge 1.1\ N_s$.

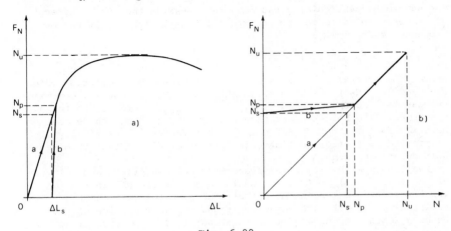

Fig. 6.22

Fig. 6.22a shows the relation between the applied external load F_N and bolt elongation ΔL. Fig. 6.22b shows the relation between F_N and the axial force N acting in the bolt shank. If the bolt has not been tightened, as F_N increases, N increases by an equal amount (Fig. 6.22b, curve a) and once the elastic limit has been reached, the bolt enters the plastic range until failure occurs at a load N_u. If the bolt is tightened, when the external applied load is zero a tensile force N_s acts on the shank causing an elongation ΔL_s. As the external load F_N increases, the tension N in the shank increases very slowly until $F_N = N_p \simeq 1.1\ N_s$, i.e. until the value entailing plate decompression (curve b). For $F_N > N_p$, the tensile force N is again equal to applied load F_N until collapse.

The load N_p to which the loss of prestressing corresponds, can therefore be assumed as a serviceability reference value if plate separation is to be eliminated to avoid dangers of atmospheric pollution. Ultimate load for bolt tension failure is, on the contrary, independent of tightening effects.

199

6.2.2.1 Serviceability limit state

The main difference between the example considered above and actual joint behaviour is that the load F_N is not applied to the bolt head but is transmitted to it through bent plates: thus complete separation is never possible. Therefore the value $N_D = 1.1\ N_s$ deduced above represents the upper limit of the force acting on the bolt before joint separation and it is always on the safe side to assume the decompression force equal to preload N_s due to tightening.

From the point of view of building practice it is obvious that if the possibility of plate separation is to be eliminated, tightening should be performed as indicated in 6.1.5 and high strength bolts should be used where necessary to avoid loss of tightening due to material relaxation.

6.2.2.2 Ultimate limit state

One should be particularly cautious in assessing the design strength of a bolted joint loaded by an external tensile force. It can be expressed as follows:

$$N_{d,o} = f_{k,N}\ A_{res}/\gamma_N$$

where $f_{k,N}$ is the characteristic strength of the bolt material (Fig. 6.2) and A_{res} is the stress resistant area. The factor γ_N, comprises two partial factors, γ_1 and γ_2 ($\gamma_N = \gamma_1 \times \gamma_2$) related to the following phenomena. The bolt head is obtained by cold rolling: therefore the bolt may fail by head detachment. Such a danger is practically eliminated if controlled tightening is performed: in fact it makes bolt defects evident.

Due to compatibility of displacements between plates and bolt head or nut (Fig. 6.23) it is rare to find a bolt perfectly in tension; more frequently a bending moment also acts on the bolt shank diminishing the ultimate capacity of the bolt. The value of the bending moment depends on plate rotation: it is more significant the lower the tensile strength of the bolt and it is negligible when tightening of the bolts reduces the deformability of the plate.

Therefore, realistic values adopted in actual codes for the overall coefficient γ_N are:

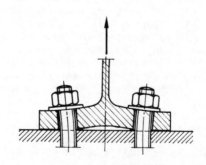

Fig. 6.23

1.25 for tightened high strength bolts
1.33 for untightened bolts, normal or high strength
1.00 for tightened normal bolts when bending stresses on the shank may
be neglected

6.2.3 *Joints under tension and shear*

In many joints tension and shear act simultaneously on bolts |8|.
According to the various limit states different interaction formulae
are applied to assess bolt strength.

6.2.3.1 Serviceability limit state

The slipping load V_f is proportional to the preload between the
surfaces in contact. The strength domain at serviceability limit
state is therefore as shown in Fig. 6.24 and is expressed by the
relation $V_f = V_{f,o}(1 - N/N_s)$, where:

$V_{f,o}$ is the friction force, when no tension is present
N_s is the preload due to tightening

The validity of the above formula should be limited to values of
$N = 0.8 \ N_s$, to provide an adequate margin with reference to plate
decompression.

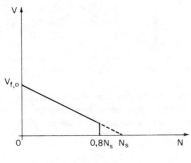

Fig. 6.24

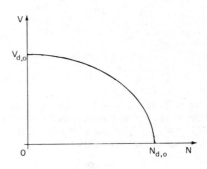

Fig. 6.25

6.2.3.2 Ultimate limit state

From the many experiments that have been carried out, the ultimate
domain for a bolt under the combined effects of an external axial
force N and a shear force V can be assumed as an ellipse (Fig. 6.25).
As a result, the following interaction formula may be used:

$$\left\{\frac{V}{V_{d,o}}\right\}^2 + \left\{\frac{N}{N_{d,o}}\right\}^2 \leq 1$$

This gives safe results whenever the plates and the bolt come into
contact along the shank and not along the threaded part. In fact, the
effects of the axial and shear forces must be combined only in the
shank section: the ultimate limit state can thus be expressed more
realistically be the following relation:

$$\left\{\frac{V}{V_{d,o}}\right\}^2 + \left\{\frac{N}{N_{d,o}} \ \frac{A_{res}}{A}\right\}^2 \leq 1$$

6.3 EFFECTS OF EXTERNAL FORCES AND STRENGTH VERIFICATIONS

Generally speaking bolted joints can be loaded in two ways:

Shear and torsion acting in the plane of the plates connected by the bolts which are loaded in shear (Fig. 6.26a)

Axial and bending loads acting in the direction of the bolts which are loaded in tension (Fig. 6.26b)

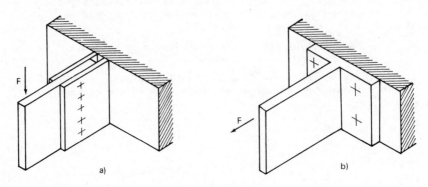

a)   b)

Fig. 6.26

The effects of such loads are distributed between the various bolts on the basis of conventional methods, supported by experimental results. A survey of the most common methods and of the assumptions on which they are based follows.

6.3.1 *Shear and torsional forces*

Referring to Fig. 6.27a the twisting moment has to be referred to a point somehow related to the geometry of the joint. This point represents the instantaneous centre of rotation of the joint. It is not invariable in space as the external loads increase   the irregular distribution of friction forces, the material elastic behaviour and the hole/bolt clearance modify its position |9-12|.

It is therefore preferable to adopt the following simplifying assumptions that have always proved to be on the safe side: the joint plates are supposed infinitely stiff and the bolts perfectly elastic.

According to this hypothesis, the relative displacement of each bolt is:

Constant when due to shear force

Proportional to its distance from the centre of gravity of the bolt group, when due to twisting moment

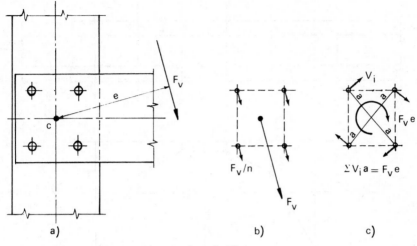

Fig. 6.27

From a qualitative point of view (Fig. 6.27) for working out the effects on bolts one has to proceed as follows:

The external force is referred to bolt centre of gravity, figuring out shear and twisting components (Fig. 6.27a)

Shear force is divided equally between the bolts in the same direction (Fig. 6.27b)

The twisting moment is divided into forces acting on the bolts normal to the line joining the bolt to the centre of gravity and proportional to the distance between the two (Fig. 6.27c)

It should be pointed out that this hypothesis does not take into consideration any force redistribution from the most loaded bolts to the less loaded ones, but it does take into account local plasticity around the holes to distribute shear stress to the bolts. In fact, it is admitted that all the bolts work in contact with the plates: this does not mean that all the bolts slip simultaneously by an amount equal to the hole/bolt clearance, but it does mean admitting a local ovalization of the holes where contact first occurs. This is why it is advisable to limit hole/bolt clearance as indicated in 6.1.3.
The joint shown in Fig. 6.28 demonstrates the role of hole/bolt clearance and of hole ovalizations. If a perfectly elastic behaviour of the plates and of the bolts were admitted and any hole/bolt clearance excluded, the force distribution on the bolts would not have a constant trend, but a hyperbolic cosinusoidal one: the end bolts would be the most loaded ones. Due to the existence of the hole/bolt clearance, it is more realistic to distribute the external force between all the bolts, provided the joint is not too long: this implies limiting bolt spacing as indicated in 6.1.4 and reducing joint length in the direction parallel to the applied force F_v.

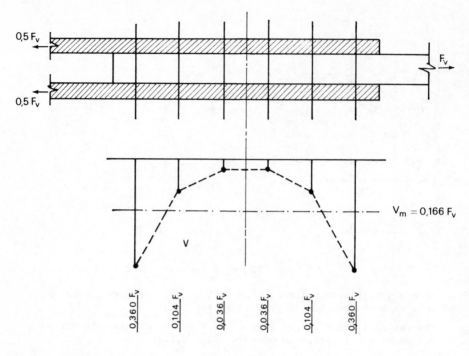

Fig. 6.28

6.3.1.1 Shear forces distribution

As a consequence of the above hypothesis, shear forces can be considered equally distributed between all the bolts (Fig. 6.29a)

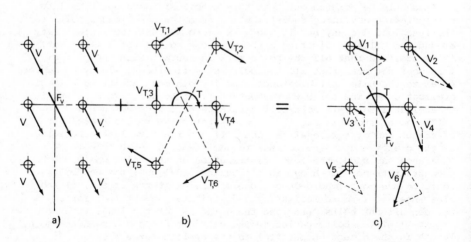

Fig. 6.29

The load acting on each face is:

$$V = \frac{F_V}{n \, n_V}$$

where:

n is the number of bolts
n_V is the number of sections in each bolt resisting the shear force V.

Such a distribution is valid for joints in which the distance between the first and the last bolt, measured in the direction of the shear component, has a length $L \leq 15 \, d$, where d is the nominal diameter of the bolt. If $L > 15 \, d$, the assumption that all the bolts carry equal shares of the shear force is no longer valid. The results of many experiments carried out by CECM-ECCS have shown that the force on the most loaded bolt can be expressed by:

$$V = V_o \, \beta$$

where:

V_o is the force calculated on the basis of all bolts carrying equal shears

$\beta \geq 1$ is a coefficient that can be assumed equal to:

1	for	$L \leq 15 \, d$
$1 + 0.33 \dfrac{L - 15 \, d}{50 \, d}$	for	$15 \, d \leq L \leq 65 \, d$
1.33	for	$L \geq 65 \, d$

6.3.1.2 Distribution of the twisting moment
The twisting moment is distributed between the bolts according to their distance from the centre of gravity. Thus on any bolt i (Fig. 6.29b):

$$V_{T,i} = k \, a_i$$

a_i being the distance between bolt centre and the centre of gravity of the bolt group. From equilibrium to rotation:

$$T = n_V \sum_{1}^{n} V_{T,i} \, a_i$$

Whence:

$$V_{T,i} = \frac{T \, a_i}{n_V \, \Sigma \, a_i^2}$$

6.3.1.3 Combined effects
By vectorially summing (Fig. 6.29c) forces V and $V_{T,i}$, one obtains the force V_i acting on one section of the bolt i. For calculation purposes, it is most convenient to assume an x-y coordinate system

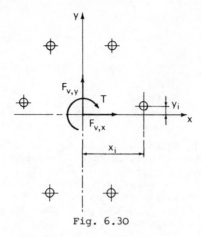

Fig. 6.30

(Fig. 6.30). Then, resolving the shear components in the direction of the axes, one obtains:

$$V_x = \frac{F_{V,x}}{n_V \, n} \quad ; \quad V_y = \frac{F_{V,y}}{n_V \, n}$$

$$V_{T,i,x} = \frac{T y i}{n_V \, \Sigma \, (x_i^2 + y_i^2)} \quad ; \quad V_{T,i,y} = \frac{T x i}{n_V \, \Sigma \, (x_i^2 + y_i^2)}$$

$$V_i = \sqrt{(V_x + V_{T,i,x})^2 + (V_y + V_{T,i,y})^2}$$

6.3.2 *Tension and bending load*

It is more difficult to define distribution of axial and bending loads on a bolted joint: it essentially depends on the stiffness of the flange through which the external load is applied.

In order to analyse the problem from a qualitative point of view refer to Fig. 6.31 which illustrates the simplest joint in tension. If the flange is sufficiently stiff, its deformation can be disregarded and the bolts can be assumed to be in pure tension (Fig. 6.31a): failure of the joint would be due to failure of the bolts. If, however, the flange is more flexible, prying forces Q are produced and the bolts, in order to follow flange deformation, are bent (Fig. 6.31b). Prying forces Q depend on the stiffness of both flanges and bolts and on the applied load; failure of the joint may be due to:

Failure of the bolt loaded by force $F_N = F + Q$ and bent

Plastic collapse of the flange

The above brief remarks show that two different methods can be followed to analyse the distribution of tension and bending.
(a) Flange is deformable and prying forces Q reduce bending in the flange. In this case force distribution between the bolts depends

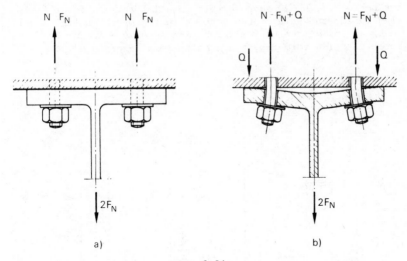

Fig. 6.31

both on joint geometry and on flange stiffness. Design strength
$f_{k,N}/\gamma_N$ of the bolts must be evaluated taking into account secondary
bending in the shank and thus assuming the highest value for γ_N
(see 6.2.2.2).
Unfortunately the experimental studies carried out in this field do
not permit a general method to be formulated for the correct
assessment of interaction between flange thickness and bolt behaviour.
As the number of rows of bolts in tension increases, the influence of
flange flexibility becomes more significant: for this reason, rational
methods have been established for the design of joints in tension and
bending only in the case of two rows of bolts set symmetrically about
the tension flange of the connected section (see 7.3.7).
(b) Flange deformability is disregarded. The behaviour of the section
may be considered similar to that of a reinforced concrete section:
tension is absorbed by the bolts, and compression by contact between
the flanges. Force distribution between the bolts thus depends on
joint geometry.
Design strength of the bolts may be evaluated disregarding secondary
bending of their shanks; therefore the most favourable values for γ_N
may be assumed. Flange thickness must be sufficient to guarantee
compliance with the assumption on which the calculation is based. It
is therefore advisable to avoid exceeding the elastic limit at any
point of the flanges: stress redistribution due to plasticity entails
an increase in deformability, which would be unacceptable. The method
is based on the assumption that bolt behaviour is independent of
flange deformation. In practice, such a statement is not true and
that is why calculation forecasts may be contradicted by experimental
evidence. Nevertheless the following design methods are widly
accepted for joints loaded by axial force and mono axial bending
moment formed by rectangular flanges and bolts of one diameter; they

can also be applied by analogy to more complex cases.
Referring to Fig. 6.32, suppose that an axial tensile force F_N is
applied inside the kern of the section formed by the bolts alone. The
force N_i acting on the bolt i can be assessed by assuming that the
resistant section formed by the n bolts remains plane. Then

$$N_i = \frac{F_N}{n} + \frac{F_N e}{n \sum\limits_1^2 i \, y_i^2} y_i$$

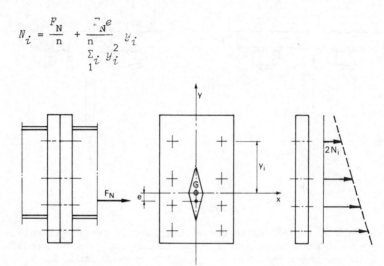

Fig. 6.32

where:

e is applied force eccentricity with reference to the centre of
gravity
y_i is the distance of the controidal axis from the bolt i.

When tension is applied externally to the inertia core of the section
formed by the bolts alone, or when compression is applied externally
to the inertia core of the rectangular section formed by the end plate,
the end plate is partially compressed and the bolts external to the
compressed zone are in tension (Fig. 6.33). Holes are generally disregarded, so that the compressed part may be considered as rectangular.
If the compressed area of the endplate is extended and unstiffened in
its most external areas, a linear distribution of strain ε and stress
σ is commonly assumed. In this case, section rotates around the axis
passing through point C: the force acting on the bolts and maximum
compression can be expressed by the following relations:

$$N_i = A_i \, k \, (y_i - y_c) ; \quad \sigma_c = k \, y_c$$

k being a proportionality constant and A_i the area of a single
bolt.

By imposing equilibrium to section rotation and translation, the
following equations are obtained, determining the position of the
neutral axis, the values of maximum pressure σ_c and of the axial
forces acting on the bolts:

208

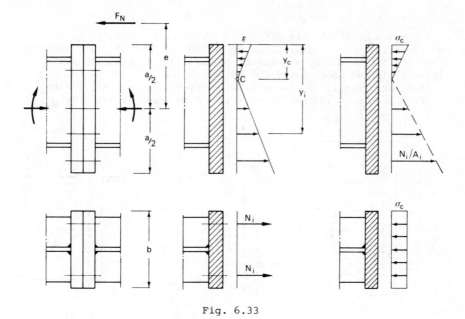

Fig. 6.33

Simple bending ($N = 0$):

$$y_c^2 \frac{b}{2} + y_c \sum_1^n A_i - \sum_1^n A_i y_i = 0$$

$$\sigma_c = My_c/I \quad \text{with:} \quad I = \frac{by_c^3}{3} + \sum_1^n A_i (y_i - y_c)^2$$

$$N_i = \frac{M}{I} (y_i - y_c)$$

Bending and axial force:

$$\frac{y_c^3 b}{6} + y_c^2 \frac{b}{2} (e - \frac{a}{2}) + y_c \sum_1^n A_i (e - \frac{a}{2} + y_i) - \sum_1^n A_i y_i$$

$$(e - \frac{a}{2} + y_i) = 0$$

$$\sigma_c = \frac{y_c}{y_c^2 \frac{b}{2} - \sum A_i (y_i - y_c)} |F_N|$$

$$N_i = |\sigma_c| \frac{A_i (y_i - y_c)}{y_c}$$

with $e > 0$ if N is compression
$e < 0$ if   is tension

The above equations are extended only to bolts in tension: if the neutral axis thus determined has an ordinate y_c greater than ordinate y_1 of the first bolt considered in tension, the calculation must be repeated without considering the corresponding bolt in the equation.

When the extension of the area in contact is presumably limited or the flange is stiffened (Fig. 6.34) it is more realistic to

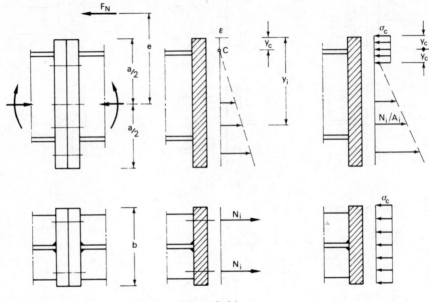

Fig. 6.34

concentrate contact pressure resultant at a reasonable point C and to impose rotation equilibrium around this point, assuming a linear distribution of bolt forces. Therefore, y_c being determined a priori and F_N being positive if it is a compression, one has:

$$N_i = k A_i (y_i - y_c)$$

$$\sum_1^n N_i (y_i - y_c) = M - F_N \left(\frac{a}{2} - y_c\right)$$

and thus:

$$N_i = \frac{M - F_N \left(\frac{a}{2} - y_c\right)}{\sum_1^n A_i (y_i - y_c)^2} A_i (y_i - y_c)$$

210

By imposing equilibrium in the direction of the acting forces, the value of contact pressure resultant R is obtained. It is determined by:

$$R = \sum_{1}^{n} N_i + F_N$$

This resultant can reasonably be assumed to be uniformly distributed on a rectangular area the sides of which are b and $2y_c$ with its centre of gravity at point C. Therefore:

$$\sigma_c = \frac{\sum_{1}^{n} N_i + F_N}{2y_c b}$$

It is also possible to evaluate joint ultimate performance. In this case, a distribution of the type illustrated in Fig. 6.35 may be assumed.

All the bolts are loaded by design axial forces N and the contact pressure has a design value f_d equal to the design strength of the flange material. The only unknown quantity is the position of the neutral axis. It is defined by the equilibrium condition in the direction of the acting forces. Assuming axial compression to be positive it is:

$$- n\, N_{d,o} + f_d\, y_c b = F_N$$

n being number of bolts in tension. Therefore:

$$y_c = \frac{F_N + n\, N_{d,o}}{f_d b}$$

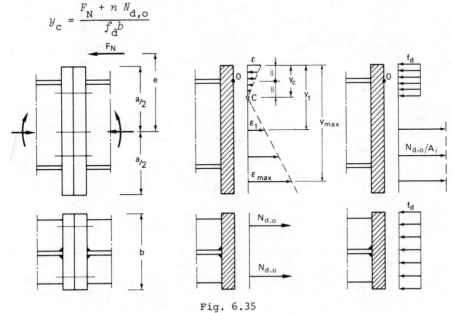

Fig. 6.35

211

Once y_c is known, it is possible to evaluate the ultimate moment, concurrent with axial force F_N. For equilibrium at rotation around O, one has:

$$M_u = F_N e = N_{d,o} \sum_1^n {}_i (y_i - \frac{y_c}{2}) + F_N (\frac{a}{2} - \frac{y_c}{2})$$

However, this moment cannot always be assumed as the ultimate one. For this to be true, the closest bolt to the neutral axis must be able to carry its maximum load without the farthest bolt having an elongation equal to the ultimate value ε_t.
Referring to Fig. 6.35 therefore $\varepsilon_{max} < \varepsilon_t$.
The value of ε_{max} may be found assuming a strain $\varepsilon_1 = f_{d,N}/E$ for the closest bolt to the neutral axis considered in the calculation. Therefore:

$$\varepsilon_{max} = \varepsilon_1 \frac{y_{max} - 2y_c}{y_1 - 2y_c} = \frac{f_{d,N}}{E} \frac{y_{max} - 2y_c}{y_1 - 2y_c} < \varepsilon_t$$

The design value for ultimate elongation of the bolts ε_t should be assumed cautiously and in any case should not exceed the minimum values indicated in the table of Fig. 6.2.

REFERENCES

1. CECM-ECCS (1978) *European Recommendations for Steel Construction*, ECCS, Brussels
2. AISC (1976) *Specification for Structural Joints using ASTM Bolts*, American Institute of Steel Construction, Chicago
3. Gaylord, E. H. and Gaylord, C. H. (1968) *Structural Engineering Handbook*, McGraw Hill, New York
4. Bowman, L. P. and Van Douwen, A. A. (1979) *The Tightening of Bolts*, Stevin Laboratory Report, Delft University of Technology
5. McGuire, W. (1968) *Steel Structures*, Prentice Hall, New York
6. Fisher, J. W. and Struik, J. H. A. (1974) *Guide to Design Criteria for Bolted and Riveted Joints*, John Wiley, New York
7. Demol, L., Lambert, J. C. and Viatour, G. (1969) *Les Assemblages par Boulons à Haute Résistance*, Report of CRIF (Centre des Recherches Scientifiques et Techniques de l'Industrie des Fabrications Métalliques)
8. Khalil, H. Shakir and Ho, C. M. (1974) Black bolts under combined tension and shear, *The Structural Engineer*, 57B, 4, 69-76
9. Abolitz, A. L. (1966) Plastic design of eccentrically loaded fasteners, *Engineering Journal*, AISC, 3, n. 3, 122-132
10. Crawford, S. F. and Kulak, G. L. (1971) Eccentrically loaded bolted connections, *Journal of Structural Division*, ASCE, 97, ST3, 765-783
11. Kulak, G. L. (1975) Eccentrically loaded slip resistant connections, *Engineering Journal*, AISC, 2, n. 2, 52-55
12. Aribert, J. M. and Machaly, E. S. (1975) Comportement à la rupture et dimensionnement optimal d'assemblages concentriques par boulons à haute résistance, *Construction Metallique*, 1, 5-19

C H A P T E R   S E V E N

Connections

7.1 GENERAL

A connection may be entirely welded or completely bolted or may have welds together with bolts.

It is an extremely difficult subject to synthetize organically mainly because it is continually evolving. It is also influenced by the fact that a joint is conceived according to the type of equipment and processing in the fabrication shop and to the destination, transportation and erection of the structure.

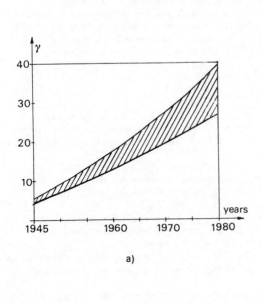

a)

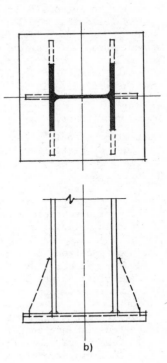

b)

Fig. 7.1

Fig. 7.1a shows an example of the substantial conditioning due to technology and economy |1|. It demonstrates the evolution that has taken place over the last 35 years in Western Europe in terms of γ the cost of one hour labour inclusive of social security taxes per cost of 1 kg of steel, prior to its transformation in the workshop. The obvious increase of this ratio might nowadays make it more attractive to adopt simpler structural details, even though they are heavier. For example, the column base illustrated in Fig. 7.1b by a continuous line is more expensive in material costs than the stiffened solution in the dashed drawing, but the latter requires more work and therefore more hours of labour. Depending on the value of the ratio γ, either one or the other solution may be more advantageous.

At present, many designers and researchers are increasingly developing and studying simpler connections, in order to eliminate technical details that influence the cost of a joint without being a factor for its strength.

This is one of the most difficult studies, as, by definition, a joint is a detail in which there is a concentration of forces: its behaviour cannot be interpreted on the basis of the elastic theory hypothesis. A joint can be modelled only on the basis of plastic design, seeking for equilibrated solutions that are consistent with strength criteria. Therefore, the evaluation of joint strength and deformability can be carried out only by means of experimental analyses or finite element techniques in the elastic-plastic field: the results so obtained make it possible to define the simplified calculation methods commonly used in design practice.

The following calculation methods can therefore be considered as safe criteria that have been judged as acceptable by means of experimental or numerical analyses, occasionally generalized by means of theoretical considerations. They must therefore be considered as conventional design criteria fit to provide a safe assessment of the strength of the joint.

7.1.1 *Joint ductility*

According to their strength, connections can be classed as full strength or partial strength joints. Full strength joints allow the ultimate load carrying capacity of the weakest connected element to be transferred.

Partial strength joints allow only a partial amount of load carrying capacity of the weakest connected element to be transferred.

According to their moment rotation behaviour connections can be divided into rigid, semi-rigid or simple joints.

Rigid joints may be considered those with sufficient rigidity to hold the angles between connected members virtually unchanged until the ultimate load carrying capacity has been reached.

Semi-rigid joints are flexible joints usually allowing a transfer of only a partial amount of bending capacity of the weakest connected element.

Simple joints transfer such a small amount of bending capacity of connected elements that it can be neglected.

In design practice rigid joints are generally full strength joints, while semi-rigid joints should be regarded as partial strength joints.

214

Simple joints are similar to articulated joints from a statical point of view, but they need local plasticization to allow relative rotation between the connected members.

For full strength and partial strength joints one must also assess 'ductility', i.e. their capability to develop their strength in the plastic range without premature collapse due to excessive strain.

Consider, for example, the behaviour of three typical tension connections illustrated in Fig. 7.2:

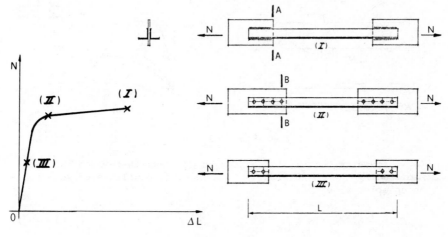

Fig. 7.2

The welded joint (I) is a full strength joint. By measuring extension ΔL of the bar it may be observed that each section reaches the plastic range.

Collapse usually takes place at $A-A$, showing that the joint guarantees a good performance also as far as ductility requirements are concerned.

The bolted joint (II) is also a full strength joint, as it reaches a collapse load equal to the connected bar yielding load. However, its collapse behaviour is brittle, due to net section $B-B$. Such behaviour offers no warranty of ductility: the structural element (the bar plus the joints) so connected remains practically elastic until rupture and cannot lengthen in the plastic field.

Therefore it cannot be relied upon to redistribute forces if the structure is designed plastically or to dissipate energy in the case of possible dynamic or seismic behaviour.

The bolted joint (III) is a partial strength joint and is therefore a weak point of the structure: the connected bar is elastic whenever the joint collapses. It can thus be adopted with the same limitations set forth for connection (II).

A further example is offered by the analysis of the behaviour of a beam connected to columns (Fig. 7.3a).

It can be schematically represented as in Fig. 7.3b, in which the maximum positive moment is M_A and the moment at the supports is $M = M_B = M_C$.

215

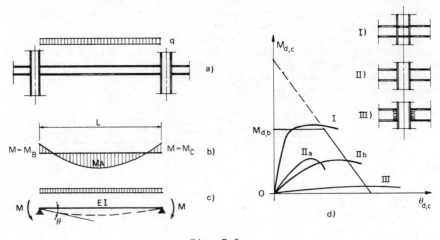

Fig. 7.3

Let the design moment of the beam be $M_{d,b}$ and that which can be transmitted by the connection, $M_{d,c}$. Then any equilibrated solution is acceptable, provided:

$$M_A = \frac{1}{8} qL^2 - M \leq M_{d,b} \tag{7.1}$$

$$M \leq M_{d,c} \tag{7.2}$$

Furthermore, at collapse load, that is when a plastic hinge forms at mid span of the beam, ductility requires that rotation θ of the beam end be less than the maximum rotation $\theta_{d,c}$ allowed by the connection. Hence, from Fig. 7.3c:

$$\theta = \frac{qL^3}{24EI} - \frac{ML}{2EI} \leq \theta_{d,c}$$

For design conditions, equations (7.1) and (7.2) should be read with the equality sign. By substituting from (7.2) in (7.1) and (7.3) and then from (7.1) in (7.3) one obtains:

$$\theta = \frac{L}{6EI} (2M_{d,b} - M_{d,c}) \leq \theta_{d,c} \tag{7.4}$$

The straight line (7.4) defines in plane $M_{d,c} - \theta_{d,c}$ the minimum rotational capacity of the joint. Obviously, the validity range of the expression is limited by the conditions relating to the joint that is not capable of transmitting moment ($M_{d,c} = 0$) and by the rigid one ($M_{d,c} = M_{d,b}$):

$$M_{d,b} \geq M_{d,c} \geq 0$$

Fig. 7.3d illustrates three typical joints: (I) rigid, (II) semi-rigid and (III) simple, the last being capable of transmitting so slight a moment that it can be considered negligible, i.e. $M_{d,c} \simeq 0$.

The same figure also indicates their behaviours and compares them with condition (7.4). The rigid and simple joints (curves I and III) have sufficient rotational capacity. The semi-rigid one can, depending on dimensional ratios, have either a sufficient rotational capacity (curve IIb) or an insufficient one (curve IIa). In the former case the connection is acceptable, but in the latter case the connection would fail prematurely.

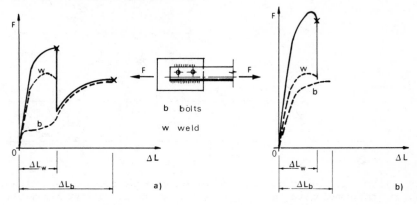

Fig. 7.4

Fig. 7.4, illustrating a joint using a combination of welding and bolting, provides a further example.

If load extension diagrams are of the type shown in Fig. 7.4a, the effects of the welding and of the bolting cannot be added.

In fact, the bolts have a greater extension (ΔL_b) under collapse compared to that of the weld seam (ΔL_w), so that the resulting overall behaviour is collapse of the one part before the other has given its ultimate performance. On the other hand, both the bolts and the weld can be considered resistant whenever the behaviour of the bolted joint is as shown in Fig. 7.4b. This is the result of reducing the hole bolt clearance so that ΔL_b is limited and becomes comparable to ΔL_w.

7.1.2 *Lamellar tearing*

Lamellar tearing is a cracking phenomenon that can take place in rolled materials when they are loaded perpendicular to their rolling plane |2|. It might therefore be dangerous with reference to the strength of some connections, and most frequently in welded joints (Fig. 7.5a) in which welding shrinkage increases strain and stress concentration.

Tearing in the material is due to localized deformations. Fig. 7.5b shows equal value lines of percentage strains as assessed by photo-elastic techniques in a full penetration butt welded joint using K preparations between thick plates |3|.

At a distance of a few millimetres from the welding seam, strains are of the order of magnitude of 2%; at a distance of 7 to 10 mm, they

217

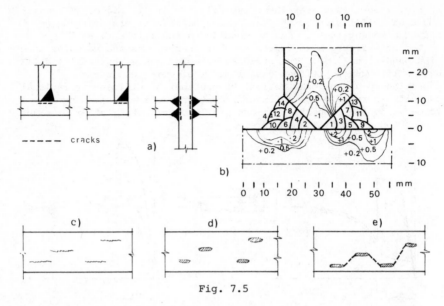

Fig. 7.5

are still of the order of 0.5% and therefore many times greater than those corresponding to the elastic limit of steel. If any non-metallic micro-inclusions are present in the material (Fig. 7.5c), they open due to the weld shrinkage (Fig. 7.5d); if material ductility is insufficient, cracks develop along the dotted lines of Fig. 7.5e. Rupture during fabrication or erection of the structure is thus possible.

Still from a qualitative point of view, the phenomenon is most likely to occur with thick components of higher strength steels (in steel, as strength increases, strain at failure and therefore ductility decrease). Generally the possibility of lamellar tearing should be considered in plates thicker than 40 mm, even though is has sometimes arisen for thicknesses in the order 25 to 30 mm.

To prevent the risk of tearing, structural details should be designed minimizing:

Welding shrinkage (Fig. 7.6a)
Quantity of parent material (Fig. 7.6b)
Strain across to the plate in which tearing is expected (Fig. 7.6c)

In some cases it is also possible to consider alternative details using parts of rolled shapes (Fig. 7.6d).

During fabrication welding shrinkage can be limited by increasing the number of weld passes and by planning their succession in such a way that the parent material at risk is buttered (Fig. 7.7).

Ultrasonic examination of steel plate makes the extent of possible inclusions evident: it reduces the danger, but does not exclude it.

Therefore, it is in any case advisable to prevent lamellar tearing by conceiving suitable structural details and planning correct fabrication techniques.

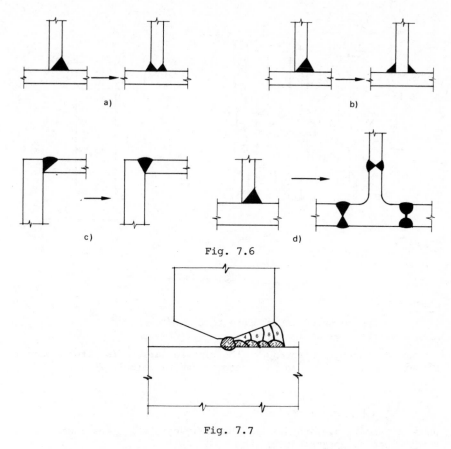

Fig. 7.6

Fig. 7.7

To control material fitness to avoid lamellar tearing due to welding, the so-called 'transverse ductility' test can be applied to material having thickness above 10 mm |4,5|.

If thickness t is below 40 mm, a cylindrical specimen must be prepared following the sequence illustrated in Fig. 7.8a. The specimen diameter d must be 6 mm if $t \leq 16$ mm or 10 mm if $t > 16$ mm.

If thickness t is 40 mm or more (Fig. 7.8b), a cylindrical specimen can be prepared directly from within the plate. The specimen diameter must be 10 mm and it must have 6 mm long threaded anchorage ends to allow gripping in the test machine by means of special extensions.

The test can be be considered positive if the percentage reduction in cross-sectional area of the specimen is such that:

$$\frac{A_o - A}{A_o} \; 100 \geq 20\%$$

219

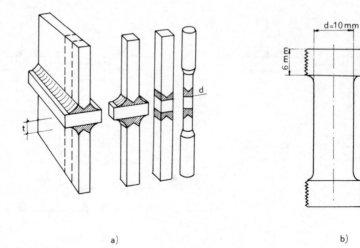

a)   b)

Fig. 7.8

where:

A_o is the measured area before test
A is the area at failure.

Due to material anisotropy, the rupture surface is often elliptical
rather than circular. In this case, A can be calculated as:

$$A = \frac{\pi}{4}\left\{\frac{d_1 + d_2}{2}\right\}^2$$

where d_1 and d_2 are the axes of the ellipse to which the rupture
surface can be approximated.

7.2 ARTICULATED JOINTS

Articulated joints were common up to the beginning of this century:
structural design strictly followed elastic theory and the constraint
conditions on which calculations were based were complied with as
faithfully as possible. As an example Fig. 7.9 illustrates the
springing and crown hinges of the roof arches of Milan, railway station,
built in 1920.

When plastic theory was developed and it became clear that each
equilibrated calculation model was in favour of safety, provided
localized failures did not take place, the importance of constraint
compliance with their model was retrenched. Nowadays a column base
with four bolts (Fig. 7.10) may be considered as a hinge or a fixed
support depending on the design assumptions.

Obviously, such a base can be considered as a hinge at the cost
of modest plasticizations: this is therefore correct within the range
of ultimate limit state design (conventional elastic or plastic design).

220

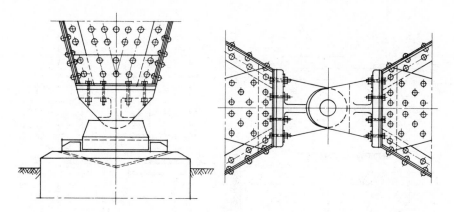

Fig. 7.9

On the other hand, in some cases it is advisable to provide for
actual movement with reference to service limit states. This is not
usually necessary for civil and industrial building structures, but
it is more typical of bridge supports and of structures supporting
machines or moving equipment.

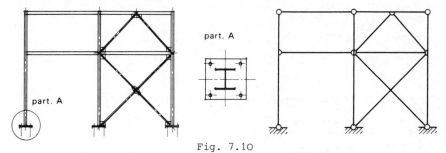

Fig. 7.10

7.2.1 *Pinned connections*

Fig. 7.11a,b show two typical connections for elements in tension.
Fig. 7.11c shows a more complex hinge, which is useful to connect
elements in tension or compression or at the foundations.

It is extremely complex to find a rigorous solution to the elastic
behaviour of pin connections. It entails problems concerning contact
between surfaces, stress diffusion in the pinhole, bending and shear
in the pin which, for its dimension, does not comply with the basic
hypotheses of beam theory. Literature |6| and recommendations |7,8|
offer some suggestions providing a correct geometry of the connection
and conventional design methods which tests have proved to be on the
safe side.

In a merely indicative context, the following dimensional ratios
are suggested for the connections shown in Figs 7.11a,b for the forged
eyebar (Fig. 7.11a):

221

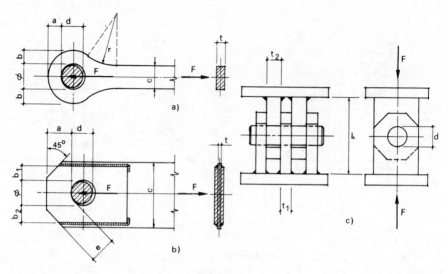

Fig. 7.11

Gross section area: $A = ct$
Net section area: $A_n = 2\,bt$
Bearing area: $A_b = t\,d$

$50 \text{ mm} \geq t \geq 12 \text{ mm};\quad c \leq 8t;\quad a \geq b;$

$1.33c \leq 2b \leq 1.50\ ;\quad d \geq \frac{7}{8}c;\quad r \geq (\phi + 2b);$

$d \leq 5t;\quad \phi - d \leq 0.8 \text{ mm}.$

For the reinforced eyebar (Fig. 7.11b):

Gross section area: $A = ct$
Net section area: $A_g = (b_1 + b_2)\,t$
Bearing area: $A_b = d\,t$

$50 \text{ mm} \geq t \geq 12 \text{ mm};\quad b_1 \leq 4t;\quad b_2 \leq 4t;\quad d \geq t;$

$a \geq \frac{2}{3}(b_1 + b_2);\quad e \geq a;\quad \phi - d \leq 0.8 \text{ mm}.$

The dimensional ratios indicated for the reinforced eyebar of Fig. 7.11b can, obviously, be followed also for the pin connecting plates illustrated in Fig. 7.11c, by interpreting t as the thickness of a single resistant plate (t_1 or t_2 in Fig. 7.11c).

For designing the plates, conventional values must be assessed for stresses in gross and net sections, bearing stresses and buckling stresses (in cases of compression).

If the external applied load is F and design strength is f_d, in the gross section having area A, $F/A \leq f_d$.

In the net section it is advisable to take greater precautions, as the edges surrounding the hole are loaded by parasitic bending (Fig. 7.12a). For this reason, recommendations impose a limit on tensile strength equal to $\alpha\,f_d$ with $\alpha=0.70$ to 0.75. Thus, $F/A_n \leq \alpha\,f_d$.

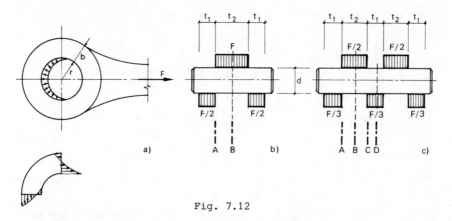

Fig. 7.12

For bearing, the conventional reference area is obtained by multiplying pin diameter by the sum Σt of the thicknesses of the connecting plates. It is advisable to use a lower limit strength than for bolted joints (see 6.2.1.3) in order to avoid deformations of the plates or of the pin, which may render the hinge unusable. A limit of 1.35 f_d is usually accepted so that $F/A_b \leq 1.35\ f_d$.

Plates in compression must be checked against buckling. In this case it is advisable to make the joint as compact as possible in order to limit their slenderness and to ensure that the mean stress calculated with reference to gross area is less than the critical stress σ_c (see chapter 9).

It is more difficult to find a calculation method for the pin: it cannot be considered as a beam, in as much as its transverse dimensions are greater than the calculation span.

By analogy with bolted joints, one can consider that the external load is uniformly distributed in each connecting plate (Figs 7.12b, c) even when the pin has more than two supports. The pin is thus designed considering only shear forces and a uniform distribution of shear stresses in the section can be assumed so that $\tau = V/A$.

The need to design against bending is more controversial.

Bending stresses can occur only if there is bending deformation: the latter can be prevented by the plates, thanks to the limited extent of hole/pin clearance. Therefore, according to the required performance of the connection, either of the two following criteria can be adopted:
(a) The pin is designed disregarding bending. The pin is thus considered as a bolt and it is not possible to guarantee that, when it bends, it doesn't interfere with the connecting plates, forcing them and therefore disturbing the required rotation capability
(b) The pin is designed under bending and shear as a bent beam. This method is on the safe side and guarantees hinge functionality. In the case illustrated in Fig. 7.12b, for example, one has:

Section A: $V = F/2$; $M = \dfrac{F}{4}\,t_1$

Section B: $V = 0$; $M = \dfrac{F}{4}\,(t_1 + t_2/2)$

223

In the case illustrated in Fig. 7.12c, one has: instead:

Section A: $V = F/3$; $M = \frac{F}{6} t_1$

Section B: $V = \frac{F}{12}$; $M = \frac{F}{6} (t_1 + \frac{5}{8} t_2)$

Section C: $V = - \frac{F}{6}$; $M = \frac{F}{6} (t_1 + \frac{1}{2} t_2)$

Section D: $V = 0$; $M = \frac{F}{6} (\frac{3}{4} t_1 + \frac{1}{2} t_2)$

7.2.2 *Articulated bearing joints*

Articulated joints obtained by direct contact between metal parts
possibly having rounded surfaces can be substantially divided into two
different types according to whether:
(a) Contact takes place between surfaces, at least one of which is
cylindrical or spherical (Figs 7.13 a,b)
(b) Contact is concentrated between a plate and a knife plate (Figs
7.13 c,d)
 In the first class of connections, point contact is achieved if
the hinge is of a spherical type and linear contact if the hinge is

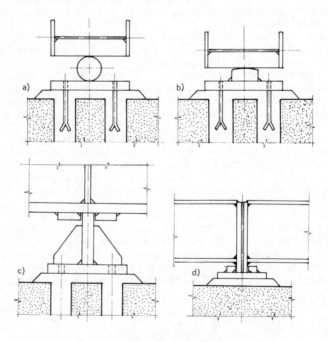

Fig. 7.13

cylindrical. Appropriate guides or wedges must be provided to
control surface rolling and to transmit possible transverse forces H
(Fig. 7.14).

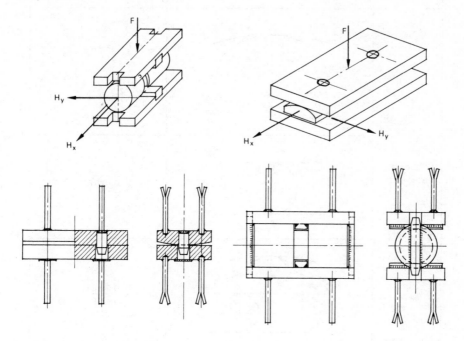

Fig. 7.14

For design purposes a conventional value of stress in the elastic
field is calculated according to the Hertz formulae given in Fig. 7.15.
It must be such that:

$$\sigma \leq \alpha \, f_d$$

α being a coefficient exceeding one, to take into account the fact
that, in the contact area, a three-dimensional stress state,
particularly favourable to strength, is present. Usually:

   $\alpha = 4.0$ for cylindrical hinges
   $\alpha = 5.5$ for spherical hinges

Articulated joints of the types illustrated in Figs 7.13 c,d are
used for small span bridges and often for the supports of important
crane runway beams. Excessive contact pressure between the plates
must be avoided so as not to deform the surfaces. This can be
achieved by ensuring that the mean contact pressure σ obtained by
dividing the load by the contact surface area, is such that:

$$\sigma \leq \alpha \, f_d$$

type of contact	shape of contact surfaces	contact stresses
cylindrical hinge of length b	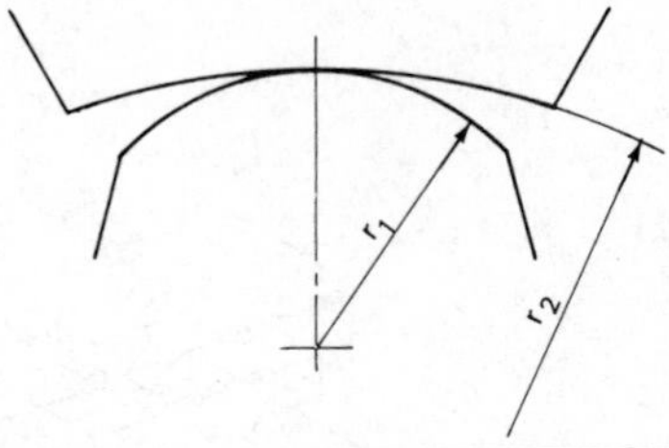	$\sigma = \sqrt{0.18\,E\,F\,\dfrac{r_2 - r_1}{r_1\,r_2\,b}}$ for $\dfrac{r_2}{r_1} \geqq 2$
	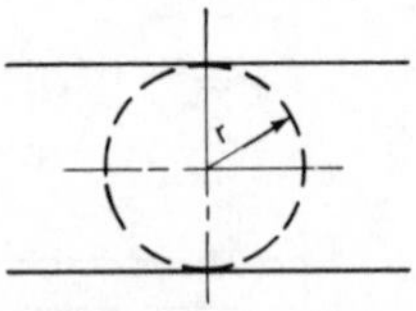	$\sigma = \sqrt{\dfrac{0.18\,E\,F}{r\,b}}$
	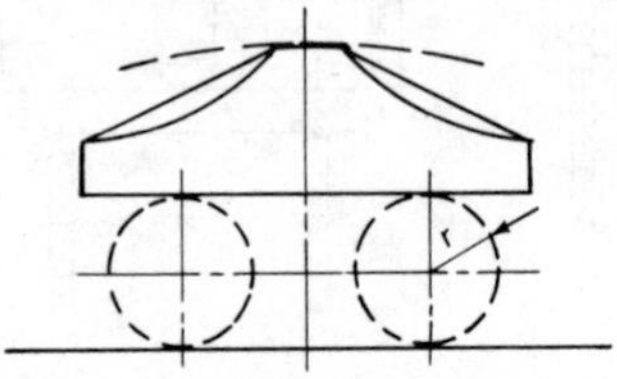	$\sigma = \sqrt{\dfrac{0.20\,E\,F}{2\,r\,b}}$
	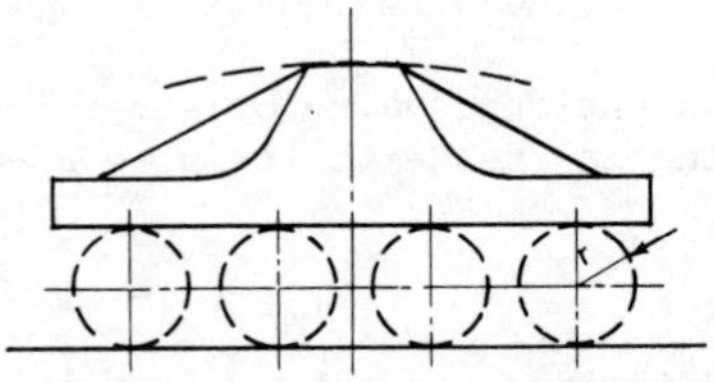	$\sigma = \sqrt{\dfrac{0.24\,E\,F}{n\,r\,b}}$ n = number of rollers
spherical hinge	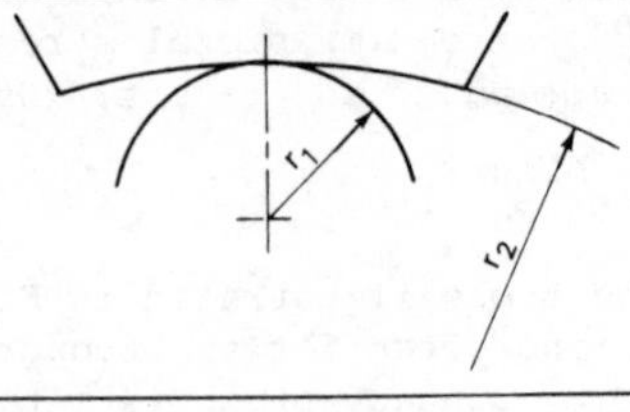	$\sigma = \sqrt[3]{\dfrac{0.06\,E^2\,F\,(r_2 - r_1)^2}{r_1^2\,r_2^2}}$
	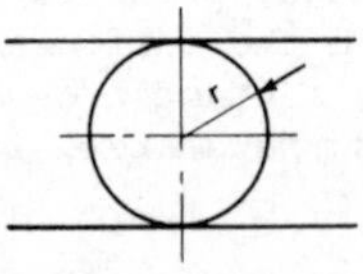	$\sigma = \sqrt[3]{\dfrac{0.06\,E^2\,F}{r^2}}$

Fig. 7.15.

226

where α is a coefficient exceeding one. This coefficient must be smaller than that adopted for bearing in bolted joints, to prevent surface deformations ($\alpha = 1.30$ to 1.50) is usually adopted.

7.2.3 *Elastomeric bearing pads*

Bearings can be formed by interposing a layer of hard rubber (Fig. 7.16a) between metal parts. Such a constraint allows relative sliding and rotations between the parts.

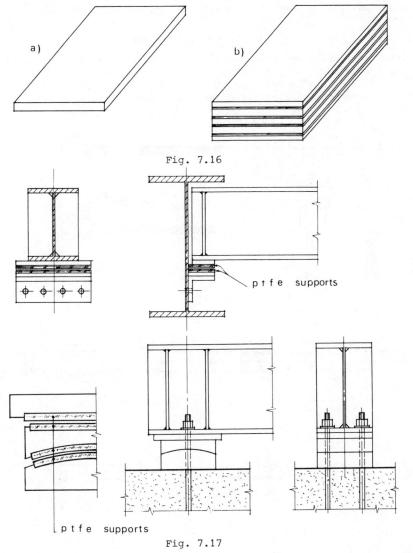

Fig. 7.16

p t f e supports

p t f e supports

Fig. 7.17

Their design essentially depends on rubber hardness and dimensional characteristics of the bearing device. For multi-layer sandwich of steel plate and rubber (Fig. 7.16b) it also depends on the number of layers.

Another particularly efficient type of expansion bearing joint can be obtained by gluing a layer of synthetic material with a base of polytetrafluoroethylene (ptfe) on to the metal surfaces (Fig. 7.17). Such a joint has an extremely low degree of friction and can excellently solve some particular structural problems. Static controls in this case essentially depend on the type of material and are generally indicated in manufacturers' specifications.

7.3 JOINTS IN TENSION

Joints for tension members essentially depend on the type of elements to be connected.

Figs 7.18a,b show two typical connections for steel cables. The first is created by fraying the cable end, pouring metal material into

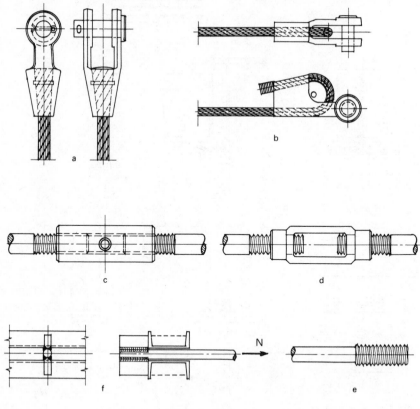

Fig. 7.18

it; the second forms a self-jamming device. Both are full strength joints. In fact, they are usually prepared and tested to ensure that they can carry a greater collapse load than that of the connected cable.

Figs 7.18c,d illustrate some typical joints for round rods.

In this case, the joint is a partial strength joint, as the threaded section is obtained by removing bar material: the connecting rod must thus be designed with reference to the thread resistant area, the weak point of the structure. For this reason, when important structures are concerned or if one wishes to take into account plastic redistribution of internal forces in the structure, it is preferable to use connecting round bars with rolled corrugated thread (Fig. 7.18e) because the gross section can be entirely exploited.

Fig. 7.18f illustrates a 'hammer' connection for a round bar in tension. It is formed by two plates welded to the rod which transmit the tension to the two channels. While the strength of the welded joint between plates and bar is a straightforward matter the trasmission of force to the two channel profiles is a more complex problem, which also includes the question as to whether it is or is not necessary to provide stiffeners (dotted lines in Fig. 7.18f) to transmit the load to the webs of the channels. Calculation methods for such problems are indicated in 7.3.1 below. Such a joint can be

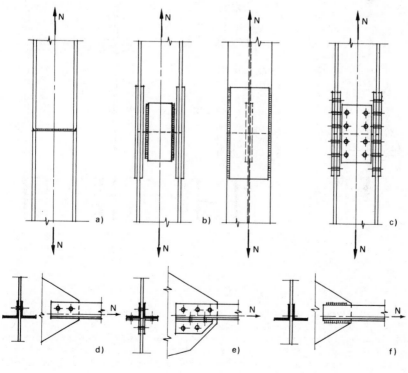

Fig. 7.19

a full or partial strength joint according to welding strength and to the capability of the channels to resist the effects of the external load N.

Fig. 7.19 shows some typical joints between profiles in tension.

They can be executed by butt welds (Fig. 7.19a), fillet welds (Fig. 7.19b) or bolted splices (Fig. 7.19c).

The first joint, if the welding is first-class, is a full strength joint and no calculation is necessary. The spliced joints must be more accurately designed: they are discussed in 7.3.1 below.

Channels, angles and tees are often used in steel structures.

They can be connected by means of bolted joints involving either a part of the section (Fig. 7.19d) or the entire section (Fig. 7.19e) or by means of welded joints, generally involving only a part of the section (Fig. 7.19f). Such connections may produce excentricities that influence profile design (see 8.2.1); the strength of the joint is analysed in 7.3.2.

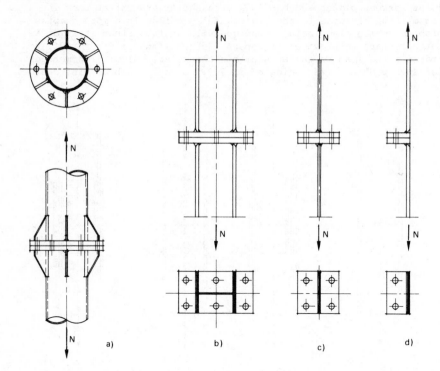

Fig. 7.20

Finally, tension members can be connected by means of end plates.

Figs 7.20a,b,c show such connections between tubes, rolled sections and plates respectively. Fig. 7.20d is an eccentric flanged connection between plates. Loads on bolts of such joints depend on end plate stiffness and some calculation methods are discussed in 7.3.4 and 7.3.5.

230

7.3.1 *Spliced joints*

It is advisable to distribute the various splices so as to reduce
stress concentration. For profiles in tension, splices should be
distributed in proportion to the elemental areas of the profile. With

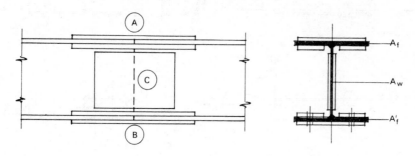

Fig. 7.21

reference to Fig. 7.21, N being the value of axial load and
$A = A_f + A_f' + A_w$ = total area of profile section, joints A, B, C
should respectively be designed for forces equal to:

$$N_A = N A_f/A; \qquad N_B = N A_f'/A; \qquad N_C = N A_w/A$$

The joints may be either welded or bolted. Full strength welded
joints, (i.e., $N = F_d A$), are quite reliable as to their ductility:
failure can take place outside the joint and for a load value leading
to the elastic limit being reached across the entire connected girder.
Full strength bolted joints, however, may have a brittle behaviour;
they may collapse by net section rupture (see 8.2.1) before plastic
deformations occur in the other sections of the connected element (see
7.3.6).

7.3.2 *Spliced joints for angles and channels*

Angles and channels are often used as bracing members in trusses.
    Their end connection raises some problems, essentially concerning
the difficulty of designing a joint for the force to be transmitted
along the centroidal axis of the profile. This implies some
eccentricities that must be taken into consideration when designing
both the profile and the joint.
    In the most general case (Fig. 7.22a) there are two possible
eccentricities, e_x and e_y. The former depends on disalignment between
the line of application of the external force and the centroidal axis
of the profile. It causes a secondary bending moment Ne_x in the
profile, which can be designed following the methods indicated in 8.2.1.
    The latter is caused by a possible difference between the
centroidal axes of the profile and of the connection.
    The profile remains simply in tension, but the connection is
subjected to a secondary twisting moment $T = Ne_y$ due to the
eccentricity.
    If the connection is welded, it is possible to eliminate

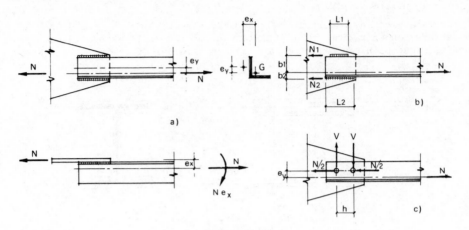

Fig. 7.22

eccentricity e_y: it is sufficient to design welds so that their centre of gravity coincides with the centroidal axis of the section (Fig. 7.22b).

For example $R_1 = t_\| \, a_1 \, L_1$ and $R_2 = t_\| \, a_2 \, L_2$ are the resistances of the two welds (see 5.3.2 and 5.3.3); therefore:

$$N_1 = \frac{N \, b_2}{(b_1 + b_2)} < R_1; \qquad N_2 = \frac{N \, b_1}{(b_1 + b_2)} < R_2$$

It is not usually possible to eliminate eccentricity e_y in bolted joints. The inner radius of the section and the dimensions of bolts and washers do not allow the bolt line to coincide with the centroidal axis.

In this case, the bolts are loaded both by parallel forces N (Fig. 7.22c) and perpendicular forces V, produced by parasitic twisting moment $T = Ne_y$ (see 6.3.1).

The joint plate must allow adequate diffusion of stress. Many theoretical and experimental researches have shown that a 30° diffusion angle is safe. Using this in the case shown in Fig. 7.23a,

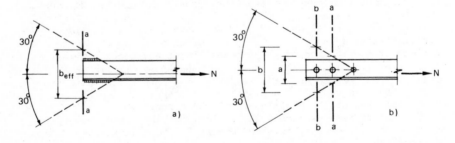

Fig. 7.23

the plate must have a thickness t such that $N/tb_{eff} \leq f_d$ to avoid failure in the weakest section a-a. Such a condition is characteristic of a partial strength joint. If one wishes the gusset to have at least the same strength as the connected element of area A, then $tb_{eff} \geq A$.

The above considerations can be applied to the bolted gusset plate of Fig. 7.23b. If the applied force is equally distributed among the bolts, the axial force in the plate is equal to $2N/3$ in section a-a and N in section b-b. It therefore follows that:

$$\frac{N}{(b - \Phi)\ t} \leq f_d; \qquad \frac{\frac{2}{3}\ N}{(a - \Phi)\ t} \leq f_d$$

Such conditions make it possible to design plate thickness t and to give it a rational shape.

7.3.3 *Ductility of bars with spliced joints in tension*

To design full strength joints in tension, one must increase the number of bolts, reinforcing the net section and avoiding eccentricities.

This is necessary whenever ductility must be guaranteed (e.g. bracing in a seismic area).

A connection may be considered full strength and ductile when the joint rupture load exceeds the yield load of the connected element. Hence:

$$\alpha\ f_t\ A_n > f_y\ A$$

where:

f_y, f_t are material yield and ultimate strength
A_n, A are net and gross sectional areas of the connected bar
α is a coefficient equal to 1 for symmetrical layouts of the joint and less than 1, and to be determined experimentally, for connections in which secondary bending might take place (see 7.3.2)

As an example, the connection of two angle bars to a gusset plate can be executed in one of the ways illustrated in Fig. 7.24. Their behaviour can vary considerably |9|. If the connection is properly welded (I), the joint is full strength and ductile. If it is bolted, it may be partial strength (II) or full strength (III, IV, V).

Secondary bending due to only one leg being connected entails a coefficient $\alpha < 1$ (experimentally, $\alpha = 0.7$ to 0.8). Therefore, connection (III) may not be ductile. To obtain ductility, either the net section must be reinforced (IV) or the connection must be rendered symmetrical (V).

233

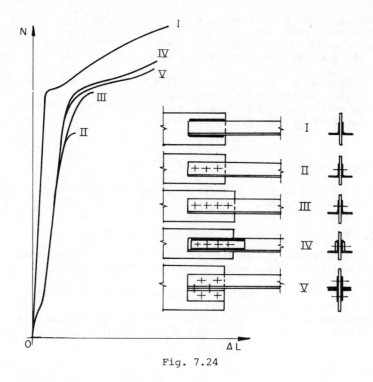

Fig. 7.24

7.3.4 *Symmetrical end plate joints*

Fig. 7.25 shows an end plate having two bolts in tension and its three possible collapse mechanisms |10|.

If plate bending deformations are small compared to bolt elongation, the collapse mechanism is as illustrated in Fig. 7.25a. The load on the bolts is $N = F/2$, secondary bending in the bolts is practically absent and the plate must resist a bending moment $M_2 = aF/2$.

If plate bending deformations are of the same order of magnitude as bolt elongation, the collapse mechanism is as illustrated in Fig. 7.25b. A prying force Q acting in the same direction as F will be produced. The bolts will be loaded by a force $N = Q + F/2$ and a not negligible bending stress will act on the shank (see 6.2.2.2.). The plate will be bent by opposite bending moments having maximum values M_1 and M_2.

The prying force Q is indeterminate a priori: an equilibrated solution must be found in compliance with the strength criterion. If $M_{d,A}$ is the resistant moment of section A-A (net of hole), $M_{d,B}$ that of section B-B and $N_{d,o}$ the maximum force allowed for the bolt, it follows that |11|:

$$N = \frac{F}{2} + Q \leq N_{d,o}$$

$$M_1 = Qc \leq M_{d,A}$$

$$M_2 = Na - Q (c - a) < M_{d,B}$$

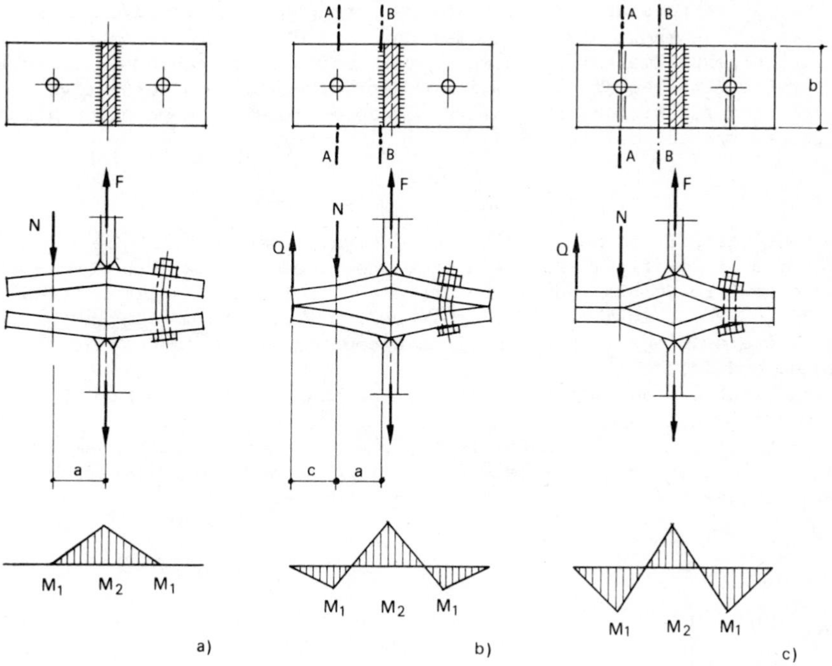

Fig. 7.25

The value of force Q can be chosen arbitrarily so as to satisfy all the inequalities. The grade of bolt and its diameter, and therefore $N_{d,o}$, can be fixed operatively. It ensues that, at the most, $Q \pm N_{d,o} - F/2$. Plate thickness will then have to satisfy the following limitations:

$$(N_{d,o} - \frac{F}{2}) \, c \leq M_{d,A}; \qquad \frac{F}{2} (a + c) - N_{d,o} \cdot c \leq M_{d,B}$$

Finally, when plate deformations are great compared to that of the bolt, the collapse mechanism is as illustrated in Fig. 7.25c: two plastic hinges form in sections A-A and B-B. It thus results that:

$$N = \frac{F}{2} + Q; \qquad Qc = M_{d,A}; \qquad \frac{F}{2} \, a - Qc = M_{d,B}$$

whence:

235

$$F = 2\,\frac{M_{d,A} + M_{d,B}}{a} \; ; \qquad Q = \frac{M_{d,A}}{c} \; ;$$

$$N = \frac{M_{d,A} + M_{d,B}}{a} + \frac{M_{d,A}}{c} \le N_{d,o}$$

Design can contemplate one of the three foregoing possibilities.
Plate thickness progressively decreases, while bolt size increases.
The above considerations, however, presuppose that plate strength values M_d are known. This problem has not yet received a satisfactory solution: experimental and numerical researches are needed in order to determine satisfactory values for M_d and, therefore, of effective width as a function of the various types of plate . Some calculation criteria are set out in 7.3.7.

7.3.5 *Eccentric end plate joints*

In some cases it is not possible to design symmetrical end plate joints, as bolts are difficult to reach from one side. One is then forced to use eccentric joints, an elementary example of which is illustrated in Fig. 7.26a. In this case, it is only possible to obtain an equilibrated diagram by taking into account a prying force Q at the flange ends.

The static diagram for the equivalent beam is given in Fig. 7.26b. Then $N = F\,(a + c)/c$.

If the plate is not sufficiently rigid to be able to transmit bending forces due to load eccentricity, considerable transverse deformations take place, due to second order effects.

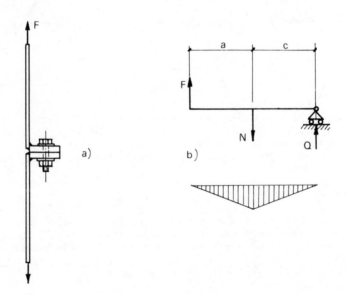

Fig. 7.26

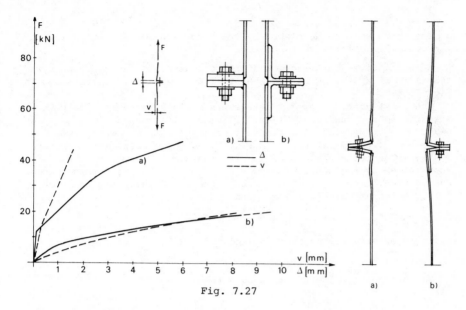

Fig. 7.27

This phenomenon is illustrated in Fig. 7.27, showing the behaviour of joints having varying degrees of rigidity |12,13|.

Joint (a) has flange and bolt designed according to the equilibrated solution of Fig. 7.26b while joint (b) has plate and bolt designed disregarding bending moment due to eccentricity.

In a first phase, due to friction, joint (a) practically does not elongate, but then evolves in a linear phase with an extremely small transversal displacement v: the joint thus only slightly departs from its initial configuration. Joint (b), however, does not display a linear phase. Plasticizations begin immediately in the plate and load can increase only provided the bending moment is maintained constant: this is possible if the lever arm decreases, i.e. if transverse displacement v becomes of the same order of magnitude as bolt eccentricity.

Specimen shape at the end of the tests clearly proves the above.

Joint (a) is still practically straight, while joint (b) has deformed transversely until the bolt axis practically coincides with the external load line.

7.3.6 *Tightening of bolts*

Bolt tightening does not affect the ultimate static load of a joint. On the other hand, it is of considerable importance whenever external loads are variable in time and thus cause fatigue in the bolts. Bolt strength under fatigue is much lower than the static strength and it is therefore advisable to avoid bolts being loaded by alternating load cycles, even if external loads vary cyclically.

This can be achieved |14| provided that:

237

The clamping force developed by the tightening of the bolts is of sufficient magnitude
The contact force developed by the tightening of the bolts is located in a favourable position

A qualitative justification of these statements can be found by a schematic study, using simple models of the two ideal joints shown in Fig. 7.28.

The first joint (Fig. 7.28a) has its contact force in line with the external force, the second (Fig. 7.28b) has its contact force at the plate ends. Consider the first joint using a model having two degrees of freedom, as shown in Fig. 7.28c. The bolts are represented by two extensional springs having stiffness k. The plates are represented by rigid girders and elastic hinges having stiffness h. Due to pre-tension $F_{N,O}$ (Fig. 7.28d), the following plate contact force, deformations and moment occur:

$$Q_O = F_{N,O} ; \qquad v = F_{N,O}/k ; \qquad \phi = v/a ; \qquad M = h\phi$$

If an external force $2F$ is applied (Fig. 7.28e):

$$Y = -F ; \qquad X = 0 ; \qquad v = 0 ; \qquad \phi = 0$$

By superposing effects, it ensues that:

$$F_N = F_{N,O} + X = F_{N,O}$$

$$Q = Q_O - Y = F_{N,O} - F$$

$$v = F_{N,O}/k ; \qquad \phi = v/a ; \qquad M = h\phi$$

Such relations are shown in Fig. 7.28f.

The obvious conclusion is that, while $F < F_{N,O}$, i.e. as long as the external force is less than the tightening force in the bolts, stresses in the plate and in the bolts do not change.

Therefore the bolts are not subjected to fatigue phenomena even if the external force F is time dependent.

Next consider the model shown in Fig. 7.28g, representing the joint of Fig. 7.28b. On tightening the bolts, the flange is bent as in Fig. 7.28h and

$$Q_O = F_{N,O} ; \qquad v = F_{N,O}/k ; \qquad \phi_1 = v/a ; \qquad M = h\phi$$

If an external load is applied (Fig. 7.28i):

$$X = F \frac{e/a}{(1+h/kba)} ; \qquad Y = F(1 - \frac{e/a}{1+h/kba})$$

$$v = X/k ; \qquad \phi = X/kb \qquad \phi_1 = Ya/h$$

Superposing the effects, gives:

$$F_N = F_{N,O} + X = F_{N,O} + \frac{F\,e/a}{1+h/kba})$$

$$Q = Q_O - Y = F_{N,O} - F(1 - \frac{e/a}{1+h/kba})$$

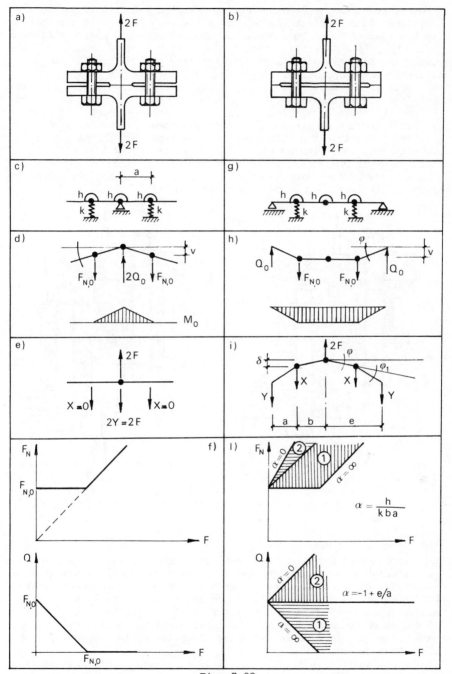

Fig. 7.28

Contrary to the proceeding case, the external force causes variations in bolt force ans bends the flange. Tension in the bolts and the contact force thus depend on applied load and plate stiffness.

If the plates are very rigid compared to bolt deformability (Fig. 7.281) ($h/kba \to \infty$) one obtains qualitatively similar results to those of the preceeding case. If plates are very deformable ($h/kba \to 0$), the force in bolts and the prying force increase simultaneously. If $h/kba = e/a - 1$, the prying force is unvaried.

From the above considerations it ensues that tightening of bolts can offer great advantages in a joint subjected to repeated loads, provided the joint is designed so as to guarantee a behaviour such as shown in Fig. 7.28a and avoiding that shown in Fig. 7.28b.

For example Fig. 7.29, shows a bearing slewing ring of a crane.

The bolts of the joint shown in Fig. 7.29 b are certainly more subjected to fatigue phenomena than those of Fig. 7.29a |14|.

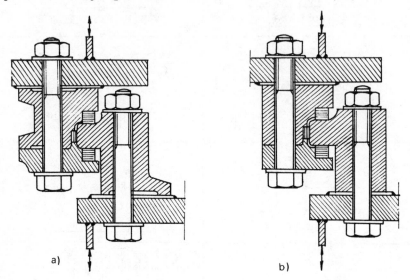

Fig. 7.29

7.3.7 *End plate design*

End plate design can be carried out, as a first approximation, by establishing an equivalence between the bi-dimensional behaviour of a plate and the uni-dimensional behaviour of a beam having appropriate stiffness: it is thus possible to provide a conventional calculation model, which is simple and useful for design purposes.

The equivalence between the real system and the model can be established in the elastic field and at collapse, independent of the limit state at stake.

The equivalence in the elastic field is always on the safe side: it under-estimates flange resistance and thus limits its deformation. It is therefore particularly appropriate whenever prying forces must be excluded (Fig. 7.25a).

The equivalence in the plastic field offers a realistic assessment of flange strength, as it considers its plastic behaviour.

It can take prying forces into account and thus it is fit to survey the mechanisms of Fig. 7.25c. On the other hand, it is inadvisable to apply such an equivalence to the calculation of flanges with many rows of bolts.

7.3.7.1 Elastic design
In order to establish the equivalence in the elastic field, consider an infinite plate, built-in along one edge and loaded by a load F distributed over a rectangular area the sides of which are c_x and c_y (Fig. 7.30).

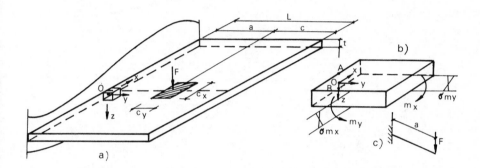

Fig. 7.30

If the load is concentrated ($c_x = c_y = 0$) and located along the free edge ($a = L$; $c = 0$), maximum values of the bending moments per unit length are as follows (Fig. 7.30b) |15|:

$$m_x = 0.509 \, F \; ; \qquad m_y = \nu \, m_x = 0.1527 \, F$$

If one now calculates the load diffusion angle and therefore the equivalent width b_{eff} of a cantilever beam having a length L, so that the same stress $\sigma_{m,y}$ as for the plate arises at the built-in joint (Fig. 7.30c), an equivalence is obtained from which it follows that:

$$m_x \, b_{eff} = Fa; \qquad b_{eff}/a = F/m_x \qquad \qquad \text{a)}$$

The load diffusion angle is therefore:

$$\alpha = \text{arc tan} \, (b_{eff}/2a) \simeq 45^{\circ}$$

A recent study |16| aimed at generalizing the above results by using finite elments. The parameters c_x and c_y were varied over the ranges:

241

$$0 \le c_x/L \le 0.40 \; ; \qquad 0 \le c_y/L \le 0.20$$

and the following observations made. Bending moment m_x at O and therefore the equivalent width b_{eff} practically does not depend on the value of c_y, but only on the position of load application.

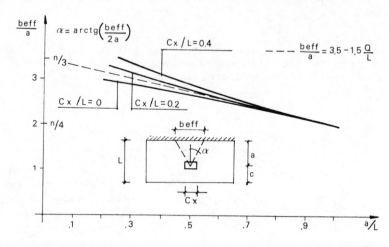

Fig. 7.31

The diffusion angle α depends in c_x/L only for small values of a/L (Fig. 7.31) and can be expressed by the following approximate formula (valid for $0.30 \le a/L \le L$):

$$\alpha = \pi (1 - a/4L)/3$$

Consequently, the moments at point O can be assessed by means of the following expressions:

$$m_x = Fa/b_{eff} \; ; \qquad m_y = \nu \, m_x \qquad \qquad \text{b)}$$

with $b_{eff} = 2\tan \alpha = 3.5 - 1.5 \, a/L$.

The above considerations justify the use of calculation models comparing plate behaviour to that of a cantilever or of a continuous beam on a number of supports, having an appropriate width. Some examples are illustrated in Fig. 7.32. The unstiffened plate shown in Fig. 7.32a can be considered as a cantilever having a width equal to the effective width $b_{eff} = 3.5 - 1.5 \, a/L$ as above.

The stiffened plates of Figs 7.32b,c can be considered as a system formed by a cantilever '1' and a beam '2' on two or more supports, perpendicular to the cantilever and connected to it: the equality of the vertical displacemnt component defines quotas F_1 and F_2 loading the cantilever and the beam. It is obvious that in the case of Fig. 7.32b the contribution of the cantilever-like behaviour is all the more negligible compared to the beam-like one, the closer the stiffeners are to each other and the farther the bolt is from the axis of symmetry.

242

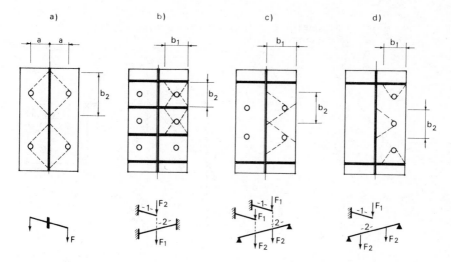

Fig. 7.32

On the contrary, in the flange of Fig. 7.32d, for the sake of simplicity, it is possible to subdivide the overall behaviour between the two directions by taking one load on the cantilever and the remaining two on the beam between the stiffeners.

The above considerations can obviously be extended also to more complex cases than those depicted in Fig. 7.32, provided analogous calculation models are formulated.

Solutions thus obtained are on the safe side: torsional moment contribution and stress redistribution in the elastic-plastic phase are neglected.

7.3.7.2 Plastic design
In order to establish the equivalence in the elastic plastic field, consider the infinite plate of Fig. 7.30, loaded by a force F concentrated ($c_x = c_y = 0$) on the free edge ($a = L$; $c = 0$).

The collapse mechanism is illustrated in Fig. 7.33a. Its limiting load can be assessed starting from that applicable to circular plates, for which many solutions are known as demonstrated in the following.

If Tresca's yield criterion is assumed as the strength criterion (Fig. 7.33b), the ultimate value $F_{u,o}$ of the concentrated load at the plate centre is $|17|$:

$$F_{u,o} = 2\pi \, m_{lim};$$

whether the plate is simply supported or its edges are clamped.

Such a result is not surprising. In fact, if the plate is simply supported (Fig. 7.33c), the exact solution gives the following bending moment distribution at collapse for both hexagonal and square resistance domains (point A of Fig. 7.33b):

243

$$m_r = 0; \qquad m_\theta = m_{lim}$$

The collapse mechanism may thus be interpreted as formed by concentric rings supporting one another.

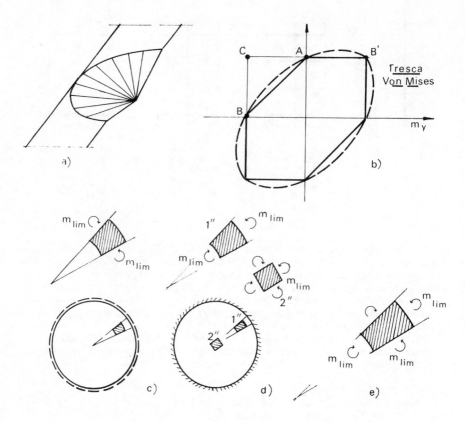

Fig. 7.33

If the plate is clamped (Fig. 7.33d) and the Tresca criterion adopted, at collapse (point B of Fig. 7.33b) we have:

$$m_r = - m_{lim}; \qquad m_\theta = 0$$

everywhere in the plate except at the centre where a discontinuity is present (point B' of Fig. 7.33b) and:

$$m_r = m_{lim}; \qquad m_\theta = m_{lim}$$

The collapse mechanism may thus be interpreted as formed by radial cantilevers, each independent of the others.

Finally, if the plate is clamped and a square resistance domain is used (Fig. 7.33e), the collapse bending moment distribution is (point C of Fig. 7.33b):

244

$$m_r = -m_{lim}; \qquad m_\theta = m_{lim}$$

The collapse mechanism thus appears as the superimposition of the previous ones and the ultimate load is |18|:

$$F_{u,o} = 4\pi \, m_{lim}$$

Such a result is used in various other works |19-21| aimed at establishing the ultimate load of steel plates subjected to a concentrated load and with various edge conditions. All these works contradict what is proved in |22|: if Tresca's criterion is adopted, a plate of any shape, undergoing a concentrated load in any position, has an ultimate load F_u that is always lower than $F_{u,o} = 2\pi \, m_{lim}$. From a physical point of view, such a result is obvious. The collapse load of a circular plate is independent of the radius r_e of the plate. Therefore, in any plate it is always possible to form around the load an analogous mechanism to that for a circular plate, and thus:

$$F_u \le F_{u,o} = 2\pi \, m_{lim}$$

Such a mechanism corresponds to the punching of a plate and $F_{u,o}$ will hereafter be referred to as punching load. Its importance is obvious. It is the upper bound to the collapse load of a plate under concentrated loading.

In practical applications, the following effects are also found:
(a) The most correct strength domain is Von Mises' ellipse (Fig. 7.33b)
(b) The load is not concentrated, but distributed over a finite area having radius c
(c) The plate has a hole having a radius r_i.

Fig 7.34 compares the values of the collapse load for simply supported or clamped circular plates, obtained by applying the Von Mises criterion as radius a of the loaded zone varies |17|.

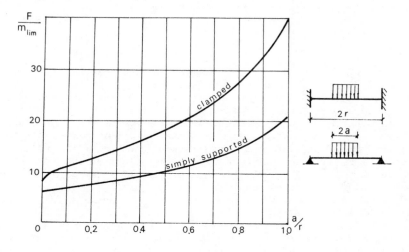

Fig. 7.34

It should be pointed out that, if the load is concentrated $(a/r = 0)$,

$$F_{u,o} = 2\pi\, m_{lim} \qquad \text{for a supported plate}$$

$$F_{u,o} = 7.26\, m_{lim} \qquad \text{for a clamped plate}$$

The Von Mises criterion thus leads to the same result as Tresca's for supported plates, but it offers an advantage of about 15% for clamped ones.

From Fig. 7.34 it also appears that if the load is distributed over an area of moderate dimensions $(a/r_e < 0.10)$, there is no advantage if the plate is simply supported but there is an advantage of about 50% if the plate is clamped. This explains why the value $F_u = 4\pi\, m_{lim}$, although incorrect from a theoretical point of view, can be considered as a realistic upper bound when it is compared with experimental results $|20,21|$.

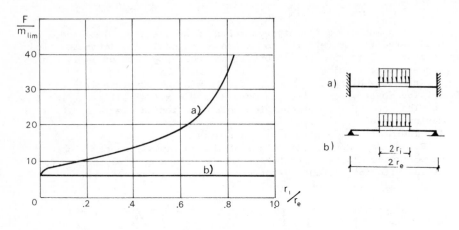

Fig. 7.35

The influence of a hole is evidenced in Fig. 7.35. It illustrates $|17|$ the analytical results obtained following Tresca's criterion in the case of a load distributed around the edge of the hole in both a clamped and a supported plate.

In the simply supported plate, the existence of the hole balances the fact that the load distance from the support decreases. In the clamped plate, however, the closer proximity of the load to the constraint more than outweighs the adverse effect of the hole.

The influence of the simultaneous existence of the hole and of the extension of the loaded area is illustrated in Fig. 7.36.

It was obtained by a finite element approach using the Von Mises criterion.

If the plate is clamped (Fig. 7.36a), it is reasonable to consider that the hole has no effect on the results obtained from an extension of the loaded area without the hole. If the plate is simply supported (Fig. 7.36b), there is, however, a distinct disadvantage and it would

246

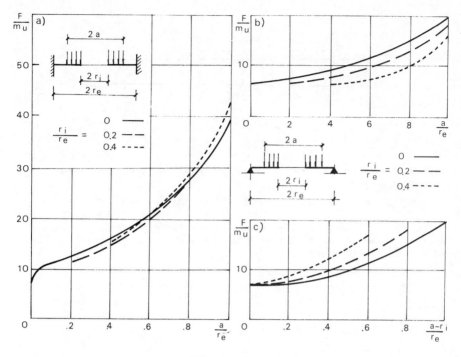

Fig. 7.36

appear cautious to refer to a fictitious load extension equal to $(a - r_i)$ (Fig. 7.36c) at least for $a/r_e < 0.60$.

The values given in Fig. 7.36 can be considered as the best estimate of the punching load for any plate. Obviously, depending on the constraints, if numerical or experimental results are lacking, one must decide whether the real collapse mechanism is closer to that of a simply supported circular plate or of a clamped one.

If the plate of Fig. 7.33a is now re-examined, it is obvious that:

c) $$3.14 \leq \frac{F_{lim}}{m_{lim}} = \frac{b_{eff}}{a} \leq 3.63$$

because the edge has an intermediate degree of constraint between the simply supported and the built-in ones. If this equivalent width is compared with that based on the elastic equivalence (Fig. 7.31), it will be seen that it is remarkably higher. This proves that there is a plastic redistribution both of the stresses in the thickness of the plate and of the bending moments surrounding the most loaded point.

If the load is not applied on the free edge, according to the limit analysis for rigid plastic materials, the two mechanisms of Figs 7.37a, b must be examined. In reality, the collapse mechanism is the incomplete one of Fig. 7.37c, made possible by the elastic-plastic nature of the material. To take this into account the following

247

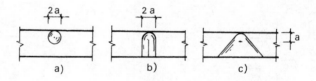

Fig. 7.37

empirical expression, derived from experimental work, has been suggested |23|:

$$b_{eff}/a = 4 + 1.25 \ c/a = 2.75 + 1.25 \ L/a$$

However, for $a = L$, this formula contradicts the results obtained above. Recent studies |13,16| have shown that, at least in the field of practical application

$$\frac{b_{eff}}{a} = - k_1 \frac{a}{L} + k_2$$

with

$$k_1 = 6 + 15 \ r/L$$

$$k_2 = 9 + 20 \ r \ L$$

$$r = \sqrt{C_x \ C_y/\pi}$$

Naturally, for plates having a finite length, $b_{eff} \leq b$.

To extend the results obtained to plates having a number of bolts in a single row, the following approach can be adopted.

If bolt spacing p is great compared to distance a, the effect of each bolt must be assessed as if it were isolated (Fig. 7.38a).

For closer bolts, the mechanism of Fig. 7.38b can be considered, or that of Fig. 7.38c if plate effect is negligible.

One can therefore assume:

$$b_{eff,n} = (n - 1) \ p + b_{eff} \leq b$$

b_{eff} being the effective width for only one bolt.

In the case of a collapse mechanism of the types illustrated in Figs 7.38b,c, plate performance can be increased by providing stiffeners. Then the mechanism of Fig. 7.39a must also be examined. It always provides upper bound values if the three sides are clamped |21| and therefore punching load values (Fig. 7.36) can be assumed as ultimate load and those of Fig. 7.36a as the most realistic ones.

Also, for a plate clamped on two sides, the mechanism of Fig. 7.39b provided higher values than the punching loads, at least for $a \leq c$. In

248

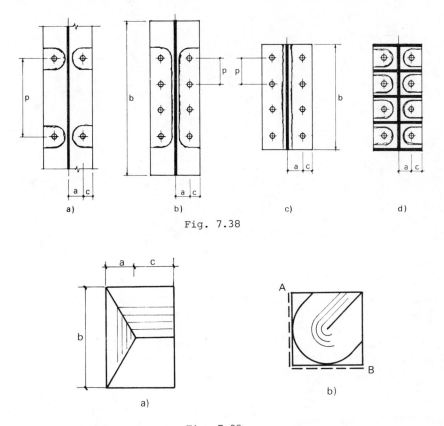

Fig. 7.38

Fig. 7.39

this case, however, the values of Figs 7.36b,c appear to be the most realistic ones.

Analogous considerations can be repeated with reference to the equivalence criterion in the plastic field for plates having a high deformability compared with that of the bolt, the ensuing behaviour of which is schematically illustrated in Fig. 7.25c. For them the possibility of using stiffeners is excluded: such reinforcements substantially diminish flange deformability and therefore prevent the development of prying forces. Consider the flange illustrated in Fig. 7.40a: its ultimate behaviour can be represented (Fig. 7.40b) as a plate constrained:

Along one edge, with a sliding-block allowing translation in the direction of the external force
Along the opposite edge with unilateral supports, i.e. capable of providing reactions only by contact
In connection with the bolt, with an elastic support simulating the deformability of the bolt itself

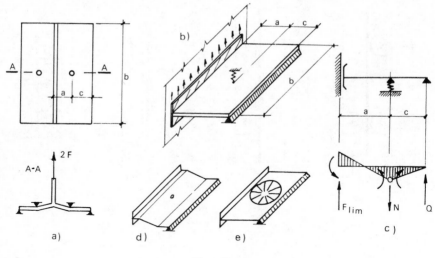

Fig. 7.40

The plate can thus be considered as the rigid-plastic beam of Fig. 7.40c. If m_{lim} is the limiting moment per unit width:

$$F_{lim} \frac{a}{2} = m_{lim} \, b_{eff};$$

whence:

$$F_{lim} = 2m_{lim} \, b_{eff}/a$$

b_{eff} being the width to be deduced by the equivalence between the beam and the collapse behaviour of the plate. Still for the beam, prying force Q and axial force in the bolt N are given by:

$$Q = F_{lim} \, a/2c \; ; \qquad N = F_{lim} \, (c + a/2)/a$$

If geometric width b is slight, the collapse mechanism of the plate is identical to that of the beam (Fig. 7.40d).

Therefore $b_{eff} = b$.

If geometric width b is considerable, the mechanism of Fig. 7.40e can take place. In this case, the punching value (Fig. 7.36) can be assumed as the ultimate load value, bearing in mind that the type of constraint makes it advisable to adopt lower values (Fig. 7.36c).

7.4 JOINTS IN COMPRESSION

The joints analysed in this section connect members which are predominantly in compression, i.e. struts, for which buckling phenomena (chapter 9) are the determinant in design. Connections between members which are predominantly in bending are examined in 7.5.

Whenever stability phenomena condition the design, the limits imposed by material strength cannot be attained. The concept of a full

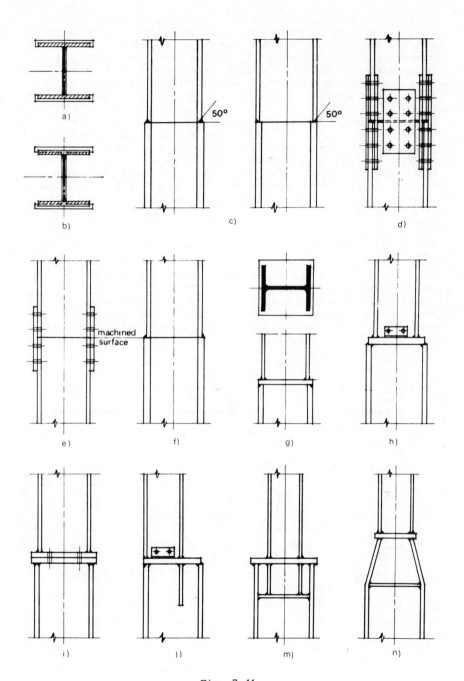

Fig. 7.41

strength joint thus becomes void of any importance: a connection is not a weak point of the connected element, provided it is designed to resist the maximum load carrying capacity of the element itself.

Connections often concern profiles having the same section or a slightly different one (Fig. 7.41a); in this case, joints are extremely simple and can be fully welded (Fig. 7.41c), bolted (Fig. 7.41d) or pure bearing joints with additional bolts (Fig. 7.41e) or welded (Fig. 7.41f).

If the dimensions of the two connected strut sections vary moderately (Fig. 7.41b), it is necessary to interpose a plate of appropriate thickness between them: the joint can then be made with butt (Fig. 7.41g) or fillet welds (Fig. 7.41h) or it can be of the bearing type (Fig. 7.41i). If there is a considerable variation in the section sizes, other measures must be adopted (Figs 7.41l,m,n).

Calculation criteria for the design of the tapered welded part are considered in 7.4.4.

7.4.1 *Welded joints*

Joints to be site welded should be so located as to allow the connection to be effected from a comfortable position (50 to 100 cm above the floor or platform).

Upper element surfaces must be prepared, while temporary locating pieces (angles, channels, etc) are welded or bolted to a part of the lower element to retain the upper element in its correct position during welding (Fig. 7.42a,b,c).

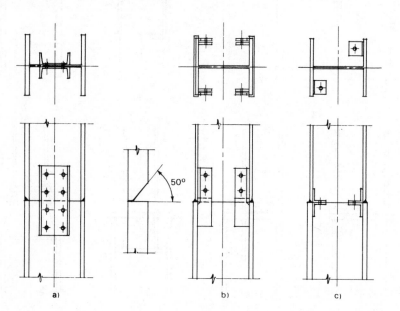

Fig. 7.42

If element sections have moderately varying dimensions (Fig. 7.41b), an intermediate plate must be provided. Joints can be butt welded, after adequate preparation, or fillet welded. In the latter case, the design of welds is based on the came criteria as for bolted or welded tension joints (see 7.3.1).

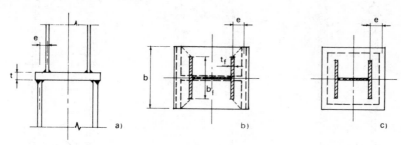

Fig. 7.43

The design of the separating plate where the sections vary considerably is more complex (Fig. 7.43a) as it is subject to bending. In the connection joint of two I-shapes (Fig. 7.43b), only the quota of load absorbed by the flanges has to be transferred, as the webs are aligned. If the mean stress in the profile is σ, plate design strength is f_d and $A_f = b_f\, t_f$ is the sectional area of a flange of the upper profile, then:

$$N_f = \sigma\, b_f\, t_f \; ; \qquad M = N_f\, e = \sigma\, b_f\, t_f\, e$$

At the elastic limit, therefore:

$$\frac{1}{6}\, b\, t^2\, f_d \geq \sigma\, b_f\, t_f\, e$$

Thus the plate thickness must satisfy the following inequality:

$$\left(\frac{t}{t_f}\right)^2 \geq 6\, \frac{b_f}{b}\, \frac{e}{t_f}\, \frac{\sigma}{f_f}$$

To put this in quantitative terms, let $b_f/b = 1$ and let $\sigma/f_d = 0.70$: it follows that, for an eccentricity e equal to upper column flange thickness t_f, one must provide a plate twice as thick as the column flange. It is thus obvious that, with this type of connection and with reasonable plate thicknesses, only moderate offsets of column flanges can be absorbed. The above calculation method can be extrapolated even when an I-beam is connected to a hollow section (Fig. 7.43c): if the web contribution is neglected, load N can be assumed to be equally distributed on the column flanges.

7.4.2 *Bolted joints*

Bolted joints can have a double cover plate (Fig. 7.44a) or a single cover plate (Fig. 7.44c). With I-sections sometimes only the flanges are connected (Fig. 7.44b).

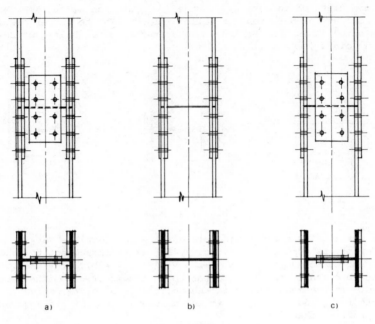

Fig. 7.44

If both flanges and web are spliced, it is advisable to distribute forces between the various joints in proportion to the areas of the several parts (see 7.3.1).

For I-sections and channels however, it is preferable for compression elements to be joined at their flanges rather than their webs: flanges, in fact, are more severely loaded by possible bending effects.

For joints of important columns it is furthermore often advisable to use a friction joint with high strength bolts in order to avoid joint slipping at service loads (see 6.2.2.1): joint slipping can entail permanent deformations of the structure, inconsistent with its use.

7.4.3 Bearing joints

When profile thickness becomes considerable, it is no longer possible to adopt welded or bolted joints: weld dimensions would become too great or the number of bolts unacceptable (see 6.3.1). It is therefore preferable to rely on bearing between surfaces to transfer load.

This can be achieved |8|:

Directly through the cross section of the column with additional light welds (Fig. 7.45a) or a small number of bolts (Fig. 7.45b)
By interposing a welded plate (Fig. 7.45c)

254

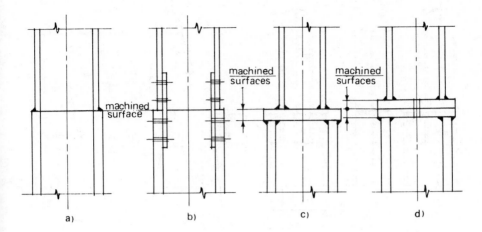

Fig. 7.45

By interposing two plates bolted to each other (Fig. 7.45d)
Contact must be ensured for the entire surface and not only for
some of its parts. In particular, squareness must be guaranteed
between the bearing surface and the column axes: if the ends are not
provided with plates , they must be sawn and finished with a planing
machine. Plates are usually flattened by press for thicknesses under
50 mm; for thicknesses above 100 mm, they must be worked by a planing
machine; for intermediate thicknesses, either a press or a planing
machine can be used, according to dimensions.

Finally, bearing joints must comply with the following
requirements |7| :

Absorb all shear and tension forces
Not be a weak point with reference to buckling

In order to comply with the first of these conditions , additional
welds or bolts must be checked:

For entire value of shear force
For tension occurring during particular loading combinations

To comply with the second condition, joints must be located as
close as possible to constraints and, in any case, at a distance not
exceeding 20% of the buckling length assumed in design.

7.4.4. *Section variations*

Fig. 7.46 indicates three typical solutions for joints between columns
with sensible section variations.

In the joint shown in Fig. 7.46a the plate welded in the shop to
the lower element is useful for easy positioning and site welding the
upper element. The web stiffeners receive the load transmitted by the

255

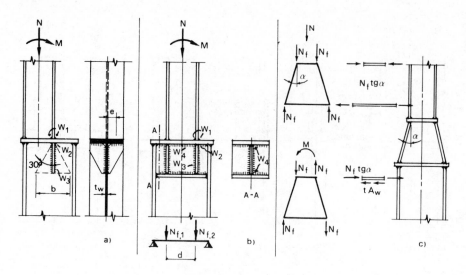

Fig. 7.46

upper flange and transfer it to the lower web, which in turn
redistributes it in the underlying part. Neglecting perpendicular
stress variations in thickness, value of force acting in the flange
is:

$$N_f = \sigma \, A_f$$

where:

A_f is the flange area, $\sigma = \dfrac{N}{A} + \dfrac{M}{W}$ is maximum stress.

Force N_f acts transverse on welds $'W_1'$ and $'W_2'$ and longitudinally
on welds $'W_3'$. The web must be able to resist stress concentration,
which means that assuming a diffusion angle of 30°:

$$N_f/bt_w \leq f_d$$

This cannot be complied with if the loads acting on the columns are
considerable. Then an analogous detail to that illustrated in Fig.
7.46b can be adopted: an I-beam loaded by two concentrated forces is
built-up.

$$N_{f,1} = \frac{N}{2} - \frac{M}{d} \; ; \qquad N_{f,2} = \frac{N}{2} + \frac{M}{d}$$

These forces transversely load welds $'W_1'$ and $'W_2'$ and
longitudinally load welds $'W_3'$. The welds $'W_4'$ absorb the longitudinal
shear between the web and the flanges of the I-beam (see 7.7).

Another possibility is illustrated in fig. 7.46c. The joint is
tapered and the stress resultant in each flange is deviated by means
of the two welded plates   that operate as tension and compression
stiffeners. If only an axial force N is acting the tension and

256

compression in the stiffeners is $\pm\, N_f\, \tan\alpha$, with $N_f = NA_f/A$. An external bending moment M produces shear stresses in the web, the maximum value of which can be assessed as $\tau = 2N_f\, \tan\alpha/A_w$, where A_w is web area and $N_f = MA_f/W$.

The web panel strength must be checked against the combined bending and axial effects. That is:

$$\sqrt{\sigma_w^2 + 3\,\tau^2} \leq f_d$$

with

$$\sigma_w = N_f/A$$

7.4.5 *Base joints*

In discussing connections between compression elements, it is natural to illustrate static problems involved in connections between concrete and steel: the column base is a typical case. In practice, a base can

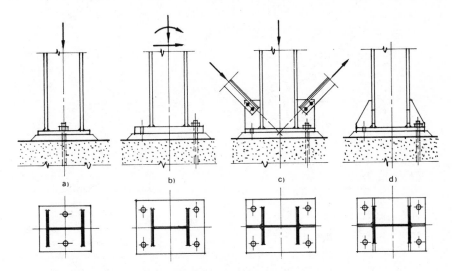

Fig. 7.47

be simply compressed (Fig. 7.47a) or be subject also to bending and shear (Fig. 7.47b). Finally, the bases of a bracing truss may be in tension or compression together with shear (Fig. 7.47c).

Bases such as that shown in Fig. 7.47d may be considered as full or partial constraints as well as hinges, depending on design assumptions (see 1.4). They are certainly capable of resisting bending moment to some extent, but their ductility provides the possibility of considering them as hinges if this leads to a more convenient and economical design.

Typical problems concerning a base are:
(a) Choice of geometrical dimensions for the base plate
(b) Transmission of shear forces to concrete
(c) Choice of plate thickness, depending also on stiffener position
(d) Anchor bolt design

7.4.5.1 Base plate geometry

Plate dimensions must be based on design values of axial force and bending moment. The section must be considered reacting only under compression and a linear distribution of stress can be assumed. Possible distributions are illustrated in Fig. 7.48.

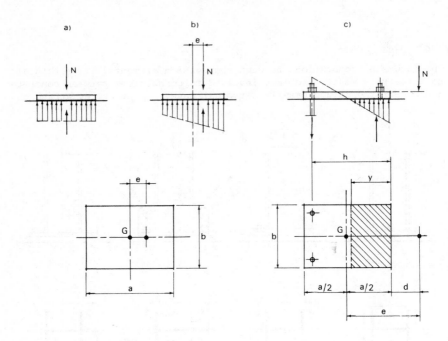

Fig. 7.48

If load eccentricity is $e = M/N$, the maximum concrete stress is:

$$\sigma = \frac{N}{ba} \qquad , \qquad \text{for } e = 0 \qquad \text{(Fig. 7.48a)}$$

$$\sigma_{max} = \frac{N}{ba}\left(1 + \frac{6e}{a}\right), \qquad \text{for } 0 \le e \le \frac{a}{6} \qquad \text{(Fig. 7.48b)}$$

If $e \ge a/6$, the section may be designed as a concrete section (Fig. 7.48c), reinforced only in tension and disregarding the bolt holes in the compressed area.

Design values for concrete strength may be assumed as for normal reinforced concrete structures. However, to enable the base plate to be adjusted to its correct level, it is always necessary to leave a

258

clearance of about 5 cm between concrete block and the steel plate
(Fig. 7.47). This space is subsequently filled with mortar, if
possible of an expansive type. Therefore, one cannot permit
compressive stresses of the order of those usually allowed for
concentrated loads on three-dimensional concrete blocks.

7.4.5.2 Shear forces
There are three possible ways for transmitting shear forces:

By means of anchor bolts
By friction between steel and concrete
By applying appropriate devices

If anchor bolts are taken into account, they must be designed
according to 6.2.3. Thus, shear action must be equally divided
between all the bolts. Then for the most loaded bolt:

$$(N_1/N_{d,o})^2 + (V_1/V_{d,o})^2 \leq 1$$

where:

N_1, V_1 respectively are axial and shear forces applied to the
anchor bolt
$N_{d,o}$ is design axial force for $V_1 = 0$
$V_{d,o}$ is design shear force for $N_1 = 0$

Some recommendations do not consider that anchor bolts are able to
transmit shear force, because of fear that permanent unacceptable
strain will occur in the superficial concrete strata in contact with
the bolts.

Therefore they indicate the possibility of transmitting shear
forces by friction between plate and concrete, accepting a friction

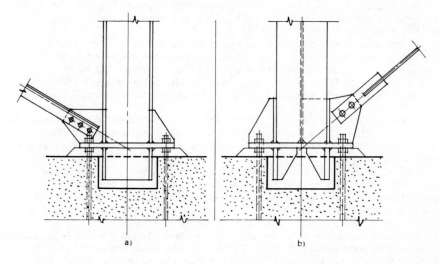

a)   b)

Fig. 7.49

coefficient $\mu = 0.40$ |24|. Control therefore becomes:

$$|V/N| \leq 0.40$$

N and V, respectively, being overall axial and shear forces applied to the base.

If the ratio N/V is small, the shear force must be resisted by direct contact between steel and concrete. This may be achieved by welding to the underside of the plate a short length of profile, identical to that above (Fig. 7.49a), or to use some stiffened plates (Fig. 7.49b). Bearing pressure must be limited to the concrete design compressive strength and the welds to the base plate must be designed for the acting shear force.

7.4.5.3 Base plate thickness

Base plate thickness and possible need for stiffeners depend on bending behaviour due to the effects of concrete pressure and of anchor bolt concentrated loads.

If the plate it not stiffened (Fig. 7.50a) and is loaded by concrete pressure, the thickness can be determined using a method adopted in US practice |8|.

The overhanging parts of the plate are considered as two cantilevers as follows:

(i) Span $a = (h_1 - 0.95h)/2$ and width b_1
(ii) Span $c = (b_1 - 0.80b)/2$ and width h_1

If the bearing pressure on the concrete is p, the thickness is the greater derived from:

$$t^2 \geq 6\,\frac{pa^2/2}{f_d}\,; \qquad t^2 \geq \frac{pc^2/2}{f_d}$$

If thicknesses become excessive, the plate must be stiffened in an increasingly complex manner as its dimensions grow (Figs 7.50b,c,d,e). In some cases conditions are such as to indicate a beam-like behaviour of the base plate (Figs 7.50b,d).

If the dimension of the sides tend to be similar, a bi-dimensional behaviour may be taken into account, using the results offered by the theory of thin plates supported or built-in on four, three or two sides. Plate thickness calculation must be followed by stiffener design. Stiffeners are usually deep beams and thus are not much affected by bending effects.

Shear effects are more significant and it is these which determine the weld sizes. Figs 7.50d,e show the load distribution of the various stiffeners and resultant actions on welds between stiffeners and column.

The effects of concentrated loads due to anchor bolts must be analysed according to the methods contained in 7.3, but disregarding prying forces. Contact between steel and concrete is not able to produce sufficiently high specific pressures to allow a prying force acting on a surface practically reduced to a line.

To diminish plate thicknesses, anchor bolt loads can be distributed over a more extended area (Fig. 7.50b) or the solution indicated in Fig. 7.50f can be adopted: it is particularly effective when anchor bolts are distant from the column.

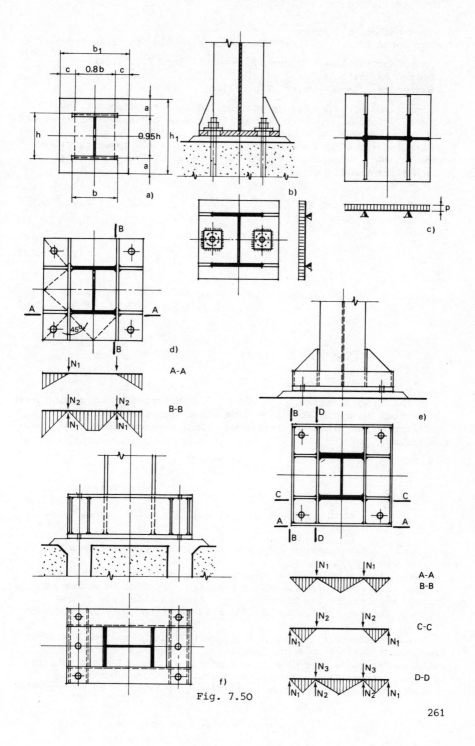

Fig. 7.50

261

7.4.5.4 Anchor bolt design

Anchor bolts (Fig. 7.51) can be classed according to three types:

Anchor bolts placed before concrete pouring (Fig. 7.51a)
Hooked anchor bolts (Fig. 7.51c)
Hammer head anchor bolts (Fig. 7.51d)

Anchor bolts placed before pouring the concrete transmit load actions to foundations by bonding. As they do not allow any clearance, they need very accurate positioning.

With reference to the notes of Fig. 7.51b, their capacity can be expressed by the following formulae, inspired by those indicated by French recommendations $|24|$:

$$N_1 = \frac{f_{b,d}}{(1 + \phi/a)^2} \; \pi \; \phi \; L \qquad\qquad \text{(Fig. 7.50b}_1)$$

$$N_1 = \frac{f_{b,d}}{(1 + \phi/a)^2} \; \pi \; \phi \; (L + 6.4 \; r + 3.5 \; L_1) \qquad\qquad \text{(Fig. 7.50b}_2)$$

$$N_1 = \frac{f_{b,d}}{(1 + \phi/a)^2} \; \pi \; \phi \; L \quad f_{c,r} \; \alpha \; \pi \; r^2 \qquad\qquad \text{(Fig. 7.50b}_3)$$

with

$$\alpha = \begin{cases} 1 - r/L & \text{for } L \le a \\ 1 - r/a & \text{for } L > a \end{cases}$$

In the above formulae, $f_{b,d}$ and $f_{c,d}$, respectively, are design bond and bearing strengths for three-dimensional concrete solids.

Following CEB (Comité Européen du Beton) recommendations $|25|$, one can assume:

$$f_{b,d} = \frac{0.36}{\gamma_c} \; \sqrt{f_{c,k}} = 0.24 \; \sqrt{f_{c,k}} \qquad\qquad (\text{N mm}^{-2})$$

$$f_{c,d} = \frac{f_{c,k}}{\gamma_c} \frac{a}{r} \quad \text{but not greater than } 3.3 \; \frac{f_{c,k}}{\gamma_c} \qquad (\text{N mm}^{-2})$$

$f_{c,k}$ being characteristic strength of concrete, measured on a cylinder (N mm^2) and γ_c = 1.5.

Hooked anchor bolts allow clearance for their positioning (Fig. 7.51c). The lower pin encased in concrete is useful to anchor the bolt during erection, before hole filling. Also in this case stress transmission takes place by bonding between anchor bolt and concrete filling.

Bonding between filling and concrete wall is guaranteed whenever the contact surface is sufficiently rugous, such as obtained when using wooden boxes or polystyrene blocks. If the holes are obtained by other methods (e.g. using plastic or steel forms) bonding is considerably lower and can be quantified only by means of experimental

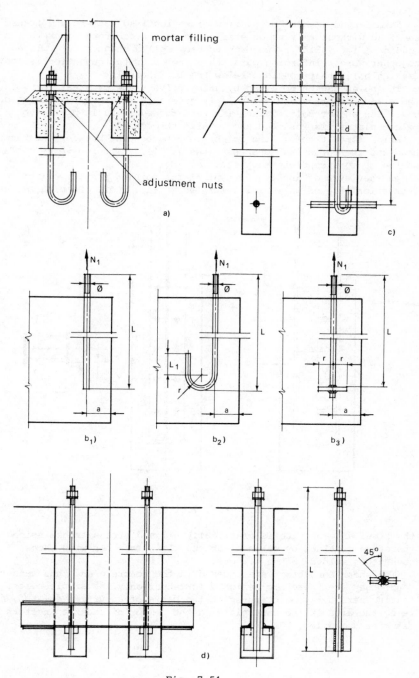

mortar filling

adjustment nuts

a)

c)

b₁)

b₂)

b₃)

d)

Fig. 7.51

tests.

Hammer head anchor bolts allow dimensional adjustment and transmit the load without relying on steel/concrete or concrete/concrete bonding (Fig. 7.51d). Two back to back rolled channels are placed together with reinforcing bars; then the concrete foundation is poured, leaving holes that continue below the two channels.

The holes can be obtained by using polystyrene or steel formwork; the extension below the channels is generally obtained by welding on a box. The anchor bolt, having its head forged or built up by two welded plates, is inserted and rotated through 90°.

Axial force transmission is obtained by bearing between the hammer head and the horizontal channels (see 7.3.1), which behave as a beam on two or more concrete supports.

Another type of anchorage, typical for joining a steel column to a foundation pile or wall is illustrated in Fig. 7.52. Bonding between

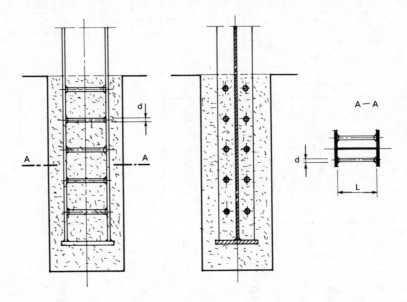

Fig. 7.52

the steel and concrete is practically nil and stress transmission therefore obtained by means of an end plate together with shear connectors welded to the flanges.

The shear connectors are encased in the concrete and thus cannot bend: they are therefore designed for shear only, as simple bolts. Bearing stress may be computed on the basis of a surface $A + ndL$, A being the area of the base plate, n the number of shear connectors of diameter d and length L.

7.5. JOINTS IN BENDING

When analysing the behaviour of joints connecting members which are predominantly subjected to bending (beams, for example), two preliminary point s should be made.

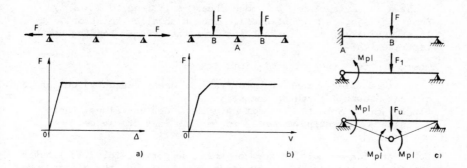

Fig. 7.53

Point 1. For tension or compression members (Fig. 7.53a), all sections simultaneously become plastic as external load increases; in beams (Fig. 7.53b) plastic hinges usually form in a well determined sequence, collapse occurring when hinges are sufficiently numerous to transform the beam into a mechanism.

Ductility needed to allow a redistribution of internal forces within the element or the structure is thus related only to the deforming capacity of the sections that evolve in the plastic field. In the case shown in Fig. 7.53c (identical to that of Fig. 7.53b) for example, hinge A forms when:

$$M_A = \frac{3}{16} F_1 L = M_{pl} \quad ; \quad M_B = \frac{5}{32} F_1 L = \frac{5}{6} M_{pl}$$

M_{pl} being the plastic moment of the section. For $F > F_1$ the beam evolves as if it were simply supported, until ultimate load F_u is reached and a hinge forms at B:

$$F_u = F_1 + (M_{pl} - \frac{5}{6} M_{pl}) \, 4/L = \quad F_1 + \frac{2}{3} M_{pl}/L = 6 \, M_{pl}/L$$

As F increases from F_1 to F_u, the hinge at A must be able to rotate by:

$$\theta_A = (F_u - F_1) \, \frac{L^2}{16EI} = \frac{1}{24} \frac{M_{pl}L}{EI}$$

without a brittle rupture due to excessive deformation.

Point 2. Shear forces V often affect section strength in a negligible way and they may be disregarded if $V < V_{pl}/3$, V_{pl} being the ultimate shear strength of the section (see 8.3.2.1).

Furthermore, they can always be neglected in beam displacement and rotation calculations.

265

From the above points it follows that:

Rigid joints maintaining the entire bending and shearing strength of the connected girder can be located at any section of the structure

Rigid joints maintaining the entire bending strength only can be located in any section, provided shear is limited to $V \leq V_{pl}/3$

Semi-rigid joints must be ductile enough to allow rotations, at least equivalent to those ensuing from the assumed bending moment distribution

There are two types of semi-rigid joint:

Semi-rigid frame joints able to transmit shear design values and a certain amount of bending moment

Simple joints able to transmit only shear design values, their resistance to bending moment being so low that it can be disregarded

Consider, for example, a beam on three supports (Fig. 7.54a) having a constant section and uniformly loaded: according to the design model illustrated in 1.4.5, various equilibrated bending moment distributions can be assumed (all are on the safe side).

Let $M_{i,u} = \alpha \, q_u \, L^2$ be the moment at the intermediate support and $M_{max,u}$ the consequent maximum positive moment corresponding to the load q_u allowed by ultimate limit state design. As the load increases, the beam behaves as a continuous beam on three supports until load $q_1 = 8 \, M_{i,u}/L^2$ is attained.

If $q > q_1$, the girder behaves as two contiguous simply supported beams loaded by $q - q_1$. Their end rotation is:

$$q_1 = 8 \, M_{i,u}/L^2 = \alpha \, 8 \, q_u$$

$$\text{for} \quad q \leq q_1 \quad , \quad \theta = 0$$

$$\text{for} \quad q \geq q_1 \quad \theta = \frac{1}{24EI} \, (q - q_1) \, L^3$$

Values of rotation θ_u and of moments in span and at support corresponding to the ultimate load q_u are:

$$\theta_u = \frac{1}{24EI} \, (q_u - q_1) \, L^3 = \frac{1}{24EI} \, (q_u - \frac{8 \, M_{i,u}}{L^2}) L^3$$

$$= \frac{1}{24EI} \, (1 - 8 \, \alpha) \, q_u \, L^3$$

$$M_{max,u} = \{0.125 - 0.5 \, \alpha \, (1 - \alpha)\} \, q_u \, L^2; \qquad M_{i,u} = \alpha \, q_u \, L^2$$

Fig. 7.54b shows diagramatically the value of θ_u as a function of the support moment $M_{i,u}$.

Various possibilities ensue for designing the beam and its connections, depending on the choice of value $M_{i,u} = \alpha \, q_u \, L^2$.

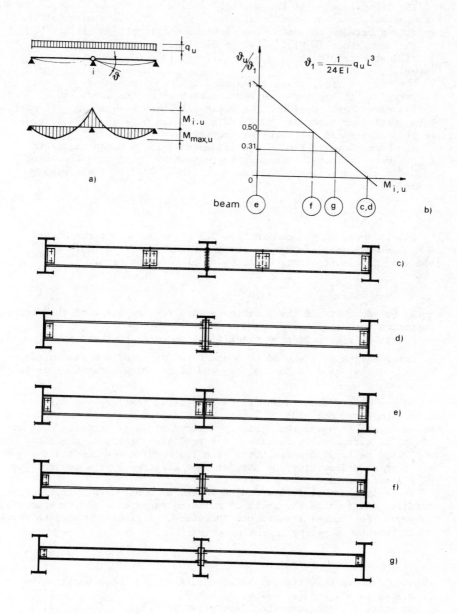

a)

$\vartheta_1 = \dfrac{1}{24\,E\,I}\,q_u\,L^3$

beam

Fig. 7.54

The beam shown in Fig. 7.54c, for example, is designed for $M_{i,u} = q_u L^2/8$.

The joint located at the point of zero moment can be a simple joint; it does not need any rotational capacity, as there are no relative rotations between the connected sections until collapse is reached.

The beam shown in Fig. 7.54d is designed as the foregoing one: the joint must be rigid ($M_{i,u} = q_u L^2/8$), but it needs no rotational capacity.

The beam of Fig. 7.54e is designed for $M_{i,u} = 0$ and $M_{max,u} = q_u L^2/8$. Therefore it has the same section as that of the above cases.

A simple joint, designed for shear only, may be used, but it must have sufficient ductility to allow a rotation $\theta_u = q_u L^3/24EI$, equal to the rotation of the simply supported beam.

The beam shown in Fig. 7.54f is designed for a moment equal to $M_{max,u} = q_u L^2/10.44$. The negative moment $M_{i,u} = q_u L^2/16$ must be provided with a semi-rigid frame joint ($M_{i,u} < M_{max,u}$) allowing a rotation:

$$\theta_u = \frac{1}{24EI} (1 - \frac{1}{16}) q_u L^2$$

Finally, the beam shown in Fig. 7.54g is designed for a moment $M_{max,u} = q_u L^2/11.66 = M_{i,u}$. Its joint must be rigid ($M_{i,u} = M_{max,u}$) and must have sufficient ductility to allow a rotation:

$$\theta_u = \frac{1}{24EI} (1 - \frac{1}{11.66}) q_u L^3$$

equal to about 30% of the rotation needed for the joint of the simply supported beam.

Joints will hereafter be classified according to their position as:

Intermediate joints used to connect two parts of the same girder
End joints, used to connect the end of one beam to another or to a column

Such a distinction is justified by reasons that are not only typological: at the ends of a beam it is always possible that plastic hinges might form and thus joints, whether rigid or semi-rigid, must always be sufficiently ductile. Intermediate joints, on the other hand, can often be located at sections with low bending moments and they do not then need ductility, as they are not affected by the possibility of plastic hinges.

The examples that follow can, however, be considered sufficiently ductile, at least on the basis of building experience and experimental surveys. For normal structures, therefore, it is unnecessary to check their rotation capacity in the plastic field.

7.5.1 *Intermediate joints*

Fig. 7.55 shows some typical intermediate joints whose qualitative behaviour is illustrated below.

Quantitative considerations are given in 7.5.3, 7.5.4 and 7.5.5.

The I-beam shown in Fig. 7.55a has both flanges and web butt welded. Therefore the joint is rigid, both with reference to bending and to shearing behaviour.

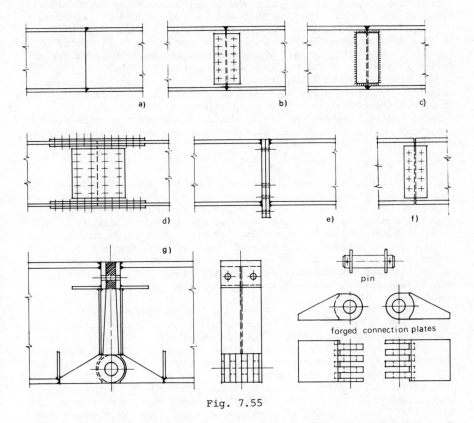

Fig. 7.55

The I-beam of Figs 7.55b,c has its flanges butt welded, and the web connected by welded or bolted splices. Thus the flanges are rigid, while the web connection is designed on the basis of the actual shear force: the joint can therefore be considered rigid with reference to bending if the web is designed for shear force V and for the portion of bending moment $M_w = f_d t_w h_w^2/6$ acting on the web. On the contrary, if the web connection is designed only for shear, the joint is semi-rigid with reference to bending: it is, in fact, capable of bearing a moment $M_f = f_d A_f d_f$, where A_f is the smallest flange area and d_f is the distance between flange centroids. Methods to be applied for web connection design are illustrated in 7.5.3.

Fig. 7.55d illustrates an entirely bolted joint. If forces are distributed between the various parts of the connection in proportion to the strength of the individual connected parts (see 7.5.3), the joint can be considered rigid with reference to bending strength.

Various tests have proved that, notwithstanding the handicap due to the holes, beam collapse takes place outside the connected area since the formation of a plastic hinge entails buckling of the compression of flange and, at the joint, the flange is stiffened by the cover plates.

269

Fig. 7.55e shows an end plate web connection: it can be considered rigid or semi-rigid according to plate and bolt strength. Problems connected with their design are illustrated in 7.5.4.

In the joint shown in Fig. 7.55f, the webs only are connected by bolted (or welded) splices. The joint is thus semi-rigid and its design is similar to the web connection of Figs 7.55b,c.

Finally, the joint shown in Fig. 7.55g is an example of a typical semi-rigid connection for rapid assembling and dismantling of temporary structures. Compression in the upper flange is transmitted by bearing (see 7.4.3), while the lower flange tension and shear act on a forged, cast or built-up pin connection (see 7.2.1).

7.5.2 *End joints*

Fig. 7.56 illustrates some solutions concerning joints between secondary and principal beams.

The welded joint shown in Fig. 7.56a is rigid with reference to both the bending and shear strengths of the secondary beam.

If site welding is to be avoided, the solutions shown in Figs 7.56b,c can be adopted: stub ends of the secondary beam are welded to the principal beam in the workshop and site connected by bolted joints to the intermediate part of the secondary beam. This is obviously an expensive solution, due both to its fabrication and erection techniques.

In the joint shown in Fig. 7.56d the lower compressed flange is connected by a bearing joint, whereas the tension flange has a bolted splice. Shear is transmitted by angle end cleats (see 7.5.5). Such a joint is capable of transferring only the design value of shear and the portion M_f of bending moment absorbed by the flanges. It may be considered rigid with respect to bending only if the upper flange connection is designed for entire bending strength M_d of the beam, that is for a force $N = M_d/d$ d being the distance between flange centroids. In this case, it is compulsory to taper the flange plate at its ends.

Fig. 7.56e shows a variant of the preceding joint: shear is transmitted by seating brackets which support the secondary beams. Design of the seats is demonstrated in 7.6.4.

Fig. 7.56f shows an end plate joint, wich obviously is semi-rigid: it is capable of transmitting the design value of shear and a bending moment $M > 0$, depending on bolt and flange design (see 7.5.3).

Fig. 7.56g illustrates the classical solution of an entirely bolted connection using end angle cleats. Design usually takes account only of shear and therefore the joint may be considered as a hinge. Fig. 7.56h shows a welded variant and Figs 7.56i,l,m show other variants with a bolted connection to welded stiffenes. All these solutions offer a good guarantee of ductility, at least for normal beam support rotations (i.e. for beams complying with the deformability limitations suggested in chapter 8). They are therefore used without further precautions, provided the beams are considered as simply supported and the joints are designed according to the criteria stated in 7.5.5.

Finally, Fig. 7.56n illustrates a further variant of the preceding joint: the beam is supported by a seat formed by a bolted angle (a

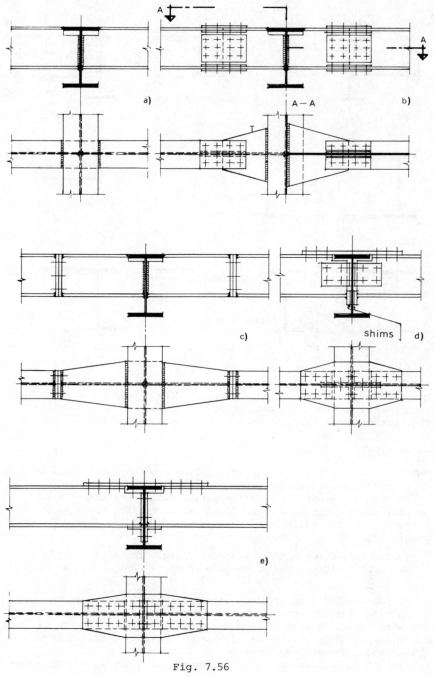

a)

b)

A — A

c)

shims d)

e)

Fig. 7.56

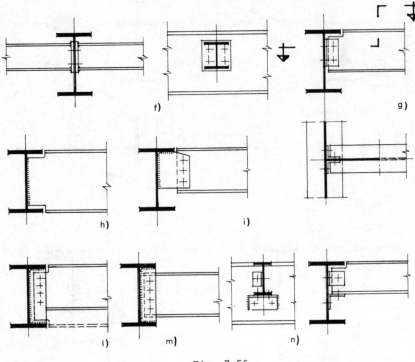

Fig. 7.56

built-up welded seat may also be used). The single web angle cleat
is provided for the sole purpose of preventing the beam overturning.
This connection is similar to a beam/column joint discussed in 7.6.

7.5.3 *Joints with cover plates*

In the case of a rigid joint, the applied bending moment must be
distributed between flange and web connections.

In the most general case of an unsymmetrical section, let the
smaller flange area be A_f, the distance between the two flange
centroids, d, the maximum section moment, M_d and the flange design
moment, $M_f = A_f f_d d$. One must design:

Each splice connection between the flanges for a force $N = A_f f_d$
Web splice connection, assuming a moment $M_w = M_d - M_f$ and design
shear V_d

For semi-rigid joints, it is often economical to design the flange
connections for the entire bending moment M_d and the web connection
for shear only. Design of flange connections (splice thickness, bolts
or welds) is identical to the elementary method for bolted or welded
joints under shear (see chapters 5 or 6).

The web connection is loaded by the design shear V_d and by a moment

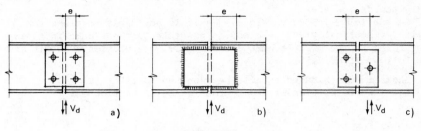

Fig. 7.57

$M_w \geq 0$.

The effects of force V_d must be determined adopting an equilibrated solution: e.g., the force can be considered as applied at the junction of the two parts (Figs 7.57a,b) or at the section passing through the centroid of the joint of one of the two connected parts (Fig. 7.57c).

The web splice must thus be designed for a shear force $V = V_d$ and a twisting moment: $T = V_d e \pm M_w$.

7.5.4 *End plate joints*

Shear force V_d is divided equally between the bolts, while bending moment is distributed between the bolts according to one of the criteria set out in 6.3.2. Flange thickness, needed to distribute the concentrated load transmitted by the bolts to the profile, can be calculated by determining effective widths according to one of the criteria set forth in 7.3.7.

End plate rigid joints are in many cases expensive, due to plate thickness and fabrication processes. For this reason, it may be preferable to design semi-rigid joints, intermediate joints located at lightly loaded sections and end joints being considered as plastic hinges capable of transmitting a smaller moment than that of the connected girder.

Following such a criterion, two classes of flanged semi-rigid joints can be distinguished; they are both capable of adapting plastically, certainly in a sufficient manner if the beams comply with the commonly accepted deformability limitations (see 8.1).

· A semi-rigid frame joint (Fig. 7.58a) is capable of transmitting the design shear V_d, together with some bending moment. Simple joints

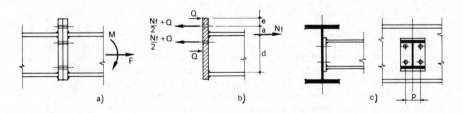

Fig. 7.58

273

(Fig. 7.58c) are typical of end plate joints and are fit to transmit only the design shear.

In the plate shown in Fig. 7.58a it is advisable to place the bolts symmetrically and as close as possible to the flange in tension. According to the ECCS-recommendations |11|, calculation criteria are as follows (Fig. 7.58b):

Shear is distributed equally between the bolts
Axial force N_f transmitted by the flange in tension is:

$$N_f = \frac{F}{2} + \frac{M}{d}$$

Maximum force that can be transmitted by each bolt including shear is (see 6.2.3):

$$N = N_{d,o}\sqrt{1 - (V/V_{d,o})^2}$$

Bending strength of the plate end plate is controlled, assuming an arbitrary value for the prying force, so that:

$$Qe = (nN - \frac{N_f}{2})\ e \leq M_{lim,2}$$

$$N_f \frac{a}{2} = (nN - \frac{N_f}{2})\ e \leq M_{lim,1}$$

where

$M_{lim,1}$ is the limiting moment of the gross section of the plate
         plate
$M_{lim,2}$ is limiting moment of the net section of the plate
n is the number of bolts in one row

The simple end plate joint of Fig. 7.58c must be fabricated taking precautions with reference to dimensional tolerances in respect of both beam length and surface squareness: it therefore requires more accurate workmanship than joints with angle cleats or seated joints (see 7.6.4). To guarantee a sufficient rotation capacity |8|, plate thickness should be between 6 and 10 mm. The welds connecting the plate to the beam and the bolts are designed on the basis of shear only. Plate thickness may be evaluated according to bolt bearing strength and to a conventional value of bending moment $M = V_D/4$.

7.5.5 *Joints with angle splices*

Joints with angle splices (Fig. 7.56g) are simple joints and allow rotation of the connected girder. To avoid cracking of the finishing of the slabs supported by the beams, joints should be placed as close as possible to the upper flange of the principal beam so that its displacement due to secondary beam rotation is minimized.

Once the position of beam reactions R is determined (see 1.4.1), joint design must take into account the two eccentricities e_1 and e_2 (Fig. 7.59a).

With reference to the symmetrical joint of Fig. 7.59b, bolts

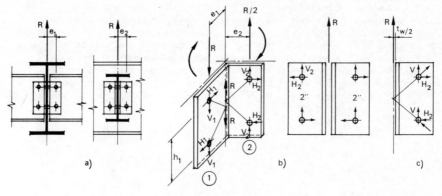

Fig. 7.59

connecting the beam web to web legs '1' must be designed for a shear force $V = R$ and a twisting moment $T_1 = Ve_1$. If there are only two bolts, they produce a vertical shear component $V_1 = V/2$ and a horizontal one $H = Ve_1/d_1$ that load the bolts on two sections by their resultant $R_1 = (V_1^2 + H_1^2)^{1/2}$.

Bolts connecting the outstanding legs '2' must be designed for a shear force $V/2$ and a twisting moment $T_2 e_2 V/2$. If there are only two bolts in each outstanding leg, they produce a vertical component $V_2 = V/4$ and a horizontal one $H_2 = (V/2) e_2/d_1$ that load the bolts on one only section by their resultant $R_2 = (V_2^2 + H_2^2)^{1/2}$.

Shear on the bolts and bearing for the holes of both principal and secondary beams must be checked against the resultant forces R_1 and R_2. The design of angle splice connections is conventional and standard dimensions are mostly used. If not, they must be checked under bolt bearing and as cantilevers built in the bolted section and loaded by a force V. Their thickness t must therefore comply with the following inequalities:

For flange type '1':

$$R_1 \leq 2td\, f_b$$

$$Ve_1 \leq (2t\, h^2/6) f_d$$

$$V \leq 2th\, f_d/\sqrt{3}$$

For flange type '2':

$$R_2 \leq td\, f_b$$

$$\frac{V}{2} e_2 \leq (th^2/6) f_d$$

$$V/2 \leq th\, f_d/\sqrt{3}$$

275

A similar design may be followed for the asymmetrical joint of Fig. 7.59. The same forces acting in the foregoing case act also on the bolts connecting the web leg, but they load only one bolt section.

The outstanding leg is loaded by forces having twice the value of those considered in the foregoing case. However, parasitic effects take place in the web leg as a consequence of eccentricity $t_w/2$.

Following U.S. common practice for non-symmetrical joints |8|, angles must have a greater thickness than half the bolt diameter.

Therefore, if standard legs are not used, the following limitations must be satisfied:

$$R_1 \leq td\, f_b \qquad\qquad R_2 \leq td\, f_b$$

$$Ve_1 \leq (th^2/6)\, f_d \qquad\qquad Ve_2 \leq (th^2/6)\, f_d$$

$$V \leq th\, f_d/\sqrt{3} \qquad\qquad t \geq d/2$$

7.6 BEAM-COLUMN JOINTS

Joints connecting beams to columns are not substantially different from joints in bending discussed in 7.5, at least as far as their function and behaviour are concerned. They are considered separately because their structural type differs from beam-beam joints and because they raise some typical design problems, such as the influence of shear and of axial force.

Beam-column joints are typical positions for plastic hinges in a possible collapse mechanism of the structure.

Between beam and column, the latter is usually the weaker element, due both to its section and to buckling phenomena.

The shear forces are transmitted from the girders to the column; negative bending moments may be transmitted from a beam to the contiguous one or to the column or to both.

Thus, it is also possible in the case of beam-column joints to maintain the distinction between rigid and semi-rigid joints, subject to the following consideration. If a joint connects a single beam to a column, the joint is rigid if its strength is not less than that of the weaker connected section. If two beams connect to a column on opposite sides, the joint is rigid if its strength is not less than that of the weaker of the two beams.

In beam-column joints, the semi-rigid solution is frequently adopted; pin-ended structures may be seen as an extreme solution adopting semi-rigid connections the bending strength of which is so low that it may be disregarded (simple joint). As with joints at beam ends in beam-beam connections (see 7.5.2), ductility as well as joint strength, is very important: it must allow the rotations of the connected beams.

Beam-column joints designed for high bending moments are typical of framed structures. In such cases, joints can be |26|:

Rigid
Rigid with reference to bending capacity alone
Semi-rigid, with sufficient rotational capacity

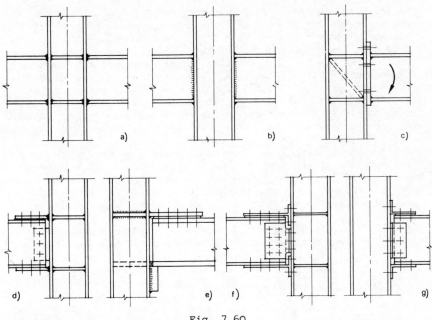

Fig. 7.60

The most common types are illustrated in Fig. 7.60.

The most natural rigid joint is the entirely welded one shown in Fig. 7.60a. Intermediate stiffeners are welded in the workshop and site welding is required between beams and column.

This has the disadvantages associated with site welding (it is uneconomical and dependent on weather conditions), but on the other hand it reduces overall structural dimensions (it avoids splices of flange plates) and guarantees a good ductility. A greater site cost is thus offset by economy in material and in fabrication. This consideration, together with the fact that beam-column joints are not numerous ·in a building, has allowed such a solution to be used especially in seismic areas.

The joint shown in Fig. 7.60b has the column unstiffened: it can provide a rigid or semi-rigid joint with reference to beam bending capacity, depending on beam and column dimensions: its rotational capacity might, however, be insufficient. Its behaviour can be assessed by the methods given in 7.6.1.

For both the joints of Figs 7.60a,b the danger of lamellar tearing in the column flange material (see 7.1.2) must be taken into account. If site welding must be excluded, the other solutions shown in Fig. 7.60 may be adopted.

In end plate joints (Fig. 7.60c), the beam end design is identical to that for elements in tension (see 7.3.4) or for elements in bending (see 7.5.4). The sole difficulty may arise from column flange thickness which might be insufficient to resist the effects of concentrated loads transmitted by bolts: it must be of the same order

277

of magnitude as that required by the end plate.

The end plate joint can be replaced by solutions using splices: they may be rigid (see 7.5.3 for considerations concerning hole effects) or, more commonly, semi-rigid. In the joint shown in Fig. 7.60d, splices are welded to the column and bolted to the beam flanges, while angle splices connect the beam web. This solution can completely restore bending capacity. It has the same disadvantage as welded joints with regard to lamellar tearing, but it offers a rational example of an entirely site bolted joint. A criticism, however is that the unprotected plates attached to the flange of the column might be bent during transportation of the column. Plate and angle splices may be designed according to 7.5.3 and 7.5.5.

An improvement of the above solution is shown in Fig. 7.60e: the web angle splice is replaced by a seat connection (see 7.6.3) which forms a bearing joint (see 7.4.3) and obviates the need for the bottom flange cover plate. To make transportation easier, it is possible to provide splices bolted both to the beam and to the column (Fig. 7.60f,g).

Fig. 7.61 shows some typical solutions for simple beam-column joints for pin-ended structures. These joints can be compared to a

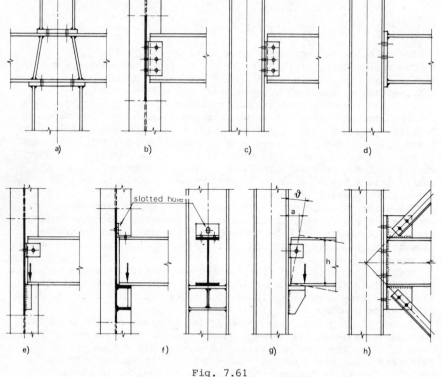

Fig. 7.61

hinge: they are able to transmit such limited bending moments that they may be disregarded in the design. On the other hand, tests and common practice have shown that their ductility is great enough to allow rotations of connected girders, complying with the usual deformation limitations (see 8.1.2.).

Pin-ended structures usually have continuous columns: at the floors, girders must therefore be considered simply supported.

Sometimes columns are interrupted at floors: it is then possible to use continuous girders with joints of the type illustrated in Fig. 7.61a, which is perfectly similar to that joining compression elements, discussed in 7.4 (Figs 7.41i,n).

The latter solution is often applied where the depth of main beams has to be limited: although it achieves economy in material, fabrication and erection are more complicated. This is why simply supported beams are often preferable: they may be connected to the columns by simple joints, that are conceptually analogous to beam end joints (see 7.5.2).

Figs 7.61b,c show angle splice joints; for their calculation, see 7.5.5. The joint shown in Fig. 7.61d is similar to those discussed in 7.5.4 with reference to plates having bolts designed for shear only: flange thickness limitations allowing a sufficient rotational capacity are identical.

Figs 7.61e,f,g illustrate three typical seat joints. An angle connecting beam web or upper flange prevents the girder from overturning. If the beam is connected to the column web, the seat must be as close as possible to the column axis, to avoid parasitic bending in the web.

If the bearing area given by the solution of Fig. 7.61e is too limited, the detail of Fig. 7.61f must be adopted. Eccentricity effects are transferred to the column flanges by means of the upper and lower plates.

If the seat is on the column flange (Fig. 7.61g), the parasitic effects of eccentricity do not usually affect the design of the column.

While seat design it dealt with in 7.6.4, attention is drawn here to the position of the hinge centre C. With a seat joint, the beam rotates about a point located near its lower flange. This (Fig. 7.61g) produces a horizontal displacement $a = \theta\, h$ of the upper flange, compared to $a = \theta\, h/2$ with the angle splice joint (Fig. 7.61c). Inconvenience might ensue in floor systems with concrete slabs: in fact localization of a crack is favoured.

Finally, Fig. 7.61h illustrates a typical solution for a beam-column joint with bracing diagonals. An analysis of stress distribution in the various connections forming the joint is set out in 7.6.4.

Connecting steel beams to concrete elements is rather a technological problem than a statical one: dimensional tolerances of concrete structures (of the order of some cm) must be complied with, together with those of steel structures (of the order of 1 to 2 mm). The entirely bolted solution (Fig. 7.62a) makes adjustment easy but it is complex: for this reason, it is sometimes advisable to adopt the site-welded solution, in which adjustments can also be made (Fig. 7.62b). The reinforcing bars may be connected directly to the encased plate (Figs 7.62c,d) rather than to a stiffener (Fig. 7.62b) by means of fillet welds.

279

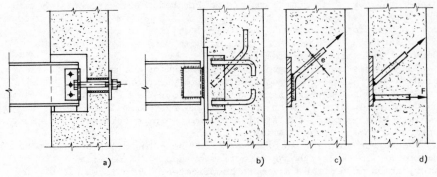

Fig. 7.62

For the horizontal bar having diameter ϕ, size b of fillet weld side must be such that:

$$\tau = \pi \, \phi \, b/\sqrt{2} \geq F = f_d \, \pi \, \phi^2/4$$

For $\tau = 0.70 \, f_d$, $b \geq \phi/2$

A complete penetration butt weld is advisable for connecting the bent bar to the plate (Fig. 7.62d). If fillet welds are used (Fig. 7.62c), the strength of the joint may be only 50 to 70% of that of the connected bar due to parasitic effects of eccentricity e.

The joints shown in Fig. 7.62 must be regarded as simple joints.

In order to transfer bending moments to the wall, rigid connections may be used but they must be considered cautiously, to avoid the danger of brittle fracture |27|.

7.6.1 *Column controls*

In the case of rigid and semi-rigid joints, one of the most delicate controls concerns the column which might fail in one of the following ways (see Fig. 7.63):

In compression (Fig. 7.63a): the web might fail due to yielding or to local buckling
In tension (Fig. 7.63b): the column flange might bend or the web might become detached from the flange
The panel of web between stiffeners and column flanges might fail due to shear (Fig. 7.63c)

In order to evaluate such collapse mechanisms the following equilibrium relations are given (Fig. 7.64a):

$$M_3 = M_2 - M_1 \quad ; \quad N_3 = V_1 + V_2 \quad ; \quad V_3 = N_2 - N_1$$

Usually axial forces N_1 and N_2 have only a slight influence on stresses, compared to that of bending moments. When analysing such a joint, the effects of bending can thus be considered concentrated in the beam flanges, while axial and shear forces can be considered acting in the webs. It ensues that the beam flanges are loaded by

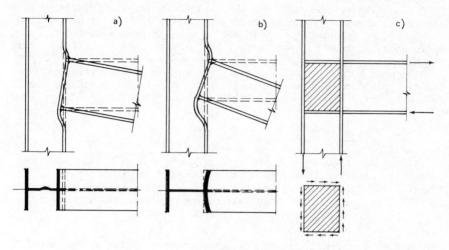

Fig. 7.63

tension and compression forces M/d (Fig. 7.64b).

Such forces must be transferred by the joint if the moments on the two beams are unbalanced, e.g. if $M_2 > M_1$; the joint is loaded by the superposition of:

Forces applied to the two sides of the column (Fig. 7.64c) produced by moment M_1 that are transferred from one beam to the other
Forces applied only on the side of beam '2', produced by moment $M_3 = M_2 - M_1$ that are transferred to the column (Fig. 7.64d)

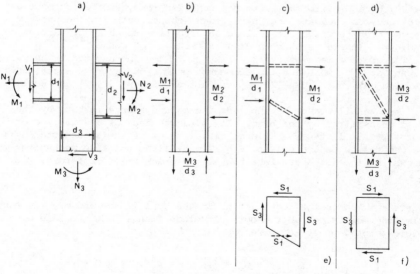

Fig. 7.64

One must therefore check whether it is necessary to provide
stiffeners ideally extending the beams (dotted in Figs 7.64c,d).

The column web can also be loaded by shear stresses due to
differences in beam heights $(d_1 \neq d_2)$ and differences in applied moments
$(M_1 \neq M_2)$.

Also when $M_1 = M_2$ (Fig. 7.64c), if beam heights are different, the
forces applied to the two sides of the column are different and the
following shear forces occur in the web (Fig. 7.64e):

$$S_1 = M_1 \left(\frac{1}{d_1} - \frac{1}{d_2}\right); \qquad S_3 = S_1 \frac{(d_1 + d_2)/2}{d_3}$$

If $M_2 \geq M_1$ (Fig. 7.64d), the difference $M_3 = M_2 - M_1$ produces
forces that must deviate their line of application in order to
transfer the bending effect to the column. This entails shear forces
S_1 and S_3 in column web, which must balance the effects of moment M_3:

$$S_1 = (M_2 - M_1)/d_2 ; \qquad S_3 = (M_2 - M_1)/d_3$$

Strength of the column may be evaluated conventionally as indicated
below.

7.6.1.1 Column strength

In the compressed area, an unstiffened web might buckle before
reaching its strength limit. Such a phenomenon is prevented by
checking that web thickness t_w complies with the following formula,
which interpolates, in favour of safety, the results of U.S.
experimental studies |28|:

$$t_w \geq \frac{h_w}{30} \sqrt{235/f_d}$$

If the above limitation is not complied with, stiffeners must be
provided. To evaluate the unstiffened web strength, the compressive
force F can be distributed along an effective width b_{eff}, varying as
indicated in Fig. 7.65 according to the type of connection.

Therefore a mean stress $\sigma = F/(b_{eff} \, t_w)$ may be assumed. The values
given for b_{eff} were derived from the results of numerous experiments
carried out in the USA |29,30|, in Great Britain |31,32| and Holland
|23|.

If a tensile force is applied, the column flange is bent by the
load transmitted by the beam. Its strength depends on the type of
connection. If the beam flange is welded or if it is connected by a
welded splice to the column (Fig. 7.66a)), the collapse mechanism
shown in Fig. 7.66b can take place. In order to prevent this:

$$F \leq 24 \, m_{lim}$$

where $F = M/d$ is the force acting on the beam flange and m_{lim} is the
limiting moment per unit length of column flange. In the case of
rigid joints, under collapse one has:

$$F_u = A_f \, f_{y,b} ; \quad m_{lim} = m_u = \frac{1}{4} f_{y,c} \cdot t_f^2 ; \quad t_f \geq 0.4 \sqrt{k \, A_f}$$

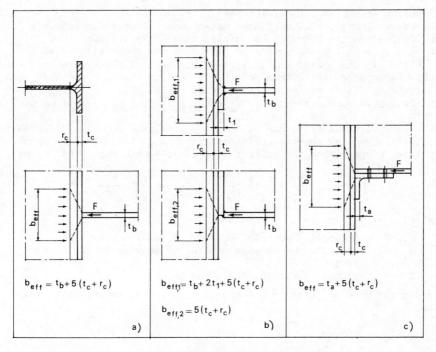

Fig. 7.65

$k = f_{y,b}/f_{u,c}$ being the ratio between beam and column yield stresses

Furthermore, experimental evidence has proved that weld length can be considered effective only for the following value:

$$b_{eff} = 2 t_w + 7 t_f$$

If the connection is end plated or has bolted splices, the column flange behaves as a symmetrical flange. Effective width may be evaluated on the basis of the collapse mechanism shown in Figs

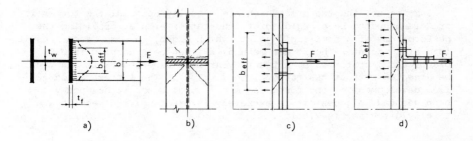

Fig. 7.66

283

7.38a,b. For flange controls, see 7.3.7.

Web yielding can cause flange detachment. If the connection is welded, a tensile force can be distributed along a length $b_{eff} = t_b + 5 (t_c + z_c)$ equal to that used for compression (Fig. 7.65a). In end plated and spliced joints (Figs 7.66c,d), behaviour is more favourable: it is thus possible to assume the same width as used for the column flange calculation (Figs 7.38a,b).

In any case, a mean stress $\sigma = F/(b_{eff}\ t_w)$ may be assumed in the column web.

Shear stresses in column web panel can be produced by a difference either in applied moments or in beam heights at the two sides of the column. Superposing the effects of Fig. 7.64a and Fig. 7.64f gives:

$$\tau_w = \left\{ \frac{M_3}{d_2} - M_1 \left(\frac{1}{d_1} - \frac{1}{d_2} \right) \right\} \frac{1}{t_w h_w} = \left\{ \frac{M_2}{d_2} - \frac{M_1}{d_1} \right\} \frac{1}{t_w h_w}$$

where

M_1, M_2 are applied moments ($M_2 \geq M_1$; $M_3 = M_2 - M_1$)
d_1, d_2 are distances between centres of gravity of the beam flanges
t_w, h_w are column web thickness and height

In the case of joints connecting a single beam to one side of a column:

$$\tau_w = \frac{M}{d} \frac{1}{t_w h_w}$$

Overall column performances expected must be decided to determine whether stiffeners are necessary, taking into account the above considerations. There are two possible approaches.

A rigid joint is, in any case, chosen. The need for stiffeners is thus independent of stresses in the connected beams, but depends only on the geometrical characteristics of the beams and columns and on possible differences in material

The unstiffened joint is analysed as a function of the stress resultants due to the external loads. The need for stiffeners thus depends on actual stress values and on possible insufficient rotational capacity of the joint compared to that required by beam deformability

Summing up, the two procedures are mutually opposed: the former is predetermined to be a rigid joint; the latter aims at providing the performances of a semi-rigid joint, which might, in some cases, result in a rigid joint. The former method |28| has been adopted in U.S. practice and confirmed by a twenty-year-long experience: it can thus be accepted without hesitation (see 7.6.1.2). The latter method |23| is a development of the foregoing one and, although it has not yet been adopted in codes, it is extremely interesting: its results are briefly discussed in 7.6.1.3.

7.6.1.2 Rigid joints

For a rigid unstiffened joint, the following conditions must be complied with in order to avoid collapse mechanisms of the types illustrated in Figs 7.63a,b. Let:

A_f = beam flange area
t_w = column web thickness
b_{eff} = equivalent width as defined above
$t_{f,b}$ = beam flange thickness
$t_{f,c}$ = column flange thickness
h_c = web height
k_1 = $f_{y,b}/f_{y,c}$ = ratio of yield strengths of beam to column
k_2 = $f_{y,c}/f_{y,s}$ = ratio of yield strengths of column to possible stiffener
$f_{y,c}$ = yield strength of column in N mm^{-2}

In a tension area, the following must be satisfied:

For column web strength: $\quad t_w \geq k_1 A_f/b_{eff}$ $\hspace{3cm}$ (7.5)

For column flange strength: $\quad t_{f,c} \geq 0.40 \sqrt{k_1 A_f}$ $\hspace{2cm}$ (7.6)

In a compression area, the following must be satisfied:

For column web strength: $\quad t_w \geq k_1 A_f/b_{eff}$, $\hspace{2.5cm}$ (7.5)

For column web stability: $\quad t_w \geq \dfrac{h_w}{30} \sqrt{235/f_{y,c}}$ $\hspace{2cm}$ (7.7)

If either one or both of the foregoing relations in compression tension area are not complied with, a stiffener must be provided, corresponding to the compression tension beam flange. It must absorb that part of the force that the web alone is incapable of bearing. Stiffener area A_s must therefore be:

$$A_s \geq (k_1 A_f - t_w b_{eff}) k_2 \hspace{3cm} (7.8)$$

Such a relation is inspired by considerations concerning web strength. If conditions (7.5) are complied with, then (7.8) is ineffective but stiffeners can be required by conditions (7.6),(7.8) to guarantee web stability or flange strength: it is then sufficient to provide stiffeners having a thickness related to the thickness of the element that must be stabilized (e.g. of the order of 50%).

It is further necessary to ensure that shear forces S due either to lack of symmetry of the beams or of the bending moments (Figs 7.64e,f) do not cause web collapse.

It there are beams on both sides of the column (+-joint), the column web must be designed for the effects of possible section differences between the two beams. If it is assumed that both will be loaded to their maximum capacity, it follows that:

$$A_{f,2} f_{y,b} - A_{f,1} f_{y,b} \leq t_w h_w f_{y,c}/\sqrt{3}$$

where:

$A_{f,2}$ $A_{f,1}$ = flange areas of the two beams '1', '2' respectively
$f_{y,b}$ = yield strength of the two beams
t_w = column web thickness
h^w = column web height
$f_{y,c}^w$ = yield strength of column

Furthermore, possible differences in stresses due to different beam loads or spans must be taken into account.

If M_1 and M_2 are calculated values of beam end bending moments

$$\frac{M_1}{d_1} - \frac{M_2}{d_2} \leq t_w \, h_w \, f_{y,c}/\sqrt{3}$$

If there is only one beam the column web must absorb the flow of forces transmitted by the beam (Fig. 7.67a).
Therefore:

$$h_w \frac{f_{y,c}}{\sqrt{3}} \, t_w \geq f_{y,b} \, A_f \quad \text{whence} \quad t_w \geq k_1 \, A_f \, \frac{\sqrt{3}}{h_w}$$

If this condition is not complied with the web panel must be reinforced by means of welded plates (Fig. 7.67b) or other stiffeners. As the web is capable of absorbing force:

$$F_w = f_{y,c} \, t_w \, h_w/\sqrt{3}$$

stiffeners must be designed for a force:

$$F = f_{y,b} \, A_f - F_w = (k_1 \, A_f - t_w \, h_w/\sqrt{3}) \, f_{y,c}$$

If a pair of diagonal stiffeners (one each side of the web) is used, the area of stiffening required must be (Fig. 7.67c):

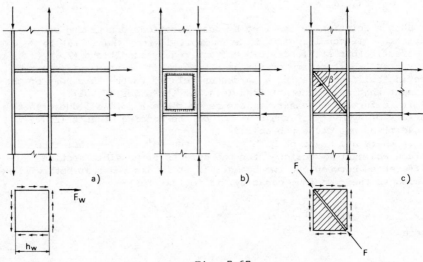

Fig. 7.67

$$A_s \geq \frac{k_1 A_f - t_w h_w/\sqrt{3}}{\cos \beta} \frac{f_{y,c}}{f_{y,s}}$$

$f_{y,s}$ being the yield strength of the stiffener.

7.6.1.3 Unstiffened rigid or semi-rigid joints

Instead of designing a column joint for maximum loads consistent with beam strength, it might be interesting to assess whether it is capable of absorbing stress induced by design values of applied loads and to check its rotational capacity.

The criteria discussed here |23| are the results of tests carried out in Holland as part of the activities of ECCS. They are relatively recent and must be considered rather cautiously, especially as far as deformation capacity is concerned. Symmetrical (Fig. 7.68a) asymmetrical (Fig. 7.68b) cross joints, ⊢(Fig. 7.68c) and knee (Fig. 7.68d) joints were all considered. Steel with a yield strength below 260 $N.mm^2$ was used and the effects of axial load in the column were disregarded.

Once the bending moments and forces applied to the column flanges have been evaluated (see Fig. 7.64b) the following checks must be performed on the basis of the considerations set out in 7.6.1:

(a) Web thickness must be:

$$t_w \geq \frac{h_w}{30} \sqrt{235/f_d}$$

to avoid sensible influences of axial forces on web strength

(b) Normal stress σ and shear stress τ in the web, calculated as in 7.6.1.1, at the connection of the tension and compression flanges of the beams must be such that:

$$\sqrt{\sigma^2 + 3\tau^2} \leq f_d$$

(c) If the beam flanges in tension are welded, welds must be designed for a bending moment equal to 1.40 times the design value, but not exceeding a value corresponding to beam ultimate moment.
This is to avoid brittle collapse of the welds. If butt welds are not

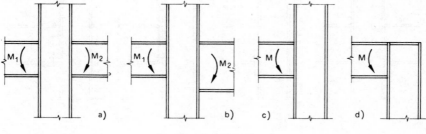

Fig. 7.68

used, fillet welds must be symmetrical and their throat section must be greater than half the beam flange thickness. For evaluating effective weld length b_{eff}, see Fig. 7.66a and comments thereon

(d) If end plated joints are used (Figs 7.69d,e), plate and bolts must be designed as in the case of connections in bending (see 7.5.4) and the column flange as a symmetrical flange in tension (see 7.3.4)

(e) If the beams have bolted angle splices (Fig. 7.69f) they must be designed as in 7.6.3; for column flanges, see 7.3.4.
      In addition to strength, rotational capacity must be checked and the effective moment of inertia of the joint must be evaluated to calculate the structure deformability under service loads (Fig. 7.69a).
      Both rotational capacity and joint stiffness depend on connection type.
      In the case of the welded joints of Fig. 7.69c, rotational capacity can be considered sufficient if:

$$M_{res}/M_u \geq 0.80 - 22.8 \,\frac{h_c}{h_b}\,\frac{h_b}{h_c}$$

where:

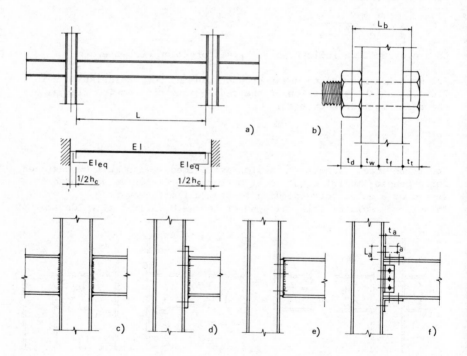

Fig. 7.69

$$M_{res} = f_d \, b_{eff} \, t_w \, d$$ = maximum moment for column web when $\tau = 0$

$$M_{res} = f_d \, A_f$$ = maximum moment for the beam

h_b, h_c = beam and column height respectively

d = distance between centroids of beam flanges

L = beam span

The equivalent moment of inertia of the joint to be used in evaluating structure deformability with reference to Fig. 7.69a, can be assumed to be:

$$I_{eq} = 0.20 \, t_w \, h_c \, h_b^2 \qquad \text{in the case of cross joints (Figs 7.68a,b)}$$

$$I_{eq} = 0.075 \, t_w \, h_c \, h_b^2 \qquad \text{in the case of} \vdash \text{and knee joints} \\ \text{(Figs 7.68c,d).}$$

The flanged joints shown in Figs 7.69d,e can provide a sufficient rotational capacity only if collapse takes place in the column.

For this reason, bolts and flange thickness must be designed for an applied bending moment of:

$$M = 1.4 \, M_{res} \qquad \text{for the joint of Fig. 7.69d}$$

$$M = 1.1 \, M_{res} \qquad \text{for the joint of Fig. 7.69e.}$$

If $M_{res} > M_u$, one assumes $M = M_u$. The equivalent moment of inertia of the joint is:

$$I_{eq} = \frac{h_c \, d^2}{4} \left/ \left\{ \frac{1}{0.4 \, t_w} + \frac{L_b}{n \, A_{res}} + \frac{m^2}{t_w^3} \right\} \right. \qquad \text{for cross joints}$$

$$I_{eq} = \frac{h_c \, d^2}{4} \left/ \left\{ \frac{1}{0.15 \, t_w} + \frac{L_b}{n \, A_{res}} + \frac{m^2}{t_w^3} \right\} \right. \qquad \text{for } T \text{ and knee joints}$$

where

d = flange lever arm

n = number of bolts in tension

A_{res} = resistant area of each bolt

m = $y + r_c/5$

y = bolt distance from beginning of column flange/web fillet

r_c = column flange/web fillet radius

L_b = $t_w + t_f + (t_d - t_t)/2$ = bolt effective length. Equal to the sum of connected plate thicknesses $(t_w + t_f)$ and half the sum of head and nut thicknesses (Fig. 7.69b).

The joints with angle splices (Fig. 7.69f) can provide a sufficient rotational capacity only if the column collapses under compression or the tension angle splice fails due to bending. Consequently, the bolts and the column flange under tension must be designed for a moment equal to 1.10 times design value.

Joint deformability can be determined according to:

$$I_{eq} = \frac{h_c \, d^2}{4} \left/ \left\{ \frac{1}{0.4 \; t_w} + \frac{L_b}{n \, A_{res}} + \frac{8a^3}{L_a \, t_a^3} + \frac{m^2}{t_w^3} \right\} \right. \qquad \text{for cross joints}$$

$$I_{eq} = \frac{h_c \, d^2}{4} \left/ \left\{ \frac{1}{0.15 \; t_w} + \frac{L_b}{n_b \, A_{res}} + \frac{8a^3}{L_a \, t_a^3} + \frac{m^2}{t_w^3} \right\} \right. \qquad \text{for} \vdash \text{and knee joints}$$

Where, in addition to the notation above (see Fig. 7.69f),

a = bolt distance from angle edge
L_a = length of leg in contact with column
t_a = angle thickness.

7.6.2 *Splice connections to columns*

Splice connections to columns can be welded (Fig. 7.70a), bolted with stiffened (Fig. 7.70b) or unstiffened ⊥ sections (Fig. 7.70c) or using angles (Fig. 7.70d).

The above solutions, in the order they are mentioned, are adopted for progressively less resistant joints.

In welded joints the column flange is loaded by tension forces through its thickness and therefore there is a danger of lamellar

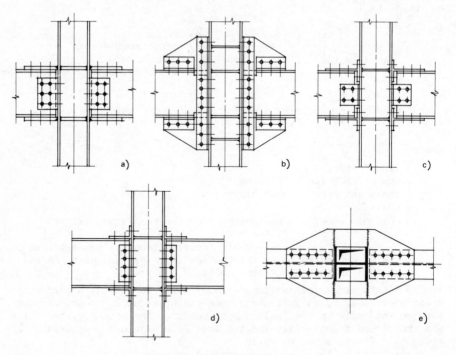

Fig. 7.70

tearing. If the flange is thick, it is advisable to test it
ultrasonically. Alternatively a different solution (i.e. as shown in
Fig. 7.70e) may be adopted, although more cumbersome and expensive.

T splices are symmetrical flanges in tension: they must be designed
as in 7.3.4. European shapes do not include ⊥ profiles with a
sufficient flange thickness and they must often be obtained by welding
two plates together or by cutting I beams.

Angle splices are unsymmetrically loaded flanges and they may offer
weak frame connections. In any case, their ductility is appreciable
because leg deformability is great.

Bolt and angle thickness can be designed according to the ultimate
mechanism shown in Fig. 7.71. The point with zero bending moment may
be assumed at half-way between the lowest bolt and the lower flange
extrados. Thus a moment $M = F$e acts on a section with a width equal
to angle length.

Bolts connecting the upper flange of the beam must be designed for
a shear force F and a bending moment $M = F$ $(e + t_a)$.

Upper bolts can be considered in tension, due to a force F in each
row, as a consequence of the prying force at the upper edge.

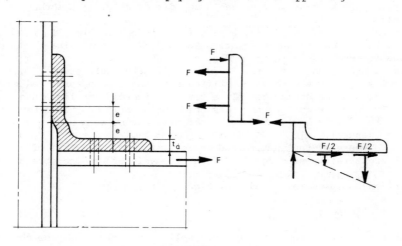

Fig. 7.71

7.6.3 *Seat connections*

Different seat connections are shown in Fig. 7.72. Specific pressure
on the beam web and seat strength must be evaluated.

For pressure on the beam web, usually a limit of 1.3 f_d is assumed.
Referring to Fig. 7.72d, the condition that

$$R \leq 1.30 \ f_d \ (b + c) \ t_w$$

makes it possible to check that reaction R does not exceed that allowed
for a web having thickness t_w, in the case of a stiffened seat. The
same condition, written with the sign of equality, can be used to
determine the position of the reaction applied to an unstiffened seat

291

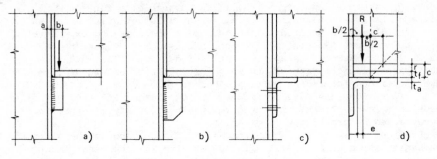

Fig. 7.72

thus:

$$b = \frac{R}{1.30 \, f_d \, t_w} - c$$

Design of the seat must be performed differently for the three types shown in Fig. 7.72.

In the case of Fig. 7.72a, eccentricity is negligible. It is thus sufficient to choose an adequate size for the welds joining the plate to the column, disregarding bending effects.

For a stiffened seat (Fig. 7.72b), it is often sufficient to design the welds joining the vertical stiffener to the column, provided the stiffener and the horizontal plate have a thickness not less than beam web thickness. Welds are thus designed for shear R and for a bending moment $M = R \, (a + b/2)$, where a is the distance between beam end and column (generally, $10 \leq a \leq 20$ mm).

In the case of the unstiffened angle seat, one must first of all check angle thickness (Fig. 7.72d) on the basis of bending moment $M = Re$.

For an angle width b_a and leg thickness t_a;

$$t_a^2 \geq 6 \, Re/b_a \, f_d$$

Then the connection between angle splice and column must be checked.

It must be either welded or bolted: in any case, joints must be designed on the basis of a shear force R and a bending moment $M = R \, (a + b/2)$.

7.6.4 *Bracing connections*

In pin-ended structures some beam/column joints are more complex due to the existence of bracing diagonals. Their shape and type varies according to the steel profiles used for columns and beams and there is the problem of distributing the forces between the various connections forming the joint, so as to comply with the assumed truss diagram.

Consider, for example the joint shown in Fig. 7.73a, which is typical of a bracing designed using members in tension only.

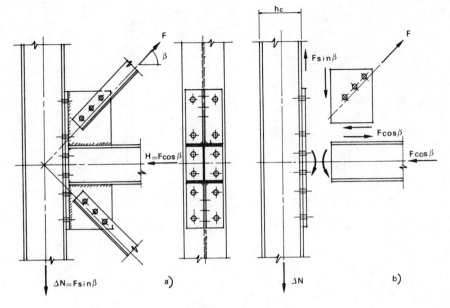

Fig. 7.73

Diagonal force F is balanced by a compression $H = F \cos \beta$ in the beam and by a decrease (or an increase for the lower diagonal) of compression $\Delta N = F \sin \beta$ in the column.

From the above consideration it is immediately possible to deduce the forces acting on the separate connections (Fig. 7.73b).

Welds joining the gusset plate to the beam and to the flange are loaded by parallel shear stresses. Flange bolts must be designed for:

A shear force V equal to the sum of the vertical component $F \sin \beta$ and of the effects of possible vertical loads acting on the beam

A bending moment $M = V\, h_c/2$, wherein h_c is column section height

In order to eliminate moment M in the bolt, the axes of the bracings can be arranged to converge at the point of intersection of the beam and bolted flange axes. The bolts thus are only in shear, while eccentricity effects bend the column.

7.6.5 *Strength of bracket loaded webs*

Another typical problem concerns the strength of a bracket at a column web (Fig. 7.74).

Collapse load calculation by the kinematic approach leads to assessing the resistant moment by means of the formula |33|:

$$M_{lim} = k\, m_{lim}\, L \qquad \text{where:} \qquad k = 2\,\frac{a}{L} + 2\,\frac{L}{a} + 2\,\sqrt{7}$$

A recent experimental survey |34| has confirmed this result, provided a limitation to joint slenderness is considered:

293

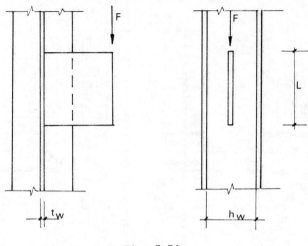

Fig. 7.74

$$a/t_w < 90 \qquad ; \qquad L/t_w < 90$$

In case of greater slendernesses, in fact, the theory based on collapse lines becomes meaningless, as membrane behaviour becomes important.

7.7 BUILT UP SECTIONS

It is often useful (and sometimes absolutely essential) to provide girders having a built up section. Symmetrical I-beams (Fig. 7.75a) are often used for important crane girders and unsymmetrical ones (Fig. 7.75b) for steel/concrete composite beams.

Box sections (Fig. 7.75c) are useful for long span bridges, for important crane girders or heavy section columns. Finally, sections as illustrated in Fig. 7.75d can provide a useful solution for tranverse loads.

In welded connections fillet welds (Fig. 7.75e) and complete (Fig. 7.75f) or partial penetration (Fig. 7.75g) butt welds may be used.

Riveted connections (Fig. 7.75h) were used in the past and bolting is now adopted mainly if there are special requirements concerning shipping, assembling or recovery of the individual parts for re-use.

Apart from technological requirements which may often determine connection sizes, load effects may be classified as:

Local effects, most often due to concentrated loads, perpendicular to the connection axis
Overall effects, due to shear forces between the connected parts

Local effects (Fig. 7.76a) may be very important for fillet welded or bolted connections. A crane wheel, for example, induces stresses $\tau_\perp = F/2a\ b$, perpedicular to the weld and it is not possible in such
294

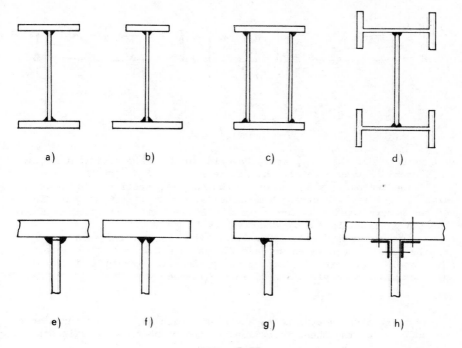

Fig. 7.75

a case to rely on the contact between web and flange being maintained: it can be guaranteed only by a complete penetration butt weld (Fig. 7.76b).

To understand the reason for the presence of shear forces between the various parts of the section, consider a bent beam built up of two superposed rectangular sections. If the two parts are not connected (Fig. 7.77a), the resulting overall section does not remain plane as

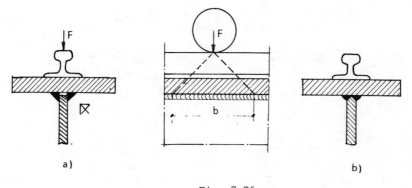

Fig. 7.76

295

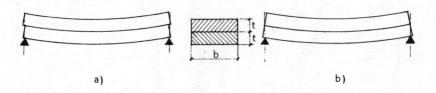

Fig. 7.77

the beam bends: the two parts behave in parallel and resistant modulus W is equal to twice that of the single section: $W = 2\ bt^2/6$.

If the two parts are connected, the built-up section behaves as one (Fig. 7.77b) and its strength modulus is: $W = b\ (2t^2)/6$ or twice that of the superposed but separate parts. However the connection between the two parts is loaded by shear forces.

Now consider an I-beam (Fig. 7.78), the longitudinal shear stress at the connection between web and flange is equal to the web shear τ_w. The total shear force dV_L acting on an elemental length dx is equal to the axial force variation dN in the flange

$$dV_L = dN\ \tau\ t_w\ dx.$$

If this value is applied to a finite length ΔL, one obtains a mean value $V_{L,m}$ for the shear force loading the connection for a length ΔL:

$$V_{L,m} = \tau_w\ t_w\ \Delta L$$

For a symmetrical fillet weld connection having a throat size a, the shear stress along the weld axis is:

$$\tau = V_{L,m}/2a$$

It must be combined with shear stress $\tau_\perp$, if induced by the above mentioned local effects.

On the other hand, for one bolt, shear stress in the shank is:

$$\tau_b = \frac{V_{L,m}}{n\ A_{res}} = \frac{V_{L,m}\ S_x}{I_x}\ p\ \frac{1}{n\ A_{res}} \simeq t_w\ p/n\ A_{res}$$

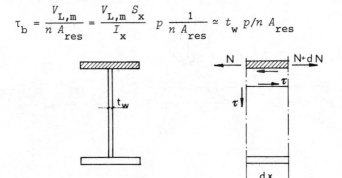

Fig. 7.78

where:

p = bolt spacing

S_x = moment of the total flange area about neutral axis of the whole girder for the flange angle splice connection; it is the 1st moment of the flange plate plus angle splices for the angle splice web connection

I_x = inertia moment about the neutral axis of the girder

V_m = mean shear force in the length p

n = number of resistant sections (=2 in both cases)

A_{res} = shear resistant area of the bolts

7.8 JOINTS IN STRUCTURAL HOLLOW SECTION STRUCTURES

Joints in structural hollow section (SHS) structures can, as those in open profile structures, be either rigid or semi-rigid; the latter are more usual, one reason being that, by using tubes in welded structures, it is possible to form the joints by direct interference of the tubes without introducing stiffeners or gusset plates. Considerable economy in fabrication often associated with a measure of aesthetic elegance, is thus obtained.

On the other hand, in the biggest tubes, local effects and consequential appreciable deformations occur. For this reason, also, semi-rigid joints often have considerable ductility.

Connections between elements in tension, compression or bending can be either welded (Figs. 7.79a,b) or bolted with flanges (Fig. 7.79c). Bolted splice connections (Fig. 7.79d) can be obtained if special bolts that may be tightened only from the outside are used. For this purpose, it is possible to apply expansion type bolts, such as that illustrated in Fig. 7.80 (patented by Dalmine SpA, Italy). The bolt essentially consists of a threaded shank, a head having the shape of

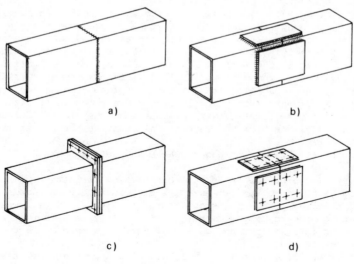

a)   b)

c)   d)

Fig. 7.79

297

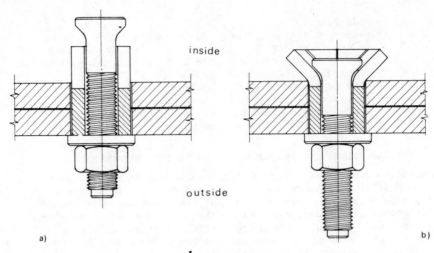

inside

outside

a)

b)

Fig. 7.80

a truncated cone, a hollow cylinder of external diameter d with
longitudinal partial cuts, a washer and a nut. The bolt head and
shank with its surrounding cylinder are inserted into the hole from
the accessible side (Fig. 7.80a); then the nut is tightened, keeping
the shank fixed; the truncated cone head is drawn into the cylinder
which is forced open (Fig. 7.80b), creating a good bearing surface.
If shank and cylinder thicknesses are properly proportioned, the bolt
has a shear strength that can be assessed on the basis of the nominal
external diameter d and a tensile strength that can be deduced on the
basis of shank area.

In buildings, although square or rectangular hollow sections may
be used as columns, the beams are normally of open profile. A
rectangular hollow section column has the advantage over an open
section of optimising the performance of the steel in compression
because the radii of gyration, and hence its slenderness ratios, are
of the same order of magnitude in all directions. However it is
heavier than an I-profile with the same in-plane bending performance.

The beam column joints shown in Figs 7.81a,b may be considered as
hinges with a sufficient rotational capacity; the elastic deformability
of RHS walls allows beam end rotation at least in the normal range
without danger.

In order to obtain continuous beams, the detail shown in Figs
7.81c,d,e may be used.

A typical use of SHS is in the construction of lattice members:
Fig. 7.82 shows some typical examples of joints between chords and
diagonals. Figs 7.82a,b,c show a joint obtained by direct welding.
Depending on the relative dimensions of the lattice member, the
tubular sections may be spaced (Fig. 7.82a), in contact (Fig. 7.82b)
or superposed (Fig. 7.82c).

Another building technique, which is more common in the case of
tubes, is to provide a gusset plate welded centrally along over which
298

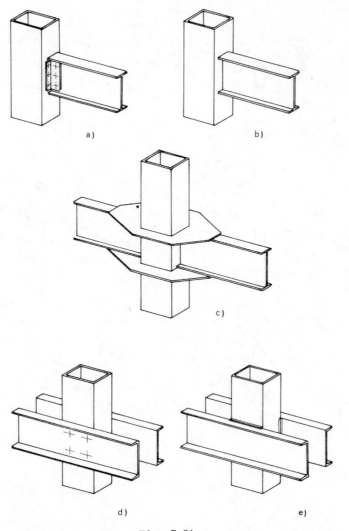

Fig. 7.81

the slotted diagonals are fixed (Fig. 7.82d).

This is a particularly efficient detail as the axial component which is perpendicular to the chord is transmitted from one diagonal to the other through the gusset plate.

The chord is thus loaded only by the tangential component that is parallel to its axis and therefore its wall is not appreciably bent. As an alternative, the same joint configuration can be obtained without cutting the tubes forming the diagonals, by dividing the gusset plate into three parts.

299

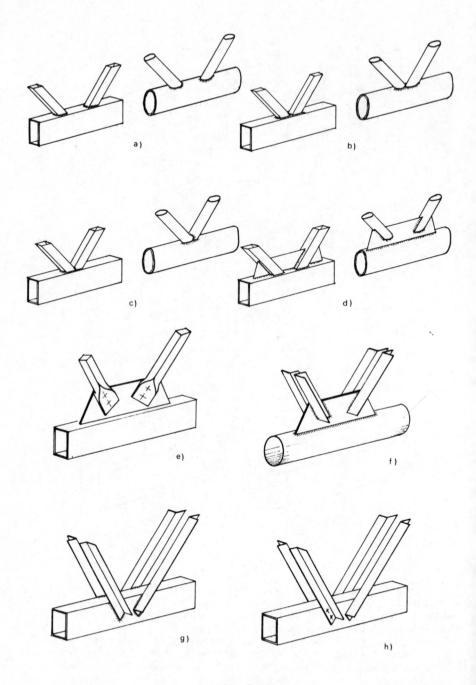

Fig. 7.82

By using gusset plates welded to the chords, it is also possible to bolt or weld diagonal open profiles (Fig. 7.82f) or flattened tubes (Fig. 7.82e).

Rectangular hollow sections can also be used as chords for lattice structures with open sections having vertical and diagonal members set on the two faces: the details are as shown in Figs. 7.82g,h.

It is useless to discuss rigid joint design: one must either use a local situation free from bending or use proper stiffeners to eliminate local effects.

It is however essential to provide an acceptable method for designing semi-rigid joints. In the past there was no great incentive provided by practical problems: the welded joint obtained by superposing tubes had always been accepted without calculation in common building practice and its validity was verified by practical experience. But since the use of hollow sections in common structures has grown considerably and great off-shore structures have been developed, it has become necessary to evolve operative calculation methods.

The problem, from a general point of view, is illustrated by Fig. 7.83 showing the forces acting on a joint of a lattice structure. If

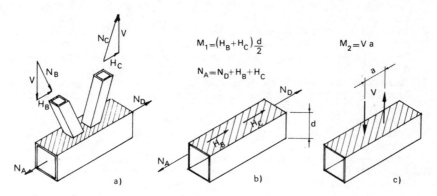

$$M_1 = (H_B + H_C) \frac{d}{2} \qquad\qquad M_2 = V a$$

$$N_A = N_D + H_B + H_C$$

Fig. 7.83

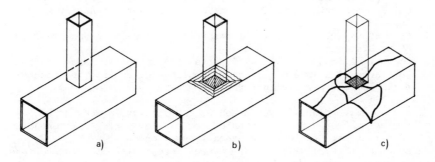

Fig. 7.84

301

it is assumed that eccentricities are negligible, $M_1 \simeq M_2$ and joint behaviour can be studied by superposing the effects of Figs 7.83b,c.

Forces H_B and H_C, which are parallel to the centre lines of the faces, create a stress resultant acting in the plane of the shaded face: it may be neglected since it does not bend the walls of the tube.

The effect of forces V, instead, is what determines joint strength: this has naturally led to a study of the behaviour of a connection between two *RHS* set at right angles (Fig. 7.84a).

Many experimental and theoretical studies |35-40| have been done on this subject and some practical design formulae suggested. |38-41| Most recently, CIDECT has performed a survey of many tests carried out on tubular joints and has proposed some design formulae correlating experimental results |42|. On the other hand, further numerical and experimental studies are probably needed in order to state final design criteria, especially with respect to the interaction between local effects and overall axial stress.

REFERENCES

1. Van Douwen, A.A. (1981), *Design for Economy in Bolted and Welded Constructions*, Proc. Int. Conf. on Joints in Structural Steelwork, Pentech Press, London
2. Chapeau, W. (1978) Lamellar tearing, *IABSE Periodica*, 3, 56, 1-16
3. Cargill, J. (1968) *Measuring Strains under Welds using the Moire Fringe Technique*, Naval Construction Research Establishment Report R 551
4. Granström, A. (1978) *The Relevance of Test Methods for Lamellar Tearing*, IIW Doc. IX-1086-78, Report of Swedish Institute of Steel Construction, Stockholm
5. Grandström, A. (1977) *Methods of Assessing the Risk for Lamellar Tearing*, Swedish Institute of Steel Construction, Publication 58, Stockholm
6. McGuire, W. (1968) *Steel Structures*, Prentice Hall, New Jersey
7. AISC (1980) *Specification for the Design, Fabrication and Erection of Structural Steel for Buildings*, AISC, Chicago
8. AISC (1981) *Manual of Steel Construction*, 8th edn, AISC, Chicago
9. Zanon, P. (1980) Résistance et ductilité des cornières boulonnées tendues, *Construction Métallique*, 3, 45-64
10. Kato, B. and McGuire, W. (1973) Analysis of T stub flange to column connections, *ASCE Journal of the Structural Division*, 99, 865-888
11. CECM ECCS (1978) *European Recommendations for Steel Construction*, ECCS, Brussels
12. Gobetti, A. and Zanon, P. (1978) Valutazioni della larghezza collaborante di flange per giunzioni bullonate: analisi sperimentale e simulazione per elementi finiti, *Costruzioni Metalliche*, 4, 162-171
13. Ballio, G., Poggi, C. and Zanon P. (1981) *Elastic Bending of Plates subjected to Concentrated Loads*, Proc. Int. Conf. on Joints in Structural Steelwork, Pentech Press, London
14. Bouwman, L. P. (1981) *The Structural Design of Bolted Connections, Dynamically Loaded in Tension*, Proc. Int. Conf. on Joints in Structural Steelwork, Pentech Press, London

15. Timoshenko, S. (1959) *Theory of Plates and Shells*, McGraw Hill, New York
16. Ballio, G., Poggi, C. and Zanon P. (1981) Notes on the behaviour of crane girders when loads are concentrated on the lower flange, *Costruzioni Metalliche*, $\underline{1}$, 3-12
17. Massonnet, C. and Save, M. (1972) *Plastic Analysis and Design of Plates, Shells and Disks*, North Holland Publ., Amsterdam
18. Mansfield, E. H. (1957) Studies in collapse analysis of a rigid plastic plate with a square yield diagram, *Proceedings of the Royal Society, London*, $\underline{241}$, A, August, 311-338
19. Zoetemeijer, P. (1974) A design method for the tension side of statically loaded bolted beam to column connections, *Heron*, $\underline{20}$, 1
20. Delesques, R. (1976) Résistance à une charge concentrée d'une plaque rectangulaire ayant un bord libre, *Construction Métallique*, 2, 59-72
21. Zoetermeijer, P. (1981) *Semi Rigid Bolted Beam to Beam-Column Connections with Stiffened Column Flanges and Flush end Plates*, Proc. Int. Conf. on Joints in Structural Steelwork, Pentech Press, London
22. Zaid, M. (1958) On the carrying capacity of plates of arbitrary shape and variable fixity under a concentrated load, *Journal of Applied Mechanics*, $\underline{12}$, 598-602
23. CECM ECCS EKS (1977) *Steifenlose Stahlskelettragwerke und Dünnwandinge Vollwandträger*, Ernst & Sons, Berlin/Munchen/Dusseldorf
24. CM (1966) *Regles de Calcul des Constructions en Acier*, Eyrolles, Paris
25. CEB (1978) *CEB FIP Model Code for Concrete Structures*, CEB Bulletin d'Information 124/125E, Paris
26. Pask, J. W. (1981) *Rationalisation of Connection Design for the UK Fabrication Industry*, Proc. Int. Conf. on Joints in Structural Steelwork, Pentech Press, London
27. Roeder, C. W. and Hawkins, A. N. M. (1981) Connections between steel frames and concrete walls, *Engineering Journal*, AISC, $\underline{1}$, 22-29
28. ASCE (1971) *Plastic Design in Steel*, A guide and commentary, Section 6.2, ASCE, New York
29. Huang, J. S., Chen, W. F. and Beedle, L. A. (1973) Behaviour and design of steel beam to column moment connections, *WRC Bulletin*, $\underline{188}$, 1-23
30. Regec, J. E., Huang, J. S. and Chen, W. F. (1973) Test of a fully welded beam to column connection, *WRC Bulletin*, $\underline{188}$, 24-35
31. Sherbourne, A. N. (1981) Bolted beam to column connections, *The Structural Engineer*, $\underline{6}$, 203-210
32. Packer, J. A. and Morris, L. J. (1977) A limit state design method for the tension region of bolted column connections, *The Structural Engineer*, $\underline{10}$, 446-458
33. Abolitz, A. L. and Warner, M. E. (1965) Bending under seated connections, *Engineering Journal AISC*, $\underline{1}$, 1-5
34. Hoptay, J. M. and Ainso, M. (1981) An experimental look at bracket loaded webs, *Engineering Journal*, AISC, $\underline{1}$, 1-7

35. Mouty, J. (1976) Calcul des charges ultimes des assemblages soudés des profilés creux, carrés et rectangulaires, *Construction Métallique*, <u>2</u>, 37-50
36. Czechowski, A. and Brodka, J. (1977) Etude de la résistance statique des assemblages soudés en croix des profilés creux rectangulaires, *Construction Métallique*, <u>3</u>, 17-25
37. Venanzi, U. (1979) Calcolo dei nodi delle strutture tubolari, *Costruzioni Metalliche*, <u>1</u>, 1-12
38. Wardemer, J. and Davies, G. (1981) *The Strength of Predominantly Statically Loaded Joints with a Square or Rectangular Hollow Section*, Report IIW Doc XV-492-81, TNO, Delft
39. Cran, J. A., Gibson, E. B. and Stadenycky, S. (1971) *Hollow Structural Sections Design Manual for Connections*, The Steel Company of Canada, Toronto
40. Mouty, J. (1980) *Manuel Assemblages: Vérification des Assemblages Soudés. Méthode Simplifiée*, Chambre Syndicale des Fabricants de Tubes d'Acier, Paris
41. Tournary, M. (1980) *Manuel Assemblages: Disposition Constructives*, Chambres Syndicale des Fabricants des Tubes d'Acier, Paris
42. Wardenier, J. (1982) *Hollow Section Joints*, Delft University Press, Delft

CHAPTER EIGHT

Strength of Structural Elements

8.1 SERVICEABILITY LIMIT STATE

8.1.1 *Influence of deformations*

Deformations in a steel structure must be contained within sufficiently low limits to avoid:
(a) Building use being in general prevented or reduced (e.g., functionality of complementary works or of plants, comfort of inhabitants)
(b) Non-bearing elements (floors, cladding, partitions and external walls) being damaged
(c) Distribution of internal forces between the various elements of the structure being altered from that anticipated when the design has been carried out assuming an undeformed scheme
(d) So called secondary stresses becoming of the same order of magnitude as principal ones, and thus no longer being negligible

While the events listed in (c) and (d) can be allowed, provided effective deformations are duly taken into account in check calculations, the dangers mentioned in (a) and (b) must be prevented by taking appropriate measures to control deformation.

Such controls, which are related to a serviceability limit state, are often determinant in the design of steel structures, particularly where bending is concerned. In general, three classes of deformation have to be considered:

Deformations due to axial forces
These must be taken into account mainly in structures formed of members subjected to axial forces only (trusses and space structures); they are usually negligible in comparison with deformations due to bending in the case of bent beams

Deformations due to bending
Such deformations usually prevail over any others and must always be taken into account. With the exception of some cases of continuous beams, the maximum deflection in a beam is located near its mid-span, independent of load distribution. It is therefore sufficient to evaluate the mid-span displacement and to compare it with the limiting values given in recommendations.

305

Deformations due to shearing effects

These are generally negligible in solid web beams, but they must be taken into consideration in beams having a latticed web or a web with holes (e.g. castellated beams)

8.1.2 *Imposed limits*

With reference to the function of a structural element in a steel framework, the various codes generally agree on permissible deformation limits, which are given in terms of the span L of the beam. In the case of cantilever beams, the same limits are assumed, but they are referred to a fictitious span L equal to double the length of the cantilever. The following is a list of the most common cases.

For beams directly loaded by walls, partitions or columns, total deflections, due to permanent loads and to accidental loads, must not exceed 1/500 of span. In the case of walls carried by steel beams belonging to a framework, this limit does not prevent the formation in the wall itself of a parabolic discharge arc (Fig. 8.1 |1|). It is assumed that height v of the arc is equal to $L/2$ and that the weight of the upper part of the wall is discharged directly to the end supports. Such beams therefore undergo bending due to the parabolic load corresponding to the part of wall under the arc and they are also in tension as a consequence of arc drift N.

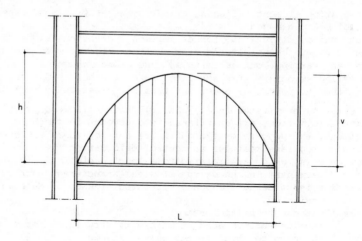

Fig. 8.1

If:

$G = \gamma \, btL$ overall weight of masonry having thickness t

$G_1 = 2vtL/3$ weight of the parabolic part directly bearing on the beam, it follows that:

$$N = (G - G_1)L/8v \quad ; \quad V_{max} = G_1/2 \quad ; \quad M_{max} = 5\,G_1L/32 \; .$$

For beams loaded by slabs some codes place a limit on deflection due to loads only of the order of 1/400 of span. Other codes, e.g. the French one |1|, impose two limits:

1/300 of span for total load (permanent + live loads)
1/500 of span for loads rapidly varying (live loads)

These limits on live loads, as well as controlling static deformation, are aimed at preventing excessive oscillations and vibrations.

For rooms used for residential purposes or as offices, it is possible to assume that rapidly varying live loads are less than half the total loads and this justifies the adoption of the French recommendation.

For secondary beams (purlins, lintels and elements of roofing and facade frames) overall deflection due to permanent load and live load together should be limited to $L/200$ to $L/150$ depending on national codes.

For beams supporting crane rails a more rigid limit is required: maximum deflection due to static loads should be from 1/700 to 1/1000 of the span.

For the overall structure of a multis-storey building, being interpreted as a solid web beam, the theoretical horizontal deflection at the top due to wind forces is often limited to 1/500 of the total height. This limit may become more strict (up to 1/1000) in the case of very high buildings due to the needs related to the comfort of the occupants. For a full discussion of this problem see |2|.

8.1.3 Deformation of beams

Deflection calculations are done according to the usual formulae for beams in bending. For example, the mid-span deflection of a simply supported beam of span L carrying a uniformly distributed load q, is expressed by:

$$v_{max} = \frac{5}{384} \cdot \frac{qL^4}{EI} \qquad (8.1)$$

Deflection control requires:

$$v_{max} < \alpha L \qquad (8.2)$$

where α = deflection limit, expressed as the ratio v/L (1/500, 1/400, etc.)

Using the allowable stress method, it is immediately possible to deduce some useful considerations for a rough design of a beam in bending. According to that method the same design loads are adopted for both strength and deflection controls. Maximum strength is given by:

$$\sigma_{max} = \frac{M}{I}\frac{h}{2} = \frac{qL^2}{8}\frac{h}{2I} \qquad (8.3)$$

This must be less than the permissible stress, i.e.:

$$\sigma_{max} < \sigma_{adm} \qquad (8.4)$$

If q/I is eliminated from (8.1) and (8.3), assuming E = 200 000 N mm^{-2} and considering $40/384 \simeq 1/10$:

307

$$\frac{v_{max}}{L} = 2.35\text{x}10^{-4} \; \frac{L}{h} \; \left\{\frac{\sigma_{max}}{235}\right\} \tag{8.5}$$

σ_{max} being expressed in N mm^{-2}.

If both deflection control (8.2) and strength control (8.4) are applied together, from (8.5) one obtains:

$$\alpha = \frac{L}{h} \; \left\{\frac{\sigma_{adm}}{235}\right\} \; 2.35\text{x}10^{-4} \tag{8.6}$$

Formula (8.6) shows that simultaneous compliance with strength and deflection does not require any particular moment of inertia value, but simply a given beam height h in relation to span. If any conditioning due to lateral buckling phenomena (see 9.3) is absent, it is therefore preferable to adopt a beam profile from the IPE series rather than from the HE series, because for a given height its weight is considerably lower (see Fig. 4.5 and 4.6).

Maximum values of L/h guaranteeing compliance with deflection limits at the permissible stress and due to live load only are given in the table of Fig. 8.2 in which g is the permanent load and q live load.

α	Max (L/h)		
	Steel Fe 360 $\sigma_{adm}=160$ N mm^{-2}	Steel Fe 430 $\sigma_{adm}=190$ N mm^{-2}	Steel Fe 510 $\sigma_{adm}=240$ N mm^{-2}
1/200	$31.4(g + q)/q$	$26.3(g + q)/q$	$20.9(g + q)/q$
1/400	$15.7(g + q)/q$	$13.2(g + q)/q$	$10.5(g + q)/q$
1/500	$12.5(g + q)/q$	$10.5(g + q)/q$	$8.4(g + q)/q$
1/700	$9.0(g + q)/q$	$7.5(g \;) q)/q$	$6.0(g + q)/q$

Fig. 8.2

In the case of large span beams, it can be useful to impose on the beam an initial deformation (or camber) equal and opposite to that due to the permanent loads.

For the simply supported scheme, as with the case of the uniformly distributed load, it is possible to obtain expressions for deflection v_o and stress σ_o at mid-span for beams with different loading conditions and having a symmetrical cross section. Thus:

For constant moment:

$$v_o = \frac{ML^2}{8EI} \qquad ; \qquad \sigma_o = \frac{Mh}{2I} \tag{8.7}$$

Hence:

$$\frac{v_o}{L} = \frac{2}{8}\frac{L}{h}(\sigma_o/E) \approx 2.9\text{x}10^{-4}\frac{L}{h}(\frac{\sigma_o}{235}) \tag{8.8}$$

For a load concentrated at mid-span

$$v_o = \frac{FL^3}{48EI} \qquad ; \qquad \sigma_o = \frac{FL}{4}\frac{h}{2I} \tag{8.9}$$

Hence:

$$\frac{v_o}{L} \approx 1.9\text{x}10^{-4}\frac{L}{h}(\frac{\sigma_o}{235}) \tag{8.10}$$

For a concentrated load at any point in the span, distant c from one, support:

$$v_o = \frac{FL^2c}{48EI}\left\{3-4\frac{c^2}{L^2}\right\} \; ; \quad \sigma_o = M_o\frac{h}{2I} \text{ with: } M_o = \frac{Fc}{2} \tag{8.11}$$

Hence:

$$\frac{v_o}{L} \approx \left\{1.2 - 1.6\frac{c^2}{L^2}\right\}(\frac{\sigma_o}{235})\frac{L}{h}2.4\text{x}10^{-4} \tag{8.12}$$

The above relation is replaced in |1| by the following approximate formula:

$$\frac{v_o}{L} \approx \left\{1.6 - 0.4\frac{M_o L}{A_m}\right\}(\frac{\sigma_o}{235})\frac{L}{h}2.4\text{x}10^{-4} \tag{8.13}$$

with an error below 3.5%, where A_m is the area of the bending, moment diagram, i.e.:

$$A_m = M_o(L - c) \tag{8.14}$$

In the case of simply supported beams, deflections always lie between those corresponding, respectively, to constant moment and to a concentrated load at mid-span. The following relation is thus conservative:

$$\frac{v_o}{L} < 3\text{x}10^{-4}(\frac{\sigma_o}{235})\frac{L}{h} \tag{8.15}$$

In the case of partially or totally fixed beams, the presence of end moments M_A and M_B, which keep the upper fibres in tension, reduce the deflection compared with that of a simply supported beam under the same loading condition so that:

$$v - v_o = \frac{(M_A + M_B)L^2}{16EI} = -\frac{\sigma_A + \sigma_B}{8E}\frac{L^2}{h} \tag{8.16}$$

where:

v is the mid-span deflection for the clamped beam
v_o is the mid-span deflection for the simply supported beam.

309

From (8.8) it follows that:

$$\frac{v}{L} \simeq \frac{v_o}{L} - \frac{(\sigma_A + \sigma_B)/2}{235} \frac{L}{h} 2.9 \times 10^{-4} \qquad (8.17)$$

and, in the case of a uniformly distributed load, it leads to:

$$\frac{v}{L} \simeq \frac{\sigma_o - 1.20(\sigma_A + \sigma_B)/2}{235} - 2.4 \times 10^{-4} \qquad (8.18)$$

where:

σ_A, σ_B are stresses corresponding to the end moments
σ_o is stress at mid-span of the corresponding simply supported beam

As stress σ at mid-span of the clamped beam can be obtained by the following relation:

$$\sigma = \sigma_o - \frac{\sigma_A + \sigma_B}{2} \qquad (8.19)$$

formula (8.18) can be expressed as follows:

$$\frac{v}{L} \simeq \frac{\sigma - 0.1(\sigma_A + \sigma_B)}{235} \frac{L}{h} 2.4 \times 10^{-4} \qquad (8.20)$$

In the case of continuous beams submitted to a uniformly distributed load on all spans, the values of L/h complying with the deflection limit for simply supported beams under similar loading (Fig. 8.2) may be increased by 60% and 30%, respectively, for intermediate and end spans.

Deflection due to shearing forces may usually be considered negligible. For example, a uniformly loaded IPE profile of height h and span L, has a deflection v_V due to shear approximately related to deflection v_M due to bending moment, as follows:

$$v_V = 10 \frac{h^2}{L^2} v_M \qquad (8.21)$$

whence it is possible to deduce that for normal ratios h/L of between 1/10 and 1/30 shearing incidence is of the order of 3% or less.

Wenever deflection limits are extremely restrictive (e.g. in the case of rails), deformation due to shearing can be evaluated on the basis of the classical formula:

$$(\frac{dy}{dz})_V = \frac{V}{GA_w} = \frac{dM}{dz} \frac{1}{GA_w} = \frac{dM}{dz} \frac{\chi}{GA} \qquad (8.22)$$

relating the inclination of the deflection of the neutral fibre to the shear therein, through area A_w of the web, in the case of I shapes, or else, in the case of full sections, of an equivalent web related to the overall section area by means of a shear factor χ. For some simple sections, shear factor values are:

1.2 for rectangular sections
1.18 for full circular sections
2 for circular tubes with thin wall

From (8.22) one deduces that the deflection due to shear may be obtained from the bending moment diagram by dividing its ordinates by GA_w. In particular, the maximum increment of mid-span deflection is equal to: $v_V = M_o / (GA_w)$.

8.1.4 *Deformation of trusses*

Deformations due to shearing effects cannot be neglected in trussed beams, as the bars belonging to the web usually have an important influence on their bending behaviour, attaining the order of magnitude of 30%.

Elastic deflection of a trussed beam can be determined by a classical method taking into account the individual deformations of all the bars of the truss.

As an alternative, but only in the case of trusses with parallel chords, two approximate methods may be adopted.

The first, which is less accurate and applicable only to trussed beams simply supported on a single span, consists in adopting an average increase of about 1/3 of the displacement due to bending moment. This substantially corresponds to evaluating the displacements in a beam formed by two concentrated masses corresponding to the chord areas and having a fictitious modulus of elasticity of approx. 150 000 N mm^{-2}.

The second approximate method (called the equivalent web method) suggests using the beam formulae, provided an equivalent web is defined having an area A_w, such that under a shear force V, a panel of the truss having length L_o undergoes a transverse deflection:

$$v_y = \frac{VL_o}{GA_w}$$

(8.23)

equal to that of an equivalent solid web beam.

Mid span displacement $v_V = M_o/GA_w$, as defined by (8.23), must be added to displacement v_M, evaluated as for a solid web beam having a moment of inertia equal to that of the cross section made of two concentrated masses formed by the chords.

Consider the general case of a simple truss (Fig. 8.3) |1|, formed by two diagonals, one having a section $A_{d,1}$, length $L_{d,1}$ and at an

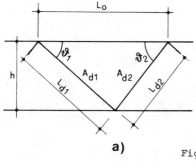

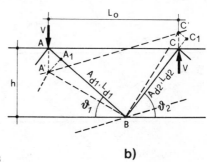

a)   Fig. 8.3   b)

angle $\theta_{d,1}$ to the horizontal, the other having analogous characteristics $A_{d,2}$, $L_{d,2}$, $\theta_{d,2}$. Further, let L_o be the length of elementary truss and h the distance between chord axes.

Now, suppose point B is fixed. Bar AB, undergoing a force $V \dfrac{L_o d_1}{h}$, shortens by:

$$AA_1 = \frac{VL_{d,1}^2}{hEA_{d,1}} \tag{8.24}$$

and moves to A'B, so that:

$$AA' = AA_1 \frac{L_{d,1}}{h} = \frac{V}{Eh^2} \frac{L_{d,1}^3}{A_{d,1}} \tag{8.25}$$

Similarly, bar BC lengthens and moves to BC', so that:

$$CC' = \frac{V}{Eh^2} \frac{L_{d,2}^3}{A_{d,2}} \tag{8.26}$$

The transverse deformation of the panel having length $AC = L_o$, is thus:

$$y_v = AA' + CC' = \frac{V}{Eh^2} \left\{ \frac{L_{d,1}^3}{A_{d,1}} \frac{L_{d,2}^3}{A_{d,2}} \right\} \tag{8.27}$$

which, assumed equal to the solid web beam deflection leads to:

$$\frac{V}{Eh^2} \left\{ \frac{L_{d,1}^3}{A_{d,1}} + \frac{L_{d,2}^3}{A_{d,2}} \right\} = \frac{VL_o}{GA_w} \tag{8.28}$$

whence the following expression is obtained for the equivalent area (for $G/E = 0.385$):

$$\frac{1}{A_w} = \frac{0.385}{L_o h^2} \left\{ \frac{L_{d,1}^3}{A_{d,1}} + \frac{L_{d,2}^3}{A_{d,2}} \right\} \tag{8.29}$$

Formula (8.29) can also be expressed as follows:

$$\frac{1}{A_w} = \frac{0.385}{\cot\theta_1 + \cot\theta_2} \left\{ \frac{1}{A_{d,1}\sin^3\theta_1} + \frac{1}{A_{d,2}\sin^3\theta_2} \right\} \tag{8.30}$$

For a symmetrical V shaped scheme (Fig. 8.4), assuming:

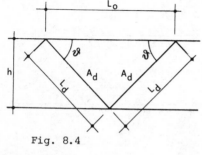

Fig. 8.4

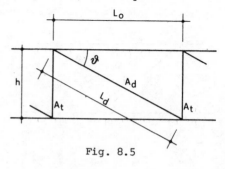

Fig. 8.5

$$L_{d,1} = L_{d,2} = L_d$$
$$A_{d,1} = A_{d,2} = A_d$$
$$\theta_1 = \theta_2 = \theta$$

formula (8.30) becomes:

$$A_w = 1.3\, A_d \frac{L_o\, h^2}{L_d^3} = 2.6\, A_d \sin^2\theta\cos\theta \qquad (8.31)$$

For an N-shaped scheme (Fig. 8.5), assuming:

$$L_{d,1} = h \quad , \quad A_{d,1} = A_t \quad , \quad \theta_1 = \frac{\pi}{2}$$
$$L_{d,2} = L_d \quad , \quad A_{d,2} = A_d \quad , \quad \theta_2 = \theta$$

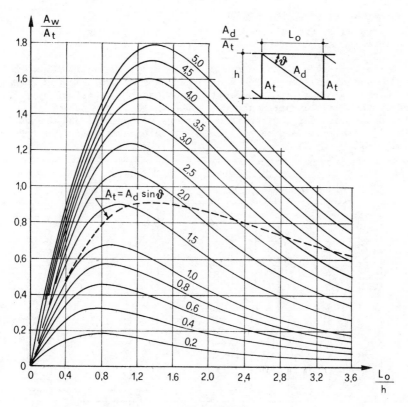

Fig. 8.6

313

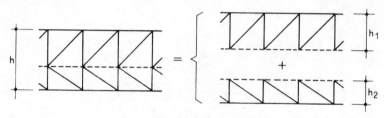

Fig. 8.7

formula (8.30) becomes:

$$A_w = \frac{2.6\, A_d \frac{L_o}{h}}{(\frac{A_d}{A_t} + \frac{L_d^3}{h^3})} = \frac{2.6\, A_d \cot\theta}{(\frac{A_d}{A_t} + \frac{1}{\sin^3\theta})} \qquad (8.32)$$

Values of A_w/A_t as a function of ratios L_o/h and A_d/A_t are provided by the curves of Fig. 8.6. Curve $A_t = A_d \sin\theta$ corresponds to the case in which axial forces are identical both in vertical and in diagonal bars.

In the case of a K-shaped scheme (Fig. 8.7), the equivalent web can be assessed as the sum of the equivalent webs of two N-shaped schemes, the heights of which are, respectively, h_1 and h_2, obtained by superposing the two parts into which the whole is divided by a fictitious chord joining the intersections of vertical and diagonal bars |1|. The same approach |1| can also be used for multiple trusses (Fig. 8.8).

MULTIPLE TRUSS	EQUIVALENT SIMPLE TRUSS	
	with compressed diagonals	diagonals only in tension
	+	o
	+ + +	+

Fig. 8.8

314

8.1.5 *Deformation of multi-cell girders*

The equivalent web method set out in 8.1.4 can also be applied to multi-cell girders. Consider, for example the Vierendel schema shown in Fig. 8.9. L_o is the vertical bar spacing, h_t is the distance between chord axes, I_t, I_1 and I_2 are the moments of inertia of the vertical bars and the upper and lower chords respectively.

When $I_1 = I_2 = I$, points of contraflexure A, B, C are at mid-span and equilibrium of the elementary system is expressed by:

$$F \frac{h}{2} = 2 \frac{V}{2} \frac{L_o}{2} \qquad\qquad F = VL_o/h \qquad\qquad (8.33)$$

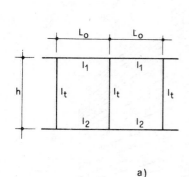

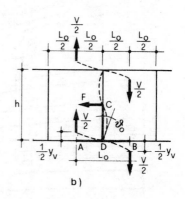

a)   b)

Fig. 8.9

Joint D rotates through an angle θ equal to that of a supported beam having span $CD = h/2$ and a bending moment distribution varying linearly from 0 at C to $Fh/2$ at D:

$$\theta = \frac{F(\frac{h}{2})^2}{3EI_t} = \frac{VL_o h}{12EI_t} \qquad\qquad (8.34)$$

The displacement of point A, expressed as half the transverse deflection of the panel is given by:

$$\frac{1}{2} y_v = \theta \frac{L_o}{2} + \frac{\frac{V}{2}(\frac{L_o}{2})^3}{3EI} = \frac{VL_o}{24E} \left\{ \frac{L_o h}{I_t} + \frac{L_o^2}{2I} \right\} \qquad\qquad (8.35)$$

Equating (8.35), multiplied by 2, to (8.23), one obtains:

$$\frac{1}{A_w} = \frac{G}{12E} \left\{ \frac{L_o^2}{2I} + \frac{L_o h}{I_t} \right\} \qquad\qquad (8.36)$$

The values of A_w can be obtained from the curves of Fig. 8.10.

When the chords have different inertias ($I_1 \neq I_2$), point C no longer falls in the middle of the vertical bar. It is then possible to determine the distribution of shear force V between the two chords and to repeat the foregoing calculation, leading to the following formula:

315

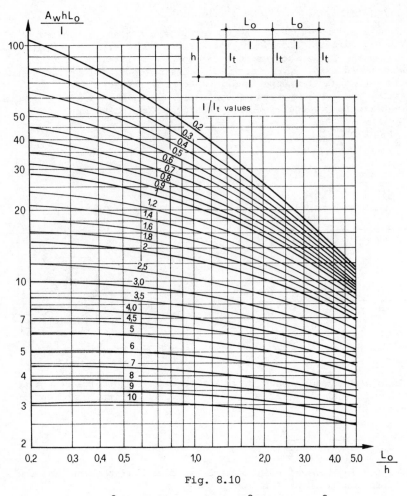

Fig. 8.10

$$\frac{1}{A_w} = \frac{G}{12E} \left\{ \frac{L_o^2}{I_1 + I_2} + \frac{L_o h}{I_t} + \frac{3L_o^2 h \left(I_1 - I_2\right)^2}{L_o I_t \left(I_1 + I_2\right)^2 + 12h I_1 I_2 \left(I_1 + I_2\right)} \right\} (8.37)$$

Usually, in practical solutions the third addendum is negligible compared to the first two and therefore the following assumption can be made, adopting $G/E = 0.385$:

$$\frac{1}{A_w} = 0.032 \left\{ \frac{L_o^2}{I_1 + I_2} + \frac{L_o h}{I_t} \right\} \qquad (8.38)$$

On this hypothesis, the curves of Fig. 8.10 can still be adopted, provided:

$$I = (I_1 + I_2)/2 .$$

8.1.6 *Deformation of bolted structures*

An important cause of deformations in beams with bolted connections and, in particular, in trussed beams formed by bars connected by normal bolts is due to the shank/hole clearance.

In general, inelastic displacements must be assessed according to the virtual work principle, by considering in the evaluation of the internal work the distortions due to the shank/hole settling in each joint.

In the case of a truss beam having diagonals bolted at their ends and jointed chords, inelastic displacement due to shank/hole effect can be determined as the sum of two factors:

$$v = v_c + v_d \qquad (8.39)$$

corresponding, respectively, to joint settling in the chords and at diagonal ends. They can be approximately evaluated by means of the following expressions:

$$v_c = \frac{n}{6}\frac{L}{h}(\phi - d) \quad ; \quad v_d = \frac{L}{p}\frac{L_d}{h}(\phi - d) \qquad (8.40)$$

where

$\phi - d$ = difference between hole diameter and bolt diameter (of the order of 1 mm)

n = $2m$, beeing m the total number of bolted joints of bearing type subjected to shear forces

L = beam span

h = beam height

p = the horizontal distance between intersections of the diagonals with the top and bottom chords

L_d = diagonal length

An evaluation of the order of magnitude of the corresponding displacements can be made by reference to the V-shaped scheme shown in Fig. 8.11, having a total of 4 joints in the chords ($n = 8$) and a joint distance of $p = L/18$.

Numerical results for various values of L/h and for $(\phi - d) = 1$ mm are tabulated in Fig. 8.12 |3|.

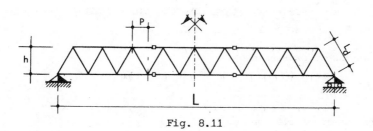

Fig. 8.11

$\dfrac{L}{h}$	v_c (cm)	v_d (cm)	v (cm)
10	1.34	2.05	3.39
15	2.00	2.33	4.33
20	2.67	2.68	5.36
25	3.34	3.08	6.42
30	4.00	3.50	7.50

Fig. 8.12

Inelastic displacements are independent of beam span L and therefore they become more important the smaller the span. Displacement/span ratios v/L as functions of L/h and L (Fig. 8.13) show that, irrespective of the span L, a height $h > 1$ m is always required in order that inelastic displacement is lower than the limit imposed for elastic displacement, i.e. not exceeding 1/400 of span.

$\dfrac{L}{h}$	v (cm)	v/L				
		L (m)				
		10	15	20	25	30
10	3.39	1/295	1/442	1/590	1/737	1/885
15	4.33	1/231	1/346	1/462	1/577	1/693
20	5.36	1/187	1/280	1/373	1/466	1/560
25	6.42	1/156	1/234	1/312	1/389	1/467
30	7.50	1/133	1/200	1/267	1/333	1/400

Fig. 8.13

From the above orders of magnitude it appears obvious that a correct evaluation of inelastic deformations is necessary for trusses having bolted joints with uncontrolled tightening, when determining imposed initial deflections.

8.2 CONVENTIONAL ELASTIC LIMIT STATE

8.2.1 *Tension*

Axial stress is expressed by:

$$\sigma = N/A_n \qquad (8.41)$$

where
N is tensile force applied to the bar
A_n is minimum net area of resistant cross section

The only problem when analysing a tension member is the definition of the net area to introduce into (8.41).

Net area means the area free from all diminutions due to the bar connecting system. Reductions in area might be caused by the existence of threaded parts or of holes for inserting bolts.

In order to determine minimum net area of a bar connected by means of bolts, various sections (straight, oblique and broken) are considered (Fig. 8.14) and for each the length of surface traversed is summed up: the smallest of these multiplied by the thickness of material is adopted as the minimum net area.

In most cases the most unfavourable section is the normal cross section; the net area is obtained by deducting the sum of empty spaces (holes) from the gross area.

Various methods for evaluating net areas are proposed by US codes |4,5|. AISC recommendations allow substracting from the gross area when non-aligned holes have to be considered the hole areas multiplied by a reduction coefficient:

$$k = 1 - \frac{a^2}{(4b + 2d)d} \qquad (8.42)$$

which can be approximated to:

$$k = 1 - \frac{a^2}{4bd}$$

where a, b and d are defined in Fig. 8.15.

Another method has been developed in |6| as part of a general study on plastic deformations. It leads to an expression for coefficient k of the following type:

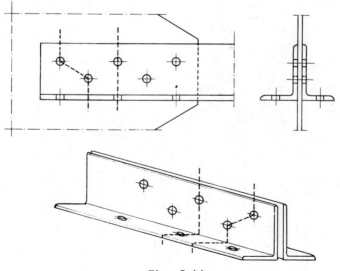

Fig. 8.14

THEORY AND DESIGN OF STEEL STRUCTURES

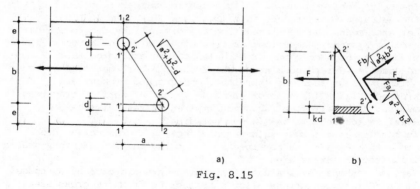

a)   b)

Fig. 8.15

$$k = \frac{a}{d} - \frac{b^2 + a^2 - d\sqrt{a^2 + b^2}}{d\sqrt{b^2 + 3a^2}}$$
(8.43)

A comparison between results deriving from the application of (8.42) and (8.43) is given in Fig. 8.16. It should be pointed out that (8.43)

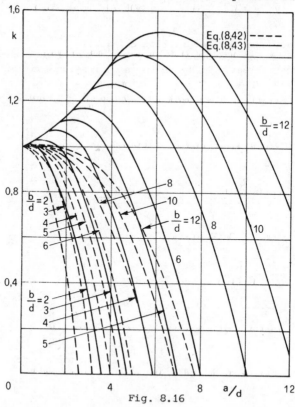

Fig. 8.16

leads to more conservative results, because, in some cases, it considers a coefficient $k > 1$ and thus an increase in conventional hole diameter to be deducted from the gross area in order to obtain the net area. Considering that the application of formula (8.42) always leads to a reduction of the area to be deducted, and thus to a bigger net area, AISC recommendations accept its results provided the net area does not exceed 85% of the gross area.

Codes usually contain special provisions for angles in tension; the resistant area for evaluating the axial stress is defined with reference to the type of connection (Fig. 8.17).

For an angle connected on both flanges, the resistant area coincides with the net area:

$$A_{res} = A_n$$

For an angle connected on only one flange (Fig. 8.17a), the resistant area is expressed by:

$$A_{res} = A_1 + \xi A_2 \tag{8.44}$$

where

A_1 is net area of connected flange
A_2 is area of non-connected flange
$\xi \triangleq 3A_1 / (3A_1 + A_2)$

Two identical angles connected to a joint plate in the flange plane (Fig. 8.17b), have a resistant area expressed by:

$$A_{res} = 2A_1 + 2\xi A_2 \tag{8.45}$$

where A_1 and A_2 have the foregoing meaning and $\xi = 5A_1 / (5A_1 + A_2)$.

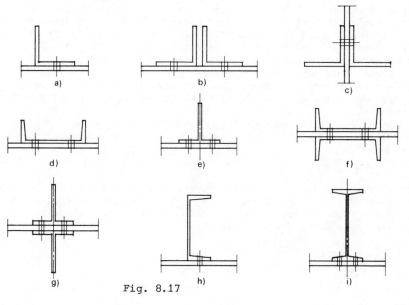

Fig. 8.17

If the two angles are connected to a joint plate between them (Fig. 8.17c), then:

$$A_{res} = 2A_n \tag{8.46}$$

where A_n is net area of each angle.

The above practical rules, derived from experimental work |7|, are adopted in British Standard BS 449, as well as in the European Recommendations |8|.

For channels connected through the web and ⊥ sections connected by their flanges (Figs 8.17d,e,f,g) the same criteria can be adopted as in the case of angles.

It is inadvisable to adopt connections fastening only one flange of a channel or I-section (Figs 8.17h,i).

8.2.2 *Compression*

A simple compression case does not often exist in steel struts, as their behaviour is usually determined by the influence of buckling phenomena (see chapter 9).

As far as material strength is concerned, steel is conventionally supposed to have the same behaviour both in tension and in compression.

Strength under compression of a cross section, neglecting buckling phenomena, can be emphasized by the stub column test (see 4.4). This test shows the lowering of the limit of proportionality f_o as a consequence of the presence of residual stresses and indicates the value of the mean tangent modulus E_{tm} referred to the entire cross section (Fig. 8.18).

8.2.3 *Bending*

Normal stresses in bent members are evaluated by considering the moment of inertia of the net cross section. This calculation can be roughly done in practice by subtracting the moment of inertia of the hole areas with respect to the centroidal axis of the gross section from the overall moment of inertia of the gross section itself.

Section modulus may be defined in different ways according to whether the member is behaving in a purely elastic manner or with yielded zones in the cross section without reaching the fully plastic condition.

Factor ψ, amplifying section modulus W, is called the plastic adaptation coefficient and the corresponding limiting moment is given by:

$$M_{lim} = \psi W f_y \tag{8.47}$$

where f_y is the yield stress of the material. This limiting moment can be related to a new kind of limit state, which can be called the plastic adaptation limit state. It is intermediate between the elastic limit state characterized by:

$$M_e = W f_y$$

and the fully plastic limit state with:

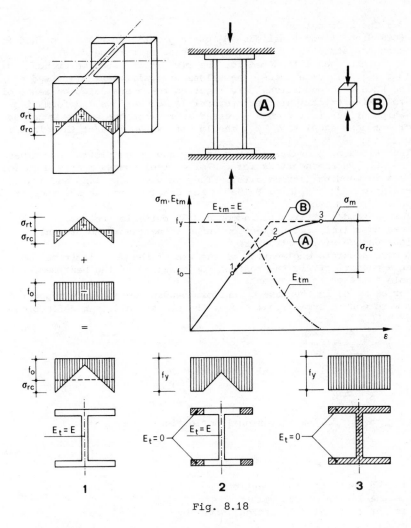

Fig. 8.18

$$M_y = \alpha M f_y$$

where α is the shape factor of the cross section (see 8.3.1.2, formula 8.97). In these terms, the plastic adaptation coefficient ψ assumes the meaning of a new reduced shape factor, thus:

$$1 < \psi < \alpha$$

This is a compromise between elastic design and plastic design and its purpose is to exploit more fully the strength properties of materials, even when designing elastically.

323

THEORY AND DESIGN OF STEEL STRUCTURES

8.2.3.1 The ECCS method

The ECCS Steel Recommendations |8| use a plastic adaptation method for checking the strength of structural members in pure bending.

This method was introduced for the first time into the French code in 1966 |1|. In both codes, the values of coefficient ψ (called η in the ECCS Recommendations) are required to be evaluated by means of the following criterion: the irreversible (or residual) deformation ε_r, which arises in the most stressed fibre when the member is unloaded after a moment equal to $\psi W f_y$, should not exceed 6.5% of the elastic deformation ($\varepsilon_e = f_y/E$).

The values of ψ evaluated in this way have been plotted as curves in the ECCS Recommendations for the mild steel shapes mostly used for members in bending (strong axis). They vary between 1.05 and 1.12, depending on depth h (Fig. 8.19a). For other common shapes they are given in Fig. 8.19b.

This method justifies the adopted 7.5% criterion because it represents a limit for deflection which is necessary to prevent serious damage of the structure.

A more accurate explanation of the use of the 7.5% limit may be given a posteriori, if the residual deformation of the bent member as a whole is taken into account.

For example, in the case of uniform bending in a simply supported beam with double symmetrical shape, the elastic mid span deflection is given by

$$v = \frac{ML^2}{8EI} = \chi \frac{L^2}{8}$$

Suppose there is residual curvature:

$$\chi_r = \frac{2\varepsilon_r}{h}$$

leading to the following residual deflection:

$$v_r = \frac{\varepsilon_r}{h} \frac{L^2}{4} \tag{8.48}$$

where

h is the depth of cross section
L is the length of the beam
ε_r is the residual deformation, given by:

$$\varepsilon_r = \frac{7.5}{100} \frac{f_y}{E} \tag{8.49}$$

Then, for a given value of ε_r, the ratio v_r/L may be computed for different kinds of steel, by varying L/h (= 10,20,30).

The corresponding values are given in the table of Fig. 8.20a, both for standard steels (Fe 360, Fe 430, Fe 510) and for high strength steels (HY 420, HY 490, HY 590, HY 690).

In the range of mild steels, it can be seen from this table that in the worst case (Fe 510, L/h = 30) the mid-span residual deflection is still contained within $L/1000$, which is universally considered as an acceptable limit for geometrical imperfections as a consequence of manufacturing tolerances for rolled steel shapes.

324

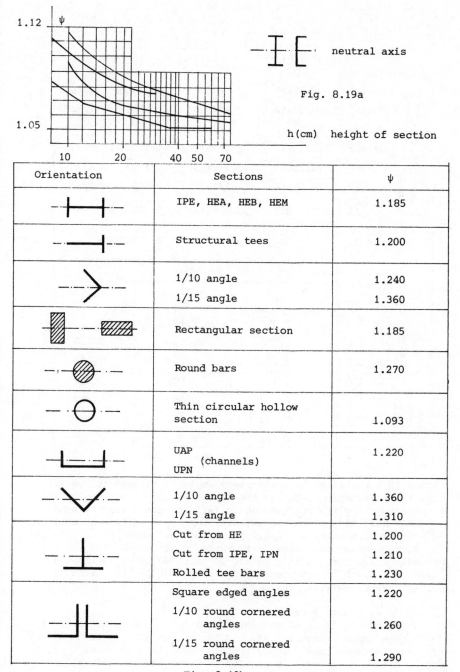

neutral axis

Fig. 8.19a

h(cm) height of section

Orientation	Sections	ψ
	IPE, HEA, HEB, HEM	1.185
	Structural tees	1.200
	1/10 angle 1/15 angle	1.240 1.360
	Rectangular section	1.185
	Round bars	1.270
	Thin circular hollow section	1.093
	UAP UPN (channels)	1.220
	1/10 angle 1/15 angle	1.360 1.310
	Cut from HE Cut from IPE, IPN Rolled tee bars	1.200 1.210 1.230
	Square edged angles 1/10 round cornered angles 1/15 round cornered angles	1.220 1.260 1.290

Fig. 8.19b

Steel		$\varepsilon_r = 7.5\% \dfrac{f_v}{E}$	v_r/L values for $\varepsilon_r = 7.5\% \varepsilon_e$		
			L/h		
			10	20	30
Mild steel	Fe 360	8.57×10^{-5}	1/4667	1/2334	1/1555
	Fe 430	10.00×10^{-5}	1/4000	1/2000	1/1333
	Fe 510	12.85×10^{-5}	1/3112	1/1556	1/1037
High strength steel	HY 420	15.30×10^{-5}	1/2614	1/1307	1/871
	HY 490	18.22×10^{-5}	1/2195	1/1907	1/731
	HY 590	21.86×10^{-5}	1/1830	1/915	1/610
	HY 690	25.51×10^{-5}	1/1568	1/784	1.523

a)

Steel		$\varepsilon_r/\varepsilon_e$ values for $v_r/L = 1/1000$		
		L/h		
		10	20	30
Mild steel	Fe 360	0.350	0.175	0.117
	Fe 430	0.300	0.150	0.100
	Fe 510	0.233	0.117	0.078
High strength steel	HY 420	0.190	0.095	0.063
	HY 490	0.160	0.080	0.053
	HY 590	0.133	0.067	0.044
	HY 690	0.114	0.057	0.038

b)

Fig. 8.20

This interpretation quantitatively explains the use of the 7.5% limit for deformations, but at the same time shows that the criterion is not always applicable, because it fails to guarantee the same residual deflection limit for all bent members of different materials with different slenderness L/h. The value 7.5% seems to be safe for mild steels, which are those mostly used in structures but it is generally inadequate if this method is extended to high strength steels, for which in many cases the residual deflection limit of $L/1000$ is exceeded (see Fig. 8.20a).

8.2.3.2 An alternative method
As already noted, the ECCS method becomes more open to criticism if its range of application is extended to include high strength steels. Its basic criterion would imply reducing the limit value of ε_r to about 3.8% ε_e, in order to guarantee in every case a residual mid-span deflection $v_r^e \leq L/1000$ (see Fig. 8.20a).

However, this approach would be extremely punitive for the lower strength steels, as well as increasing the lack of uniformity of the method in relation to the different types of steel.

In order to eliminate such incongruities, an alternative method |9| of plastic adaptation can be based on the following criterion: coefficient ψ must be computed under the condition that the residual mid span deflection does not exceed the limit of $L/1000$. As a consequence, the residual deflection becomes independent of the material being related only to the geometry of the beam through its slenderness parameter L/h.

On this basis, equation (8.48) can be written:

$$v_r = \frac{1}{1000} = \frac{\varepsilon_r}{4}\frac{L}{h} \tag{8.50}$$

Hence:

$$\varepsilon_r = \frac{4}{1000}\frac{h}{L}$$

$$\begin{array}{lll}
\varepsilon_r = 0.0004 & \text{for} & L/h = 10 \\
\varepsilon_r = 0.0002 & \text{for} & L/h = 20 \\
\varepsilon_r = 0.00013 & \text{for} & L/h = 30
\end{array}$$

This gives greater conceptual coherency compared to the previous ECCS definition in which the residual deformation was based on a constant percentage.

The values of the calculated deformations have the same meaning as the deformation $\varepsilon_r = 0.002$ conventionally used in defining the elastic limit of materials and they to not require that yielding be proportional to strength, which physically contradicts the actual characteristics of materials where reduced ductility is generally associated with higher strength. The value of the residual deformation which substitutes the constant value of 0.075 in the ECCS method, is given by:

$$\frac{\varepsilon_r}{\varepsilon_e} = \frac{4}{1000}\frac{h}{L}\frac{E}{f_y} \tag{8.51}$$

and it varies in accordance with both the material (E, f_y) and the beam slenderness (L/h).

For the same qualities of steel already considered in Fig. 8.20a, the percentage values of the residual deformation are given in the table of Fig. 8.20b. These vary from 3.8% to 35% of the elastic deformation, depending on the type of steel, and its variation is inversely proportional both to slenderness (L/h) and to strength (f_y).

The results of the application of this method to the standard rolled shapes (IPE, HEA, HEB, HEM) in structural steel Fe 360, Fe 430, Fe 510, Fe 550 are plotted in Fig. 8.21, for values of L/h of 10 and 30. For $L/h = 30$, the results for I shapes bent about the weak axis, together with other rolled shapes having a single axis of symmetry ⊏, ⊥, ∟ sections), are given in the table of Fig. 8.22, where they are compared with the results derived from ECCS Recommendations. They practically coincide, in the case of Fe 510 steel (for $L/h = 30$), as expected (see also Fig. 8.19).

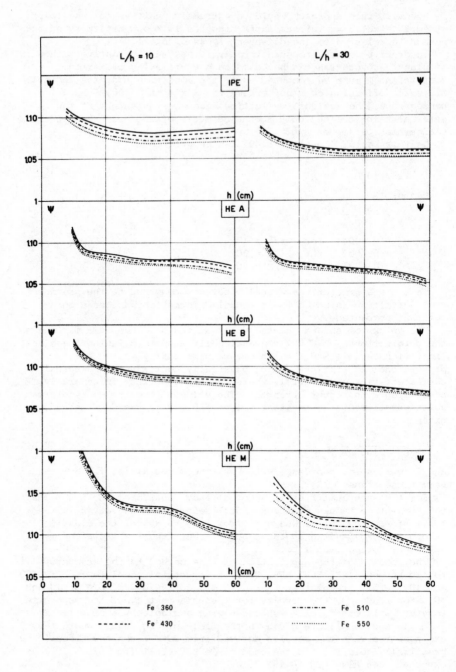

Fig. 8.21

	SHAPE	ECCS	ALTERNATIVE VALUES			
			Fe360	Fe430	Fe510	HY420
	IPE 80	1.190	1.224	1.211	1.192	1.178
	IPE 220	1.189	1.223	1.210	1.192	1.178
	HE M 300	1.189	1.223	1.210	1.192	1.177
	HE M 500	1.189	1.223	1.210	1.192	1.177
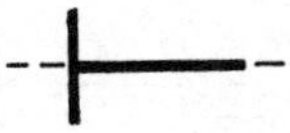	T 30	1.176	1.228	1.214	1.193	1.176
	T 50	1.176	1.226	1.213	1.193	1.176
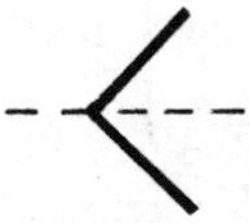	L 25x4	1.263	1.320	1.308	1.284	1.263
	L 80x10	1.250	1.297	1.283	1.265	1.250
	L 50x5	1.253	1.286	1.274	1.256	1.241
	L 100x10	1.250	1.285	1.272	1.252	1.237
	L 60x4	1.236	1.277	1.203	1.240	1.221
	L.120x8	1.232	1.277	1.261	1.236	1.215
	U 50x38	1.231	1.275	1.259	1.235	1.213
	UPN 220	1.254	1.301	1.283	1.258	1.238
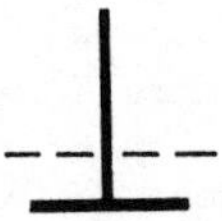	$\frac{1}{2}$ IPE 220	1.210	1.252	1.237	1.213	1.195
	$\frac{1}{2}$ HEM 600	1.196	1.241	1.225	1.200	1.180
	T 60	1.210	1.254	1.237	1.273	1.195
	T 100	1.202	1.248	1.231	1.206	1.298

Fig. 8.22

In order to codify in a simple way the results that are valid for both rolled and welded I shapes, a suitable analytical relation has been developed. It expresses the correlation between coefficient ψ and geometrical ratios h/t_f and b/t_w for each given material. For the cases involved, identified by the following variations of geometrical parameters:

$$1/\rho_w = b/t_w \text{ from 10 to 50} \quad ; \quad 1/\rho_f = h/t_f \text{ from 10 to 100}$$

and for $L/h = 30$, the expression for ψ is:

$$\psi = k_1 + k_2\rho_f + k_3\left\{\rho_f + k_u\right\}^{-k_5} \times \left\{k_6 - \log \frac{1}{\rho_w}\right\} \tag{8.52}$$

The numerical values of the constants k_1 to k_6 for the various steels are given in the table of Fig. 8.23.

Steel	k_1	k_2	k_3	k_4	k_5	k_6
Fe 360	0.9848	1.1612	0.000957	0.0150	1.275	2
Fe 430	0.9844	1.1725	0.000431	0.0220	1.573	2
Fe 510	0.9877	1.1063	0.000357	0.0200	1.552	2
HY 420	1.0042	0.9070	0.000679	0.0080	1.166	1.75
HY 490	1.0042	0.8675	0.000928	0.0040	0.994	1.75
HY 590	1.0053	0.8140	0.000414	0.0055	1.189	1.75
HY 690	1.0066	0.7575	0.000638	0.0040	1.025	1.75

Fig. 8.23

8.2.4 *Shear*

Tangential stresses due to shear force in bent bars are rigorously assessed by the following well known formula:

$$\tau = \frac{VS}{Ib} \tag{8.53}$$

For most commonly adopted sections (C , Π , $\perp$, I , L) it is usual in designing to neglect any flange contribution |8| and to assume that the shear force is totally taken by the web which, having an area A_w, is subjected to a mean stress of:

$$\tau_m = V/A_w$$

Whenever there are bolt holes in the web, the above stress is multiplied by the ratio of gross web area/net web area.

From formula (8.53) the maximum shear stress at the centroidal axis can be evaluated. In the case of some simple profiles having area A, it has the following values:

Rectangular section: $\tau_{max} = \frac{3}{2}\frac{V}{A}$

Solid circular section: $\tau_{max} = \frac{4}{3}\frac{V}{A}$

Thin walled circular hollow section: $\tau_{max} = 2\frac{V}{A}$

8.2.5 *Torsion*

Torsional behaviour of steel bars is deeply influenced by the shape of the cross section. Thin walled open profiles, which are very common in steel constructions, have the characteristic of having three geometric dimensions of different orders of magnitude, i.e. the thickness t of the several parts composing the profile (flange, web), the overall dimensions of the cross section (height, width) and the beam length L.

The classical theory of beams (de Saint Venant) under-estimates the strength of such profiles. In order to obtain more realistic results, it is necessary to adopt the so called non-uniform torsion theory [10]. When analysing the torsional behaviour of thin walled beams according to this, tangential stress flow due to torsional moment T must be divided into two parts: the classical primary flow characterizing de Saint Venant's theory, which is associated with the so called pure torsion, and the secondary flow, which is associated with the tangential stresses related to normal stresses arising from non-uniform warping of cross section, due to primary flow (warping tension).

The main characteristic of the torsional behaviour of thin walled profiles essentially consists in the fact that, as an effect of a space behaviour caused by torsional loads, deformation may take place along the longitudinal fibres of the beam, together with consequent normal stresses. Such a stress condition is called complementary or secondary. It is not taken into consideration in the classical theory of torsion of de Saint Venant, whereas it is the main object of the non-uniform torsion theory. The values of normal stresses due to warping may not be negligible, but of the same order of magnitude as those due to the bending behaviour.

Let us consider an I shaped cantilever, undergoing a horizontal force F applied to the upper flange (Fig. 8.24a). Force F is statically equivalent to two loading conditions: the first (Fig.8.24b), having two forces acting in the same direction equal to $F/2$, produces a simple bending deformation in the beam; the second (Fig. 8.24c), having two forces acting in opposite directions equal to $F/2$, produces torsion and bending deformations. Therefore, using the principle of superposition, the complete deformation condition may be obtained as the sum of a pure bending condition and of a flexural torsional condition. For the first loading condition, the cross section remains plane, whereas for the second loading condition the beam flanges bend in their own plane, in mutually opposite directions, and the cross section does not remain plane (it warps). Where warping is prevented by beam restraints, normal stresses arise, together with shear stresses due to their variation.

Normal stress resultants create a system of self balancing longitudinal forces in each cross section.

In the case of Fig. 8.24, the two opposite bending moments acting in both flanges produce a generalized force characterizing the cross

331

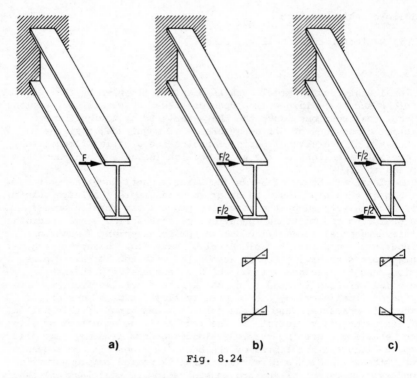

a)   b)   c)

Fig. 8.24

section, called the warping moment or *bimoment*. Bimoment can be assessed as a moment of the second order, expressed by the product of the bending moment acting in one of the flanges multiplied by the distance between the two flange planes.

8.2.5.1 Pure (or uniform) torsion

Consider a beam of constant cross section having length L, undergoing pure torsion (uniform, or de Saint Venant torsion). It is loaded by two equal and opposite torques applied at the ends. Such a loading system is self equilibrated, so that the beam does not require any constraint. The uniform torsion angle $d\theta/dz$ is constant along the beam and is proportional to the torsional moment:

$$\theta' = \frac{d\theta}{dz} = \frac{T_T}{GI_T} \tag{8.55}$$

where:

$\theta = \theta(z)$ = the relative twisting rotation between a generic section at abscissa z and the reference section ($z=0$), usually assumed at one end

G = the shear modulus

I_T = the torsional moment of inertia

Product GI_T expresses the primary torsional rigidity. In the case of a compact circular section, I_T coincides with the polar moment of

332

inertia I_o, whereas for all other shapes $I_T < I_o$. For open profiles composed of slender rectangular elements, I_T is expressed by the following formula [11]:

$$I_T = \frac{1}{3} \int_s t^3 ds$$

where:

$t = t(s)$ = profile thickness
ds = the elementary length on the mid line
$\int_s$ = the integral extended to the overall cross section

The position of the various elements of the cross section scarcely influences torsional stiffness and the value of I_T is almost independent of section shape. Consequently, in the case of a profile formed by different, straight or curved, elements, each having a constant although different thickness, it can be assumed that, approximately [11]:

$$I_T = \frac{1}{3} \sum_1^n b_i t_i^3 \qquad \text{(see p. 338)} \qquad (8.56)$$

(see p. 338)

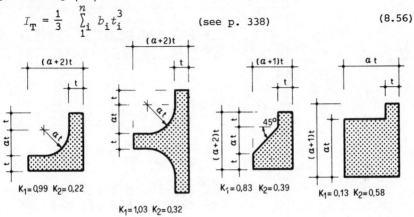

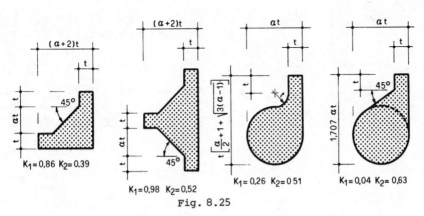

Fig. 8.25

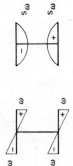

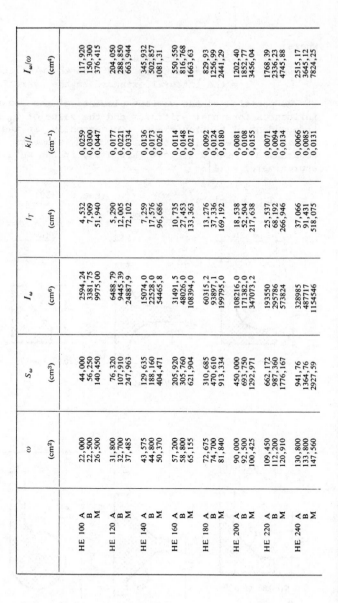

	ω	S_ω	I_ω	I_T	k/L	I_ω/ω
	(cm²)	(cm³)	(cm⁶)	(cm⁴)	(cm⁻¹)	(cm⁴)
HE 100 A	22,000	44,000	2594,24	4,532	0,0259	117,920
B	22,500	56,250	3381,75	7,909	0,0300	150,300
M	26,500	140,450	9975,00	51,940	0,0447	376,415
HE 120 A	31,800	76,320	6488,79	5,290	0,0177	204,050
B	32,700	107,910	9445,39	12,005	0,0221	288,850
M	37,485	247,963	24887,9	72,102	0,0334	663,944
HE 140 A	43,575	129,635	15074,0	7,259	0,0136	345,932
B	44,800	188,160	22528,0	17,576	0,0173	502,857
M	50,370	404,471	54465,8	96,686	0,0261	1081,31
HE 160 A	57,200	205,920	31491,5	10,735	0,0114	550,550
B	58,800	305,760	48026,0	27,453	0,0148	816,768
M	65,155	621,904	108394,0	133,363	0,0217	1663,63
HE 180 A	72,675	310,685	60315,2	13,276	0,0092	829,93
B	74,700	470,610	93897,1	37,336	0,0124	1256,99
M	81,840	913,334	199795,0	169,192	0,0180	2441,29
HE 200 A	90,000	450,000	108216,0	18,538	0,0081	1202,40
B	92,500	693,750	171382,0	52,504	0,0108	1852,77
M	100,425	1292,971	347073,2	217,638	0,0155	3456,04
HE 220 A	109,450	662,172	193550	25,537	0,0071	1768,39
B	112,200	987,360	295786	68,192	0,0094	2336,23
M	120,910	1776,167	573824	266,946	0,0134	4745,88
HE 240 A	130,800	941,76	328985	37,066	0,0066	2515,17
B	133,800	1364,76	487717	91,431	0,0085	3645,12
M	147,560	2927,59	1154546	518,075	0,0131	7824,25

HE 260	A	154,375	1254,30	517245	46,552	0,0059	3350,58
	B	157,625	1792,98	754925	110,424	0,0075	4806,16
	M	172,525	3756,73	1732085	598,516	0,0115	10039,62
HE 280	A	179,900	1637,09	786478	55,474	0,0052	4371,75
	B	183,400	2310,84	1131768	128,846	0,0066	6171,03
	M	199,440	4738,69	2524959	679,769	0,0102	12660,25
HE 300	A	207,000	2173,50	1201676	75,715	0,0049	5805,20
	B	210,750	3003,19	1690357	165,689	0,0061	8020,68
	M	233,275	7050,74	4394828	1160,967	0,0101	18839,69
HE 320	A	220,875	2567,67	1514527	97,010	0,0050	6856,94
	B	224,625	3453,61	2071851	201,011	0,0061	9223,60
	M	246,427	7614,61	5014019	1228,380	0,0097	20346,83
HE 340	A	235,125	2909,67	1827067	114,452	0,0049	7770,62
	B	238,875	3851,86	2457438	229,056	0,0060	10287,5
	M	260,332	8044,27	5596396	1233,937	0,0092	21497,1
HE 360	A	249,375	3273,05	2179893	133,919	0,0049	8741,42
	B	253,125	4271,48	2887808	259,617	0,0059	11408,6
	M	273,350	8419,18	6150650	1235,227	0,0088	22501,0
HE 400	A	278,250	3965,06	2946894	169,682	0,0047	10590,8
	B	282,000	5076,00	3823867	314,359	0,0056	13559,8
	M	330,860	9236,40	7427733	1242,382	0,0080	24688,3
HE 450	A	314,250	4949,44	4154212	218,074	0,0045	13219,4
	B	318,000	6201,00	5267886	386,012	0,0053	16565,7
	M	336,164	10320,26	9275178	1256,582	0,0072	27591,1
HE 500	A	350,250	6041,81	5652322	274,685	0,0043	16137,9
	B	354,000	7434,00	7031063	467,221	0,0050	19862,0
	M	370,259	11329,96	11217934	1266,516	0,0066	30297,4
HE 550	A	387,000	6966,00	7201559	311,191	0,0041	18608,7
	B	390,750	8498,81	8874085	519,144	0,0047	22710,4
	M	406,980	12453,59	13355434	1281,333	0,0060	33307,4
HE 600	A	423,750	7945,31	8994962	350,973	0,0039	21227,0
	B	427,500	9618,75	10989742	575,000	0,0045	25707,0
	M	442,250	13488,62	15957975	1291,884	0,0056	36083,6

Fig. 8.26

	ω (cm^2)	S_ω (cm^4)	I_ω (cm^6)	I_T (cm^4)	k/L (cm^{-1})	I_ω/ω (cm^4)
IPE 80	8,602	5,143	118,754	0,601	0,0441	13,805
IPE 100	12,966	10,162	353,476	1,031	0,0335	27,261
IPE 120	18,192	18,337	895,242	1,512	0,0255	49,210
IPE 140	24,290	30,588	1988,57	2,153	0,0204	81,865
IPE 160	31,283	47,456	3976,21	3,156	0,0175	127,104
IPE 180	39,130	71,216	7469,96	4,228	0,0147	190,901
IPE 200	47,875	101,734	13018,6	6,130	0,0134	271,929
IPE 220	57,970	146,664	22773,7	7,979	0,0116	392,854
IPE 240	69,060	203,036	37624,3	11,225	0,0107	544,806
IPE 270	87,682	301,847	70870,8	14,045	0,0087	808,266
IPE 300	108,487	435,306	126378	17,880	0,0074	1164,91
IPE 330	127,400	586,040	199841	24,758	0,0069	1568,61
IPE 360	147,602	796,684	314509	33,101	0,0064	2130,79
IPE 400	173,925	1056,59	492214	44,922	0,0059	2830,04
IPE 450	206,815	1434,26	794311	59,170	0,0053	3840,69
IPE 500	242,000	1936,00	1254440	79,353	0,0049	5183,62
IPE 550	279,720	2525,85	1893450	108,881	0,0047	6769,09
IPE 600	319,550	3339,30	2858297	146,525	0,0044	8944,76

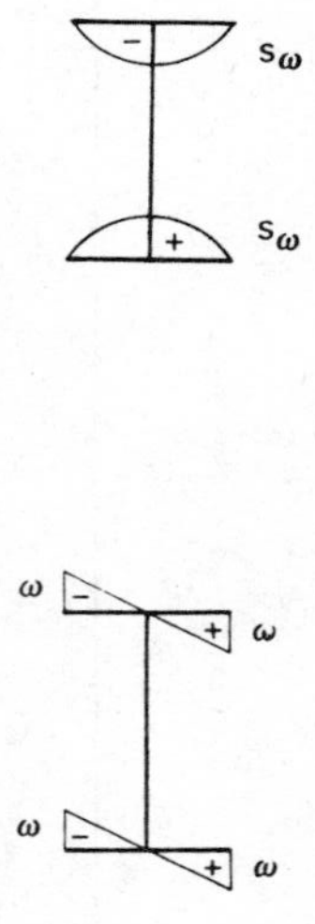

Fig. 8.26 (cont.)

Shape	Shear centre e	Primary moment I_T	Sectorial moment I_ω
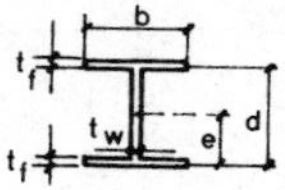	$d/2$	$\dfrac{1}{3}(2bt_f^3 + dt_w^3)$	$\dfrac{t_f\, d^2\, b^2}{24}$
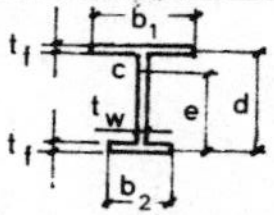	$h\,\dfrac{b_1^3}{b_1^3 + b_2^3}$	$\dfrac{1}{3}\left[(b_1+b_2)t_f^3 + dt_w^3\right]$	$\dfrac{t_f h^2}{12}\,\dfrac{b_1^3\, b_2^3}{b_1^3 + b_2^3}$
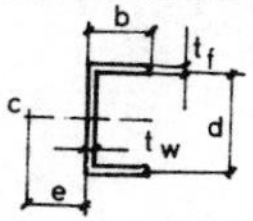	$\dfrac{3b^3 t_f}{6bt_f + dt_w}$	$\dfrac{1}{3}(2bt_f^3 + dt_w^3)$	$\dfrac{t_f b^3 h^2}{12}\,\dfrac{3bt_f + 2dt_w}{6bt_f + dt_w}$
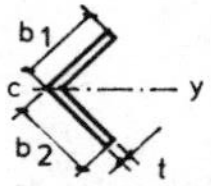	0	$\dfrac{1}{3}(b_1 + b_2)t^3$	$(b_1^3 + b_2^3)\,\dfrac{t^3}{36}$
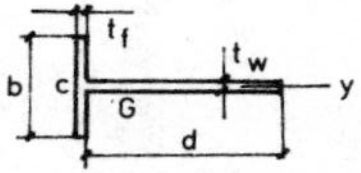	0	$\dfrac{1}{3}(bt_f^3 + ht_w^3)$	$\dfrac{t_f^3 b^3}{144} + \dfrac{t_w^3 d^3}{36}$
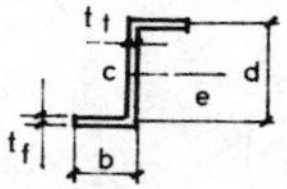	$d/2$	$\dfrac{1}{3}(2bt_f^3 + dt_w^3)$	$\dfrac{b^3 t_f d^2}{12(2b+d)^2}\,\left|\,2t_f(b^2+bd+d^2)+3t_w\, bd\,\right|$
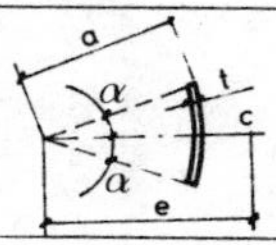	$2a\,\dfrac{\sin\alpha - \alpha\cos\alpha}{\alpha - \sin\alpha\cos\alpha}$	$\dfrac{2\alpha t^3}{3}$	$\dfrac{2ta^5}{3}\left(\alpha^3 - \dfrac{6(\sin\alpha - \alpha\cos\alpha)^2}{\alpha - \sin\alpha\cos\alpha}\right)$

Fig. 8.27

where:

b_i = the length of a single element
t_i = the mean thickness of a single element
n = the number of elements

The existence of thickened links or bulbs (Fig. 8.25) in a thin profile entails an increase of I_T, sometimes a significant one. Such an increase can be approximately assessed by means of the following expression |12|:

$$\Delta I_T = \{(k_1 + k_2 \alpha) t\}^4 \tag{8.57}$$

where:

t = the mean thickness of elements connected by the link or of the element reinforced by a bulb
α = the ratio between link or bulb height and thickness of the elements connected thereby
k_1 and k_2 = empirical constants, the numerical values of which are given in Fig. 8.25 for the most common cases

The torsional moment of inertia is found by adding ΔI_T to the value given by (8.56) for the complete section without links or bulbs.

Values of I_T, according to (8.56) and (8.57), are given in the table of Fig. 8.26, for various types of European I sections. For other shapes see Fig. 8.27.

Once I_T is known, the maximum tangential stress in each section is:

$$\tau_{T,max} = Gt \frac{d\theta}{dz} = \frac{T_T}{I_T} t \tag{8.58}$$

Torsional shear stresses corresponding to pure torsion vary linearly within the thickness of each part of the cross section (Fig. 8.28a,b).

In the case of thin walled box sections, in which thicknesses t_i are small compared to the transverse dimensions, it is an acceptable approximation to consider transverse shear stress constant within the thickness (Fig. 8.28c). This leads to the well known Bredt's formula |11|:

$$\tau_T = \frac{T_T}{2At} \tag{8.59}$$

which expresses equilibrium between internal stresses and torque according to the hydro dynamical analogy. In (8.59) A is the area contained in the mean line of the wall. It also follows that the uniform twist angle is:

$$\frac{d}{dz} = \frac{T_T}{4A^2 G} \int_s \frac{ds}{t} \tag{8.60}$$

Comparison between (8.60) and (8.55) allows the primary torsional moment of inertia I_T to be defined for closed sections |11|:

$$I_T = \frac{4A^2}{\int_s \frac{ds}{t}} \tag{8.61}$$

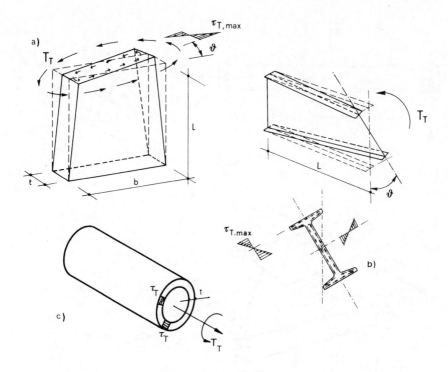

Fig. 8.28

8.2.5.2 Torsion due to warping

Consider the example of an I beam shown in Fig. 8.29 loaded at its mid-span by a concentrated torque and constrained at its ends by a torsional pinned support, which reacts with a torsional moment without preventing displacement component w parallel to beam axis z. By contrast, a torsional fixed support prevents not only the end section twisting, but also displacement w, and thus its warping.

The beam of Fig. 8.29 is torsionally pinned, symmetrical and symmetrically loaded with reference to the mid-span section, where displacement w must be nil, because they are not compatible with the symmetry of the system. Consequently, in this case, the mid-span section is the restraint preventing warping and the problem is one of mixed torsion. In fact, the torsional strength of the mid span section is due to the flange being prevented from warping, while at the ends (which are free to warp) the section resists only pure torsion. In other transverse sections of the beam, torsional behaviour is intermediate between those described above and strength reveals itself partly as resisting pure torsion and partly as preventing warping.

In the case of beams variously loaded and torsionally restrained, section by section distribution between the two ways of resisting torsion depends on the loading condition and is strictly related to the degree of clamping against warping of the end restraints.

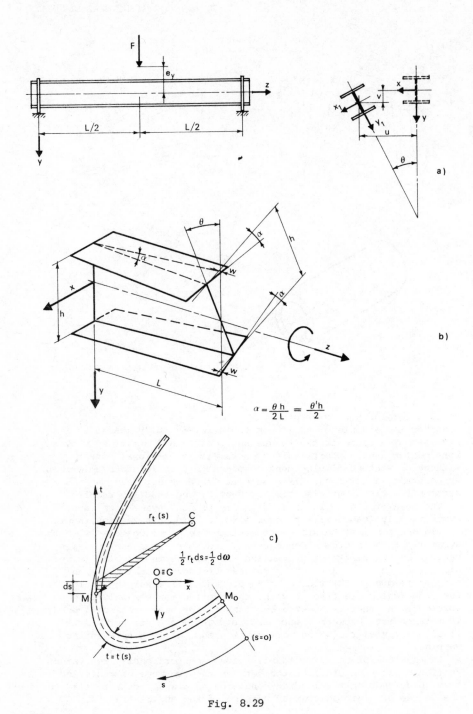

$$\alpha = \frac{\theta\,h}{2\,L} = \frac{\theta'\,h}{2}$$

$$\frac{1}{2}\,r_t\,ds = \frac{1}{2}\,d\omega$$

Fig. 8.29

Resuming the examination of Fig. 8.29, θ being the overall twist angle on length L and h being the distance between flange axes, the cross section warps for an angle equal to $\theta h/2L$ (Fig. 8.29b).

The values of w corresponding to warping represent the displacement undergone in the longitudinal direction by each point of the section, with reference to the undeformed position.

For a beam of constant section undergoing torsion, the displacement component w in relation to z (warping) is related to the uniform twist angle by the following:

$$w = \omega \frac{d\theta}{dz} \tag{8.62}$$

where $\omega = \omega(s) = \omega(x,y)$ is a function of cross sectional shape, defined as sectorial area, or uniform warping, which the deformed longitudinal configuration depends upon.

Function $\omega(s)$ is proportional to twice the surface generated by vector radius CM, where M defines the mean section line (Fig. 8.29c). If origin M of the curved abscissa s is fixed at a given point of the mean section line, the generic value of ω is expressed as follows:

$$\omega = \omega(s) = \int_0^s r_t(s)\,ds$$

$r_t(s)$ being the distance between point C and axis t tangent to the section in point (s). Point C is called the pole and M_o is called the nil sectorial point.

In the case of pure or uniform torsion, $d\theta/dz$ being constant along the beam, ω (8.62) is independent of z, whatever the value of function ω.

A beam is said to undergo non uniform torsion whenever displacements w are a function of z. This might happen essentially for two reasons:

(a) Torsional moment varies along z, and therefore the uniform twist angle, to which torsional moment is proportional

(b) Restraint conditions impose some limitations on w (prevented warping)

Whatever the cause may be of w varying along z, longitudinal components of deformation take place at every cross section:

$$\varepsilon_{z,\omega} = \frac{\partial w}{\partial z} = \omega\theta'' \tag{8.63}$$

The following normal stresses correspond to the above components:

$$\sigma_{z,\omega} = E\varepsilon_z = E\omega\theta'' \tag{8.64}$$

Beam stresses and deformations thus depend on ω. Its assessment is, therefore, absolutely necessary in order to study non-uniform torsion in the various types of structural profiles. In the case of the European series of I sections (IPE and HE) the distribution of ω and its maximum values are given in the tables of Fig. 8.26.

Prevented warping not only causes the above normal stresses, but it is also accompanied by tangential stresses τ_ω in the cross section, given by:

341

$$\tau_\omega = - \frac{ES_\omega}{t} \; \theta'''$$ (8.65)

The sectorial static moment introduced into (8.65) is defined by the following relation:

$$S_\omega(s) = \int_A \omega dA$$

Like ω , S_ω also depends only on the shape of the cross section; its distribution and its maximum values are given in the tables of Fig. 8.26.

Secondary torsional moment T_ω is defined by the sum along the mean line of the cross section of the moments of the shear forces with respect to the shear centre. Its value is expressed as follows:

$$T_\omega = - EI_\omega \theta'''$$ (8.66)

The sectorial moment of inertia I_ω is expressed as:

$$I_\omega = \int_A \omega^2 \, dA$$ (8.67)

and depends on cross sectional shape. Its numerical values are also given in the tables of Fig. 8.26, for I beams of European series |13|. Values for other shapes may be evaluated from formulae of Fig. 8.27 or may be found in the literature |14,15|.

8.2.5.3 Mixed torsion

From the foregoing consideration it ensues that a thin walled open section beam undergoing torsional loads is subject not only to torsional shear stresses τ_T due to primary torsion (see formula 8.58 or 8.59), but also to a secondary stress state characterized by normal stresses $\sigma_{z,\omega}$ (8.64) and shear stresses τ_ω (8.65). Fig. 8.30 gives an outline of all these stresses $(\tau_T, \tau_\omega, \sigma_{z,\omega})$, produced by mixed torsion in I sections and channels.

Torsional moment T thus separates, section by section, into a primary portion T_T (pure torsion), equilibrated by the primary torsional shear stresses, and a secondary portion T_ω (warping torsion), equilibrated by the combination of the secondary stresses due to prevented warping. Therefore:

$$T = T_T + T_\omega$$ (8.68)

Secondary torsional moment T_ω is nil only in cases of pure torsion. Even when there is a constant torsional moment T_T and T_ω vary complementarily along z, by keeping their sum constant.

In the case of actual beams, once torsional moment distribution has been determined as a function of loading and restraint conditions, the problem of mixed torsion is that of dividing, section by section, torsional characteristic T into primary torsional moment T_T and secondary torsional moment T_ω (Fig. 8.31).

Such a distribution is strongly influenced by cross sectional shape. It can be stated, in general, that, in the case of solid or hollow sections, portion T_ω is always negligible compared to T_T. Conversely

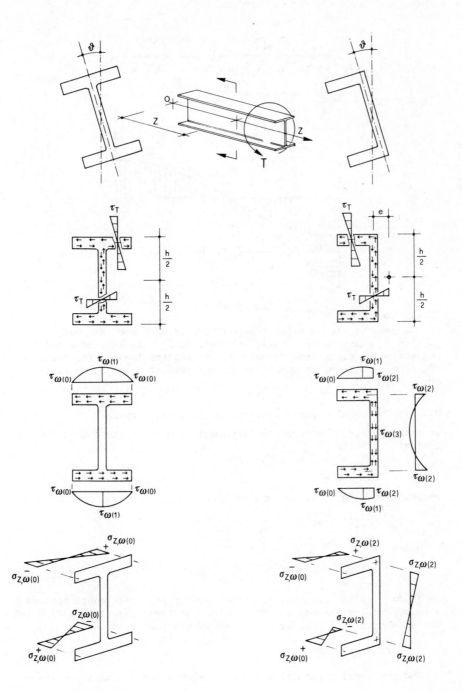

Fig. 8.30

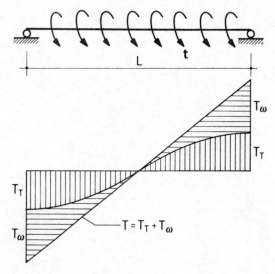

Fig. 8.31

in the case of open sections, especially if they are thin walled, portion T_{T} might be very small compared to T_ω. In this case, therefore neglecting T_ω might lead to serious underestimations of the torsional rigidity of a beam. A quantitative analysis of such a problem is set out in 8.2.5.4.

8.2.5.4 Torsional moment distribution in mixed torsion

In each section of a beam, the torsional moment is the sum of the two following effects:

$$T = T_{\mathrm{T}} + T_\omega$$

where, according to (8.55) and (8.66):

$$T_{\mathrm{T}} = GI_{\mathrm{T}}\theta' \quad \text{and} \quad T_\omega = -EI_\omega\theta'''$$

Hence:

$$T = GI_{\mathrm{T}}\theta' - EI_\omega\theta''' \qquad (8.69)$$

If $t(z)$ is the external torsional moment per unit length applied to a beam, due for example to the presence of transverse load $q(z)$ which are eccentric with respect to the cross section shear centre:

$$t(z) = q(z)e(z) \qquad (8.70)$$

The equilibrium condition for beam element dz is expressed as:

344

$$- T + t(z) + (T + \frac{dT}{dz} dz) = 0 \qquad (8.71)$$

which becomes:

$$\frac{dT}{dz} = -t(z) \qquad (8.72)$$

Substituting for T from (8.69) gives:

$$EI_\omega \theta^{IV} - GI_T \theta'' = t(z) \qquad (8.73)$$

Formula (8.73) is the fourth degree differential equation with constant coefficient for solving the problem of mixed torsion in thin walled beams. It should be pointed out that this equation is structurally identical to that for solving the problem of a beam bearing a transverse load $q(z)$ together with an axial tensile force N (Fig. 8.32), i.e.:

$$EIv^{IV} - Nv'' = q(z) \qquad (8.74)$$

This analogy becomes evident by introducing the following substitutions into (8.74):

$$I \to I_\omega \; ; \; v \to \theta \; ; \; N \to GI_T \; ; \; q(z) \to t(z) \qquad (8.75)$$

In such a context, the primary torsional stiffness of a beam assumes the same meaning as axial tensile force N. The flexural stiffening effect given by N to a bent beam corresponds to the torsional stiffening effect due to GI_T in the mixed torsion case. With position:

$$k = L\sqrt{\frac{GI_T}{EI_\omega}} \qquad (8.76)$$

where k represents beam characteristic non-dimensional length, (8.73) becomes:

$$\theta^{IV} - \frac{k^2}{L^2} \theta'' = \frac{t(z)}{EI_\omega} \qquad (8.77)$$

Its general integral can be expressed in the following form:

$$\theta = \theta_o + C_1 + C_2 \frac{z}{L} + C_3 \sinh \frac{k}{L} z + C_4 \cosh \frac{k}{L} z \qquad (8.78)$$

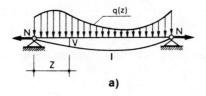 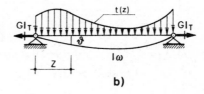

a)   b)

Fig. 8.32

or else:

$$\theta = \theta_o + C_1 + C_2 \frac{z}{L} + C_3 e^{-\frac{k}{L}z} + C_4 e^{-\frac{k}{L}z} \qquad (8.78)$$

where:

θ_o is the particular integral, depending on the distribution of $t(z)$ on the beam

C_1, C_2, C_3 and C_4 are the integration constants of the homogeneous equation, which can be determined according to edge conditions for each part of the beam

Once quantities $\theta_o, C_1, C_2, C_3, C_4$ are known, the twist angle expressed by (8.78) leads first of all to the uniform twist angle θ', to which section warping (8.62) is proportional:

$$w = \omega\theta' \qquad (8.79)$$

It is also possible to determine all magnitudes characterizing the problem, i.e. the warping moment M_ω, resulting from normal stresses due to torsion:

$$M_\omega = - EI_\omega \theta'' \qquad (8.80)$$

and the total torsional moment T, which is the sum of the primary torsional moment T_T (the resultant of the distribution of torsional shear stresses τ_T varying within the thickness of each part of the cross section) and the secondary torsional moment T_ω (the resultant of warping shear stresses τ_ω). From (8.68) and (8.69), it ensues that:

$$T = - EI\theta''' + GI_T\theta' \qquad (8.81)$$

Therefore, once function θ (8.78) and its first three derivatives are determined, formulae (8.80) and (8.81) give the internal actions for the mixed torsion problem.

Thus, recalling formulae (8.78), (8.80) and (8.81), the four fundamental magnitudes of the problem in their explicit form are expressed as follows:

$$\theta(z) = C_1 + C_2 \frac{z}{L} + C_3 \sinh \frac{k}{L}z + C_4 \cosh \frac{k}{L}z + \theta_o(z)$$

$$\theta'(z) = \frac{C_2}{L} + C_3 \frac{k}{L} \cosh \frac{k}{L}z + C_4 \frac{k}{L} \sinh \frac{k}{L}z + \theta_o'(z) \qquad (8.82)$$

$$M(z) = -GI_T\{C_3 \sinh \frac{k}{L}z + C_4 \cosh \frac{k}{L}z + \frac{L^2}{k^2}\theta_o''(z)\}$$

$$T(z) = GI_T\{C_2 + \theta_o'(z) - \frac{L^2}{k^2}\theta_o'''(z)\}$$

Integrals $\theta_o(z)$ for particular cases has the following expressions:

For concentrated torque:

$$t(z) = 0 \quad ; \quad \theta_o(z) = 0 \qquad (8.83)$$

Fig. 8.33

$$\chi = \frac{\dfrac{1-\operatorname{ch} k\zeta_1}{\operatorname{tgh} k} + \dfrac{\operatorname{ch} k\zeta_1 - 1}{\operatorname{sh} k} + \operatorname{sh} k\zeta_1 - k\zeta_1}{\dfrac{\operatorname{ch} k + \operatorname{ch} k\zeta_1\,\operatorname{ch} k - \operatorname{ch} k\zeta_1 - 1 + k(\zeta_1-1) - \operatorname{sh} k\zeta_1}{\operatorname{sh} k}}$$

#	scheme ($\zeta = \frac{z}{l}$)	boundary conditions (left)	boundary conditions (right)	range	torsional rotation
1	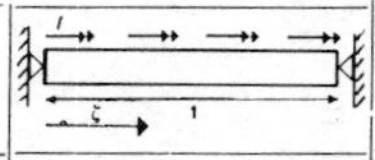	$\theta = \theta'' = 0$	$\theta = \theta'' = 0$	$0 \leqslant \zeta \leqslant 1$	$\theta = \dfrac{tl^2}{GI_T}\left\{ \dfrac{1}{8} - \dfrac{1}{k^2} - \dfrac{1}{2}\left(\zeta - \dfrac{1}{2}\right)^2 + \dfrac{\operatorname{ch} k(\zeta - 1/2)}{k^2 \operatorname{ch} k/2} \right\}$
2	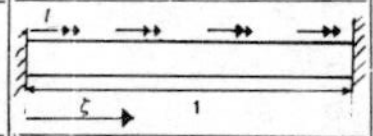	$\theta = \theta' = 0$	$\theta = \theta'' = 0$	$0 \leqslant \zeta \leqslant 1$	$\theta = \dfrac{tl^2}{GI_T}\left\{ \dfrac{1}{k^2}\left[\left(\dfrac{1+\operatorname{ch} k}{\operatorname{sh} k}\right)(\operatorname{ch} k\zeta - 1) + k\zeta(1-\zeta) - \operatorname{sh} k\zeta\right] \right\}$
3	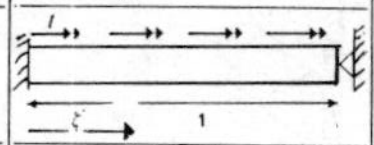	$\theta = \theta'' = 0$	$\theta = \theta'' = 0$	$0 \leqslant \zeta \leqslant 1$	$\theta = \dfrac{tl^2}{GI_T}\left\{ \dfrac{1}{k^2}\left[\left(\dfrac{k^2}{2} - 1 + \dfrac{1}{\operatorname{ch} k}\right)\dfrac{1}{\operatorname{tgh} k}k\zeta - k - (\operatorname{tgh} k - k\zeta - \zeta\,\operatorname{tgh} k\,\operatorname{ch} k + \operatorname{sh} k\zeta) + \dfrac{\operatorname{ch} k\zeta}{\operatorname{ch} k} - \dfrac{1}{\operatorname{ch} k} - \dfrac{k^2\zeta^2}{2}\right] \right\}$
4	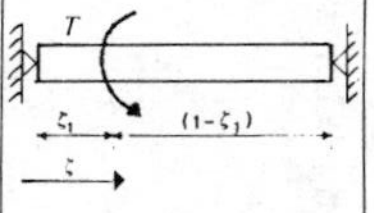	$\theta = \theta'' = 0$	$\theta = \theta'' = 0$	$0 \leqslant \zeta \leqslant \zeta_1$	$\theta = \dfrac{TL}{GI_T}\left\{ (1-\zeta_1)\zeta - \dfrac{\operatorname{sh} k(1-\zeta_1)}{k\cdot\operatorname{sh} k}\cdot\operatorname{sh} k\zeta \right\}$
				$\zeta_1 \leqslant \zeta \leqslant 1$	$\theta = \dfrac{TL}{GI_T}\left\{ \zeta_1(1-\zeta) - \dfrac{\operatorname{sh} k\zeta_1}{k\,\operatorname{sh} k}\operatorname{sh} k(1-\zeta) \right\}$
5	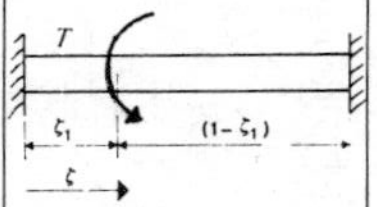	$\theta = \theta'' = 0$	$\theta = \theta'' = 0$	$0 \leqslant \zeta \leqslant \zeta_1$	$\theta = \dfrac{TL}{GI_T}\left\{ \dfrac{1}{(1+\chi)k}\left[\chi\left(\dfrac{1}{\operatorname{sh} k} + \operatorname{sh} k\zeta_1 - \dfrac{\operatorname{ch} k\zeta_1}{\operatorname{tgh} k} + \dfrac{1}{\operatorname{tgh} k}\right) + \dfrac{\operatorname{ch} k\zeta_1 - 1}{\operatorname{sh} k}\cdot(\operatorname{ch} k\zeta - 1) - \operatorname{sh} k\zeta + k\zeta + \operatorname{sh} k\zeta_1\right] \right\}$
				$\zeta_1 \leqslant \zeta \leqslant 1$	$\theta = \dfrac{TL}{GI_T}\left\{ \dfrac{1}{\left(1+\dfrac{1}{\chi}\right)k}\left[\dfrac{1}{\chi}\left(\dfrac{\operatorname{ch} k\zeta_1 - 1}{\operatorname{sh} k} - \operatorname{ch} k + k\operatorname{sh} k + \operatorname{ch} k\cdot\left(\dfrac{1-\operatorname{ch} k\zeta_1}{x\,\operatorname{tgh} k}\right) + \dfrac{1-\operatorname{ch} k\zeta_1\,\operatorname{ch} k}{\operatorname{sh} k} + \operatorname{sh} k\left(\dfrac{\operatorname{ch} k\zeta_1 - 1}{x} + \operatorname{ch} k\zeta_1 - 1\right)(\operatorname{ch} k\zeta_1 - 1) - k\zeta_1\right)\zeta\right] \right\}$
6	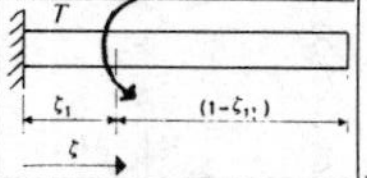	$\theta = \theta' = 0$	$\theta'' = 0$	$0 \leqslant \zeta \leqslant \zeta_1$	$\theta = \dfrac{TL}{GI_T}\left\{ \dfrac{1}{k}\left[(\operatorname{ch} k\zeta_1 - 1)\dfrac{\operatorname{sh} k - \operatorname{sh} k(1-\zeta_1)}{\operatorname{ch} k} + k\zeta - \operatorname{sh} k\zeta\right] \right\}$
				$\zeta_1 \leqslant \zeta \leqslant 1$	$\theta = \dfrac{TL}{GI_T}\left\{ \dfrac{1}{k}\left[(\operatorname{ch} k\zeta_1 - 1)\dfrac{\operatorname{sh} k - \operatorname{sh} k(1-\zeta)}{\operatorname{ch} k} - \operatorname{sh} k\zeta_1 + k\zeta_1\right] \right\}$
7	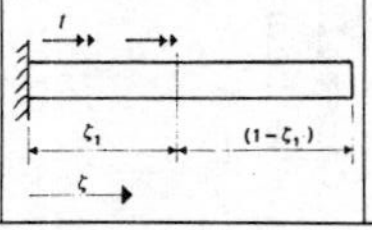	$\theta = \theta' = 0$	$\theta'' = 0$	$0 \leqslant \zeta \leqslant \zeta_1$	$\theta = \dfrac{tl^2}{GI_T}\left\{ \dfrac{1}{k^2}\left[\operatorname{tgh} k(k\zeta_1 - \operatorname{sh} k\zeta_1)(\operatorname{tgh} k - \operatorname{tgh} k\,\operatorname{ch} k\zeta + \operatorname{sh} k\zeta) + \operatorname{ch} k\zeta_1](\operatorname{ch} k\zeta - 1) - k\zeta_1\,\operatorname{sh} k\zeta + k^2\zeta\left(\zeta_1 - \dfrac{\zeta}{2}\right)\right] \right\}$
				$\zeta_1 \leqslant \zeta \leqslant 1$	$\theta = \dfrac{tl^2}{GI_T}\left\{ \dfrac{1}{k^2}\left[(\operatorname{sh} k\zeta_1 - k\zeta_1)(\operatorname{tgh} k - \operatorname{tgh} k\,\operatorname{ch} k\zeta + \operatorname{sh} k\zeta) - \operatorname{ch} k\zeta + \dfrac{k^2\zeta_1^2}{2} + 1\right] \right\}$

For uniformly distributed torque

$$t(z) = t = q e \quad ; \quad \theta_0(z) = -\frac{t}{2} \frac{z^2}{GI_T} \qquad (8.84)$$

For torque having a triangular distribution

$$t(z) = t \frac{z}{L} \quad ; \quad \theta_0(z) = -\frac{t}{6} \frac{1}{L} \frac{z^3}{GI_T} \qquad (8.85)$$

For torque having a trapezoidal distribution

$$t(z) = e(q_1 + q_2 \frac{z}{L}) \; ; \; \theta_0(z) = -\frac{e}{6GI_T} (3q_1 z^2 + q_2 \frac{z^3}{L}) \qquad (8.86)$$

For torque having a parabolic distribution

$$t(z) = t \frac{z^2}{L^2} \quad ; \quad \theta_0(z) = -\frac{t}{12} \frac{1}{GI_T} \frac{z^4}{L^2} - t \frac{EI}{(GI_T)^2} \frac{z^2}{L^2} \qquad (8.87)$$

In order to determine the 4 constants C_1, C_2, C_3, C_4 appearing in (8.82), it is possible to refer to the following conditions, corresponding to the most common end restraints:

For torsional pinned support

$$\theta = 0 \quad ; \quad \theta'' = 0 \, (M_\omega = 0) \qquad (8.88)$$

For torsional fixed support

$$\theta = 0 \quad ; \quad \theta' = 0 \, (w = 0) \qquad (8.89)$$

Free end

$$\theta'' = 0 \, (M_\omega = 0) \quad ; \quad T = 0 \qquad (8.90)$$

The table of Fig. 8.33 provides the expressions for solving some significant load and restraint cases.

8.2.5.5 Stress state in flexural torsion
By combining the results of mixed torsion with those already known for bending, one obtains the complete stress state in thin walled beams undergoing flexural torsion.

The stress state with respect to the principal generalized coordinates $x(s)$, $y(s)$ and $\omega(s)$ of the cross section of the beam, is represented by the following components:

Generalized normal stress, acting along the longitudinal fibres of the beam:

$$\sigma_z = \frac{M_x}{I_x} y + \frac{M_y}{I_y} x + \frac{M_\omega}{I_\omega} \qquad (8.91)$$

Generalized shear stress, acting along the mean line of the cross section

$$\tau(s) = \frac{1}{t(s)} \left\{ \frac{T_x}{I_x} S_x(s) + \frac{T_y}{I_y} S_y(s) + \frac{T_\omega}{I_\omega} S_\omega(s) \right\} \qquad (8.92)$$

Primary torsional shear stress, linearly variable along the thickness of each element forming the cross section:

$$T_T = \frac{T_T}{I_T} t(s) \qquad (8.93)$$

where

$t(s)$ is profile thickness

$S_x(s)$, $S_y(s)$, $S_\omega(s)$ are the static moments of the point having curved coordinate (s), in which (8.92) is evaluated

I_x, I_y, I_ω are the principal moments of inertia of the cross section
I_T is the primary torsional moment of inertia of the cross section
M_x and M_y are the bending moments
T_x and T_y are the shear forces
M_ω is the bimoment
T_T and T_ω are the primary and secondary torsional moments, respectively

For evaluation of I_T and I_ω see Figs 8.26 and 8.27.

8.2.5.6 Final remarks on torsional behaviour

The above illustrated results are valid for open sections, but with some limitations they can be extended to hollow sections. However, torsional behaviour of hollow and open sections are substantially different. The value of the coefficient $k = L\sqrt{GI_T/EI_\omega}$ characterizes the problem of torsion case by case and indicates whether pure torsion or warping effects are predominant. Open sections, for example, can be characterized by high values of secondary torsional stiffness EI_ω/L^2 compared to the primary one GI_T. In this case, the value of coefficient k (determining the law by which torsional moment is subdivided into the primary and secondary portions in the various sections of the beam) is slight. Secondary torsion increases the greater EI_ω/L^2 is compared to GI_T, or, more exactly, span L being equal, the smaller parameter k is. Also the law by which the effects of non-uniform torsion dampen depends on k. For the same span L, the damping grows as k increases. The damping of the warping and the consequent flow of the secondary stresses are therefore weak in the case of open sections for which the values of k are low and thus the corresponding deformations can be considered negligible.

In the case of hollow sections, on the contrary, primary stiffness GI_T is predominant over EI_ω/L^2, so that the torsional stress state is characterized by the flow of primary stresses. On the other hand, due to the high values of k, the damping of the warping and therefore of the flow of the secondary stresses is extremely violent. It ensues that, although the values of secondary torsional moments proportional to EI_ω are low, the great variations of $\varepsilon_{z,\omega}$ (and thus of $\sigma_{z,\omega}$) can entail high values of secondary stresses τ_ω, the existence of which, in the case of hollow sections, makes it more difficult to accept the basic hypothesis according to which their corresponding deformations are neglected compared to the primary ones. It must, however, be pointed out that, in the case of closed sections (due to the remarkable damping effect) the existence of high secondary stresses is limited to short parts corresponding to the sections in which the torque or the end restraints are applied, so that they have the meaning of a local effect, whereas in all the other parts of the beam secondary stesses are usually very small and almost always negligible compared to the primary stresses. In these cases, possible local yield reduces stress peaks, making the assumption of the primary torsion theory acceptable.

In some cases mixed open and hollow sections are adopted. A design criterion could be to reduce the damping parameter k, by means of a suitable increase in secondary stiffness EI_ω/L^2.

It is therefore useful (Fig. 8.34) to provide a classification (although necessarily an approximate one), emphasizing the torsional behaviour of the various types of thin walled profiles in terms of coefficient k:

Torsion due to warping only ($0 < k < 0.5$): thin walled cold formed sheets and open section bridge girders made of orthotropic plates

Warping prevalent ($0.5 < k < 2$): cylindrical thin shells and composite steel/concrete bridges having an open section

Warping and uniform torsion ($2 < k < 5$): hot rolled profiles (I sections and channels)

Uniform torsion prevalent ($5 < k < 20$): stocky sections (profiles for rails, jumbo shapes) and hollow sections (tubes and box sections)

Uniform torsion only, of the de Saint Venant type ($20 < k < \infty$): compact sections.

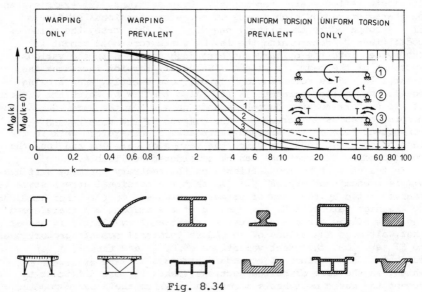

Fig. 8.34

A more detailed view of primary torsion/secondary torsion interaction in the case of rolled (HE, IPE) standard I profiles can be found by plotting L/i_y against t/h for those values of k that define the areas of characteristic torsional behaviour listed above, L/i_y being the lateral slenderness of the beam and t_f/h the flange thickness/depth ratio (Fig. 8.35).

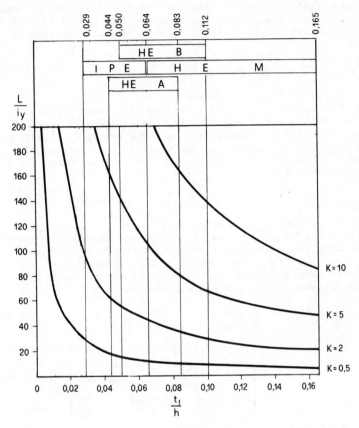

Fig. 8.35

8.3 PLASTIC LIMIT STATE

8.3.1 *Ultimate values of internal actions*

8.3.1.1 Axial force
The value of the fully plastic axial force acting in a generic section is expressed by:

$$N_{pl} = A_n \, f_y \qquad (8.94)$$

351

where:

A_n is net sectional area, i.e. free from holes or other possible weaknesses

f_y is the yield stress of the steel.

8.3.1.2 Simple bending

Consider the evolution of the stress state in a section of a beam made of elastic-perfectly plastic material (Fig. 8.36). As the load increases, the maximum elastic moment M_e is attained as soon as the stress in the most stressed fibres reaches material yielding and its value is:

$$M_e = f_y \, W \qquad (8.95)$$

W being the elastic section modulus.

If bending moment is further increased, the plasticization extends towards the centre and stress distribution is modified from a bi-triangular one to a bi-rectangular one. The full plasticization of the cross section corresponds to the formation of a so-called plastic hinge which is characterized by the maximum bending moment the section can bear, expressed by:

$$M_{pl} = f_y \, Z \qquad (8.96)$$

Z being the plastic section modulus.

Modulus Z can be assessed as the sum of the absolute values of the static moments about the plastic neutral axis of the two parts having the same area into which the section is subdivided.

The ratio between plastic and elastic moduli, corresponding also to the ratio between plastic and elastic moments is:

$$\alpha = \frac{Z}{W} = \frac{M_{pl}}{M_e} \qquad (8.97)$$

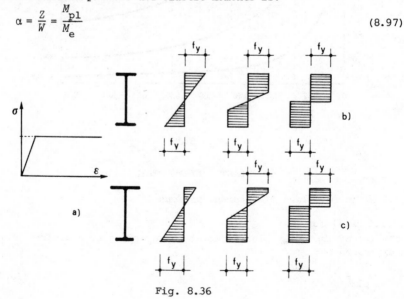

Fig. 8.36

This is called the shape factor of the section and expresses the increase in strength due to the elastic limit being exceeded.

The ratio M_{pl}/M_e can be deduced from the moment/curvature diagrams of the various sections (Fig. 8.37). As the maximum stress cannot exceed f_y, the advantage due to complete yielding is just an extension of the plastic areas near the neutral axis and this is appreciable if the section bulges towards its centre (circular and rhombic shapes). Values of shape factor α, for the most common sections, are |16|:

1.10 to 1.20 for I sections and channels
1.27   for thin walled tubes
1.50   for rectangular sections
1.70   for solid circular sections
2.00   for rhombic sections
2.37   for triangular sections

Resources due to plasticization are thus modest (10 to 20% increase in elastic limit moment) in the sections that are commonly used in steel structures (I sections, channels, tubes, etc.).

Referring to the plastic collapse limit state, the design limit value $M_{d,pl}$ for the bending moment of a cross section can be evaluated by dividing the fully plastic moment M_{pl} (8.96) by 1.12:

$$M_{d,pl} = \frac{M_{pl}}{1.12} \tag{8.98}$$

The coefficient 1.12 is considered by the European Recommendations |8| as an average shape factor for structural steel shapes. In a more general way, the design value of plastic moment is expressed as:

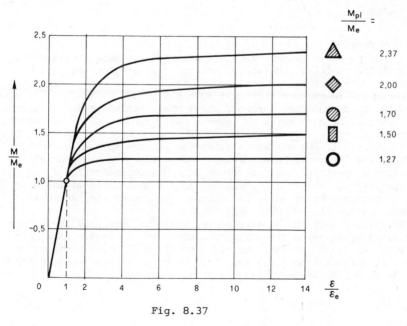

Fig. 8.37

353

$$M_{d,pl} = \frac{M_{pl}}{\gamma_{m,pl}} \qquad (8.99)$$

in which

$$\gamma_{m,pl} = \alpha/\psi \qquad (8.100)$$

where
α is mean shape factor (=1.12)
ψ is a possible plastic adaptation coefficient (see 8.2.3), which in any case must not exceed the value of α

The attainment of the ultimate value M_{pl} can be allowed only under the condition that premature local buckling does not take place in the compressed parts. This is why width/thickness ratio in such parts must be contained within the values set out in 9.6.2.2.

8.3.1.3 Shear
Under the current hypothesis that in I sections and similar shapes the shear force is entirely absorbed by the web with a uniform distribution of shear stresses expressed as:

$$\tau = V/A_{w,n}$$

$A_{w,n}$ being web area net of possible holes, the value of shear producing the complete plasticization of the web is expressed by the following formula:

$$V_{pl} = A_{w,n} \, f_y/\sqrt{3} \qquad (8.101)$$

8.3.1.4 Torsion
It is not possible to determine directly the ultimate conditions in thin walled profiles undergoing torsion. In fact, in the case of primary torsion alone, in the section there are only the tangential stresses due to primary torsional moment T_T, but, in the case of mixed torsion or of secondary torsion, three characteristics are present in the section: primary torsional moment T_T, secondary torsional moment T_ω and bimoment M_ω.
Therefore, if one wished to solve the problem in a completely general way, it would be necessary to determine the yield domain defining compatible limiting values of characteristics T_T, T_ω, M_ω.

Considering the primary torsional moment alone, its ultimate value can be assessed by assuming in collapse conditions the tangential stress distribution shown in Fig. 8.38.

By analogy with the elastic phase, ultimate torsional moment can be evaluated by assuming the collapse moments of the rectangular parts forming the cross section. In the case of an elongate rectangular section, ultimate moment is evaluated by using stress function $F(s,n)$ or the sand pile analogy, leading to:

$$T_{pl} = 2 \int_A F(s,n) \, dA \qquad (8.102)$$

354

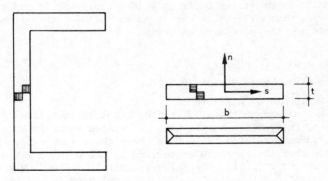

Fig. 8.38

where function F (coinciding with the surface characterizing the shape of a sand pile set on the section) is defined by two areas of constant slope meeting along the axis of the section. The constant slope value is:

$$- \frac{\partial F}{\partial n} = k \, f_y \qquad (8.103)$$

where $k f_y$ is the tangential stress limit, k being provided by the assumed yield criterion. In the case of steel, by analogy with formula (8.101), it can be assumed that $k = 1/\sqrt{3}$.

Hence, for an elongate rectangular section:

$$T_{pl} = k \, f_y \, bt^2/2 \qquad (8.104)$$

and, for a thin walled profile composed by n elongate rectangular elements:

$$T_{pl} = k \, f_y \, \sum_{i}^{n} \frac{b_i t_i^2}{2} \qquad (8.105)$$

8.3.2 *Interaction between internal actions*

In a section undergoing combined bending moment, shear force and axial force, the ultimate bending moment M_u cannot generally attain the value of the plastic moment M_{pl} expressed by formula (8.96).

The reduction of the plastic bending capacity of the section due to components N and V can be determined by applying the static theorem, i.e. by assuming a statically admissible stress distribution.

The following presentation of the influences causing a reduction of the carrying capacity in bending is mainly based on the rules set out in the European Recommendations |8,16|. These rules are applicable to I sections and their reliability is based on experimental studies.

8.3.2.1 Influence of shear force

In the case of symmetrical rolled I sections, bent about the strong axis, it can be assumed that the ultimate moment is not influenced by shear if the latter has been entirely allotted to the web.

355

For doubly symmetrical built up welded profiles, if the shear force $V > 1/3 \; V_{pl}$, where V_{pl} is as expressed in (8.101), the ultimate moment can be assumed to be reduced as follows:

$$M_u = M_{pl}(1.1 - 0.3 \; \frac{V}{V_{pl}}) \qquad (8.106)$$

M_{pl} being given by (8.98) or (8.99).

8.3.2.2 Influence of axial force

In the case of a symmetrical I section imagined to be composed of three rectangles, interaction curves have the shapes shown in Fig. 8.39.

Fig. 8.39a is related to bending about the strong axis. It is noticeable that the rectangular section is the upper limit of the interaction curves, while their lower limit is defined by the line joining points $N = N_{pl}$ and $M = M_{pl}$, corresponding to a section composed of two concentrated masses connected together by a flange having nil area.

Fig. 8.39b is related to bending about the weak axis and, in this case, the rectangular section is the lower limit.

In general, ultimate moment can be expressed as:

$$M_u = k \; M_{pl} \qquad (8.107)$$

where reduction factor k depends on the one hand on cross sectional geometry and, on the other hand, on axial load intensity. I sections used in steel structures have a particular geometry corresponding to a clearly defined band in the $M\text{-}N$ curves of Fig. 8.39. From the point of view of application therefore it is possible to assess the factor k by establishing numerical values for the geometric parameters of practical interest and finding interaction curves nearest the actual ones by means of linear or parabolic formulae.

Such an approximation is proposed in the European Recommendations, where limiting values are given for axial force, below which its influence on the plastic moment can be considered negligible. In the table of Fig. 8.40 such limits are expressed as a function of M_{pl} as expressed by (8.94).

The expressions for reduction factor k, to be applied whenever N exceeds the values set out in the table, are as follows:

Profiles of the HE series
Bending about the strong axis

$$k = 1.11(1 - \frac{N}{N_{pl}}) \qquad (8.108)$$

Bending about the weak axis

$$k = 1 - (\frac{N}{N_{pl}} - 0.2)^2 1.56 \qquad (8.109)$$

Profiles of the IPE series
Bending about the strong axis

$$k = 1.22(1 - \frac{N}{N_{pl}}) \qquad (8.110)$$

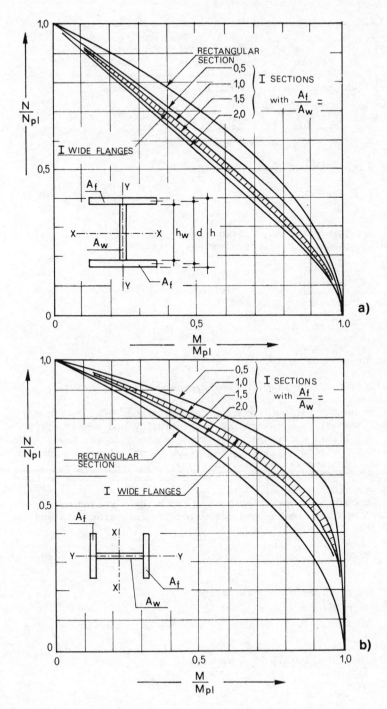

Fig. 8.39

357

Shape	Bending	
	Strong axis	Weak axis
HEA, HEB, HEM	$0.10\,N_{pl}$	$0.20\,N_{pl}$
IPE	$0.18\,N_{pl}$	$0.36\,N_{pl}$

Fig. 8.40

Bending about the weak axis

$$k = 1 - (\frac{N}{N_{pl}} - 0.36)^2 2.44 \qquad (8.111)$$

Alternatively, without any distinction as to the type of profile, the following approximate values have been proposed:

$$k = (1 - \frac{N}{N_{pl}}) \frac{1}{0.9} \quad , \qquad \text{for } N \geq 0.1 N_{pl} \qquad (8.112)$$

in the case of the strong axis and:

$$k = 1 - (\frac{N}{N_{pl}})^2 \frac{1}{0.84} \quad , \qquad \text{for } N \geq 0.4 N_{pl} \qquad (8.113)$$

in the case of the weak axis.

8.3.2.3 Influence of the simultaneous presence of axial and shear forces

The influence on the ultimate moment of axial N and shear V forces acting simultaneously can be studied theoretically and numerically, but relatively little experimental testing has been done. Some tests have shown that, if the ultimate shear value is calculated by formula (8.101) (i.e. ignoring flange contributions), its effects are practically negligible and the only reason for concern is due to the effects of the axial load.

ECCS Recommendations more cautiously permit the simultaneous presence of N and V to be ignored, provided the following relation is complied with:

$$\frac{1}{\gamma} \frac{N}{N_{pl}} + \frac{V}{V_{pl}} \leq 1 \qquad (8.114)$$

where
  $\gamma = 0.10$ for profiles of the HE series
  $\gamma = 0.18$ for profiles of the IPE series

In general, it is still possible to express an analogous relation to (8.107):

$$M_u = k\,M_{pl} \qquad \text{where} \qquad k = k(\frac{N}{N_{pl}}, \frac{V}{V_{pl}}) \leq 1$$

In the case of symmetrical I sections, the linear expressions in the table of Fig. 8.41 may be adopted for k. They lead to the domain which is also shown in the same figure and are on the safe side, provided flange contribution is neglected in the evaluation of V_{pl} by means of formula (8.101).

It should be pointed out that, for $N = 0$ or $V = 0$, the values of k obtained are in any case more conservative than those indicated, respectively, in 8.3.2.1 and 8.3.2.2.

	$V/V_{pl} \leq 1/3$	$1/3 \leq V/V_{pl} \leq 0.9$
$N \leq N_{pl}/11$	$k = 1$	$k = 1.15 - 0.45\, V/V_{pl}$
$N > N_{pl}/11$	$k = 1.10(1 - N/N_{pl})$	$k = 1.25 - 1.10\, N/N_{pl} - 0.45\, V/V_{pl}$

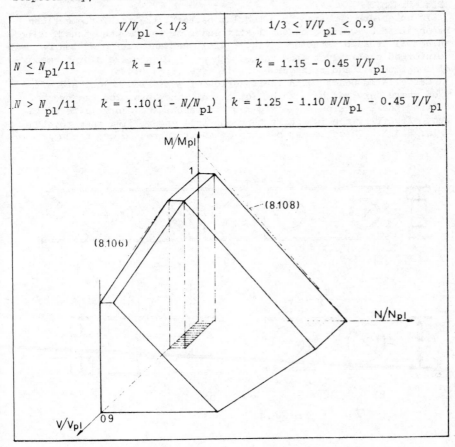

Fig. 8.41

8.4 SECTION WEAKENING DUE TO HOLES

The installation of services often requires holes (usually circular or rectangular) to be cut in the webs of girders.

It thus becomes necessary to control whether the strength reduction and the concentration of stresses due to the presence of the holes can be tolerated, or whether it is necessary to adopt special reinforcements.

In general, in the case of circular holes, if hole diameter d_h is smaller than 1/3 of the web height and if the holes are spaced at a distance greater than $3d_h$, no reinforcement is required. Otherwise, especially in zones mainly subjected to shear, appropriate reinforcing elements must be added to restore the strength of the eliminated part.

A circular hole can be reinforced, for example, by inserting into it a portion of tube welded along its perimeter (Fig. 8.42a), or by welding flat rings to the two faces of the web around the hole (Fig. 8.42b).

If holes are repeated in a regular pattern, the reinforcement can be obtained by welding diagonal stiffeners on to the web, thus forming an actual trussed structure (Fig. 8.42c). Rectangular holes can be reinforced similarly to the methods shown in Figs 8.42a,b or by means of longitudinal and transverse stiffeners (Fig. 8.42d).

Castellated beams, with their series of preformed holes, are often used where services have to be installed. They are produced by cutting the I sections on a zig-zag line about the centre line of the web, offsetting the two parts thus obtained and welding them together (Fig. 8.43). These beams, because of their increased depth compared

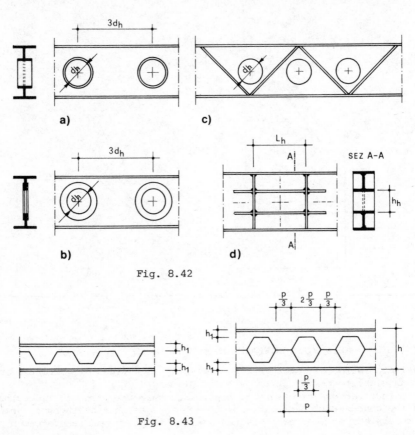

Fig. 8.42

Fig. 8.43

to the original section, also have the advantage of enhanced section modulus without increase in weight.

The problem of hole influence on static and deformative behaviour of members is further examined in the following pages with reference to the limit states previously considered |17|.

8.4.1 *Conventional elastic limit state*

The first surveys on this subject were made with reference to an infinite plate in which stresses were modified by a circular hole |11|. In the case of actual girders, the solution thus obtained is approximate, because the flanges limit the web dimensions. According to |18,19|, it leads to seriously deficient results, mainly whenever large holes and high shear to moment ratios are involved.

An alternative approach was proposed in |20|, also for central circular holes, introducing into the elementary beam theory a suitable stress concentration factor.

A later study |21| showed that, in general, the largest stress values given in |18-20| can be considered sufficiently close to those deriving from an accurate solution. Further comparisons between the two methods and studies on eccentric holes have been carried out and reported in |22|.

The qualitative distribution of the principal stresses surrounding a circular hole in a thick plate is illustrated in Fig. 8.44, together with some experimental results |19|. In the case of thin plates $(h_w/t_w \geq 200)$, the buckling effects reduce compressive stresses and increase tensile ones as shown in Fig. 8.45, derived from some experimental results |23|. The distributions of stresses around a circular hole are referred to two different load levels: '1' in the elastic field; '2' in the elastic plastic field.

In the case of rectangular holes, stresses in connection with the hole can be assessed by superposing bending effects and those due to shear acting as in a Vierendeel girder.

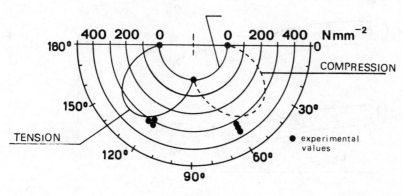

Fig. 8.44

361

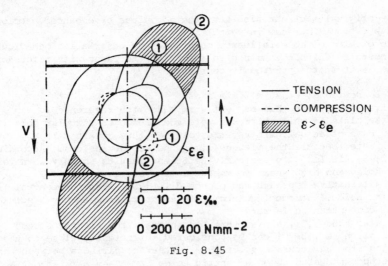

Fig. 8.45

Consider the general case of an I section having unequal flanges and an eccentric rectangular web hole (Fig. 8.46a).

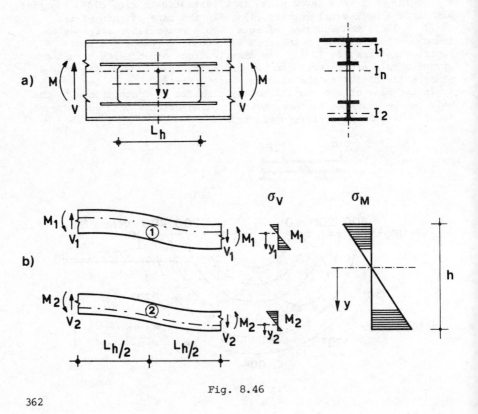

Fig. 8.46

Stresses caused by the bending moment M can be calculated on the net section, on the hypothesis that plane sections remain plane:

$$\sigma_M = \frac{My}{I_n}$$

I_n being the moment of inertia of the net cross section.
Additional stresses due to the shear force V are obtained assuming points of contraflexure in the shapes above and below the hole, according to Fig. 8.46b.

Shear V is distributed between the upper and lower shapes in the proportions V_1 and $V_2 = V - V_1$, in direct proportion to their stiffnesses, according to the following formula:

$$\frac{V_1}{V_2} = \frac{(L_n^2/12EI_2) + (1/GA_{w2})}{(L_h^2/12EI_1) + (1/GA_{w1})} \qquad (8.115)$$

where indices 1 and 2 refer to the upper and lower shape respectively, I being the moment of inertia of the shape with respect to its own centroid and A_w its area.

Secondary stresses due to the Vierendeel effect are expressed by the following formula:

$$\sigma_{V,i} = \frac{M_i}{I_i} y_i = \frac{V_i L_h/2}{I_i} y_i \qquad (i = 1,2) \qquad (8.116)$$

Normal stresses $\sigma_M(y)$ and $\sigma_{V,i}(y_i)$ must be added according to their signs.

Shear stresses in the webs, expressed as an average, are:

$$\tau_i = \frac{V_i}{A_{w,i}} \qquad (i = 1,2) \qquad (8.117)$$

8.4.2 *Ultimate limit state*

8.4.2.1 Shear

In perfectly plastic conditions, ultimate shear capacity of a web having an unreinforced hole (Fig. 8.47) can be assessed by means of the following expression |17,24,25|:

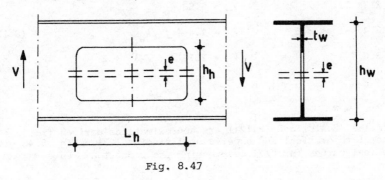

Fig. 8.47

$$V_u = (1 - \frac{h_h}{h_w}) \ V_{pl} \ \sqrt{\beta/(1 + \beta)}$$ (8.118)

where

$$\beta = 0.75 \, (h_w - h_h)^2/L_h^2 \qquad \text{for a concentric hole}$$

or else:

$$\beta = 0.75 \, \frac{(h_w - h_h)^2}{L_h^2} \left\{ 1 + \frac{4e^2}{(h_w - h_h)^2} \right\} \text{ for an eccentric hole}$$

$$V_{pl} = 0.58 \, f_y t_w h_w$$

where

t_w = web thickness
h_w = web depth (flange centroid distance)
L_h = rectangular hole length or 45% of circular hole diameter

Equation (8.118) gives a lower bound estimate of the shear capacity which is conservative compared with experimental results for rolled profiles. It is also applicable in the case of thick webs characterized by a slenderness $\lambda_w < 2.5$ where conventional web slenderness is defined by the following expression:

$$\lambda_w = \frac{h_w}{t_w} \sqrt{\frac{f_y}{E}} \simeq \frac{1}{30} \frac{h_w}{t_w} \sqrt{\frac{f_y}{235}}$$ (8.119)

In the case of thin webs ($\lambda_w > 2.5$), web buckling must be taken into account and the value of shear capacity is [17,23]:

$$V_u = \begin{cases} \{1 - (h_h - 0.31L_h)/h_w\} V_V & \text{for rectangular holes} \\ (1 - d_h/h_w) V_V & \text{for circular holes, diameter } d_h \end{cases}$$ (8.120)

where

V_V is the shear capacity with regard to web buckling for the unperforated web, related to slenderness λ_w by the following:

$\lambda_w \leq 2.4$	$V_V = 0.58 \, A_w f_y$
$2.4 < \lambda_w \leq 3.6$	$V_V = 1.4 \, A_w f_y/\lambda_w$ (8.121)
$3.6 < \lambda_w$	$V_V = 2.22 \, A_w f_y/(\lambda_w + 2)$

where

$$A_w = t_w h_w$$

Formulae (8.120) and (8.121) are approximations derived from the results of theoretical and experimental studies [23]. Fig. 8.48 gives theoretical curves for V/V_{pl} expressing the reduction of shear capacity

364

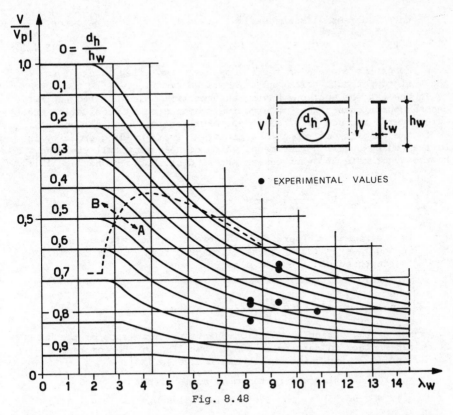

Fig. 8.48

compared to the full yielding value V_{pl} as a function of slenderness λ_w and of the ratio d_h/d_w, in the case of a circular hole. In the same figure, two validity areas of the above formulae are shown dashed, area B (low slenderness) relates to expression (8.118), while area A (mean and high slenderness) relates to (8.120).

8.4.2.2 Bending and shear
In the case of compact sections (IPE and HE types), ultimate bending capacity on the hypothesis of a fully yielded section (rectangular distribution of stresses) is $|17|$:

$$M_u = M_{pl}\left\{ 1 - \frac{(h_h + 2e)^2/h_w^2}{1 + 4\,A_f/A_w} - k_1\,\frac{V}{V_u} \right\} \qquad (8.122)$$

V_u being expressed by (8.118)
A_f and A_w being flange and web areas, respectively
V being applied shear in the section through the hole
M_{pl} being full plastic moment of the unperforated section

$$k_1 = \frac{1 + 3(L_h/h_w)(0.7 - h_h/h_w)}{1 + 4\,A_f/A_w} \qquad (8.123)$$

365

Factor k_1 agrees with suggestions given in $|24|$ for the range

$$0.25 < h_h/h_w < 0.75 \qquad (8.123)$$

even though the expression for k_1 has been simplified in order to adapt it to for both thick and thin walled profiles.

In the case of sections having an eccentric hole, bending capacity can be approximately and conservatively evaluated by considering the existence of a concentric hole that inscribes the eccentric one.

In the case of non compact sections, ultimate bending capacity is influenced by buckling phenomena, which do not allow full yielding. On the hypothesis of a triangular stress distribution, one has $|17|$:

$$M_u = k_2 k_3 M_e \left\{ 1 - \frac{t_w (h_h + 2e)^3}{12 I_x} - k_1 \frac{V}{V_u} \right\} \qquad (8.124)$$

where

I_x is the moment of inertia of the cross section without hole
V is the lower value of those calculated from (8.118) and (8.120)
M_e is the unreduced limit of elastic moment

$$k_3 = \begin{cases} 1, \text{ for } \lambda_w \leq 4.8 \\ 1 - 0.15 \dfrac{A_w}{A_f} (1 - 4.8/\lambda_w), \text{ for } \lambda_w > 4.8 \end{cases}$$

k_2 expresses the strength reduction due to the existence of the hole, which takes place in particular due to buckling of the compressed parts, the following values have been indicated $|23|$:

$k_2 = 1$ for circular and rectangular holes with $\lambda_w \leq 4.2 \, h_w/L_h$

$\quad = 1.13 - 0.03 \, {}_w L_h/h_w$ for rectangular holes with $\lambda_w > 4.2 \, h_w/L_h$

Design tables and a more deep discussion of theoretical and experimental results may be found in $|25\text{-}28|$.

8.4.3 *Serviceability limit state*

The existence of a hole in a beam causes an increase in deformability for the beam itself. It is thus necessary to take it into account when calculating the maximum displacement to be considered in checking the serviceability limit state.

The deflection of a beam with a hole can be determined by adding to the vertical displacement of the beam without holes both the contributions due to bending and shear deformations of the sections above and below the hole and those due to the locally reduced bending stiffness (Fig. 8.49).

The first effect (Fig. 8.49c) provides the following vertical displacements:

$$\Delta v(z) = - v_v \, z_1/L \qquad \text{for } 0 \leq z_1 \leq a - L_h/2$$

$$+ v_v \, z_2/L \qquad \text{for } 0 \leq z_2 \leq b - L_h/2$$

respectively to the left and to the right of the hole, as:

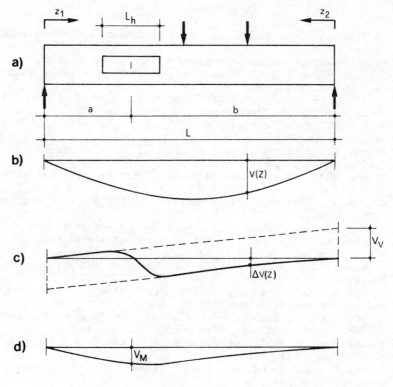

Fig. 8.49

$$v_V = V_1 \, L_h \, (\frac{L_h^2}{12EI_1} + \frac{1}{GA_{w,1}})$$ (8.125)

Formula (8.125) is strictly valid only if the hole is concentric and then the terms V_1, I_1 and $A_{w,1}$ in it can be related to either the upper or the lower section. If the hole is eccentric, the same terms can be related in close approximation to the part having the smaller area $A_{w,1}$ |17|.

The second effect (Fig. 8.49d) can be interpreted as a concentrated reduction of bending stiffness, producing an additional displacement concentrated at the centre of the hole, expressed as:

$$v_M = ML_n \, (\frac{1}{EI_n} - \frac{1}{EI}) \, \frac{ab}{L_h}$$

where

I_n and I are the moments of inertia of the sections with and without the hole respectively

M is the bending moment at the hole centre

367

REFERENCES

1. CM (1966) *Regle de Calcul des Constructions en Acier* - Eyrolles, Paris
2. Council on Tall Buildings, Group SB (1979) *Structural Design of Tall Steel Buildings*, Volume SB of Monograph on Planning and Design on Tall Buildings, ASCE, New York
3. Finzi, L. and Nova, E. (1969) *Elementi strutturali*, CISIA, Milan
4. McGuire, W. (1968) *Steel Structures*, Prentice Hall, New York
5. AISC (1980) *Specification for the Design, Fabrication and Erection of Structural Steel for Buildings*, AISC, Chicago
6. Bijlaard, P.P. (1940) Theory of local plastic deformation, *IABSE Publications*, 6, 27-44
7. Nelson, H.M. (1953) *Angles in Tension*, BCSA Pubblication No. 7, London
8. CECM ECCS (1978) *European Recommendations for Steel Structures*, ECCS, Brussels
9. De Martino, A., Faella, C. and Mazzolani, F.M. (1981) The use of plastic adaptation coefficients in the limit state design of steel shapes, *Costruzioni Metalliche*, 3, 120-132
10. Vlassov, B.Z. (1962) *Pièces Longues en Voiles Minces*, Eyrolles, Paris
11. Timoshenko, S.P. and Goodier, J.N. (1970) *Theory of Elasticity*, 3rd Ed., McGraw Hill, New York
12. Terrington, J.S. (1968) *Combined Bending and Torsion of Beams and Girders*, (1st & 2nd part) BCSA Publication No. 31
13. Mazzolani, F.M. (1972) *La torsione nei profilati e nelle travi metalliche*, CISIA, Milan
14. Roark, R.J. and Young, W.C. (1975) *Formulas for Stresses and Strain*, McGraw Hill, New York
15. Bleich, F. (1952) *Buckling Strength of Metal Structures*, McGraw Hill, New York
16. Massonnet, Ch. and Save, M. (1967) *Calcul Plastique des Constructions*, Vol. I, Centre Belgio Lux d'Inf. de l'Acier, Brussels
17. CECM, ECCS, EKS, *Steifenlose Stahlskelettragwerke und Dünnwandinge Vollwandträger*, Ernst & Sohn, Berlin, Munchen, Dusseldorf
18. Bower, J.E. (1966) Elastic stresses around web holes in wide flange beams, *ASCE Journal of the Structural Division*, 92, ST2, 85-101
19. Bower, J.E. (1966) Experimental stresses in wide flange beams with holes, *ASCE Journal of the Structural Division*, 92, ST5, 167-188
20. Sahmel, D. (1969) Konstruktive Ausbildung und Näherungsberechnung Gescheweisster Biegeträger und Torsionsstäbe mit grossen Stegausnehmungen, *Schweissen und Schneiden*, 4,
21. Chan, P.W. (1971) *Approximate Methods to Calculate Stresses around Circular Holes*, Report to Canadian Institute of Steel Construction, Project 695
22. Chan, P.W. and Redwood, R.C. (1974) Stresses in beams with circular eccentric web holes, *ASCE Journal of the Structural Division*, 100, ST1, 231-248
23. Höglund, T. (1970) *Bärförmaga hos Tunnväggig I-balk med Cirkulärt eller Rektangulärt Hal i Livet*, Meddelande No. 87, Institutionen för Byggnadsstatik, KTH, Stockholm

STRENGTH OF STRUCTURAL ELEMENTS

24. Redwood, R.G. (1969) The strength of steel beams with unreinforced web holes, *Civil Engineering and Public Works Review*, <u>64</u>, June, 559-562
25. Redwood, R.G. (1972) Tables for plastic design of beams with rectangular holes, *Engineering Journal*, AISC, 1, 2-19
26. Redwood, R.G. (1973) *Design of Beams with Web Holes. Canadian Steel Industries*, Constructional Council, Ontario
27. Redwood, R.G. (1978) Analyse et dimensionnement des poutres ayant des ouvertures dans les âmes, *Construction Métallique*, No. 3, 15-27
28. Redwood, R.G. (1978) *Dimensionnement du Renfort d'Âme des Poutres Comportant une Ouverture*, CRIF Centre de Recherches Scientifiques et Technique de l'Industrie des Fabrications Métalliques, NT 16, Liege

C H A P T E R   N I N E

Stability of Structural Elements

9.1 GENERAL REMARKS

9.1.1 *Historical introduction*

An interesting and meaningful introduction to the complex and vast
theme of the stability of structural steel elements could be
provided by a historical survey (Fig. 9.1) |1,2|, pointing out which
steps have been progressively made in this field and, at the same
time, allowing one to follow the various tendencies that have arisen in
the past. The survey, which follows has been made with reference
to the problem of the *column*, which can be considered as the
structural element that has been most accurately studied in the
history of building.

The first qualitative remarks on column strength and stability
were given by Erone of Alexandria (75 B.C.) and similar descriptions
of buckled columns were found in drawings of Leonardo da Vinci (1452-
-1519).

P. Van Musschenbroek's (1693-1761) studies are less well known.
He was the first to search for physical models for materials and
structural elements. He defined the constitutional properties of
materials (hardness, elasticity, flexibility) and, by systematic
experiments on columns, he found empirical formulae which show that
the critical load is inversely proportional to the square of the
length.

D. Bernoulli's (1700-1782) studies on the elastic line inspired
L. Euler's (1707-1783) researches. In addition to his famous formula
for the elastic critical load, Euler had a foresight of advance
issues, such as the post-critical problems in the field of large
deformations and the non-elastic behaviour of materials, which are
issues that have been recently reconsidered by modern theories on
stability.

In his first paper (1774), in fact, Euler directed his theory to
the post-critical behaviour of a structure and, as a particular case,
he determined the serviceability limit state for a column due to
large displacements under the *critical load:*

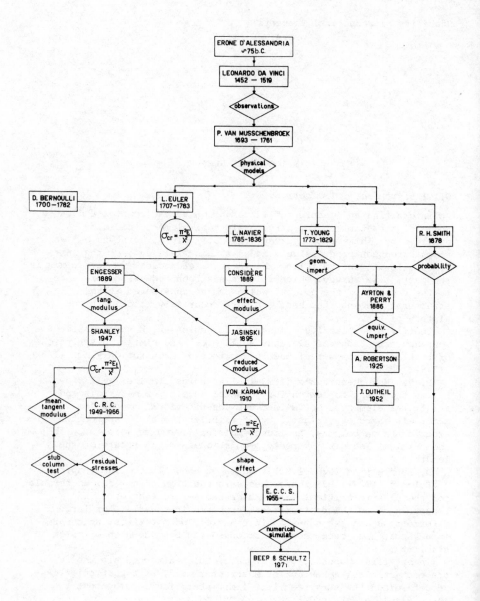

Fig. 9.1

$$N_{cr} = \frac{\pi^2 B}{L^2} \tag{9.1}$$

He defined the term B, presently known as bending stiffness (EI), first as 'absolute elasticity' and later on (1775) as 'stiffness moment' (moment de roideur) "... *because it occurs in all bodies that resist flexure, whether they are elastic or not*". Concerning the determination of stiffness EI, he proposed bending tests on a beam having identical boundary conditions as the column: this can be considered as the first attempt at formulating an inelastic theory with a constitutive equation relating moment to flexural stiffness. Unfortunately, neither Euler himself nor his contemporaries followed up this concept and about 200 years had to pass before his successors rediscovered it.

In his third paper (1778), Euler introduced the classical elastic formula:

$$N_{cr} = \frac{\pi^2 EI}{L^2} \tag{9.2}$$

where I is defined as the moment of inertia of the cross section and E as the elastic modulus of the material, the numerical values of which were derived by Euler from the experiments of Van Musschenbroek.

D. Bernoulli (1700-1736) included other restraint conditions in the problem of elastic column buckling. He also performed some buckling tests on timber columns, but he was not satisfied with the comparison between theory and test results.

Some time later L. Navier (1785-1836) analysed his contemporaries' experiments and found that Euler's critical load provided an upper bound to experimental results. So he defined the validity limit of Euler's formula for various structural materials (timber, cast iron, iron) by means of the so-called *slenderness limit* or *proportionality slenderness*, expressed by (Fig.9.2):

$$\lambda_e = \pi \sqrt{E/f_e} \tag{9.3}$$

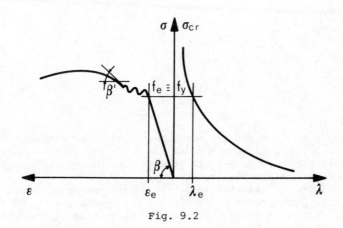

Fig. 9.2

f_e being the elastic limit of the material separating slender struts ($\lambda > \lambda_e$), from stocky struts ($\lambda < \lambda_e$), for which failure takes place mainly due to material crushing.

T. Young (1773-1829) had already pointed out for the first time the importance of *imperfections* for column strength, when he was trying to find an explanation for the differences between theory and practice. Young noted both the difficulty in applying perfectly centred loads and the causal non-homogeneity of materials: *considerable irregularities may be observed in all the experiments ... ... and there is no doubt but some of them are occasioned by the difficulty in applying the forces precisely at the extremities of the axis and others by the accidental inequalities of the substances, of which the fibres must often have been in such directions to constitute originally rather bent than straight columns.*

He then gave a rough but correct analysis of a column having an initial curve and load eccentricity. He considered a strut affected by an initial sinusoidal imperfection having amplitude v_o and bearing an axial load N. In addition, this strut is bent by a second order moment, expressed by the relation:

$$M = \frac{Nv_o}{1 - N/N_{cr}} \tag{9.4}$$

where $1/(1 - N/N_{cr})$ indicates the amplification coefficient of the mid-span deflection v_o, N_{cr} being Euler's critical load for the strut. Young postulated that the limit load N_c for such a column is attained whenever normal stress in the most loaded fibre reaches the strength limit value:

$$\frac{N_c}{A} + \frac{N_c v_o}{(1 - N_c/N_{cr})W} = f_{lim} \tag{9.5}$$

where: A and W are, respectively the area and the section modulus of the cross-section of the strut.

By considering the strength limit f_{lim} equal to the material elastic limit f_e and σ_c and σ_{cr} being, respectively, the maximum compressive stress and Euler's critical stress, (9.5) can be expressed as follows:

$$\sigma_c + \frac{\sigma_c v_o}{1 - \sigma_c/\sigma_{cr}} \frac{A}{W} = f_e \tag{9.6}$$

or else:

$$(f_e - \sigma_c)(\sigma_{cr} - \sigma_c) = \eta \sigma_c \sigma_{cr} \tag{9.7}$$

where

$$\eta = v_o \frac{A}{W} \tag{9.8}$$

indicates only the effect of the geometrical imperfection.

As one of Young's successors at the Royal Institution (Rayleigh) recognized a century later, his studies were not appreciated by his contemporaries as they deserved.

This concept of imperfections, which has been reconsidered in modern times, received a *probabilistic interpretation* thanks to R.H. Smith (1878), who noticed that strength scatterings are a matter of probability, not of 'exact' theory. Following the same direction, his successors Ayrton and Perry (1886) proposed practical expressions for *generalized or equivalent imperfections*, that were later on considered in A. Robertson's formulations (1925). Ayrton and Perry proposed adopting an expression for an imperfection parameter η of the following type:

$$\eta = k\lambda \tag{9.9}$$

defining it as proportional to strut slenderness.

Some years later J. Dutheil (1946) assumed an imperfection parameter of the following type as the basis of his theory:

$$\eta = k\lambda^2 \frac{f_e}{\pi^2 E} \tag{9.10}$$

Resuming Smith's introduction, Dutheil's formulation aimed at interpreting it in a statistical form.

Dutheil's approach was the determinant in the safety philosophy adopted by the present French code and it influenced the basis of the ECCS-CECM researches.

It should be pointed out that, from a general point of view, parameter η adopted in Young's formula (9.7) can be applied to interpret not only the effect of a geometric imperfection - such as the initial curvature that was first examined by Young - but also the effect of mechanical imperfections, such as residual stresses and elastic limit scatter through the cross section.

Thus equation (9.7) can be written in the more general modern non-dimensional form:

$$(1 - \bar{N})(1 - \bar{N}\bar{\lambda}^2) = \eta\bar{N} \tag{9.11}$$

where:

$$\bar{N} = \sigma_c/f_y \quad ; \quad \bar{\lambda} = \lambda/\lambda_e \quad ; \quad \lambda_e = \pi\sqrt{E/f_y}$$

yield stress being considered as the elastic limit. Form (9.11) is still used to numerically interpret buckling curves by choosing a suitable function η considering all the imperfections existing in the strut (see 9.2.2.3).

In parallel with such researches, different authors looked back at Euler's formula with the aim of introducing the elastic-plastic characteristics into the critical phenomenon. Engesser (1889) proposed the first method of determining critical stress in the field of slenderness ($\lambda < \lambda_e$) characterizing stocky struts (Fig. 9.2). He proposed replacing the elastic modulus E, defined as:

$$E = \sigma/\varepsilon = \tan \beta \tag{9.12}$$

by an instantaneous modulus:

$$E_t = \frac{d\sigma}{d\varepsilon} = \tan \beta_1 \tag{9.13}$$

called the *tangent modulus*, according to which the ultimate load in the elastic-plastic field is simply given by:

$$N_c = \frac{\pi^2 E_t I}{L^2} \tag{9.14}$$

375

This procedure was succesively criticised by Considère (1889) and Jasinski (1895) who pointed out the need to take into account elastic unloading by introducing an *effective modulus* intermediate between the elastic and tangent moduli. In fact, due to lateral deflection, part of the section of a strut is further compressed, and part is unloaded. The former can follow the law of the tangent modulus, while the latter during unloading will behave according to a modulus which is practically equal to the elastic one. Therefore, the over-all behaviour of the strut beyond the elastic limit, during the incipient lateral buckling, is ruled by an intermediate modulus between E (9.12) and E_t (9.13).

This concept was more precisely defined by the *reduced modulus* proposed by T.H. Von Karman (1910), which led to the following expression for ultimate load:

$$N_c = \frac{\pi^2 E_{red} I}{L^2} \tag{9.15}$$

Von Karman's theory makes it possible to determine E_{red}, the value of which, for example, is:

$$E_{red} = \frac{4E\, E_t}{(\sqrt{E} + \sqrt{E_t})^2} \qquad \text{for rectangular sections} \tag{9.16}$$

$$E_{red} = \frac{2E_t + E}{E_t + E} \qquad \text{for symmetrical I-sections} \tag{9.17}$$

The latter author pointed out the influence of the form of the cross section (*shape effect*) on the buckling phenomenon.

The use of the *tangent modulus* was later revalued by F.R. Shanley (1947), who gave it a new theoretical justification based on the observation that when the bifurcation takes place the load increment process continues without any elastic unloading and the phenomenon is essentially governed by the tangent modulus, while the reduced modulus can never be attained (Fig. 9.3).

Duberg and Wilder (1950) extended Shanley's theory to an I-section idealized into two concentrated masses.

The application of Shanley's theory, which represents the arrival point of the trend begun by Euler, appears to be justified by the possibility of comprehensively taking into account *structural imperfections* (residual stresses; non-homogeneous distribution of mechanical properties) characterizing industially produced steel profiles. This tendency was followed by the Column Research Council (CRC) |3-5| which in the period 1949-1966 undertook a vast programme of experiments in order to measure structural imperfections and inaugurated the overall compression test on steel shapes (*stub column test*) (see 4.4.5), as an experimental instrument replacing the tensile test on individual specimens, capable of giving an answer on the load versus deformations relationship concerning the whole cross section, by considering the effect of all the existing imperfections. The *average tangent modulus* obtained by that test

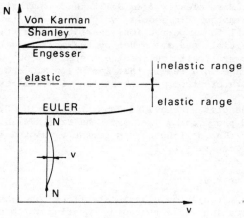

Fig. 9.3

was applied in Shanley's theory for the bifurcation study of column stability.

From the foregoing survey it is possible to recognize, even in the remote past, some trends that, developing spontaneously and independently, have been reproposed in a more modern version and investigated both by single authors and by international organizations (SSRC; ECCS) with the purpose of theoretical research or of issuing codes and recommendations |6-11|.

The concepts which have arisen indicate the main up-to-date aspects in the steel construction field: inelastic behaviour, the post buckling behaviour, geometrical imperfections, structural imperfections, the probabilistic approach. With this perspective, successive steps have led to the unquestionable abandonment of the concept of a perfectly straight *theoretical bar* made of isotropic and homogeneous material and to the adoption of the more realistic concept of an *industrial bar* affected by inevitable imperfections of an accidental nature, the characterization of which requires an analysis of a probabilistic-simulative type.

From the point of view of calculation methods, as an alternative to the various interpretations belonging to methods based on the equilibrium bifurcation, procedures of the incremental type based on numerical simulation are more and more being applied. The interest in this kind of approach is justified by the possibility of obtaining many answers to the most urgent problems which on the one hand, require the interpretation of inelastic material behaviour and hardening effects and, on the other hand, have to take into account the effects of geometrical and structural imperfections characterizing industrial bars.

With reference to such requirements, *simulation methods* allow a great versatility than the bifurcation methods. The latter in fact are not effective for the interpretation of the actual shape of the bar, while they can consider only 'as an average' the material dishomogeneity through a suitable choice of an average tangent modulus.

It will appear more clearly from 9.1.2 that simulation methods substantially use mathematic models, which allow the introduction into the bar of all the imperfections characterizing it, as input data. If such initial data are given by the statistical interpretation of systematic full scale tests, the output of the simulation procedure is a result that can be considered equivalent to that of a laboratory test, and therefore of a semi-experimental nature.

The above tendencies have guided Committee 8 of ECCS, where the simulation method proposed by Beer and Schultz (1971) has permitted the drawing up of the buckling curves recently adopted in the codes of many European Countries |12|.

9.1.2 *Simulation methods*

The development of studies on elastic stability based on the equilibrium bifurcation was very rapid, covering up to now all problems having a practical interest. Such a success is essentially due to the simplicity of the method which, once the principal equilibrated configuration is known, always leads to the search for the highest values of a linear problem. When, however, even the simplest case of stability, i.e. that concerning compressed struts, is more closely examined, one notices that the actual problem is much more complex than that appearing through a theory based on the concept of equilibrium bifurcation. A true strut is always affected by imperfections due to the load being not perfectly centred. Consequently, as load increases, equilibrium configurations progressively depart from the initial one, first in the elastic and then in the elastic-plastic phase, until equilibrium is no longer possible.

The phenomenon of divergence is closely connected with the plasticization of the most loaded area in the strut. Therefore, elastic-plastic behaviour of the cross section is extremely important in this phenomenon. Consequently, the ultimate value of the load depends not only on the elastic modulus and on strut slenderness, as in Euler's field, but also on yield stress, section shape and residual stress distribution. Determination of the quantitative effects of these factors on the load bearing capacity of a strut is essential.

The study of strut stability according to this approach is obviously very difficult, as non-linear problems are involved. On the other hand, at the present state of knowledge, it seems the only suitable way of follow in order to investigate more accurately the actual behaviour of structures.

The first level of approximation consists in reducing the problem to only one degree of freedom, assuming a priori the deflection configuration and checking the elastic-plastic equilibrium condition at mid-span.

The natural extension of this simplified approach is to consider systems with several degrees of freedom and to solve the non-linear problem according to two alternative criteria:

(a) To assume an external load multiplier as an independent variable

378

and to determine the equilibrium

(b) to assume a displacement component as an independent variable and to determine the corresponding equilibrium configuration and its external loads.

The methods assuming a displacement parameter as an independent parameter (b) have the advantage of being fit to proceed according to constant increases of such a parameter. The methods assuming axial load as an independent parameter (a) must render the increases of N adequate for the inclinations of the $N - v$ diagram in order to avoid convergence difficulties in proximity of the maximum value N_c, where the tangent to the diagram itself becomes horizontal.

According to the first type of approach (a), one initial strut configuration is known and by iterating on the elastic-plastic system stiffness matrix, one looks for the equilibrium configuration corresponding to given axial and bending external loads.

In particular, it is possible to repeat calculations by varying the external load multiplier and to assess its maximum value corresponding to the stiffness matrix no longer being defined positive.

Interaction curves for beam-columns are obtained by maintaining axial load constant and varying bending |13-15|. Load bearing capacity of a compressed strut is determined |16-17| by varying axial load, bending actions being equal.

The methods which follow this type of approach are based on determining the stiffness matrix of the structure in the elastic-plastic phase and therefore they can interpret the phenomenon only until the value of the load parameter for which the matrix becomes singular. Local and overall elastic unloading processes are thus excluded, as well as the analysis of the post-buckling phase.

The second type of approach (b) has been applied both to determine interaction curves for beam-columns and to evaluate different aspects of load bearing capacity of simply compressed struts.

Assuming the transverse component of the mid-span displacement as an independent variable, it is possible to determine the equilibrium configuration in respect of a given value of the axial load and thus to deduce eccentricity at the ends related to the values of the end moments. It is then possible to determine also the unstable branch of the load process, still assuming that local elastic unloadings are absent. One thus arrives directly at the equilibrium configuration without iterative processes |18-19|.

The method followed by Batterman and Johnston |20| was specifically set up for determining the load bearing capacity of simply compressed struts. Its essential lines are the following. Starting from a given equilibrium configuration, one supposes as a first approximation that the deformation corresponding to the next step is similar to the preceding one and defined by a displacement increase of the mid-span. This allows the determination of curve variations in connection with the various strut section considered in the calculation.

Once the curve has been chosen, the equilibrium conditions are satisfied and bending moment and axial load values are determined by iteratively assigning the neutral axis positions. Once this first series of iterations is completed, equilibrium has been satisfied

379

but for values of axial load which differ in the various strut
sections. One then determines its mean value and the resulting
deviations, in terms of which the deflection is corrected until
constant values of axial load are obtained (within the desired
tolerance). The external force is thus known and it is possible to
define the relations between the external force and the stress and
deformation parameters, always on the hypothesis that elastic
unloading is absent.

The method proposed by F. Frey and Ch.Massonet |21| follows a
substantially analogous procedure. In fact, starting from a generic
equilibrium situation, it applies a similar increase of the deflection,
consequently varying the curvature in each section of the strut.
From the newly obtained curvature one iterates on the position of
the neutral axis until equilibrium is satisfied in each section.
This entails values of N varying along the strut axis. The
condition that the axial force be constant along the strut is
satisfied by iteratively correcting the curvature. This method is
able to include any type of imperfection, as well as to take into
consideration local elastic unloadings.

A substantially different approach was developed by H. Beer and
C. Schulz (1969) |22-24| within the range of the theoretical and
experimental researches of Committee 8 of ECCS for the purpose of
determining buckling curves for compressed struts. It operates as
follows: for a given type of cross section, characterized by its
geometry and its residual stress distribution, one first of all
evaluates the relationship $M = M(\chi)$ between bending moment M and
curvature χ, for given values of axial force N acting on the section.
Once the relationship and the intial imperfection value are known,
one operates a model of a continuous strut having a fixed initial
curvature. The strut is divided lengthwise into stubs and the axial
force N is maintained constant, assuming slenderness and maximum
transverse displacement as unknown quantities. By a double
iteration procedure it is possible to determine, for a given axial
load N_c, the relationship $\lambda - v_o$ between slenderness λ and mid-span
displacement v_o. It is then possible to evaluate the maximum slenderness λ_{max} that is compatible with the pre-determined axial load N_c.
Point (N_c, λ_{max}) is therefore a point on the buckling curve. This
method is very effective because it maintains axial force N constant
during the whole calculation and so makes it possible to operate in
two distinct or independent phases: one concerning the calculation
of the moment-curvature relationship and the other concerning the
calculation of maximum slenderness compatible with a predetermined
axial load. This allows a reduction of the programme and of
calculation time. The above procedure is swift, but scarcely applicable to the study of more complex problems in the post-buckling range
that take into account local or overall unloadings. It is therefore
limited to merely technical-applicational purposes.

Ballio, Petrini and Urbano |25| proposed a general method assuming
an external load multiplier as an independent variable of the
incremental process. Such a method therefore belongs to type (a),
and it overcomes the numerical difficulties that, as already mentioned,
are found in connection with this type of procedure in proximity of

maximum load. As the system is differentiated with respect to the
load parameter, a differential equation at the initial values is
formulated and it is thus possible to find the solution by means of
a numerical integration, applying commonly used and available
techniques.

Finally, mention should be made of the procedure proposed by
Faella and Mazzolani |26| for the purpose of obtaining a calculation
instrument that, without setting aside   generalization
requirements of the Ballio-Petrini-Urbano method, offered the
advantages of the quicker convergence obtainable with the type (b)
methods, i.e. those assuming a displacement parameter as an
independent variable. Furthermore the material constitutional law
$\sigma - \varepsilon$ can be totally generic, various categories of steel can be
interpreted and the Bauschinger effect can be taken into account
in the case of cyclic loads.

9.1.3 *Present situation*

In recent times, the definition of critical load has undergone a deep
evolution, substantially due to the following remark: as the applied
load multiplier α increases, a structure or a structural element
passes from the elastic range to the elastic-plastic one until
collapse is attained (collapse being defined by the compatible and
equilibrated situation for which the load multiplier reaches its
maximum value α_u). It therefore appears that there is no longer
much sense in computing a critical multiplier α_{cr}, characterizing
- at least in the case of conservative loads - only a situation of
bifurcation of the elastic equilibrium for the structure.

There are however two classes of difficulty to be faced when
determining the collapse multiplier α_u. Such a multiplier heavily
depends on the magnitudes defining:
   Structure geometry (geometrical imperfections, load eccentricity)
   Material mechanical properties ($\sigma - \varepsilon$ law)
   Cross sectional properties (shape, residual stress and elastic
   limit distribution)
Therefore, one no longer operates on a *perfect* model, such as for
example Euler's strut (perfectly straight, having a constant section
and bearing a perfectly centred axial load) but one must consider
models *affected by imperfections*. On the other hand, such
imperfections cannot be considered as deterministic variables, i.e.
known a priori, but only as aleatory variables. Probabilistic
analysis thus becomes the only approach capable of rigorously solving
the problem.

The difficulties connected with such a method are so great that
only recently ECCS achieved some conclusions from twenty-year long
studies directed at finding a suitable solution to the problem of
determining the load bearing capacity of pin-ended compression
members made of structural steel |24,27-29|.

However, in general it cannot be expected that such methods will
shortly solve all the buckling problems characterizing structural
elements and structures as a whole. There are too many parameters

381

defining the phenomenon and sometimes the analytical approach in the
non-linear field is too complex to give practical results, even in a
deterministic way only. One must therefore accept to operate with
a certain degree of approximation, following methods that past
experience has shown to be on the safe side.

Consider, for example, the classical stability problem shown in
Fig. 9.4. Once a structural element (Fig. 9.4a) or a structure (Fig.
9.4b) is given, how is it possible to determine with sufficient
accuracy the maximum load bearing capacity F_c and its influence $N(F_c)$
on the values of the reaction of the restraints and the internal
forces in the various elements? The methods currently available for
solving such a problem are summarized in Fig. 9.5 $|30|$.

A general method is indicated in the first column. It entails a
vast experimental survey and a statistical exploitation of the results.
It is applicable only for a large organization and solves the problem
completely, as it allows a statistical evaluation of the maximum load
F_c and its effects $N(F_c)$, depending on displacement components and
therefore on initial geometrical imperfections. This method was
adopted to define the load bearing capacity of pin-ended compression
members $|31|$ and of individual angle bars of transmission towers $|32|$
(see 9.1.3.1).

The method indicated in the second column is also based on
experimental tests, but only on a limited number of them, as they
are partially replaced by the results of numerical simulation
calculations. This method therefore provides reliable results,
although they are not complete from a probabilistic point of view.
Although the mean value of F_c and of $N(F_c)$ are determined with
reasonable accuracy no information is given on their statistical
distribution (see 9.1.3.2).

In the third column and its subgroups are listed several methods
up to now indicated in the codes for computing load bearing capacity
although they are based on the determination of Euler's critical
load F_{cr} of the structural system, which is therefore assumed to be
free from imperfections and made of perfectly elastic material. Once
that value has been computed, it is possible to operate according to

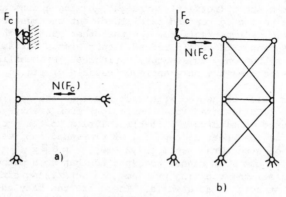

Fig. 9.4

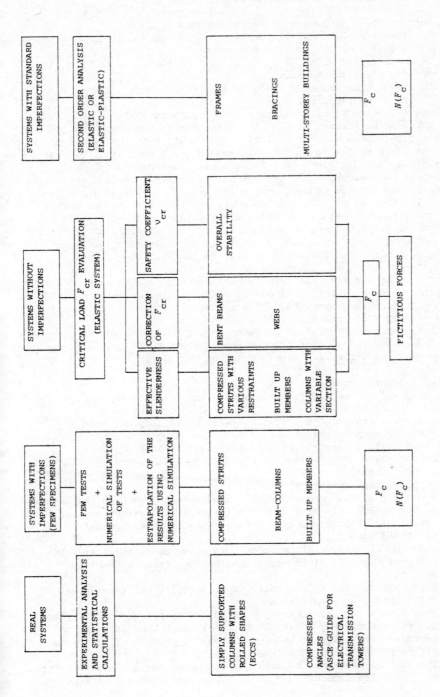

Fig. 9.5

one of the three following alternatives (see 9.1.3.3):

To determine the ideal slenderness of the system, which makes it equivalent to a simple pin-ended strut

To correct that value (F_{cr}) in order to take account of material elastic-plastic behaviour

To divide that value (F_{cr}) by an appropirate safety factor ν_{cr} chosen on an empirical basis

The foregoing methods indicate an approximate value of the load F_c that can be carried by the structure, but do not allow the evaluation of the effects $N(F)$ as a contribution of the reactions of the internal forces. For this reason, when restraints or bracing structures are designed, calculations must consider *fictitious external forces*, which are not due to the external loads, but simulate an average of the unstable effects of the actual loads (see 8.1.2). The method referred to in the fourth column can be useful in studying the overall stability of structures, even though it often entails considerable difficulties in calculation. A structure affected by imperfections of a predetermined value is calculated taking account of deformation effects in the elastic or, better, elastic-plastic field. Even though imperfection values are not confirmed by experimental results, this method determines the maximum capacity load F_c (or F_{cr}, if one is operating in the elastic field) and displacement components, which are useful in order to determine the effects $N(F)$ (see 9.1.3.4).

Some general information is given in the following sub-sections (9.1.3.1 to 9.1.3.4) on the four trends to which all operative methods used in steel structure calculations can be related.

9.1.3.1 Actual systems
The only refined method for determining the load bearing capacity of members affected by buckling phenomena is to perform an experimental analysis on a statistically reliable population of specimens. Such an investigation was planned by ECCS for evaluating the load bearing capacity of pin-ended axially loaded struts.

To emphasize all the aspects of strut behaviour, the experimental programme contained:

Buckling tests on actual columns

A measure of initial geometrical imperfections, such as the scatter of the actual axis line from the theoretical straight one and variations in cross sectional dimensions from the nominal tabulated one

A measure of structural imperfections, such as residual stresses and variations of yield stress and rupture through the cross section

Stub column tests to determine experimentally the mean value of yield stress in compression

To obtain a representative population of European products, twenty specimens of nominally identical bars for each kind of shape and slenderness were selected at random from the product of the various countries belonging to ECCS (see Fig. 4.5, 4.6).

The initial programme foresaw experimenting on: 516 profiles type IPE; 54 profiles type HE; 139 round tubes; 188 square tubes; 94

T-shaped profiles; 76 box sections; i.e. a total of 1067 specimens. The programme was later completed by approx. 500 more tests, some of which concerned very thick profiles (HEM 340). Other tests, on similar profiles to those listed above, were carried out by the individual countries to check their own products.

The characteristic values of load bearing capacity were defined as the mean value of the test results minus two standard deviations. Fig. 9.6 shows the overall experimental results for IPE profiles: it should be pointed out that the characteristic strength $f_{c,k} = f_{c,m} - 2s$ is in relation to the actual cross sectional dimensions.

In Fig. 9.7 the same results are given in non-dimensional form in relation to the mean yield stress f_y of each series of profiles, with $\lambda_e = \pi\sqrt{E/f_y}$.

Starting from the experimental points of collapse stresses defined *characteristic* as indicated above, buckling curves $\sigma_c - \lambda$ were obtained for the various profiles. An attempt was made to interpret them deterministically, i.e. by obtaining the same curves by means of numerical simulation programmes considering the bar made of elastic -plastic material, having a pre-determined initial configuration and a given residual stress distribution through its section. This appeared to be possible, provided one considered:

Initial configurations of a sinusoidal shape having amplitude v_o equal to 1/1000 of bar length (this was justified by the reason illustrated in 4.3)

Conventional distributions of residual stresses, as those illustrated in Fig.9.8 that schematically represent those

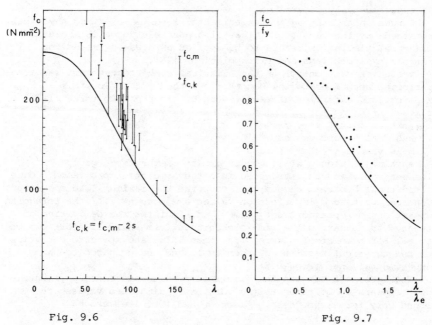

Fig. 9.6   Fig. 9.7

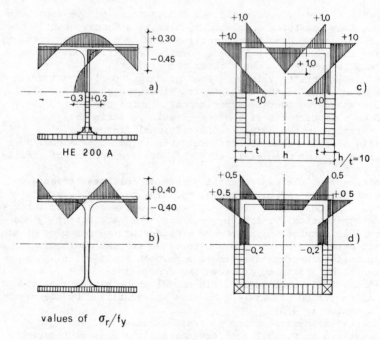

HE 200 A

values of σ_r / f_y

Fig. 9.8

experimentally determined (see 4.2)

Curves interpolating the experimental results were thus numerically obtained; at the same time it was also possible to calibrate a numerical simulation method to be applied as a useful means for further exploitations and extrapolations |23|.

A statistical control of the numerical simulation method was also performed according to the Montecarlo method |33|. For this application, experimental results can be grouped into three distinct classes referred to:

Initial curvature v_o/L
Residual stress values
Mean values of yield stress

A value is chosen at random from each of the three groups independently and is introduced into the simulation programme. Once slenderness has been chosen, one computes the maximum load the strut can bear when such imperfection values are present. If the foregoing operation is repeated twenty times, it simulates the delivering of twenty struts having the same slenderness from an ideal store common to all European steel plants. The mean value and standard deviation of the numerical results are then evaluated and compared to the experimental ones (Fig. 9.9).

The agreement between the simulated results and the experimental ones appears to be remarkably good. Lower simulated values are found only for slenderness between 80 and 120, i.e. around

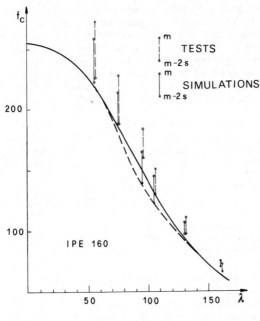

Fig. 9.9

proportional slenderness, where - as it is known - imperfection
influence is most important. Such a difference can be explained by
the hypothesis that geometrical and structural imperfections could
be mutually related in order to increase load bearing capacity of
the bar. In fact, where initial curvature is greater, residual
stresses are lower, or vice versa.

9.1.3.2 Systems with imperfections (few tests)
By this method it is possible to obtain reliable results, although
not rigorous ones from a statistical point of view, within the field
of problems that have not yet been studied at a more sophisticated
level. This method is also within the reach of small organizations
and it might therefore be very useful in solving many practical
problems.

One must, substantially, mix reasonable doses of testing and
numerical calculation in order to reduce considerably the experimental
costs without losing too much in result reliability. To simulate
experiments on a computer, the initial data concerning the unloaded
strut must be introduced. Such data are:
    Section shape
    Yield stress in each fibre
    Residual stress in each fibre
    One therefore introduces, at a deterministic level, the values
of the parameters the influence of which should be strictly considered
at a probabilistic level.

387

If however a limited number of experimental tests are performed on a sufficiently representative population, and if the values of the parameters characterizing the specimens, once they are introduced into the simulation programme, lead to numerical results that are close to the experimental ones, it is reasonable to state that simulation can be used as an extrapolation of experimental results.

This method is being applied to obtain results for:

Columns differing in material and/or shape, from those tested by ECCS

Beam-columns in mono- or bi-axial bending (see 9.4)

Compound struts (see 9.2.3)

Consider for example, the problem of determining which buckling curve given in codes is appropriate for tubular struts made of high strength steel |34|.

The procedure to adopt is the following:

A certain number of specimens, all of the same nominal slenderness, are built

For each specimen the mean yield strength over the cross section is determined by means of a stub column test

For each specimen initial curvature and cross sectional dimensions are measured

Each specimen is tested in compression to determine its maximum load experimentally (Fig. 9.10)

At this point, the experimental results are simulated by a numerical procedure and a check is made that the mean values and standard deviations of the experiments and of the calculations are close to each other (Fig. 9.11)

If this is satisfactory, it can be assumed that all the parameters influencing the phenomenon have been correctly introduced by means

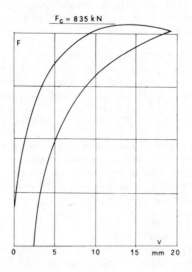

Fig. 9.10

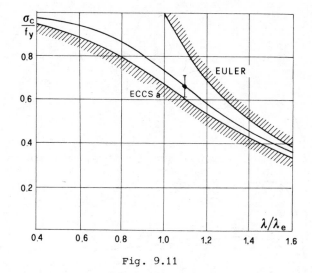

Fig. 9.11

of the simulation and such results can thus be extrapolated, with the added advantage of being able to study the influence of the various parameters acting separately

In this case, the curve obtained by numerical simulation passes through the point characterizing the mean value of the experimental results. Fig. 9.11 also shows the line corresponding to the mean value minus two deviations of the experimental results.

9.1.3.3 Systems without imperfections

All the problems that have been examined so far have been related to the compressed strut pinned at its ends and variously loaded. In steel structures, however, there are also many other problems that certainly cannot be solved by similar techniques.

When no other way presents itself, an attempt can only be made to find some criteria enabling the maximum load carrying capacity of a structure or structural element to be related to the corresponding Euler critical load. In doing this, there are essentially three available criteria to be considered:
a) Use of ideal or equivalent slenderness
b) Correction of the critical load
c) Introduction of a safety factor

(a) *Equivalent slenderness*. Consider a structural element (e.g. a compressed strut) where the applied load is being increased. If the material had a perfectly elastic behaviour, the transverse displacement would increase together with load increase, according to the dashed line of Fig. 9.12.

For the actual material, once the elastic limit is attained in the most stressed fibre of the most loaded section, there is a lowering of the diagram from the ideal behaviour. Each point of the

389

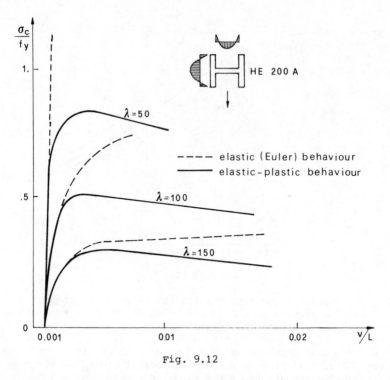

Fig. 9.12

actual curve corresponds to an equilibrum situation characterized by a load that increases to a maximum value and then decreases.

In many cases it will be recognized that the elastic-plastic phase for an increasing load has a rather limited extension compared to the elastic one. In other words, load increase in the elastic-plastic phase is a small portion of the maximum load and this obviously becomes more sensible as strut slenderness increases. In these cases, therefore, an error in the evaluation of such an increase cannot have a significant influence on the determination of strut load bearing capacity. On the basis of these considerations, the equivalent slenderness criterion can be expressed as follows:

If the Euler critical load for the system is known, it is possible to define the slenderness λ_{eq} of a simple pin-ended strut having the same Euler critical load as the system for which maximum load is being sought

Maximum load of the pinned equivalent strut having slenderness λ_{eq} is known. If the differences between the elastic-plastic phase of the pinned equivalent strut and that of the real system are neglected, maximum load for the system can be stated to be equal to that of the pinned equivalent strut

In other words, the criterion is based on the statement that, if two systems have the same Euler critical load, they also have the same maximum load bearing capacity. Obviously, this is not always

390

true, but it can be considered reasonable for:
Simple struts having other restraints than hinges (see 9.5.1)
Compound struts in which shear or bending deformability takes
place in the connection (see 9.2.3)
Struts having variable section (see 9.5.2)

(b) *Correction of the critical load*. A less sophisticated method
consists in computing the Euler critical load and then in correcting
it by means of formulae that must comply with the following
characteristics. The maximum stress σ_c corresponding to the ultimate
situation, defined as a function of the corresponding value σ_{cr} in
bifurcation conditions, must tend to σ_{cr} whenever the latter becomes
small, while it must tend to yield stress whenever σ_{cr} tends to
infinity.

This result can be obtained, for example, by a formula such as:

$$\sigma_c = \frac{\sigma_{cr}}{\sqrt[n]{\sigma_{cr}^n + f_y^n}} f_y \qquad (9.18)$$

used to interpret flexural-torsional buckling (see 9.3). A similar
criterion, but in a different formulation, is also applied in the
calculation method of web panels according to the linear theory (see
9.7.2).

The method is operatively simple, but must be supported by
experimental or numerical studies based on more rigorous approaches.
Otherwise, it might become arbitrary and void of any meaning.

(c) *Introduction of a safety factor*. For certain problems, a semi
-empirical method is sometimes proposed. It can roughly be formulated
as follows: the Euler critical multiplier for the applied load is
evaluated and checked that it is comprised between 2.5 and 5,
depending on the cases and the problems under examination.

This method can be also justified. In the elastic field, the
influence of second order geometrical effects on system deformability
can often be expressed by the following formula:

$$v = \alpha v_o \frac{1}{1 - F/F_{cr}} = \alpha v_o \frac{1}{1 - 1/\nu_{cr}}$$

where
v = the displacement of a significant point of a structure
v_o = the initial displacement of the same point
α = a numerical coefficient
F = the live load multiplier
F_{cr} = the Euler load multiplier
ν_{cr} = F_{cr}/F = the safety factor against elastic buckling
To increase the ratio F_{cr}/F in the design phase would render
structural imperfection effects less influential and therefore take
care against a non-linear result which could amplify the displacements
of the structure. On the other hand, safety factors ν_{cr} can be
defined only in an arbitrary manner. In the literature their values
depend on the problems and on the various authors. The method must
therefore be classed as a semi-empirical one, but it is unfortunately

391

necessary to apply it, especially for those problems of overall stability for which it is not possible to operate at a more qualified level, given the present state of knowledge.

9.1.3.4 System with standard imperfections

It has already been pointed out that methods based on the evaluation of the critical load of the system without imperfections have a big limitation. They do not make it possible to determine the forces acting in the restaints and, therefore, the lowering influence of load on the stress state.

To determine such fares (which is absolutely necessary to design, for example bracing structures) one must analyse systems affected by geometrical imperfections. Analyses of a statistical type have not yet been carried out in this direction. It is however possible to introduce standard values of the imperfections in a deterministic way, such as those corresponding to dimensional tolerances (see 1.5). It is not possible, for example, to evaluate the lowering effects on the bracing shown in Fig. 9.13 if the system is considered perfect. This goal is obtained if the structure is loaded by a horizontal force H or if an imperfection such as column out-of-straightness v_o is introduced. The value of such an imperfection is conventional and, in the elastic field, the unstabilizing effects are proportional to its value.

The study of single storey and multi-storey frames is considerably more complicated, because the problem shows a double non-linearity: a geometrical one, due to the so-called $P - \Delta$ effect and a structural one, as the moment curvature relationship of the cross section is not linear in the plastic range. One must furthermore take into consideration:

The possibility of elastic unloading of existing plastic hinges which tend to reduce their rotation as external load increases
The influence of the axial load on the limiting moment on the section
The possibility that a beam-column may reach its maximum carrying capacity and go into the post buckling range
The possible collapse of joints having limited ductility (as bolted bracing diagonals)

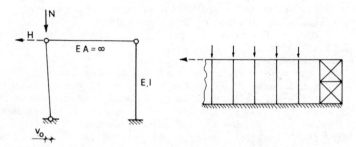

Fig. 9.13

This field is being actively explored, but the problem is extremely complex. In fact, it is difficult even to simulate the behaviour of a beam-column and it is certainly more compex to simulate framed structures.

As a principle, a calculation method to be applied in practice should be based on incremental techniques of the following kind:

For given values of the vertical and horizontal loads calculate the displacements of the system in a linear range

Once the horizontal loads have been removed, apply the vertical loads to the previously determined deformed system and evaluate the further displacements due to their unstabilizing effect

Repeat the latter operation until the difference between displacements of the two consecutive iterations is negligible

Such a method is not, however, directly applicable, as at each iteration it is necessary to correct the configuration on which one is operating, on the basis of the results of the previous iteration. An automatic calculation programme is therefore required.

The Report of the International Conference on Tall Buildings $|35|$, as well as $|3,7|$ refer to some design criteria that are easier to apply in order to obtain an approximate solution of similar problems. The most direct method is the 'fictitious shear force' one, that offers the advantage of operating by successive iterations, but always on the same configuration (see 9.5.4.3).

So that such an analysis seems to provide the most adequate solution to the problem, both in the elastic and in the elastic-plastic field, by giving sufficiently complete information on the overall stability of a structure. From this point of view it is possible to understand the reason why an empirical design practice limits the transverse displacement of a structure due to fictitious tranverse forces equal to a fraction of the vertical loads (see 8.1.2).

In this way transverse displacements can be made sufficiently slight that the unstabilizing contribution of the vertical load (so-called $P - \Delta$ effect) can be kept within acceptable limits. A few tests performed on the most usual types of structure (multi-storey buildings, mill buildings) have shown that, if displacement is limited to a value of the order of 1/500 of the height, the unstabilizing effects are always slight and produce just a small percentage increase with respect to the stress evaluated in the undeformed situation.

A numerical survey $|36|$ of approx. 600 diagrams of multi-storey steel frames has shown that a transverse deflection limit of 1/500 of the height due to imaginary horizontal forces equal to 1/80 of the vertical load provides a safety factor ν_{cr} of about 11.5 with reference to elastic buckling of Euler's type (Fig. 9.14). If the same criterion is applied to a cantilever clamped at its base, a safety factor ν_{cr} of about 5 is obtained.

An overall stability check based on similar criteria is therefore always on the safe side, but it is not yet possible to know by how much. If such a check is not satisfactory, the degree of safety of the structure must be investigated by means of more complex and sophisticated methods.

393

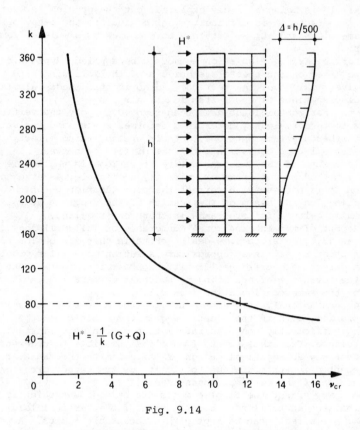

$$H^* = \frac{1}{k} (G + Q)$$

Fig. 9.14

9.2 CENTRALLY COMPRESSED MEMBERS

An example of a strut undergoing a perfectly centred axial compression
is a column in a pin-ended steel framework, in which horizontal
forces are totally entrusted to proper bracing structures (see 1.2).

With reference to cross sectional shape, a simply compressed strut
might buckle according to three different phenomena:

Plane buckling

Torsional buckling

Flexural-torsional buckling

If the increasing deflection of the strut is so controlled as to be
contained within a plane, it is a case of plane buckling. This
phenomenon is typical of sections having double symmetry when bending
takes place according to a principal inertia plane coinciding with:

The plane of minimum inertia, if restraint conditions are the same
in the two principal directions

The plane of maximum slenderness, in the case of different
restraint conditions in both directions

Sections lacking double symmetry may still be subject to plane buckling, provided the strut is prevented from twisting so as to avoid flexural-torsional buckling. Pure torsional buckling is a phenomenon that particularly concerns sections in which secondary torsional stiffness is negligible (e.g., cross-shaped sections), because of the convergence into one point only of all the elements forming the section.

Plane buckling is discussed with reference to axially loaded compression members, as simple columns in 9.2.1 and 9.2.2 and as columns composed of a certain number of profiles connected on the flanges (9.2.3). Torsional buckling is discussed in 9.2.4.

Beam-columns, which are studied in 9.4, can undergo both plane buckling and flexural-torsonal buckling.

9.2.1 *Influence of imperfections*

The definition of industrial bars has pointed out a number of reasons why its actual behaviour differs from the theoretical behaviour of perfect bars. With reference to the buckling behaviour of struts, in addition to the so-called imperfections of a geometrical (initial curvature) or mechanical nature (residual stresses, variation of yield stress) that have already been defined in chapter 4, the mean value of yield stress and the cross sectional shape must also be considered as influence parameters.

The various influences are analysed in the following by comparing the results of numerical investigations made by means of simulation methods |7,37,38|.

The comparison is related to the curves of plane buckling for simply compressed memebers, represented in the non-dimensional plane $\bar{N} - \bar{\lambda}$, where:

$$\bar{N} = \sigma_c/f_y \qquad ; \qquad \bar{\lambda} = \frac{\lambda}{\pi\sqrt{E/f_y}} \qquad\qquad (9.19)$$

assuming:

f_y = the mean yield stress of the cross section
E = the material elastic modulus.

Buckling curves for symmetrical I-sections are indicated with (min) if the weak axis buckling plane (parallel to flanges) is concerned and with (max) if the strong axis buckling plane (perpendicular to flanges) is concerned.

9.2.1.1 Mean yield stress effect

The influence of mean yield value is shown in Fig. 9.15 for a wide flange I-DIE-20 profile which buckles about the weak axis and has an initial displacement v_o equal to $L/1000$. Within the range of yield values for mild steel, the influence of mean yield on buckling can be considered negligible.

9.2.1.2 Shape effect

The shape effect of the most common rolled profiles (IPE, HEA, HEB, HEM) having 200 mm of depth, is examined in Fig. 9.16. The material is considered free from mechanical imperfections, but an initial

395

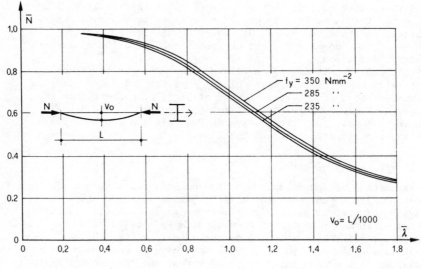

Fig. 9.15

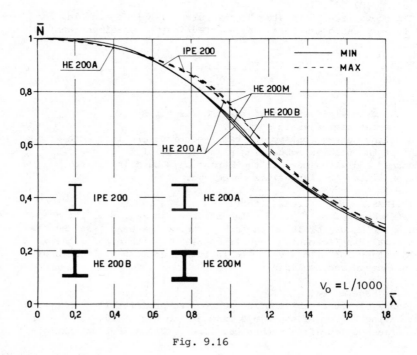

Fig. 9.16

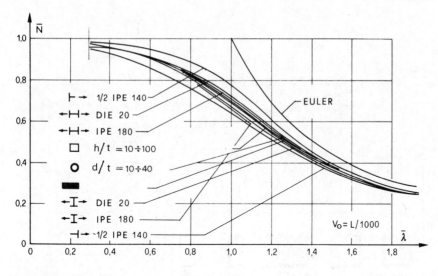

Fig. 9.17

curvature of $L/1000$ is present.

As might be expected in the case of I-sections, shape effect has practically no influence. In fact, no substantial differences between the curves for the various profiles can be noticed, with respect either to the strong axis or to the weak axis.

Greater differences are found with other section shapes (Fig. 9.17).

In particular, ⊥ sections behave in a remarkably different way from other types. In fact, they give the behavioural limit cases: an upper curve when the flange is in compression and a lower curve when the web is compressed.

For box sections, Fig.9.18 shows that even if there are considerable differences between wall thickness and depth/width ratios, the behavioural scattering is quite restricted.

It should in any case be pointed out that the shape effect is almost never the sole influence, as a profile shape and its manufacturing process produce a mechanical imperfection distribution the influence of which - as shown later - is far from negligible. Shape effect fully takes place only when special heat treatment has eliminated residual stresses in the bar.

9.2.1.3 Initial curvature effect

The geometrical imperfection that is universally chosen as representative in the analysis of struts is an initial curvature having a sinusoidal shape and a mid-span displacement $v_o = L/1000$ (see 4.3).

The effect of different values of v_o on the load bearing capacity of a column is examined in Fig. 9.19. The curves are for $L/2000$, $L/1000$ and $L/500$ for a wide flange I-section without (full line) and with (dashed line) residual stresses. A comparison shows that the

397

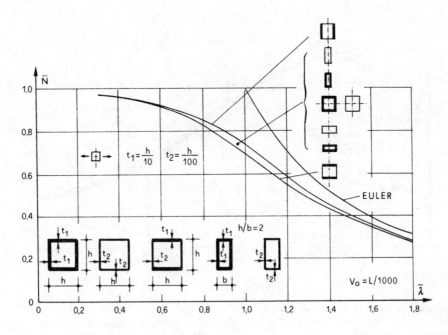

Fig. 9.18

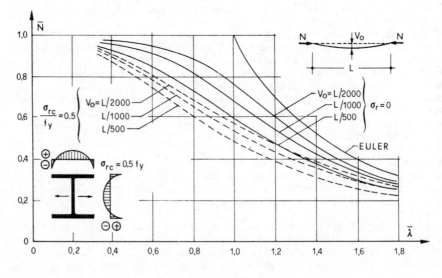

Fig. 9.19

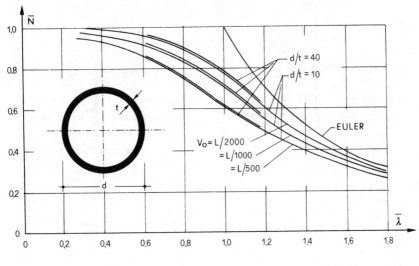

Fig. 9.20

influence of initial curvature decreases due to the presence of
residual stresses and that the dimensionless slenderness value at
which its influence is a maximum is approx.1.0 if residual stresses
are absent and approx. 1.3 if they are present.

In the case of tubes load bearing capacity reduction due to
initial curvature is practically independent of the diameter
thickness ratio (Fig. 9.20).

9.2.1.4 Yield scatter effect
The effect of yield stress distribution are in cross section has
been examined with reference to the behavioural models of Fig. 9.21,
for IPE 200, HE 200A, HE 200B, HE 200M, HE 300M shapes, the buckling
curves of which are compared in Fig. 9.22. Due to the lowering
caused by the yield stress scatter effect, in particular in the
elastic-plastic range ($\bar{\lambda}$ = 0.2 to 1.2), the curves are considerably
flattened compared to the ones corresponding to the shape effect
alone.

The severest lowering concerns strong axis buckling, with a
different amount according to the value of the ratio between mean
yield stress in the flanges and in the web: $f_{y,f}/f_{y,w}$.

This entails a superposition of the max and min curves, which
practically cancels every qualitative difference in behaviour related
to section shape and bending plane.

9.2.1.5 Residual stress effect
This effect is perhaps the one that has received most attention in
the field of studies on column stability.

The influence of residual stress distribution on a section is shown
in Fig. 9.23. Welded profiles (see 4.2.3) have high tensile residual

399

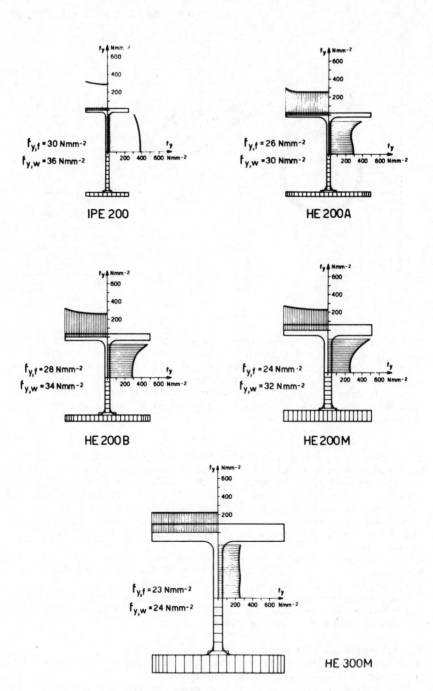

$f_{y,f} = 30\ Nmm^{-2}$
$f_{y,w} = 36\ Nmm^{-2}$

IPE 200

$f_{y,f} = 26\ Nmm^{-2}$
$f_{y,w} = 30\ Nmm^{-2}$

HE 200 A

$f_{y,f} = 28\ Nmm^{-2}$
$f_{y,w} = 34\ Nmm^{-2}$

HE 200 B

$f_{y,f} = 24\ Nmm^{-2}$
$f_{y,w} = 32\ Nmm^{-2}$

HE 200 M

$f_{y,f} = 23\ Nmm^{-2}$
$f_{y,w} = 24\ Nmm^{-2}$

HE 300 M

Fig. 9.21

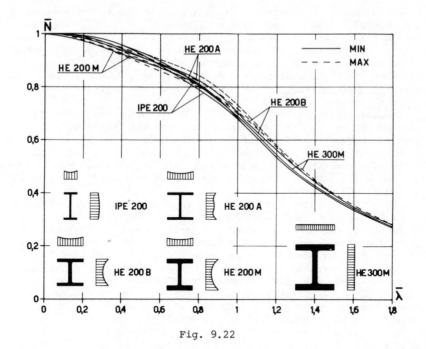

Fig. 9.22

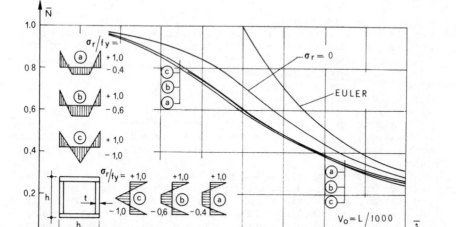

Fig. 9.23

stresses in the area near the welds which, in the case of box sections, are located at the corners; for equilibrium, therefore, compressive residual stresses are produced in the central areas of the section walls. Distribution and order of magnitude of these depend on the overall dimensions of the box section itself. Compressive residual stresses can reach approx. the same level as the tensile one, i.e. material yield stress, in small boxes, but they decrease as the overall dimensions increase.

To demostrate the foregoing, Fig. 9.23 shows three distributions related to boxes of various dimensions, but the curves do not show sensible differences in ultimate load values.

With reference to the I-sections of series IPE and HE, residual stress influence has been examined for the distribution models shown in Fig. 9.24. The corresponding results are indicated in Figs 9.25, 9.26 and 9.27.

Fig. 9.25 compares the effects of the various residual stress distributions for profile HE 200B: parabolic model '1', mean distribution '2', and rolling straightening effect '3'. It can be seen that the most conservative distribution is that producing curves '1' and this corresponds to the parabolic model shown in 4.2.1.

Different residual stress distributions on profiles of the same depth (HEA, HEB, HEM, IPE 200 mm of depth) produce the curves shown in Fig. 9.26. Again in this case, the lowest curves for both bending planes correspond to profile HEB 200, with a parabolic distribution, providing a lower bound for all the profiles examined.

In the case of I-profiles with a depth of the same order of magnitude as flange width, the residual stress effect becomes more severe as depth increases, as may be seen from the comparison (Fig. 9.27) between curves corresponding to the same series of profiles (HEM) having different depths (100, 200, 300 mm).

9.2.1.6 Combined effects

Fig.s 9.28 to 9.35 compare the dimensionless curves (for both strong and weak axes) for profiles IPE 200, HEA 200, HEA 200, HEB 200, HEM 200, under the four following conditions:
(a) with no structural imperfections
(b) with yield scatter only
(c) with residual stresses only
(d) with yield scatter and residual stresses together

In all the cases there is a geometrical imperfection $v_o = L/1000$.

The comparison shows the strong lowering role of the yield scatter with particular reference to the differences between flange and web values.

In all the examined cases, except that for the HE 200B (weak axis) (Fig. 9.33), the yield scatter effect is much more important than the residual stress effect. Excluding cases HE 200A and HE 200B (weak axis)(Figs 9.31 and 9.33), the presence of residual stresses has always reduced yield scatter influence, thus becoming a favourable effect, which increases load bearing capacity. This quite unexpected result can be explained by the comparison - for the case in which this takes place - between yield stress and residual stress distribution, by superposing them on the cross section. It can in

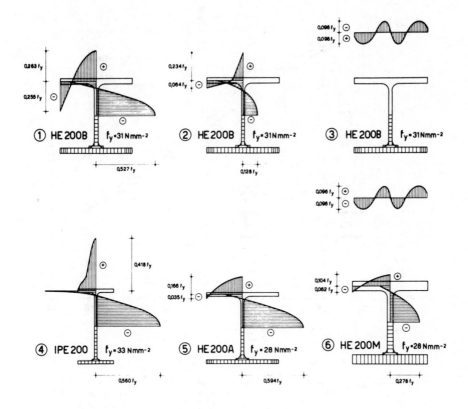

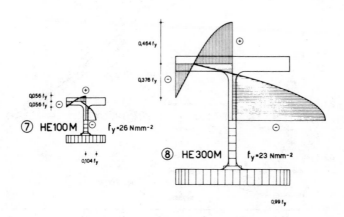

Fig. 9.24

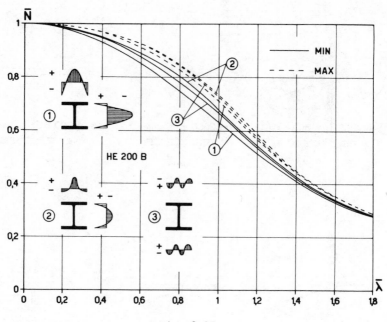

Fig. 9.25

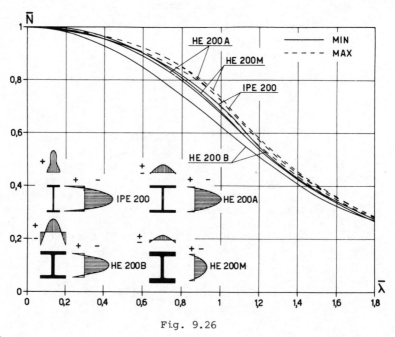

Fig. 9.26

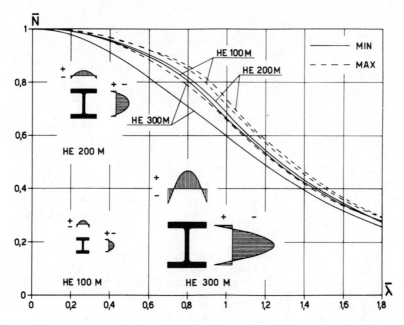

Fig. 9.27

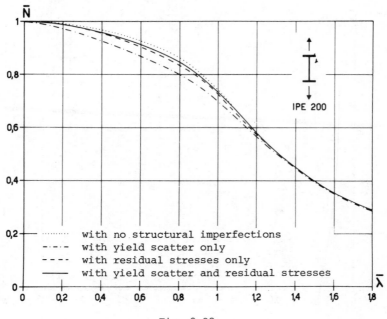

Fig. 9.28

405

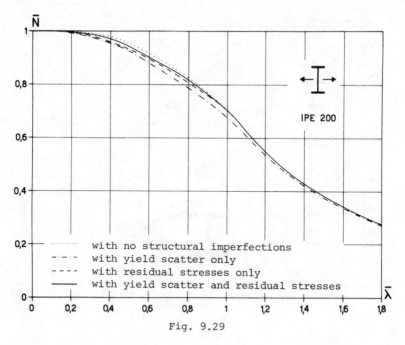

Fig. 9.29

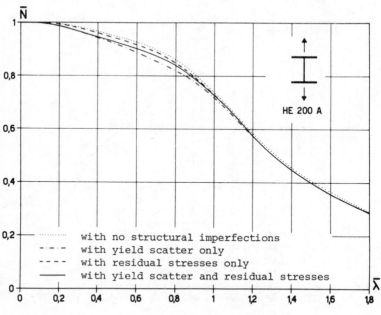

Fig. 9.30

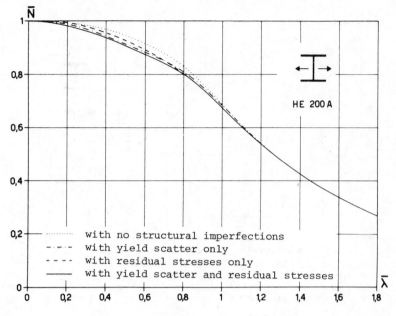

Fig. 9.31

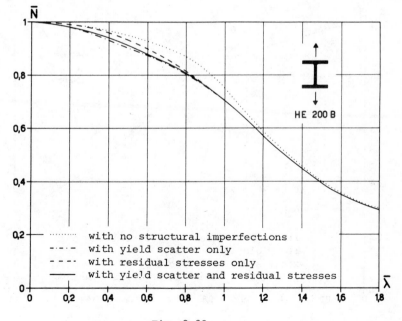

Fig. 9.32

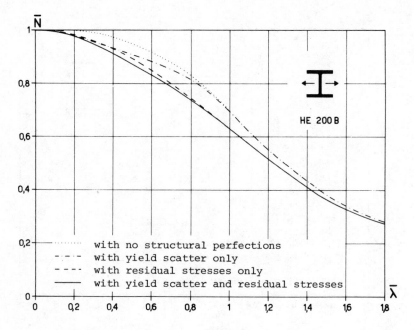

Fig. 9.33

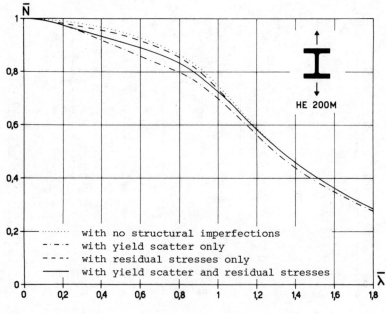

Fig.9.34

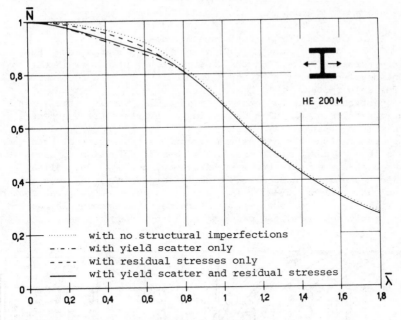

Fig. 9.35

fact be noticed that in the flanges - where yield stress values are always lower than in the web - tensile residual stresses usually prevail (all residual stresses are tensile either because of the geometrical effect of the I-shapes, or because of the relieving of compression peaks at the flange ends as a consequence of straightening by rolling). Therefore in the flanges, although they have a lower yield strength there is a compensating effect because the residual stresses create strength reserve with respect to compression. On the contrary, in the web where yield stresses are higher, the unfavourable presence of high compressive residual stresses is moderated by greater local strength.

The combination of both effects interacting in the flanges and in the web is, on the whole, favourable to column load bearing capacity, having obviously a different weight according to the bending plane.

The combination of yield scatter and residual stresses leads to lower buckling curves only in the case of the weak axis for profiles HEA 200 (Fig. 9.31) and HEB 200 (Fig. 9.33).

From the above results, it does not therefore seem conservative to neglect the yield scatter effect when studying steel column buckling.

9.2.2 *Recommended curves*

9.2.2.1 International Organizations and codes
The Column Research Council (CRC) in the United States (now SSRC) and the European Convention for Constructional Steelwork (ECCS) in Europe developed their activities in order to prepare recommendation for design purposes.

In particular, CRC has the merit of having provided the first organic settlement of rules concerning steel structures, especially in the field of stability |3|.

Several tests on mechanical imperfections in industrial bars made it possible to associate a residual stress distribution to each series of American structural steel shapes, not only for rolled profiles made of mild steel but also for high strength steels and for welded sections |39-43|. At the same time in |44,45| was showed theoretically how to take residual stresses into account in the evaluation of maximum load carrying capacity of struts.

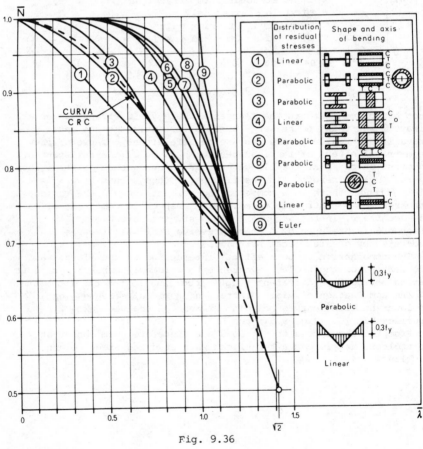

Fig. 9.36

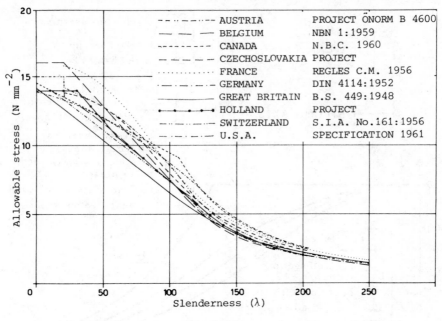

Fig. 9.37

From these results (1966), the CRC |3| established a buckling curve based on the theoretical calculation obtained on profiles having various shapes with residual stress distributions varying parabolically and linearly only in the flanges, with maximum values in compression equal to 0.3 of yield stress (Fig. 9.36). The European Convention for Constructional Steelwork (CECM, ECCS, EKS) was founded in Zurich (1955) with the fundamental purpose of carrying out a common research programme in the various member countries and of promoting the corresponding results from the point of view of application. Later, ECCS also aimed at obtaining as completely as possible a unification of technical recommendations concerning steel construction in the various European countries.

At the beginning of its activities, with reference to the problem of column stability, the fundamental buckling curves in the various national codes were found to be very different and differently conceived |46| (Fig. 9.37), showing a large scatter mainly in the most commonly used slenderness range.

Such discrepancies increased when comparing the various recommendations concerning stability of beam-columns, compound columns, struts having a variable section, beams, etc.

Faced with such a variety of practical results, Committee 8 of ECCS (*Stability problems*) had the great merit of attempting to overcome the differences.

411

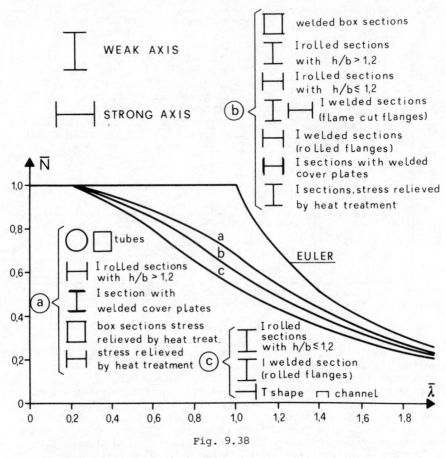

Fig. 9.38

9.2.2.2 ECCS Recommendations |31|
The systematic studies carried out by ECCS led to defining three
buckling curves for simply compressed columns |14,17-29,47|; curve
'a' is typical for tubes, curve 'b' for welded box sections and curve
'c' for wide flange I-profiles buckling about their weak axis.
    The three curves are cut horizontally at λ = 0.2, an approximation
which is conservative because it neglects the favourable effects of
strain hardening which, however, has not been taken into account in
computing the curves.
    The three curves are shown in Fig. 9.38 in the non dimensional
plane $\bar{N} - \bar{\lambda}$, where

$$\bar{N} = N_c/N_{pl} = \sigma_c/f_c \qquad ; \qquad \bar{\lambda} = \lambda/\lambda_c \qquad (9.20)$$

and'
N_c = maximum axial load
N_{pl}^c = fully plastic axial load

412

Values of σ_c/f_c for curve "a"

λ/Λ_c	0.	0.01	0.02	0.03	0.04	0.05	0.06	0.07	0.08	0.09
.0	1.0000	1.0000	1.0000	1.0000	1.0000	1.0000	1.0000	1.0000	1.0000	1.0000
.1	1.0000	1.0000	1.0000	1.0000	1.0000	1.0000	1.0000	1.0000	1.0000	1.0000
.2	1.0000	0.9981	0.9962	0.9942	0.9922	0.9900	0.9877	0.9854	0.9829	0.9805
.3	0.9780	0.9756	0.9731	0.9706	0.9682	0.9657	0.9632	0.9607	0.9582	0.9556
.4	0.9530	0.9504	0.9477	0.9449	0.9421	0.9392	0.9362	0.9330	0.9298	0.9265
.5	0.9230	0.9193	0.9156	0.9117	0.9078	0.9039	0.9000	0.8961	0.8923	0.8885
.6	0.8848	0.8810	0.8772	0.8733	0.8693	0.8652	0.8611	0.8570	0.8530	0.8489
.7	0.8447	0.8404	0.8359	0.8312	0.8264	0.8214	0.8164	0.8115	0.8055	0.8015
.8	0.7965	0.7914	0.7860	0.7806	0.7749	0.7692	0.7634	0.7575	0.7515	0.7455
.9	0.7394	0.7333	0.7270	0.7207	0.7143	0.7078	0.7013	0.6947	0.6880	0.6813
1.0	0.6746	0.6678	0.6610	0.6541	0.6473	0.6404	0.6336	0.6267	0.6198	0.6130
1.1	0.6061	0.5993	0.5925	0.5858	0.5791	0.5725	0.5660	0.5595	0.5530	0.5466
1.2	0.5403	0.5339	0.5276	0.5213	0.5151	0.5090	0.5029	0.4970	0.4911	0.4854
1.3	0.4798	0.4742	0.4687	0.4633	0.4580	0.4527	0.4475	0.4423	0.4372	0.4321
1.4	0.4271	0.4221	0.4172	0.4124	0.4077	0.4030	0.3984	0.3939	0.3894	0.3850
1.5	0.3807	0.3764	0.3722	0.3681	0.3640	0.3600	0.3560	0.3521	0.3482	0.3444
1.6	0.3406	0.3369	0.3333	0.3297	0.3262	0.3227	0.3193	0.3159	0.3126	0.3094
1.7	0.3062	0.3031	0.3000	0.2970	0.2940	0.2910	0.2881	0.2852	0.2824	0.2796
1.8	0.2768	0.2741	0.2714	0.2687	0.2661	0.2635	0.2609	0.2583	0.2557	0.2532
1.9	0.2507	0.2482	0.2458	0.2434	0.2410	0.2387	0.2364	0.2342	0.2320	0.2298
2.0	0.2277	0.2256	0.2235	0.2215	0.2194	0.2174	0.2153	0.2133	0.2113	0.2094
2.1	0.2076	0.2056	0.2041	0.2024	0.2007	0.1990	0.1973	0.1956	0.1939	0.1923
2.2	0.1906	0.1890	0.1873	0.1857	0.1842	0.1826	0.1811	0.1795	0.1780	0.1766
2.3	0.1751	0.1737	0.1723	0.1709	0.1696	0.1682	0.1668	0.1655	0.1642	0.1628
2.4	0.1615	0.1602	0.1589	0.1576	0.1563	0.1551	0.1539	0.1527	0.1515	0.1503
2.5	0.1492	0.1482	0.1471	0.1461	0.1449	0.1437	0.1425	0.1414	0.1404	0.1394
2.6	0.1384	0.1373	0.1362	0.1351	0.1341	0.1332	0.1323	0.1313	0.1303	0.1294
2.7	0.1285	0.1275	0.1266	0.1256	0.1247	0.1238	0.1229	0.1220	0.1212	0.1203
2.8	0.1195	0.1187	0.1179	0.1171	0.1163	0.1155	0.1147	0.1140	0.1132	0.1124
2.9	0.1117	0.1110	0.1103	0.1096	0.1089	0.1082	0.1075	0.1068	0.1061	0.1055
3.0	0.1048	0.1041	0.1035	0.1028	0.1022	0.1015	0.1008	0.1002	0.0995	0.0988
3.1	0.0982	0.0976	0.0970	0.0964	0.0958	0.0952	0.0945	0.0940	0.0935	0.0929
3.2	0.0923	0.0917	0.0912	0.0906	0.0901	0.0895	0.0889	0.0884	0.0878	0.0873
3.3	0.0868	0.0863	0.0858	0.0854	0.0849	0.0844	0.0839	0.0834	0.0829	0.0824
3.4	0.0819	0.0814	0.0810	0.0806	0.0801	0.0797	0.0793	0.0788	0.0784	0.0779
3.5	0.0775	0.0771	0.0766	0.0762	0.0758	0.0754	0.0750	0.0746	0.0742	0.0738
3.6	0.0734									

Fig. 9.39a

Values of σ_c/f_c for curve "b"

λ/λ_c	0.	0.01	0.02	0.03	0.04	0.05	0.06	0.07	0.08	0.09
.0	1.0000	1.0000	1.0000	1.0000	1.0000	1.0000	1.0000	1.0000	1.0000	1.0000
.1	1.0000	1.0000	1.0000	1.0000	1.0000	1.0000	1.0000	1.0000	1.0000	1.0000
.2	1.0000	0.9967	0.9933	0.9899	0.9865	0.9830	0.9795	0.9760	0.9724	0.9687
.3	0.9650	0.9612	0.9573	0.9533	0.9493	0.9453	0.9412	0.9372	0.9331	0.9291
.4	0.9250	0.9211	0.9171	0.9132	0.9093	0.9054	0.9014	0.8974	0.8933	0.8892
.5	0.8850	0.8807	0.8762	0.8717	0.8671	0.8624	0.8577	0.8529	0.8480	0.8430
.6	0.8380	0.8329	0.8278	0.8227	0.8174	0.8122	0.8068	0.8015	0.7960	0.7905
.7	0.7850	0.7794	0.7738	0.7681	0.7624	0.7566	0.7508	0.7449	0.7390	0.7330
.8	0.7270	0.7210	0.7148	0.7087	0.7024	0.6961	0.6897	0.6832	0.6766	0.6700
.9	0.6633	0.6566	0.6500	0.6434	0.6369	0.6305	0.6241	0.6177	0.6114	0.6051
1.0	0.5987	0.5924	0.5861	0.5799	0.5737	0.5675	0.5615	0.5554	0.5495	0.5435
1.1	0.5376	0.5318	0.5260	0.5202	0.5145	0.5088	0.5031	0.4975	0.4919	0.4864
1.2	0.4809	0.4754	0.4700	0.4647	0.4593	0.4541	0.4489	0.4438	0.4387	0.4337
1.3	0.4288	0.4240	0.4192	0.4145	0.4098	0.4052	0.4007	0.3962	0.3918	0.3874
1.4	0.3831	0.3788	0.3746	0.3704	0.3663	0.3622	0.3582	0.3542	0.3503	0.3464
1.5	0.3426	0.3389	0.3352	0.3317	0.3281	0.3246	0.3212	0.3178	0.3144	0.3111
1.6	0.3078	0.3046	0.3014	0.2982	0.2950	0.2919	0.2888	0.2857	0.2826	0.2796
1.7	0.2766	0.2737	0.2709	0.2681	0.2654	0.2617	0.2601	0.2576	0.2551	0.2526
1.8	0.2502	0.2478	0.2455	0.2431	0.2408	0.2385	0.2362	0.2340	0.2317	0.2295
1.9	0.2273	0.2251	0.2230	0.2208	0.2188	0.2167	0.2147	0.2127	0.2108	0.2089
2.0	0.2070	0.2052	0.2034	0.2016	0.1999	0.1982	0.1965	0.1948	0.1931	0.1914
2.1	0.1897	0.1880	0.1864	0.1848	0.1833	0.1818	0.1804	0.1790	0.1776	0.1761
2.2	0.1746	0.1730	0.1715	0.1701	0.1688	0.1675	0.1662	0.1648	0.1635	0.1621
2.3	0.1607	0.1594	0.1580	0.1567	0.1555	0.1542	0.1530	0.1518	0.1506	0.1494
2.4	0.1483	0.1471	0.1460	0.1449	0.1438	0.1427	0.1417	0.1407	0.1397	0.1387
2.5	0.1377	0.1366	0.1356	0.1346	0.1336	0.1327	0.1319	0.1311	0.1303	0.1293
2.6	0.1283	0.1273	0.1263	0.1253	0.1244	0.1237	0.1230	0.1222	0.1214	0.1206
2.7	0.1198	0.1190	0.1182	0.1174	0.1166	0.1158	0.1150	0.1142	0.1134	0.1127
2.8	0.1119	0.1111	0.1104	0.1096	0.1088	0.1081	0.1074	0.1066	0.1059	0.1052
2.9	0.1045	0.1038	0.1031	0.1024	0.1017	0.1010	0.1003	0.0997	0.0990	0.0983
3.0	0.0977	0.0971	0.0964	0.0958	0.0951	0.0945	0.0939	0.0932	0.0926	0.0920
3.1	0.0914	0.0908	0.0902	0.0896	0.0891	0.0885	0.0879	0.0874	0.0868	0.0863
3.2	0.0857	0.0852	0.0846	0.0841	0.0835	0.0830	0.0825	0.0819	0.0814	0.0809
3.3	0.0804	0.0799	0.0794	0.0789	0.0784	0.0779	0.0774	0.0769	0.0764	0.0760
3.4	0.0755	0.0750	0.0746	0.0742	0.0737	0.0733	0.0729	0.0724	0.0720	0.0716
3.5	0.0712	0.0708	0.0704	0.0700	0.0697	0.0693	0.0689	0.0686	0.0682	0.0679
3.6	0.0675									

Fig. 9.39 b

Values of σ_c/f_c for curve "c"

λ/λ_c	0.	0.01	0.02	0.03	0.04	0.05	0.06	0.07	0.08	0.09
.0	1.0000	1.0000	1.0000	1.0000	1.0000	1.0000	1.0000	1.0000	1.0000	1.0000
.1	1.0000	1.0000	1.0000	1.0000	1.0000	1.0000	1.0000	1.0000	1.0000	1.0000
.2	1.0000	0.9949	0.9899	0.9849	0.9799	0.9750	0.9702	0.9654	0.9606	0.9558
.3	0.9510	0.9461	0.9412	0.9362	0.9312	0.9261	0.9210	0.9158	0.9106	0.9053
.4	0.9000	0.8947	0.8893	0.8838	0.8783	0.8727	0.8671	0.8613	0.8555	0.8496
.5	0.8436	0.8376	0.8316	0.8256	0.8196	0.8136	0.8076	0.8015	0.7954	0.7892
.6	0.7829	0.7766	0.7701	0.7636	0.7571	0.7506	0.7441	0.7377	0.7314	0.7250
.7	0.7187	0.7124	0.7060	0.6997	0.6933	0.6869	0.6804	0.6738	0.6673	0.6608
.8	0.6543	0.6478	0.6416	0.6353	0.6292	0.6232	0.6171	0.6111	0.6051	0.5991
.9	0.5931	0.5871	0.5812	0.5754	0.5696	0.5640	0.5584	0.5529	0.5474	0.5421
1.0	0.5368	0.5315	0.5263	0.5211	0.5159	0.5108	0.5057	0.5006	0.4956	0.4906
1.1	0.4856	0.4807	0.4758	0.4710	0.4662	0.4614	0.4567	0.4521	0.4474	0.4428
1.2	0.4383	0.4338	0.4293	0.4249	0.4205	0.4162	0.4119	0.4076	0.4034	0.3993
1.3	0.3952	0.3911	0.3871	0.3832	0.3792	0.3754	0.3715	0.3678	0.3640	0.3604
1.4	0.3567	0.3532	0.3496	0.3462	0.3427	0.3393	0.3360	0.3328	0.3295	0.3263
1.5	0.3232	0.3211	0.3170	0.3139	0.3109	0.3078	0.3048	0.3018	0.2989	0.2959
1.6	0.2930	0.2900	0.2871	0.2842	0.2813	0.2785	0.2758	0.2731	0.2704	0.2678
1.7	0.2652	0.2626	0.2600	0.2575	0.2550	0.2525	0.2501	0.2478	0.2455	0.2432
1.8	0.2410	0.2388	0.2366	0.2345	0.2324	0.2303	0.2282	0.2262	0.2242	0.2222
1.9	0.2203	0.2184	0.2165	0.2146	0.2128	0.2110	0.2092	0.2075	0.2058	0.2041
2.0	0.2024	0.2007	0.1991	0.1974	0.1958	0.1942	0.1926	0.1910	0.1895	0.1879
2.1	0.1864	0.1850	0.1837	0.1823	0.1807	0.1790	0.1774	0.1759	0.1745	0.1731
2.2	0.1718	0.1703	0.1688	0.1674	0.1662	0.1650	0.1637	0.1624	0.1611	0.1598
2.3	0.1585	0.1572	0.1560	0.1548	0.1536	0.1524	0.1512	0.1501	0.1489	0.1478
2.4	0.1467	0.1456	0.1445	0.1435	0.1424	0.1414	0.1404	0.1394	0.1385	0.1375
2.5	0.1366	0.1357	0.1347	0.1337	0.1328	0.1318	0.1308	0.1300	0.1292	0.1283
2.6	0.1273	0.1261	0.1250	0.1244	0.1237	0.1230	0.1222	0.1214	0.1205	0.1196
2.7	0.1188	0.1181	0.1173	0.1165	0.1158	0.1150	0.1142	0.1135	0.1128	0.1120
2.8	0.1113	0.1106	0.1098	0.1091	0.1084	0.1077	0.1070	0.1063	0.1056	0.1050
2.9	0.1043	0.1036	0.1030	0.1023	0.1017	0.1010	0.1003	0.0997	0.0990	0.0984
3.0	0.0977	0.0971	0.0964	0.0958	0.0951	0.0945	0.0939	0.0932	0.0926	0.0920
3.1	0.0914	0.0908	0.0902	0.0896	0.0891	0.0885	0.0879	0.0874	0.0868	0.0863
3.2	0.0857	0.0852	0.0846	0.0841	0.0835	0.0830	0.0825	0.0819	0.0814	0.0809
3.3	0.0804	0.0799	0.0794	0.0789	0.0784	0.0779	0.0774	0.0769	0.0764	0.0760
3.4	0.0755	0.0750	0.0746	0.0742	0.0737	0.0733	0.0729	0.0724	0.0720	0.0716
3.5	0.0712	0.0708	0.0704	0.0700	0.0697	0.0693	0.0689	0.0686	0.0682	0.0679
3.6	0.0675									

Fig. 9.39c

Thickness	k_2	k_1	f_c	Dimensional curve
$t \leq 20$	1.06	1	$1.06\,f_y$	A1, B1, C1
		0.94	$1.00\,f_y$	A2, B2, C2
$20 < t \leq 30$	1	1		
		0.94	$0.94\,f_y$	A3, B3, C3
$30 < t \leq 40$	0.94	1		
		0.94	$0.88\,f_y$	A4, B4, C4

Fig. 9.40

$\sigma_c = N_c/A$

$f_c = k f_y$ = mean compressive strength

$k = k_1 \times k_2$ = coefficient, k_1 being a reduction coefficient for some welded profiles and k_2 being a function of thickness

λ = strut slenderness

$\lambda_c = \pi\sqrt{E/f_c}$ = conventional proportionality slenderness

Numerical values of the dimensionless ratio σ_c/f_c are given in the tables of Fig. 9.39.

In ECCS Recommendations, the dimensionless curves 'a', 'b', 'c' $(\bar{N} - \bar{\lambda})$ are transformed into dimensional curves 'A', 'B', 'C' $(\sigma_c - \lambda)$ where:

$$\sigma_c = f_c \bar{N} \quad ; \quad \lambda = \lambda_c \bar{\lambda} \quad\quad (9.21)$$

To evaluate the compressive strength F_c, values of coefficients k_1 and k_2 must be stated.

The coefficient k_1 may be assumed equal to one for all shapes except for some welded ones. The welding process causes deviations of cross sectional geometry from nominal dimensions. Such imperfections have a lowering effect that may be taken into account by a fictitious reduction of yield stress.

The coefficient k_2 takes into account the variations of compressive strength due to thickness. The following three intervals are considered:

$$t \leq 20 \text{ mm} \quad ; \quad 20 < t \leq 30 \text{ mm} \quad ; \quad 30 < t \leq 40 \text{ mm}$$

The intermediate interval has a yield stress equal to the nominal one $(k_2 = 1)$; in the case of thicknesses below 20 mm, a 6% increase is accepted $(k_2 \cong 1.06)$, while for thicknesses exceeding 30 mm a 6% reduction is imposed on the yield stress $(k_2 \cong 0.94)$. These values have been justified by probabilistic considerations based on several stub column tests.

416

Therefore, for each grade of steel one passes from one class to the immediately lower one by multiplying yield stress by 0.94 and 4 dimensional curves may be worked out, according to the following values for f_c:

$$f_c = 1.06f_y \quad ; \quad f_c = 1.0f_y \quad ; \quad f_c = 0.94f_y \quad ; \quad f_c = 0.94^2 f_y$$

The table of Fig. 9.40 shows both coefficients k_2 and k_1 and gives all the possible values of strength f_c for all combinations of profiles and thicknesses.

Compressed sections can thus be related to their corresponding dimensional curves 'A', 'B' and 'C' by applying the table of Fig. 9.41. There are altogether 36 curves (3 for 4 strength variations of materials).

Rolled tubes use the most favourable curve 'a' (A1, A2, A3) whereas welded hot-worked tubes are slightly penalized by factor k_1 = 0.94 (A2,A3,A4).

Welded box sections use curve 'b' in the version reduced by factor k_1 = 0.94 (B2, B3, B4) except for butt welded thick wall sections having h/t < 30 (see Fig. 9.42)

Rolled I-sections use all three curves 'a', 'b' and 'c', according to depth width ratio (h/b) and to the buckling plane (strong or weak axis). The distinction between profiles having narrow flanges $(h/b > 1.2)$ or wide flanges $(h/b \leq 1.2)$ reflects the previously mentioned dependence of residual stress intensity and distribution on the cross sectional shape and the corresponding variation in its influence on their load carrying capacity.

In the case of welded I-profiles, a distinction is made not only with reference to the buckling plane, but also according to whether rolled or flame cut plates are used as flanges. As flame cutting produces tensile residual stresses at the ends of the plate, the corresponding effect is favourable in buckling about the weak axis and thus it allows curve 'b' to be used instead of 'c', as imposed in the case of the rolled flange. For buckling about the strong axis, however, curve 'b' is applicable to both flange forms. In any case, k_1 = 0.94.

Rolled I-sections with plates welded to their flanges are favourably affected due to the welding which acts in the same way as flame cutting. Tensile residual stresses arise in the welds located at the flange edges, increasing the column strength mainly in buckling about the weak axis. Curves to be applied are therefore 'a' for the weak axis and 'b' for the strong axis, always with k_1 = 0.94.

Tees and channels are allotted to curve 'c', as they are particularly penalized by their shape effect (see 9.2.1.2).

When welded profiles are subjected to heat treatment to relieve residual stresses, more favourable curves can be applied, assuming k_1 = 1.

Thick wall $(h/t < 30)$ butt welded box sections are assigned to dimensional curve 'B' or 'C' depending on whether column slenderness is below or above a limit of slenderness λ_{lim}, given in the table of Fig. 9.42 for European qualities of steel.

In addition to curves 'a', 'b' and 'c', two other curves have been successively introduced: 'a$_o$' upper bound above 'a' and 'd' a lower

417

SHAPE OF SECTION		NON-DIMENSIONAL CURVE	k_1	DIMENSIONAL CURVE		
				$t \leq 20$	$20 < t \leq 30$	$30 < t < 40$
ROLLED TUBES-WELDED TUBES 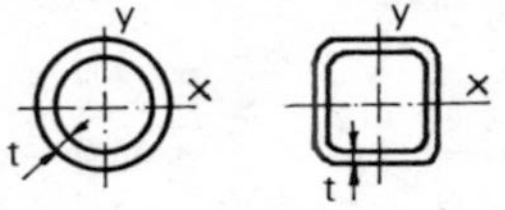	Rolled tubes	a	1.00	A1	A2	A3
	Welded tubes (hot finished)	a	0.94	A2	A3	A4
WELDED BOX SECTIONS 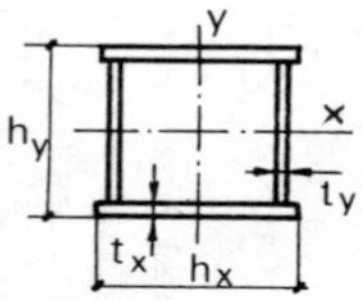	– Buckling about x–x: h_x, t_x; y–y: h_y, t_y. Not valid for heavy welds (full penetration) when $h/t < 30$	b	0.94	B2	B3	B4
I AND H ROLLED SECTIONS	– Buckling about x–x: $h/b > 1.2$	a	1.00	A1	A2	A3
	$h/b \leq 1.2$	b		B1	B2	B3
	– Buckling about y–y: $h/b > 1.2$	b	1.00	B1	B2	B3
	$h/b \leq 1.2$	c		C1	C2	C3
I AND H WELDED SECTIONS	– Buckling about x–x: a) flame cut flanges	b	0.94	B2	B3	B4
	b) rolled flanges	b		B2	B3	B4
	– Buckling about y–y: a) flame cut flanges	b	0.94	B2	B3	B4
	b) rolled flanges	c		C2	C3	C4

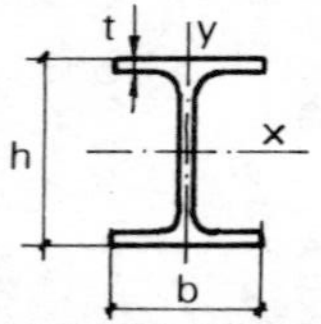

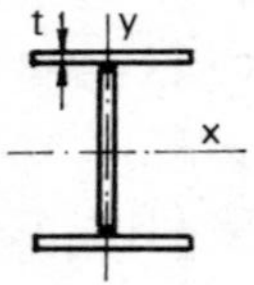

Fig. 9.41a

SHAPE OF SECTION	NON-DIMENSIONAL CURVE	k_1	DIMENSIONAL CURVE		
			$t \leq 20$	$20 < t \leq 30$	$30 < t \leq 40$
I AND H SECTIONS WITH WELDED FLANGE COVER PLATES 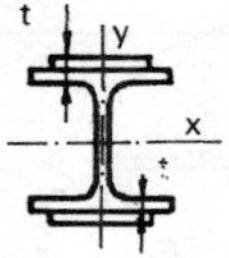 t – Buckling about $y-y$ – Buckling about $x-x$ $t = t_{imax}$	b a	0.94	B2 A2	B3 A3	B4 A4
BOX SECTIONS STRESS RELIEVED BY HEAT TREATMENT 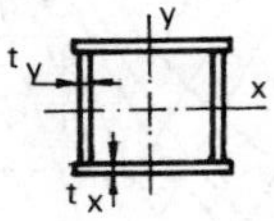 – Buckling about $x-x$: $t = t_x$ $y-y$: $t = t_y$	a	1.00	A1	A2	A3
I AND H SECTIONS, STRESS RELIEVED BY HEAT TREATMENT 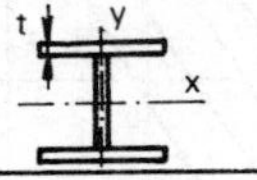 – Buckling about $x-x$ – Buckling about $y-y$	a b	1.00	A1 B1	A2 B2	A3 B3
T-SECTION – HALF I SECTIONS 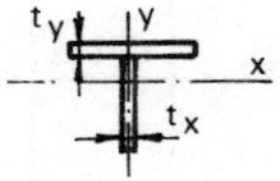 – Buckling about $x-x$: $t = t_x$ $y-y$: $t = t_y$	c	1.00	C1	C2	C3
CHANNELS 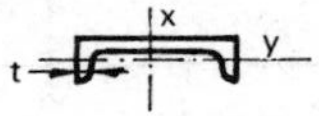	c	1.00	C1	C2	C3

Fig. 9.41b

Thickness		$t \leq 20$ mm	$20 < t \leq 30$	$30 < t \leq 40$
Fe 360	λ_{lim}	47	50	53
	$\lambda \leq \lambda_{lim}$	B2-36	B3-36	B4-36
	$\lambda > \lambda_{lim}$	C1-36	C2-36	C3-36
Fe 430	λ_{lim}	43	45	48
	$\lambda \leq \lambda_{lim}$	B3-43	B3-43	B4-43
	$\lambda > \lambda_{lim}$	C1-43	C2-43	C3-43
Fe 510	λ_{lim}	37	39	41
	$\lambda \leq \lambda_{lim}$	B2-51	B3-51	B4-51
	$\lambda > \lambda_{lim}$	C1-51	C2-51	C3-51

Fig. 9.42

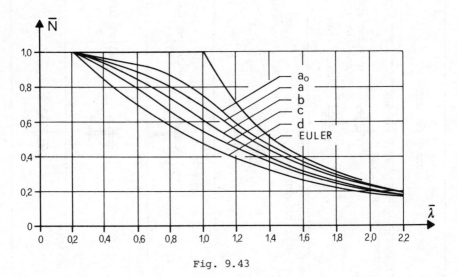

Fig. 9.43

$\lambda \Lambda_c$	0.	0.01	0.02	0.03	0.04	0.05	0.06	0.07	0.08	0.09
.0	1.0000	1.0000	1.0000	1.0000	1.0000	1.0000	1.0000	1.0000	1.0000	1.0000
.1	1.0000	1.0000	1.0000	1.0000	1.0000	1.0000	1.0000	1.0000	1.0000	1.0000
.2	1.0000	0.9983	0.9966	0.9948	0.9930	0.9910	0.9891	0.9872	0.9852	0.9833
.3	0.9813	0.9794	0.9775	0.9756	0.9737	0.9719	0.9700	0.9682	0.9664	0.9645
.4	0.9627	0.9608	0.9590	0.9571	0.9552	0.9533	0.9515	0.9496	0.9477	0.9459
.5	0.9440	0.9421	0.9403	0.9384	0.9366	0.9346	0.9327	0.9308	0.9288	0.9269
.6	0.9249	0.9229	0.9208	0.9188	0.9168	0.9148	0.9129	0.9108	0.9087	0.9065
.7	0.9040	0.9013	0.8982	0.8949	0.8914	0.8876	0.8836	0.8794	0.8751	0.8708
.8	0.8659	0.8610	0.8560	0.8509	0.8456	0.8401	0.8345	0.8267	0.8228	0.8166
.9	0.8103	0.8039	0.7973	0.7905	0.7838	0.7765	0.7692	0.7618	0.7543	0.7467
1.0	0.7390	0.7313	0.7235	0.7157	0.7078	0.6999	0.6920	0.6840	0.6761	0.6681
1.1	0.6601	0.6522	0.6443	0.6364	0.6286	0.6208	0.6131	0.6055	0.5979	0.5904
1.2	0.5831	0.5758	0.5685	0.5614	0.5543	0.5473	0.5404	0.5336	0.5268	0.5202
1.3	0.5136	0.5071	0.5007	0.4944	0.4882	0.4820	0.4760	0.4701	0.4643	0.4586
1.4	0.4529	0.4474	0.4419	0.4366	0.4313	0.4261	0.4209	0.4159	0.4109	0.4060
1.5	0.4011	0.3964	0.3917	0.3871	0.3828	0.3781	0.3737	0.3694	0.3651	0.3610
1.6	0.3569	0.3528	0.3488	0.3449	0.3410	0.3372	0.3335	0.3298	0.3262	0.3226
1.7	0.3191	0.3156	0.3122	0.3089	0.3056	0.3023	0.2991	0.2959	0.2928	0.2898
1.8	0.2868	0.2838	0.2809	0.2780	0.2752	0.2724	0.2696	0.2669	0.2642	0.2618
1.9	0.2590	0.2564	0.2539	0.2514	0.2489	0.2465	0.2441	0.2418	0.2395	0.2372
2.0	0.2349	0.2327	0.2305	0.2284	0.2262	0.2241	0.2220	0.2200	0.2180	0.2160
2.1	0.2140	0.2121	0.2102	0.2083	0.2064	0.2046	0.2028	0.2010	0.1992	0.1974
2.2	0.1957	0.1940	0.1923	0.1907	0.1891	0.1875	0.1859	0.1843	0.1827	0.1812
2.3	0.1797	0.1782	0.1767	0.1753	0.1738	0.1724	0.1710	0.1696	0.1683	0.1669
2.4	0.1656	0.1642	0.1629	0.1616	0.1603	0.1591	0.1578	0.1566	0.1554	0.1542
2.5	0.1530	0.1518	0.1506	0.1495	0.1483	0.1472	0.1461	0.1450	0.1439	0.1428
2.6	0.1417	0.1407	0.1396	0.1386	0.1376	0.1366	0.1356	0.1346	0.1336	0.1326
2.7	0.1317	0.1307	0.1298	0.1289	0.1279	0.1270	0.1261	0.1253	0.1244	0.1235
2.8	0.1227	0.1216	0.1210	0.1201	0.1193	0.1185	0.1177	0.1169	0.1161	0.1153
2.9	0.1145	0.1138	0.1130	0.1123	0.1115	0.1108	0.1100	0.1093	0.1086	0.1079
3.0	0.1072	0.1065	0.1058	0.1051	0.1045	0.1038	0.1031	0.1025	0.1018	0.1012
3.1	0.1005	0.0999	0.0993	0.0987	0.0981	0.0975	0.0969	0.0963	0.0957	0.0951
3.2	0.0945	0.0939	0.0934	0.0928	0.0922	0.0917	0.0911	0.0906	0.0901	0.0895
3.3	0.0890	0.0885	0.0880	0.0874	0.0869	0.0864	0.0859	0.0854	0.0849	0.0844
3.4	0.0839	0.0834	0.0830	0.0825	0.0820	0.0815	0.0811	0.0806	0.0802	0.0797
3.5	0.0793	0.0788	0.0784	0.0779	0.0775	0.0771	0.0767	0.0762	0.0758	0.0754
3.6	0.0750									

Values of σ_c / f_c for curve "a"

Fig. 9.44

SHAPE OF SECTION			COLUMN CURVE	k_1
TUBES		Rolled	a	1.00
		Welded (not finished)	a	0.95
WELDED BOX SECTIONS		Buckling about x-x: h_x, t_x y-y: h_y, t_y Not valid for heavy welds (full penetration) when h/t < 30	a	0.90
ROLLED I-SECTION		Buckling about x-x	a	1.00
		Buckling about x-x	b	1.00
WELDED I-SECTION		Buckling about x-x		
		- flame cut flange	a	0.90
		- rolled flange	a	0.90
		Buckling about y-y		
		- flame cut flange	a	0.90
		- rolled flange	b	0.90
ROLLED I-SECTIONS WITH WELDED FLANGE PLATE		Buckling about x-x Buckling about y-y	a	0.90
HEAT TREATED BOX SECTIONS			a_o	1.00
HEAT TREATED I-SECTIONS		- Buckling about x-x	a_o	1.00
		Buckling about	a	

bound below 'c' (Fig. 9.43). The purpose is to cover a larger range of cases, including the use of high strength steel ($f_y > 430$ Nmm^{-2}) on the one hand and of thick sections (*jumbo profiles*) on the other.

The behaviour of high strength steel profiles in more favourable than that of mild steel ones. In fact, steel quality does not at all modify the residual stress distribution, the influence of which on the pattern of the buckling curve depends only on the ratio between maximum residual stress and yield stress. Therefore, as yield stress increases, the unfavourable effect of residual stresses decreases. These intuitive considerations have been confirmed by simulation calculations that led to qauntifying curve 'a$_0$', the dimensionless values of which σ_c/f_c are given in the table of Fig. 9.44.

It provides an upper bound limit for the behaviour of high strength steel profiles in the particularly favourable case of sections which have undergone heat treatment to relieve residual stresses.

In order to cover all types of high strength steel section, curves 'a' and 'b' must also be considered, according to the table of Fig. 9.45. Also in this case yield stress is fictitiously reduced by means of coefficient k_1 set out in the table, which must be applied to evaluate both stress f_c and slenderness λ_c whenever one enters into the dimensionless curve. ECCS Recommendations are not explicit with reference to the use of coefficient k_2 but it seems reasonable to use the values of k_2 given in Fig. 9.40 as a function of thickness.

TYPE OF CROSS SECTION 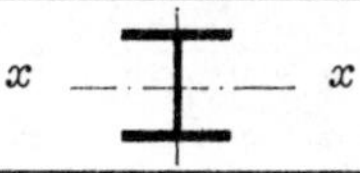 x —·—x	COLUMN CURVE
Rolled H-Shapes	
- Buckling about x-x	d
- Buckling about y-y	d
Welded H-Shapes built-up from rolled plates	
- Buckling about x-x	c
- Buckling about y-y	d
Welded H-Shapes built-up from flame-cut plates	
- Buckling about x-x	
- Buckling about y-y	c

Fig. 9.46

423

Values of σ_c/f_c for curve "d"

λ/λ_c	0.	0.01	0.02	0.03	0.04	0.05	0.06	0.07	0.08	0.09
.0	1.0000	1.0000	1.0000	1.0000	1.0000	1.0000	1.0000	1.0000	1.0000	1.0000
.1	1.0000	1.0000	1.0000	1.0000	1.0000	1.0000	1.0000	1.0000	1.0000	1.0000
.2	1.0000	0.9916	0.9829	0.9742	0.9656	0.9570	0.9487	0.9405	0.9325	0.9247
.3	0.9170	0.9093	0.9017	0.8941	0.8866	0.8790	0.8713	0.8637	0.8560	0.8483
.4	0.8407	0.8332	0.8259	0.8187	0.8115	0.8044	0.7974	0.7903	0.7833	0.7762
.5	0.7691	0.7620	0.7549	0.7478	0.7407	0.7336	0.7266	0.7196	0.7126	0.7057
.6	0.6989	0.6921	0.6853	0.6786	0.6719	0.6653	0.6587	0.6522	0.6457	0.6393
.7	0.6329	0.6265	0.6202	0.6140	0.6078	0.6017	0.5957	0.5897	0.5837	0.5776
.8	0.5720	0.5662	0.5605	0.5549	0.5493	0.5438	0.5383	0.5329	0.5276	0.5223
.9	0.5171	0.5119	0.5067	0.5018	0.4988	0.4916	0.4870	0.4821	0.4774	0.4727
1.0	0.4681	0.4635	0.4589	0.4544	0.4500	0.4456	0.4413	0.4370	0.4328	0.4286
1.1	0.4244	0.4204	0.4163	0.4123	0.4084	0.4045	0.4006	0.3968	0.3930	0.3892
1.2	0.3855	0.3819	0.3782	0.3746	0.3711	0.3676	0.3641	0.3606	0.3572	0.3538
1.3	0.3505	0.3472	0.3439	0.3407	0.3375	0.3343	0.3312	0.3281	0.3250	0.3219
1.4	0.3189	0.3159	0.3130	0.3101	0.3072	0.3043	0.3016	0.2987	0.2959	0.2932
1.5	0.2905	0.2878	0.2862	0.2826	0.2800	0.2774	0.2749	0.2724	0.2700	0.2675
1.6	0.2651	0.2627	0.2603	0.2580	0.2557	0.2534	0.2511	0.2489	0.2467	0.2445
1.7	0.2423	0.2402	0.2381	0.2360	0.2339	0.2319	0.2299	0.2279	0.2259	0.2239
1.8	0.2220	0.2201	0.2182	0.2163	0.2145	0.2126	0.2100	0.2090	0.2073	0.2055
1.9	0.2038	0.2021	0.2004	0.1988	0.1971	0.1955	0.1939	0.1923	0.1907	0.1891
2.0	0.1876	0.1861	0.1846	0.1831	0.1816	0.1802	0.1787	0.1773	0.1759	0.1745
2.1	0.1731	0.1717	0.1704	0.1691	0.1677	0.1664	0.1651	0.1639	0.1626	0.1614
2.2	0.1601	0.1589	0.1577	0.1565	0.1553	0.1542	0.1530	0.1519	0.1507	0.1496
2.3	0.1485	0.1474	0.1463	0.1452	0.1442	0.1431	0.1421	0.1410	0.1400	0.1390
2.4	0.1380	0.1370	0.1361	0.1351	0.1341	0.1332	0.1322	0.1313	0.1304	0.1295
2.5	0.1286	0.1277	0.1268	0.1259	0.1251	0.1242	0.1234	0.1225	0.1217	0.1209
2.6	0.1201	0.1193	0.1185	0.1177	0.1169	0.1161	0.1153	0.1146	0.1138	0.1131
2.7	0.1123	0.1116	0.1109	0.1101	0.1094	0.1087	0.1080	0.1073	0.1066	0.1059
2.8	0.1052	0.1045	0.1038	0.1031	0.1024	0.1018	0.1011	0.1004	0.0998	0.0991
2.9	0.0985	0.0978	0.0972	0.0965	0.0959	0.0952	0.0946	0.0940	0.0934	0.0927
3.0	0.0921									

Fig. 9.47

The second integrative curve, 'd', has been added in order to take account of particularly unfavourable residual stress distributions in thick profiles, i.e. where thickness exceeds 40 mm.

In fact experiments carried out in the USA on both rolled and welded *jumbo profiles* have shown that compressive residual stresses attain yield value at the flange ends (see 4.2). This is extremely penalizing, specially in the field of low slendernesses, which is exactly where thick profiles are prevailingly used. The introduction of curve 'd' was necessary.

Both curves 'c' and 'd' are used to check columns of this type, according to the table of Fig. 9.46. Values of σ_c/f_c for curve 'd' are given in the table of Fig. 9.47. In order to make them dimensional, ECCS Recommendations suggest using the minimum guaranteed yield stress deriving from stub column tests, with $k_1 = k_2 = 1$.

9.2.2.3 Approximate formulae for buckling curves
As shown in the foregoing paragraphs, the curves proposed in the ECCS Recommendations, which have been adopted in several national codes, originate from systematic simulation calculations, they are therefore simulation curves which are not directly derivable from an analytical formulation of a definite form.

This deficiency was overcome by Maquoi and Rondal $|48,49|$ who examined the possibility of traslating the ECCS buckling curves into formulae.

They started from the modern interpretation (formula (9.11) in 9.1.1) of Young's equation $|50|$:

$$(1 - \bar{N})(1 - \bar{N}\bar{\lambda}^2) = \eta\bar{N} \qquad (9.22)$$

later also adopted by other authors such as Ayrton, Perry, Robertson and Dutheil (see Fig. 9.1).

The problem consists in correctly choosing the dimensionless imperfection factor η so as to generate the European curves characterized by a 'plateau' $\sigma_c/f_y = 1$ in the area of low slenderness $0 \le \lambda \le 0.2$.

The following expressions for η have been taken into consideration:

$$\eta_1 = \alpha_1(\bar{\lambda} - 0.2)$$
$$\eta_2 = \alpha_2\sqrt{\bar{\lambda}^2 - 0.04}$$
$$\eta_3 = \alpha_3(\bar{\lambda} - 0.2)^2 \qquad (9.23)$$
$$\eta_4 = \alpha_4(\bar{\lambda}^2 - 0.04)$$

The value of parameter α for each European curve was determined so as to minimize the sum of scatters $(\Delta\bar{N})^2$ between the theoretical values computed by means of the analytical relation and those given in the ECCS Recommendations, for the range of slendernesses $\bar{\lambda}$ between 0.2 and 3.61, in increments of 0.1.

It has been pointed out that the first and second expressions η_1, η_2 offer the best results. Thus two alternative analytical expressions for the European curves may be assumed:

$$\bar{N} = \frac{1 + \alpha(\bar{\lambda} - 0.20) + \bar{\lambda}^2}{2\bar{\lambda}^2} - \frac{1}{2\bar{\lambda}^2}\sqrt{(1 + \alpha(\bar{\lambda} - 0.20) + \bar{\lambda}^2)^2 - 4\bar{\lambda}^2}$$

(9.24)

$$\bar{N} = \frac{1 + \alpha\sqrt{\bar{\lambda}^2 - 0.04} + \bar{\lambda}^2}{2\bar{\lambda}^2} + \frac{1}{2\bar{\lambda}^2}\sqrt{\left\{1 + \alpha\sqrt{\bar{\lambda}^2 - 0.04} + \bar{\lambda}^2\right\}^2 - 4\bar{\lambda}^2}$$

(9.25)

The values of parameter α are given in the table of Fig. 9.48 for the five characteristic European curves for both (9.24) and (9.25).

Formula 9.24		
Curve	α	Max error (%) ($\bar{\lambda} \leq 3.0$)
a_o	0.125	$-0.96\,(\bar{\lambda} = 0.5)$ $+1.24\,(\bar{\lambda} = 1.0)$
a	0.206	$-0.79\,(\bar{\lambda} = 0.6)$ $+2.15\,(\bar{\lambda} = 1.8)$
b	0.339	$-1.80\,(\bar{\lambda} = 3.0)$ $+0.51\,(\bar{\lambda} = 1.2)$
c	0.489	$-1.26\,(\bar{\lambda} = 0.8)$ $+3.23\,(\bar{\lambda} = 2.1)$
d	0.756	$-1.78\,(\bar{\lambda} = 0.7)$ $+5.71\,(\bar{\lambda} = 2.0)$

Formula 9.25			
Curve	α	Error (%) ($0.6 \leq \lambda \leq 2.1$)	Max error (%)
a_o	0.093	$+0.49\,(\bar{\lambda} = 0.8)$ $-0.53\,(\bar{\lambda} = 1.2)$	$-0.53\,(\bar{\lambda} = 1.2)$
a	0.158	$+0.48\,(\bar{\lambda} = 1.8)$ $-0.39\,(\bar{\lambda} = 0.8)$	$-1.56\,(\bar{\lambda} = 0.3)$
b	0.281	$+2.60\,(\bar{\lambda} = 0.6)$ $-2.58\,(\bar{\lambda} = 2.1)$	$-5.51\,(\bar{\lambda} = 3.5)$
c	0.384	$+1.77\,(\bar{\lambda} = 0.6)$ $-1.81\,(\bar{\lambda} = 1.0)$	$-3.84\,(\bar{\lambda} = 0.3)$
d	0.587	$+1.89\,(\bar{\lambda} = 2.1)$ $-1.87\,(\bar{\lambda} = 0.9)$	$-4.55\,(\bar{\lambda} = 0.3)$

Fig. 9.48

In the range of greatest application of curves $(0.6 \leq \bar{\lambda} \leq 2.1)$ formula (9.25) is more satisfactory on the other hand (9.24) minimizes the errors in the entire field of definition. For this reason the aim of some new codes is to adopt (9.24) in order to offer considerable advantages to designers, especially if a computer is systematically used in carrying out stability checks.

A completely different approach was proposed in |51| as an attempt at simplifying the multiple curve method and further complicated by the introduction of coefficients k_1 and k_2. According to that proposal, it is possible to refer to only one basic curve (e.g. curve 'a$_o$'), provided imaginary slenderness $\lambda*$ is used. This slenderness can be assessed on the basis of imaginary radius of gyration $i*$, that in turn is a function both of the type of the ECCS curve to which the profile is referred and of the considered slenderness range. One thus obtains a change in scale of the abscissa in the non-dimensional plane $\bar{N} - \bar{\lambda}$, variable with the abscissa itself, that refers each curve to the sole reference curve 'a$_o$'. It is thus possible to build a standard set of profiles, introducing into it, in addition to the geometrical and static properties of the various cross sections, three numbers - i_1*, i_2*, $\lambda*$ - which are useful for checking stability.

If the slenderness range is divided into two areas, each characterized by a different change in scale, and the imaginary slenderness $\lambda*$ is expressed by the formulae and values given in Fig. 9.49, errors do not exceed 3% for $\bar{\lambda} \leq 1.6$.

9.2.3 *Built-up members*

This section concerns struts made up of two or more equally cross sectioned and parallel chords interconnected either by lacing or by batten plates.

According to the type of connection between the chords, built-up members can be classed as:

Laced members (Fig. 9.50a)
Struts with batten plates (Fig. 9.50b)
Buttoned struts (Fig. 9.50c)

	$0 \leq \lambda \leq 0.5$ $\lambda = L/i_1^*$	$0.5 < \lambda \leq 1.6$ $\lambda = L/i_2^* + \lambda_1 \, 235/f_y$	
Curve	i_1^*/i	i_2^*/i	λ_1^*
a$_o$	1.00	1.00	0
a	0.76	1.15	21
b	0.65	1.18	32
c	0.58	1.23	43
d	0.51	1.27	56

Fig. 9.49

427

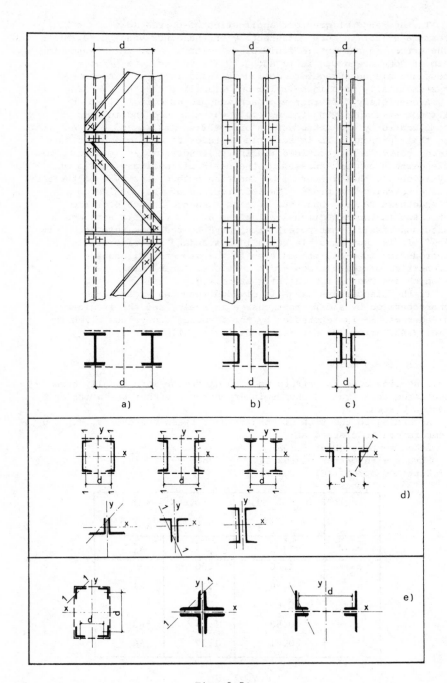

a)   b)   c)

d)

e)

Fig. 9.50

In general, such struts can be considered either as a simple or a
built-up strut depending on the plane of bending. A strut having
a cross section of the type illustrated in Fig. 9.50d, for example
must be considered as a simple strut if it bends about the x axis
(and therefore moment of inertia is assessed about axis x). On
the contrary, it must be considered as a built-up strut if it bends
about the y axis in the plane that cuts the profiles forming the
strut.

Struts having a cross section of type illustrated in Fig. 9.50e,
behave as built-up struts both in direction x and in direction y.

The behaviour of built-up members does not depend only on their
bending performance, it is also influenced by the deformability of
the connections which must absorb any sliding action between the
profiles forming the section. The connection deformation increases
the overall and lateral deflection of the memeber and thus the
unstabilizing effects of vertical loads.

In order to evaluate the load carrying capacity of built-up
members, it is necessary to analyse:

The overall behaviour of member

The local behaviour of each chord

The forces loading the connections

Overall behaviour is influenced by deformabilities due to bending
and shear, on which the lateral deflection of the member depends,
due to the presence of initial geometrical imperfections.
Bending deformability is related to the moment of inertia of the
overall cross section, the value of which, if the section is composed
by two chords, is:

$$I = 2I_1 + 2A_1 \ d^2/4 \qquad\qquad (9.26)$$

where

I_1 = moment of inertia of the chord

A_1 = cross sectional area of the chord

d^1 = distance between the centroids of the two chords

Shear deformability is related to the deformability of battens
and chords.

In the case of laced members (Fig. 9.51a), it depends on the axial
deformability of the lattice members. In struts with batten plates
(Fig. 9.51b), it is influenced by the Vierendeel beam behaviour of
the unit and therefore is a function of the flexural deformability of
chords and batten plates. Finally, in buttoned struts (Fig. 9.51c),
it depends on the flexural deformability of the chords and, if the
joint is bolted, on bolt slip due to hole-bolt clearance.

The behaviour of each chord between one joint and the next depends
on the type of connection. In general, in the case of laced struts
(Fig. 9.51a) each chord can be considered as a simple strut, having
a buckling length equal to the joint spacing. In the case of struts
with batten plates (Fig. 9.51b), the chords are compressed and bent,
with a practically bi-triangular diagram with opposed bending moments.
In fact, the bending moments can be assessed in close approximation
with reference to the undeformed situation, while the vertical load
buckling effect can be considered balanced by equal and opposite

429

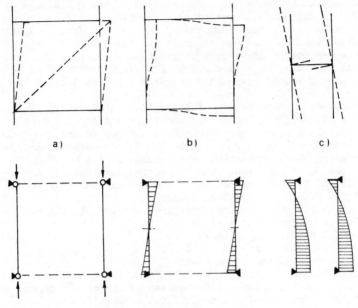

Fig. 9.51

axial forces in the chords. In buttoned struts, the single chord
is compressed and bent with a bending moment distribution that cannot
be considered bi-triangular (Fig. 9.51c). In this case the overall
lateral bending contribution becomes significant compared to the
local one of the single chord so that bending moment values must be
computed with reference to the deformed configuration.

The influence of local behaviour on the overall performance of
a strut is difficult to quantify. It is preferable to ignore such
interaction but, at the same time, to set dimensional limitations to
built-up member geometry which if complied with, garantee that the
overall behaviour of the strut is practically independent of local
behaviour of any single chord.

Forces loading connections depend on:
(a) Transverse external loads which may be acting
(b) Strut deformability due to initial imperfections
(c) Vertical load eccentricity

Condition a) occurs in built-up beam-columns and it must be
superposed on conditions b) and c).

The importance of b) grows as slenderness increases, while c)
has a greater influence at low slendernesses. These facts are
considered by the various codes in different ways.

Fig. 9.52 shows the ratio V_i/N_c as a function of slenderness,
assuming:

V = maximum shear on the basis of which the force acting on the
connection is assessed

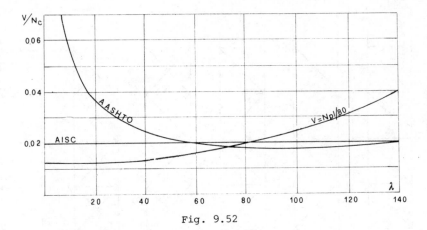

Fig. 9.52

N_c = maximum load the strut can bear under compression

The AASHTO-AREA Specification |52,53| gives more importance to c). Recommendations attributed to DIN 4114 |54| are more cautious in considering b), by requiring a value of V as a fraction of the maximum compressive load N_{pl} which can be borne when buckling phenomena are absent, while they neglect c). Finally, the AISC |55| rules combine the two contributions, by suggesting a constant ratio V/N, which is certainly the simplest solution.

The load carrying capacity of built-up members is, however, determined on the basis of the equivalent slenderness criterion (see 9.1.3.3). It also depends on the inertia radius of gyration of the built-up section:

$$i^2 = i_1^2 + d^2/4 \eqno(9.27)$$

where:

i_1 = radius of gyration of the chord cross section in the direction of the considered deflection plane

d = distance between chord centroids

On the basis of the contribution of i_1, built-up members can thus be divided into two classes, independent of connection type:

Struts with spaced chords $(d > 6i_1)$
Struts with close chords $(d < 3i_1)$

For the first class it is possible to neglect the bending contribution of single chord with respect to the overall bending contribution of the cross section assumed to be made of two concentrated masses. For the second class this is no longer correct, as it would entail a too significant penalty. It is only in the last few years the ECCS has developed a unified approach to the problem indicating a calculation method that is independent of the chord position.

Some design criteria for struts having either spaced or close chords are given in the following as well as the criteria based on

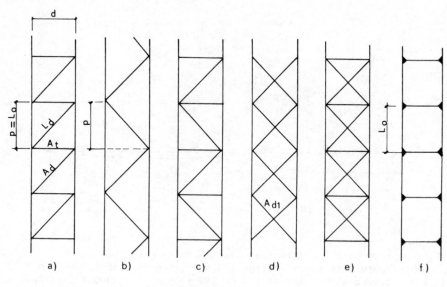

Fig. 9.53

the ECCS method.

9.2.3.1 Built-up members with spaced chords

The most commonly adopted schemes for struts composed of two spaced chords are illustrated in Fig. 9.53. The following symbols will be adopted:

L = overall length of strut

A_1 = area of the individual chord

A = $2A_1$ overall cross sectional area

d = distance between the centroids of the chords

I_1 = moment of inertia of the individual chord

I = $2A_1 \, d^2/4 + 2I_1$ = moment of inertia of overall cross section

i_1 = $\sqrt{I_1/A_1}$ = radius of gyration of the individual chord

i = $\sqrt{I/A}$ = radius of gyration of overall cross section

λ = L/i = strut slenderness

For laced struts(Fig.s 9.53a to e) the following will also be used:

L_d = diagonal length

L_o = chord length between two successive joints

p = distance between two successive joints

L_t = d = distance between centroids of chords

A_d = cross sectional area of diagonal (= $2A_{d,1}$ for Fig. 9.53d,e)

A_t = cross sectional area of bar perpendicular to chords

For struts with batten plates (Fig. 9.53f) the following also apply:

L_o = batten plate distance

I_c = batten plate moment of inertia

λ_1 = L_o/i_1 = local chord slenderness between one batten plate and the next

The first studies aimed at determininig equivalent slenderness were performed around 1910 by Mann, Engesser, Wentzel and others. Around 1930, they were resumed and improved by Von Mises, Ratzersdofer, Chwalla and systematized by Bleich |56|. They are on the safe side as they neglect the moment of inertia of the individual chords $2I$ in comparison with their moment about the overall centroidal axis. The overall moment of inertia of the built-up member is therefore assumed to be:

$$I \cong 2A_1 \, d^2/4 \qquad (9.28)$$

Using this assumption, the critical load in the purely elastic field for the lattice scheme in Fig. 9.53a is:

$$N_{cr} = \frac{\pi^2 EI}{L^2} \frac{1}{1 + \dfrac{\pi^2 I}{L^2} \dfrac{1}{L_t^2 L_o} \left(L_d^3/A_d + L_t^3/A_t \right)} \qquad (9.29)$$

If one assumes $\lambda^2 = L^2/i^2 = L^2 A/I$, the corresponding critical stress becomes:

$$\sigma_{cr} = N_{cr}/A = \frac{\pi^2 E}{2} \frac{1}{1 + \dfrac{\pi^2 A}{\lambda^2 L_t^2 L_o} \left(L_d^3/A_d + L_t^3/A_t \right)} \qquad (9.30)$$

A simple strut having the same critical stress in the purely elastic field has a slenderness λ_{eq} such that:

$$\sigma_{cr} = \frac{\pi^2 E}{\lambda_{eq}^2} = \frac{\pi^2 E}{\lambda^2 + \pi^2 (A/A_d)(L_d/L_o)\left\{ L_d^2/L_t^2 + (A_d/A_t)(L_t/L_d) \right\}} \qquad (9.31)$$

whence:

$$\lambda_{eq} = \sqrt{\lambda^2 + \pi^2 (A/A_d)(L_d/L_o)\left\{ L_d^2/L_t^2 + (A_d/A_t)(L_t/L_d) \right\}} \qquad (9.32)$$

By an analogous procedure, an equivalent slenderness expression is obtained for laced struts of the types illustrated in Figs 9.53c,d,e:

$$\lambda_{eq} = \sqrt{\lambda^2 + \pi^2 \frac{A}{A_d} \frac{L_d^3}{L_o L_t^2}} \qquad (9.33)$$

Formula (9.33) is also applicable to the type shown in Fig. 9.53b, provided L_o is replaced by p.

In the case of struts with batten plates (Fig. 9.53f), an analogous procedure can also be adopted and, if shear deformability of the batten plate is neglected, one obtains

$$\lambda_{eq} = \sqrt{\lambda^2 \left(1 + \frac{\pi^2}{12} \frac{I}{I_c} \frac{L_o d}{L^2} \right) + \frac{\pi^2}{12} \lambda_1^2} \qquad (9.34)$$

433

If the strut is properly designed, the batten plate is practically undeformable ($I_c = \infty$) and therefore the term in parenthesis can be assumed equal to one; it then follows that:

$$\lambda_{eq} = \sqrt{\lambda^2 + \lambda_1^2 \, \pi^2/12} \qquad (9.35)$$

For reasons both of simplicity and safety, some codes replace the term $\pi^2/12$ in (9.35) with one.

The above formulae do not take into account possible stiffened zones at the ends of the member (Fig. 9.54a), eccentricities in the joints (Fig.9.54b) and net span of chord between batten plates (Fig. 9.54c). Such aspects have recently been taken into consideration by Johnston and others |57|, by defining a non-dimensional parameter μ characterizing the shear deformability in built-up members.

Adopting the above defined symbols and referring to Fig. 9.54b, for a laced strut with identical chords:

$$\mu = \frac{\xi_t}{1 + \xi_c} \, (d/L_a)^2 \, A/A_d \left\{ \frac{d}{\xi_c L_o} \left(1 + \frac{\xi_c L_o}{d}^2 \right)^{\frac{3}{2}} + \frac{d}{\xi_c L_o} A_d/A_t \right\} \qquad (9.36)$$

For a strut with batten plates (Fig. 9.54c)

$$\mu = \left\{ \frac{1}{L_a/i_1} + \left(\frac{d}{2L_a} \right)^2 \right\} \left\{ (A_1/A_c) \left(\frac{L_o}{6i_c^2} + \frac{5.2 \, L_o}{d} \chi_c \right) \right.$$

$$\left. + 2.6 \, \xi_c \chi_1 + \xi_c^3 (d/i_1^2)^2/12 \right\} \qquad (9.37)$$

where, in addition to the symbols defined in Fig.9.54:
A_c = area of a batten plate
i_c = radius of gyration of a batten plate in this plane
χ_i, χ_c = shear factors of chords and batten plates (equal to 1.2 for rectangular sections)

Once parameter μ has been computed, it is possible to determine the equivalent slenderness λ_{eq} function of the ratio a/L, by using the curves of Fig. 9.55.

Once equivalent slenderness has been computed as set out above, the critical stress can be evaluated. It is conservative to derive the critical stress from curve 'c' (see 9.2.2).

The local behaviour of the strut must then be checked. For laced struts, codes indicate the criterion that chords must be dimensioned so that their local slenderness λ_1 does not exceed overall strut slenderness λ.

In schemes having a hyperstatic lattice of the type illustrated in Fig 9.53e, the external load induces secondary stresses in the lattice due to the influence of axial deformability of the bars.

N_d being the axial force in the diagonals (Fig. 9.56), the chords are compressed by a force equal to $N/2 - N_\alpha \sin \alpha$, while the horizontal bar is subjected to a tensile force $N_t = 2N_d \cos \alpha$. The solution of the hyperstatic problem is then given by:

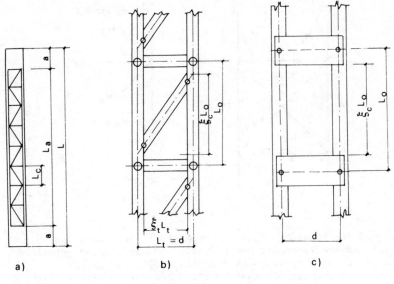

Fig. 9.54

$$N_d = \frac{F}{2} \frac{A_d}{A} \frac{\sin^2\alpha}{1 + A_d/A_t \cos^3\alpha + A_d/A \sin^3\alpha} \qquad (9.38)$$

Ratio A_d/A is usually negligible with respect to one.

Therefore, in order to render the values of N_d more favourable, diagonals should not have high inclinations ($\alpha \leq 45°$) and their area should be bigger than that of the transverse bar.

The above considerations are not applicable if the diagonals are

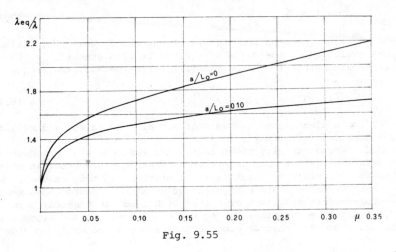

Fig. 9.55

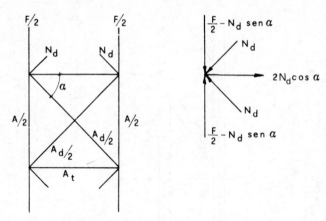

Fig. 9.56

designed for tension. Then the hyperstatic scheme of Fig. 9.53e reverts to the isostatic one of Fig. 9.53a and equivalent slenderness must be computed by means of formula (9.32).

Stricter precautions must be taken for struts with batten plates. In such struts, the chords are compressed and bent and their local behaviour might be more penalizing than the overall one.

The commonly adopted criterion is to check chord strength only on the basis of the loading components due to external loads and to ensure that interaction between local and overall behaviour is negligible.

DIN 4114 |54| limits to 50 the chord local slenderness in the buckling plane of the overall struts. A derogation of this limit is allowed only if the axial force is small or if the overall strut slenderness λ_x with respect to the x axis (see Fig. 9.50d) is high. Then $\lambda_1 = 50$ can be exceeded, provided:

$$\lambda_1 \leq (4 - 3\sigma/\sigma_c) \; \lambda_x/2 \tag{9.39}$$

$\sigma = N/A$ being the mean compressive stress and σ_c being the ultimate stress evaluated on the basis of equivalent slenderness λ_{eq}. If one operates according to the permissible stress method, the inequality becomes:

$$\lambda_1 \leq \{4 - 3(\sigma/\sigma_{adm}) \; (\sigma_{adm}/\sigma_c) \}\lambda_x/2 \tag{9.40}$$

σ being evaluated on the basis of actual load values and not increased by coefficients γ_f.

Load bearing capacity might be further reduced as a consequence of the deformability of the connection between chords and batten plates. As it is impossible to quantify it, limitation in this case also must be set in order to avoid a significant penalty. Connections must therefore be either welded or bolted with adequate tightening to prevent sensible rotations due to the clearance. As an alternative Johnston indicates that deformable joints are acceptable, provided

436

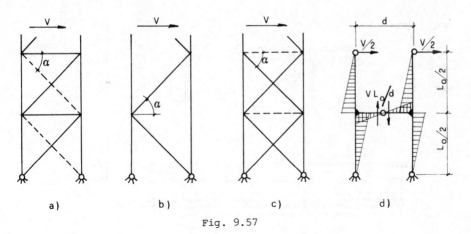

Fig. 9.57

(Fig. 9.54c):

$$\lambda_1 = L_o/i_1 \leq 14\sqrt{235/\sigma}\ \Big/\xi_c \tag{9.41}$$

$\sigma = N/A$ being the mean stress in the column, expressed in N mm^{-2}

Lattices and batten plates must be checked for the effects due to imaginary forces V illustated in Fig. 9.57.

With reference to the scheme of Fig. 9.57a, if there is only one diagonal or if, when there are two, only the tension one is considered efficient, one has:

In the diagonal $N_d = V/\cos \alpha$
In the transverse bar $N_t = V$

For the arrangement in Fig. 9.57b, the diagonal must be designed for both tension and compression for a value $N_d = V/\cos\alpha$.

In Fig. 9.57c, the diagonals must resist the force $N_d = V \cos\alpha/2$, both in tension and in compression. If a transverse bar (dashed in the figure) is added, the scheme becomes hyperstatic and, in addition to force N_d, the diagonals must resist the secondary stresses caused by the deformability of the bars. In the case of struts with batten plates (Fig. 9.57d), force V acts in two equal parts in the chords and, from equilibrium, the force acting on the batten plate is given by VL_o/d. The batten plate and its connection to the chord must be designed for the shear force VL_o/d and a moment $VL_o/2$.

The chords must be checked for the axial component $N/2$ and for a bending moment equal to $V \xi_c L_o /4$ (see Fig. 9.54c for the definition of ξ_c). As already pointed out it is sufficient to check chord strength only, provided the above mentioned dimensional limitations are complied with, as they are imposed to prevent interaction between local and overall stability phenomena.

9.2.3.2 Built-up members with close chords

Built-up members having close chords are not often used as columns, but rather to form lattice bars for truss structures.

Some typical sections are illustated in Fig. 9.58. Some are composed of two channels (Fig. 9.58a) or of two equal or unequal

437

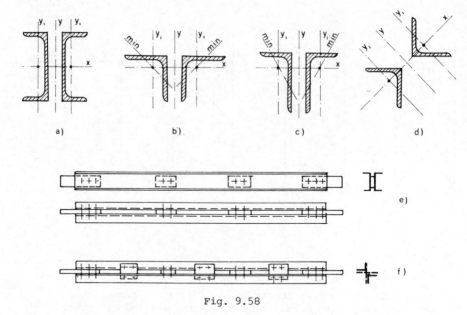

Fig. 9.58

angles (Figs 9.58b,c) connected by means of welded or bolted plates (Fig. 9.58e).

Two angles can also be arranged as a 'butterfly' or 'cruciform' section by alternately connecting their opposite sides (Fig. 9.58f).

Before facing the calculation of such built-up struts, it should be pointed out that not all struts formed by two angles (Fig.s 9.58b, c) must be considered as built-up members.

In many cases it might be useful to provide a connection between the two angles which does not resist sliding forces but must only prevent buckling of the angle in the weakest direction.

Consider, for example, Fig. 9.59: the two angles are connected by a pin-ended bar, which therefore is not capable of resisting sliding forces. The two angles cannot buckle in their weakest direction (Fig. 9.59a), because it would violate the constraint conditions. In order to avoid violating the constraint (Fig.9.59b), one angle must buckle in the weakest direction and the other in its strongest direction. It ensues that the second angle supports the first one and prevents its buckling. There remain, therefore, only the two possibilities illustrated in Figs 9.59c, d. The two angles buckle in the same direction, behaving as two single and simple struts working in parallel (Fig. 9.59 e). The connections thus have just a kinematic function without resisting sliding forces.

The case illustrated in Fig. 9.60 is different. The connection can resist sliding forces and the two angles behave as simple struts but when bending in direction x (about y), the section behaves as a built-up member (Fig. 9.60c), where the connections have a static function (Fig. 9.60c).

If angle sides are equal ($i_x = i_y$), there is no reason for

438

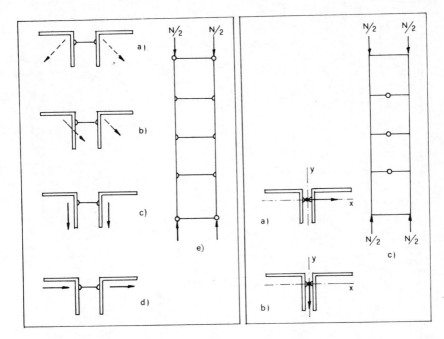

Fig. 9.59   Fig. 9.60

statically   connecting the two bars, when their buckling lengths
are equal in both planes. Conservatively, it might be economical
to connect the angles statically, only if their sides are not equal
($i_x > i_y$) or if buckling lengths in the two planes differ. However,
a suitable composition can equalize the slenderness λ_x and λ_y in
the two planes. For the same reason, it is always convenient to
compose a section of two channels if buckling lengths in the two
planes are equal. The cruciform solution is particularly efficient,
because the minimum radius of gyration i_y of the whole section (Fig.
9.58d) is equal to the maximum radius of gyration of the single angle.

Built-up struts having statically   connected close elements can
be considered as battened struts, so that it is no longer admissible
to neglect the moment of inertia of the chords.

It was shown |58,59| that the problem can be solved by numerical
simulation. In this way a behaviour substantially differing from that
of battened struts with spaced elements is obtained.

For example, analysing a compound strut divided into 5 fields by
n = 4 intermdiate connections (Fig. 9.61a) set at a distance L_0 = $50i_1$
and having overall slenderness λ = 130, one obtains mid-span
displacement over length v/L forms the abscissa, and mean stress
σ_m = $N/2A$ over yield stress (f_y = 235 N mm^{-2}) is represented on the
ordinate. The curves of Fig. 9.61c are of greater interest. They
show mean stress values σ_1 and σ_2 in the two chords, as a function of
mean stress σ_m = $N/2A$ (and therefore of external load). If there

439

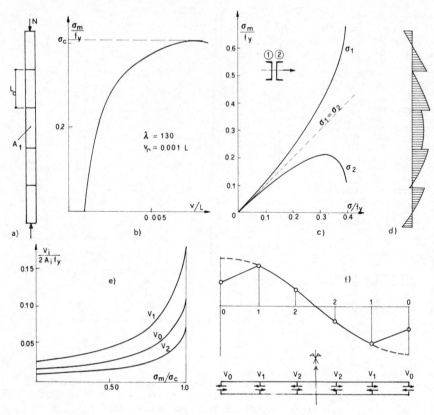

Fig. 9.61

were no bending effects, $\sigma_1 = \sigma_2 = \sigma_m$ (dashed line in the figure). On the contrary, due to lateral bending, one chord is compressed more than the other and, in particular, chord (2) tends to unload once it has reached a given maximum compression. Fig. 9.61d shows the bending moment distribution on a chord. The diagram is not bi-triangular and this demonstrates the contribution of the chords to the overall behaviour in bending. The difference $\sigma_1 - \sigma_2$ of mean stresses in the two chords is proportional to the force which the connection undergoes. Fig. 9.61e shows the ratio between the force acting on each connection (Fig. 9.61f) and the fully plastic force in the strut, as a function of the ratio σ_m/σ_c between the mean and maximum bearable stresses.

It is not surprising that the end connection (point 0) is less loaded from bending than the intermediate one (point 1): it undergoes sliding forces relative to only half a field and not an entire field. On the other hand, the end connection must transmit to the strut not only the sliding force but also the concentrated effect of the external load . Therefore in practice it is the most loaded one.

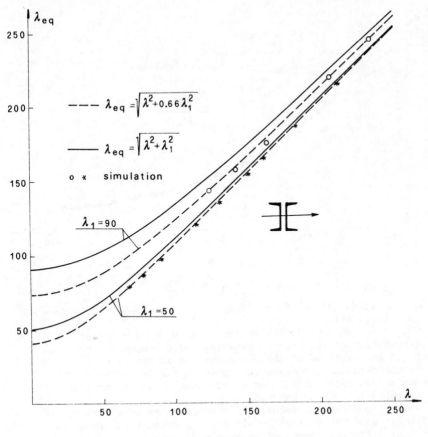

Fig. 9.62

Forces V acting on the connections are proportional to the function $\cos x/L$ (Fig.9.61f). This statement has been confirmed by many numerical checks and it is a logical consequence of the practically sinusoidal shape of a deflected strut until collapse.

From the point of view of design, it is necessary to define equivalent slenderness and forces V acting on the connections.

If an infinitely elastic behaviour of the material is assumed it is possible to evaluate the first critical load of the system N_{cr} and thence the equivalent slenderness:

$$\lambda_{eq} = \pi \sqrt{\frac{2EA_1}{N_{cr}}} \qquad (9.42)$$

Fig. 9.62 compares values of λ_{eq} obtained from this equation for two different values of local slenderness of chords (50 and 90) and those obtained using the approximate expressions:

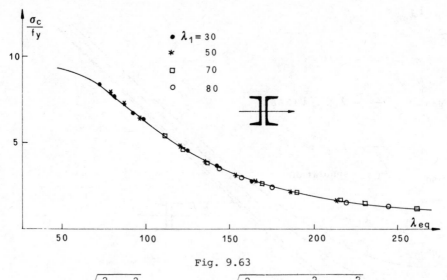

Fig. 9.63

$$\lambda_{eq} = \sqrt{\lambda^2 + \lambda_1^2} \quad \text{and} \quad \lambda_{eq} = \sqrt{\lambda^2 + 0.80 \ (\pi^2/12) \ \lambda_1^2} \qquad (9.43)$$

where the slenderness λ is evaluated on the basis of the total moment of inertia (9.26).

The dashed curve from the second expression is a better interpretation of the elastic behaviour of a compound strut. One must not be surprised at the 80% reduction in the effects of strut local slenderness λ_1 compared to those in struts having spaced elements. In fact, chord bending stiffness significantly contributes to overall behaviour.

Fig. 9.63 shows the results of a check aimed at assessing the reliability of the equivalent slenderness criterion. It gives for four different values of local slenderness λ_1, values representing maximum load carrying capacity as equivalent slenderness varies. The results were obtained by simulating the behaviour in the elastic-plastic field of compound struts having close elements and they are compared with the bending curve of a simple strut having a section of the same shape. They confirm and fully justify this type of approach, at least for compound struts having close elements and for high values of local slenderness λ_1. A penalty on load carrying capacity due to interaction between local and overall stability was noticed only for local slenderness values $\lambda_1 > \lambda_e$.

If a sinusoidal shape is assumed for the deflection line, it is possible to deduce the resultant V_{tot} of forces V_i loading the connections located in one of the two strut halves. Referring to Fig. 9.61f:

$$V_{tot} = V_o + \overset{n/2}{\underset{1}{\Sigma_i}} V_i = V_o + 2V_o \overset{n/2}{\underset{1}{\Sigma_i}} \cos i \ \pi/m \qquad (9.44)$$

where:
V_o = force loading the end connection

442

n	m	k_o	k_1	k_2	k_3	k_4
2	3	0.50	0.50			
4	5	0.31	0.50	0.19		
6	7	0.22	0.40	0.28	0.10	
8	9	0.17	0.33	0.27	0.17	0.06

Fig. 9.64

V_i = force loading the i-th intermediate connection
m = $n + 1$ = number of fields of equal length into which the strut is
  divided.
  Therefore:

$$V_o = k_o V_{tot} \quad ; \quad V_i = k_i V_{tot} \tag{9.45}$$

where $1/k_o = 1 + 2 \cos i\pi/m$: $k_i = 2k_o \cos i\,\pi/m$
  Values of k_o and k_i are given in the table of Fig. 9.64 for struts
having 3,5, 7, 9 fields.
  The value of resultant V_{tot} depends on the applied load N, the
ratio d/i, strut slenderness λ and the mid-length deflection v_o and
as a close approximation it can be expressed by:

$$V_{tot} = N\,\alpha\,\frac{d}{i}\,\frac{v_o/L}{1 - N/N_{cr}}\,\lambda \tag{9.46}$$

$N_{cr} = \pi^2 EA/\lambda^2$ being the elastic critical load and α being a numerical
coefficient. To obtain slightly excessive and therefore safe values,
one can assume $\alpha = 0.25$.
  Intermediate connections are usually identical to each other and
therefore only the first needs checking. The end connection must
usually also transmit the applied load N, in addition to the sliding
force V_o.
  In the case of built-up members with bolted connections, in
order to apply the above formulae, the hole clearance effect must
be avoided. The possible inelastic sliding of the joints causes an
increase - even a considerable one - in strut deflection. It thus
appears advisable either to reduce working tolerances or to tighten
the bolts so as to allow transmission by friction of forces V_i in
connection with the axial force corresponding to the design load (or
with value νN if the design follows the permissible stress method).

9.2.3.3 ECCS Recommendations

ECCS Reccommendations |31| indicate a general method for designing
built-up members independent of the type of connection and the
distance between the chords. It can be summed up as follows. The
critical load is evaluated in the purely elastic field on the basis
of a reduced moment of inertia I_{red}:

$$I_{red} = 2A_1 \, d^2/4 + 2kI_1 \tag{9.47}$$

The reduction coefficient $k < 1$ is assumed to have the following values:

For laced struts $k = 0$

For battened struts:

$$k = \begin{cases} 1 & \text{, for } \lambda = 75 \\ 2 - \lambda/75 & \text{, for } 75 \le \lambda \le 150 \\ 0 & \text{, for } 150 \le \lambda \end{cases}$$

where the slenderness $\overline{\lambda}$ is evaluated on the basis of the total moment of inertia (9.26).

Equating the critical load N_{cr} to that of the equivalent strut

$$N_{cr} = \frac{\pi^2 EI_{red}}{L^2} \; \frac{1}{1 + \dfrac{\pi^2 EI}{L^2 GA_w}} = \frac{\pi^2 EA}{\lambda_{eq}^2} \tag{9.48}$$

the equivalent slenderness may be defined:

$$\lambda_{eq} = \sqrt{\lambda_{red}^2 + \pi^2 \frac{EA}{GA_w}} \tag{9.49}$$

Shear stiffness in the elastic field GA_w is defined as the ratio between shear force V and the corresponding shear deformation Δ/L_o of a field of the strut (Fig. 9.65a,b). In the case of battened struts, it must be reduced by a factor $\pi^2/12$, as a precaution against interaction between local and overall stability.

Equivalent slenderness can also be assessed without taking shear stiffness GA_w into consideration. It is sufficient to adopt the appropriate formula from (9.32) to (9.37), provided the value of λ is replaced by the reduced value λ_{red} and in the case of battened struts, without applying the reduction coefficient $\pi^2/12$ to local slenderness λ_1.

Therefore, as an effect of applied load N, the strut is assumed to bend according to the following relation (Fig. 9.65c):

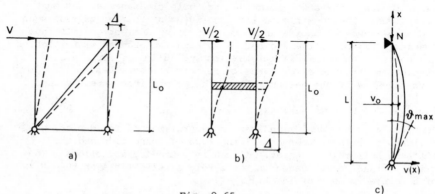

a)   b)   c)

Fig. 9.65

$$v(x) = v_o \frac{1}{1 - N/N_{cr}} \sin \frac{\pi x}{L} \tag{9.50}$$

The value of v_o includes two effects ($v_o = v_1 + v_2$). The first corresponds to initial out-of-straightness of the strut and the second takes into account possible load eccentricity and residual stresses. On the whole, it can be assumed $v_o = 0.002\ L$, i.e. twice the geometrical imperfection that is considered in simple strut calculations.

The strut must be checked at the middle and end sections, where bending and shear forces, respectively, are greatest. At the mid-section the strut must be checked for the effect of the axial force N and bending moment $M = Nv_o (1/1 - N/N_{cr})$.

The mean compression in the most loaded chord is:

$$\sigma_m = \frac{N}{2A_1} + \frac{M}{I_{red}}\ d/2 \tag{9.51}$$

This must be less than the ultimate stress σ_c related to the chord slenderness λ_1.

At the end section axial force is N and shear force $V = N \tan \theta_{max}$
The value of $\tan \theta_{max}$ is derived from formula (9.50). It is:

$$\tan \theta_{max} = \left\{ \frac{dv}{dx} \right\}_{x = 0} = \pi \frac{v_o}{L} \frac{1}{1 - N/N_{cr}}$$

For laced struts, it is sufficient to check the diagonal and transverse bars with respect to axial forces produced by shear V. By applying this method, V is directly evaluated and therefore assumed definitions of its value (Fig. 9.52) are not necessary.

In the case of battened strut, both effects caused by V and N are present in the chord. Each chord is thus loaded by axial force $N_1 = N/2$ and bending moment $M = V\xi_c L/4$ (see 9.2.3.1).

It is necessary to check that $M^c \leq M_u(N_1)$ where $M_u(N_1)$ is the ultimate moment of the section, compatible with the axial load. If the chord sections are not symmetrical, ultimate moments in the two chords are different. In this case, it is possible to refer to the mean value of the ultimate moment between the two planes of the section.

Finally, connections between batten plates and chords must be checked, according to the method set out in 9.3.2.1, and always on the basis of shear force V as computed above.

The main advantages of the ECCS method lie in the fact that its approach is general and shear force V can be directly evaluated. At the same time, as in battened strut design a distribution with opposite bending moments is considered (Fig. 9.51b), this method leads to safe results in defining shear V for struts with close elements.

Confirmation of this statement is given by Fig. 9.66, showing the ratio between value of V computed by numerical simulation and that obtained by the ECCS method. The latter results overestimate the shear force, particularly when the number of intermediate connections is high and the ratio d/i is low

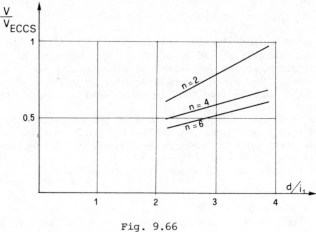

Fig. 9.66

9.2.4 *Torsional buckling*

A strut might also become unstable due to torsional twist. On the hypothesis that the shear centre coincides with the centre of gravity of the cross section, the phenomenon is independent of that related to flexural buckling and therefore can be analysed separately.

In the elastic field, the differential equation governing the phenomenon is:

$$EI_\omega \theta^{IV} + (I_o N/A + GI_T)\theta'' = 0 \qquad\qquad (9.52)$$

where I_ω and I_T are the sectional and torsional moments of inertia of the cross section, as defined in 8.2.5, and I_o is its polar moment of inertia.

Obviously, the first term becomes significant for sections with two flanges (Fig. 9.67a). For cruciform sections (Fig.9.67b), torsion due to warping can be neglected and the equation becomes a second order one.

With the edge conditions $\theta = \theta'' = 0$ (8.88) corresponding to end

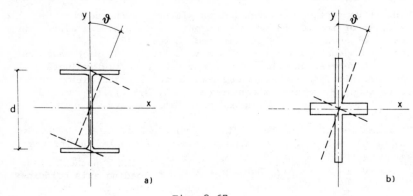

Fig. 9.67

torsional supports set at a distance L, i.e. preventing twist but allowing possible warping, the following critical load is obtained:

$$N_{cr,\theta} = \frac{\pi^2 EA}{L^2}\left\{I_\omega/I_o + \frac{L^2}{\pi^2}\frac{GI_T}{EI_o}\right\} = \frac{\pi^2 EA}{\lambda_{eq}^2} \tag{9.53}$$

It is thus possible to define an equivalent slenderness:

$$\lambda_{eq} = L/i_{eq} \quad \text{with} \quad i_{eq} = \sqrt{\frac{I_\omega}{I_o} + \frac{L^2}{\pi^2}\frac{GI_T}{EI_o}} \tag{9.54}$$

As a close approximation, one can assume $|56|$:
$I_\omega = d^2/4\, I_y$ for I-sections
$I_\omega = 0$ for cruciform sections
$I_T = \Sigma t_i^3 b_i/3$ for sections composed of a number of rectangles having dimensions $b_i t_i$.
For I-sections, therefore (Fig. 9.67a):

$$i_{eq} = \sqrt{\frac{d^2}{4}\frac{I_y}{I_x + I_y} + 0.0390\, L^2 \frac{\Sigma t_i b_i/3}{I_x + I_y}} \tag{9.55}$$

and, for cruciform sections in which $d = 0$ (Fig. 9.67b):

$$i_{eq} = 0.197\, L\sqrt{\frac{\Sigma t_i^3 b_i/3}{I_x + I_y}} \tag{9.56}$$

Further the equivalent radius of gyration depends on the buckling length of the strut. In practice, the buckling length in relation to pure torsional buckling is not greater than that for flexural buckling. It ensues that, in the case of I-profiles, torsional buckling phenomena may be neglected. It is not the same for cruciform profiles, for which a check is in any case advisable.

Once equivalent slenderness λ_{eq} has been defined, it is possible to check that:

$$\sigma = N/A \le \sigma_c(\lambda_{eq}) \tag{9.57}$$

σ_c being the buckling stress conservatively deduced from curve 'c' (Fig. 9.39) for slenderness λ_{eq}.

9.3 BENT MEMBERS

9.3.1 *Analysis of the problem*

From a physical point of view, the problem of flexural-torsional stability consists in the fact that a beam can simultaneously bend in the lateral plane and twist, until it reaches collapse without having displayed all the bending resources that would have appeared available in a safety check ignoring such a buckling phenomenon. Fig. 9.68 illustrates the phenomenon with reference to a cantilever loaded at its free end, so as to allow a simple and intuitive determination of characteristic parameters, on the hypothesis that the beam has a constant section and that the loading axis coincides with the principal inertia axis according to which the section can

447

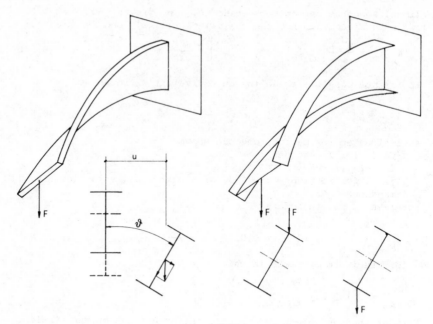

Fig. 9.68

provide its maximum bending stiffness. Static and geometrical
parameters influencing stability are: the bending and torsional
stiffnesses of the cross section and the position of the load
application point in relation to the centre of gravity, together
with beam length, load and constraint distribution.

From a geometrical point of view, it is also apparent that, when
describing the buckled configuration of beam within the above
hypothesis, two mutually dependent functions are necessary, to
represent lateral displacement u and rotation θ of the cross section
in relation to the abscissa x of the beam axis.

In the case of industrial beams, checks for flexural-torsional
stability must be performed to determine the length L at which
restraints must be provided to prevent lateral buckling. Such restraints
can be provided by the end connections or by specially installed
intermediate members or by transverse beams properly connected to
the beam being checked (Fig. 9.69).

In other cases there is a continuous constraint along the beam axis
due to the presence of a concrete slab or profiled steel sheet welded
or bolted to the beam.

The position of the load relative to the centre of gravity has a
rather important influence. If the load is applied to the upper
flange, its effect is unstabilizing, whereas if it is applied to the
lower flange, it produces a stabilizing effect that increases
critical load.

The interest in this problem is evidenced by the vast technical

448

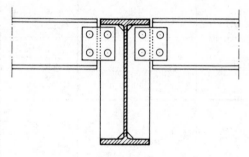

Fig. 9.69

literature that, in the field of steel construction, has covered the subject from both the theoretical and experimental aspects as well as with reference to design recommendations.

Flexural-torsional buckling of beams was first studied, in the elastic field and for rectangular sections, by Prandtl and, simultaneously, by Mitchell (1899)

For I-section steel beams, the first elastic solution was proposed by Timoshenko (1910), who introduced the warping effect into the buckling phenomenon in pure bending. This approach was extended by Timoshenko |60| and Bleich |56| to more general cases.

Starting from these theories, many authors tried to provide approximate solutions, often for practical purposes, taking into account various load and constraint conditions. The result of such studies in the elastic field is synthetically summarized in the following expression for the critical moment for wide flange I-beams:

$$M_{cr} = \psi_1 \left(\pi / L_{c,h} \right) \sqrt{EI_y GI_T} \sqrt{1 + \pi^2/k^2} \qquad (9.58)$$

where:

$$k = L_{c,h} \sqrt{\frac{GI_T}{EI}}$$

ψ_1 = a coefficient depending on load distribution and on contraint conditions

$L_{c,h}$ = lateral buckling length
EI_y = lateral bending stiffness
GI_T = primary torsional stiffness
EI_ω = secondary torsional stiffness (see 8.2.5)

Bucking length $L_{c,h}$ is equal to distance L between the constraints, if they are simple torsional supports. It is equal to 0.5 L if the constraints also prevent warping.

For ψ_1, the following expression was proposed |61|:

$$\psi_1 = 1.75 - 1.05 (M_B/M_A) + 0.3 \ (M_A/M_B)^2 \qquad (9.59)$$

applicable also to the case of a beam loaded at its ends by two couples M_A and M_B with $M_B < M_A$.

The effect of load position relative to the centre of gravity is taken into consideration by introducing the following coefficient:

	Loading and constraint condition	Bending moment	Torsional Constraint	ψ_1	ψ_2
①	Q	$\frac{QL}{8}$	Supported	1.13	0.45
			Clamped	0.97	0.29
②	Q	$\frac{QL}{24}$	Supported	1.30	1.55
			Clamped	0.86	0.82
③	F	$\frac{FL}{4}$	Supported	1.35	0.55
			Clamped	1.07	0.42
④	F	$\frac{FL}{8}$	Supported	1.70	1.42
			Clamped	1.04	0.84
⑤	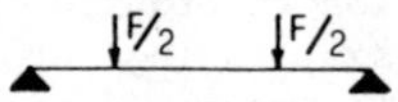 F/2 F/2	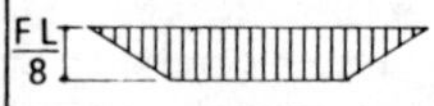 $\frac{FL}{8}$	Supported	1.40	0.42
⑥	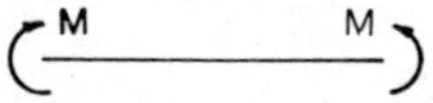 M M	 M	Supported	1.0	0.00

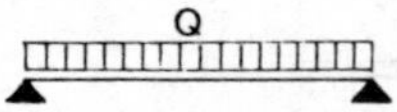

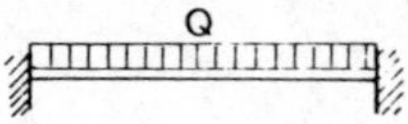

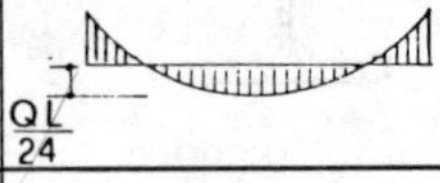

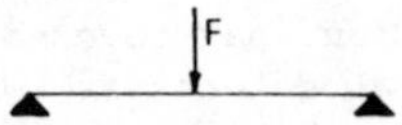

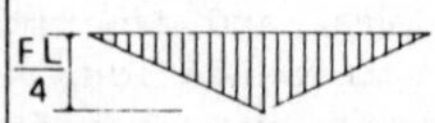

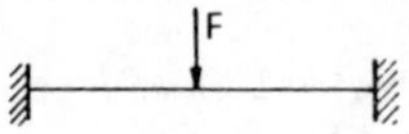

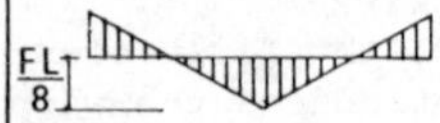

Fig. 9.70

$$\psi_4 = \psi_1 \; \pi \left\{ \sqrt{1 + (\pi^2/k^2)(\psi_2^2 + 1)} \pm \psi_2 \; \pi/k \right\} \qquad (9.60)$$

ψ_2 being a function of the vertical position of the load ($\psi_2 = 0$, if the load is applied at the centre of gravity) and the sign $\pm$ corresponding to a load applied to the lower or upper flange respectively.

Numerical values for ψ_1 and ψ_2, based on results given by Clark and Hill |62|, relating to the mid-span bending moment are set out in Fig. 9.70 for various load and constraint conditions. By introducing the coefficient ψ_4, the critical moment becomes:

$$M_{cr} = \psi_4 (1/L_{c,h}) \sqrt{I_y I_T} \sqrt{EG} \qquad (9.61)$$

This formula has been assumed as the basis of the American code |3|. Kerensky |63| studied a vast range of symmetrical I-sections and,

on the basis of formula (9.58), obtained a simplified formula relating
critical stress to lateral slenderness $L_{c,h}/i_y$ by means of ratio t/h:

$$\sigma_{cr,D} = M_{cr}/W_x = \frac{13E}{(L_{c,h}/i_y)^2} \sqrt{1 + \{(L_{c,h}/i_y)(t/h)\}^2/20} \qquad (9.62)$$

This approximate expression, together with an analogous one for
I-beams with unequal flanges, was used in British Standards 153
(1958) and 449 (1959), for beams and girders for bridges and
buildings, respectively.

At the same time, other authors sought on the one hand to derive
simpler solutions to the problem and, on the other, to extend it to the
elastic-plastic field, by introducing also the effects of geometrical
and structural imperfections.

To take into account the effect of geometrical imperfections on
the critical phenomenon in the elastic field, Flint |64| derived a
formulation on the assumption of an initial imperfection corresponding
to a bending curve or to a torsional rotation. By making reasonable
assumptions concerning cross-sectional shape and imperfection
intensity, limiting stress σ_D can be obtained by solving the following
equation:

$$\sigma_D^3 - \sigma_D^2(f_y + 385\, i_y/L_{c,h})$$
$$- \sigma_D \sigma_{cr,D}(\sigma_{cr,D} + 385\, i_y/L_{c,h}) + f_y \sigma_{cr,D}^2 = 0 \qquad (9.63)$$

where
$\sigma_{cr,D} = M_{cr}/W_x$
M_{cr} being provided by formulae (9.58) and (9.61).

In order to extend the elastic results to the elastic-plastic
field, Timoshenko |60| suggested reducing lateral flexural and
torsional stiffnesses, on the basis of the tangent modulus obtained
by the tension test.

Bleich |56| pointed out that a substitution of the tangent modulus
in elastic formulations provides a lower bound to critical stress
in the elastic-plastic field. A comparison between theoretical results
based on such an extension and the results of many flexural-torsional
buckling experiments on steel beams has shown that the tangent modulus
can be satisfactoruly applied in the case of constant moment, whereas,
in the case of a loading condition producing a variable moment,
experimental values have exceeded theoretical ones by as much as 40%.

The effects of yielding on torsional stiffness were examined by
Neal and Horne. Neal |65| proved that, in the case of rectangular
sections, primary torsonal stiffness at the beginning of lateral
deflection is not influenced by yielding due to bending. Horne |66|
evaluated lateral bending stiffness and secondary torsional stiffness
for various degrees of yielding in I-beams.

The influence of residual stresses on flexural-torsional buckling
of steel beams in pure bending conditions was first studied by
Galambos |67|. His formulation is an extension of the results of
the classical elastic theory, based on the introduction of corrective
coefficients β_1 and β_2 for lateral bending stiffness and secondary
torsional stiffness, which are functions of the stress state at

bifurcation and therefore functions of M_D.

Galambos' relation implicitly defining M_D is as follows:

$$M_D^2 (L_{c,h}/i_y)^4 - (\pi^2 EG \ AI_T \beta_1)(L_{c,h}/i_y)^2 +$$
$$- (\pi^2 E)^2 (A(h - t_f))^2 \ \beta_1 \beta_2 = 0 \tag{9.64}$$

The connections between applied moment and β_1, β_2 were obtained for various I-sections, assuming a bi-triangular residual stress distribution in the flanges with a maximum compression at the ends equal to $0.3 \ f_y$ and a constant distribution in the web (see Fig. 4.12).

The problem of flexural-torsional buckling of a beam due to pure bending when residual stresses are present, was investigated by Como and Mazzolani |68,69|. Their analysis led to an expression for the following implicit form:

$$M_D = F(\alpha, \beta, \gamma, \rho_o, \varepsilon_e) \ M_{cr} \tag{9.65}$$

where

$$F( ) = \left\{ - \beta/2 + \sqrt{\alpha^2 - (2/\pi^2)(L_{c,h}^2/h^2) \ \alpha \gamma \rho_o \varepsilon_e} \right\}$$
$$\times \left\{ 1 - (2/\pi^2)(L_{c,h}^2/h^2) \ \rho_o \varepsilon_e \right\}^{-\frac{1}{2}} \tag{9.66}$$

is a dimensionless function taking into account the presence of residual stresses bi-triangularly distributed in the flanges with $\sigma_{rc} = \sigma_{rt} = \sigma_r$ and with a field intensity defined by the ratio between maximum residual stress and yielding stress (Fig. 9.71):

$$\rho_o = \sigma_r/f_y \tag{9.67}$$

Quantities α, β, γ are functions of deformation and equilibrium bifurcation and thus of M_D. Term α expresses secondary flexural and torsional stiffness reduction due to yielding. In addition to α, which was considered also in Galambos' formulation, terms β and γ have been introduced. Term β corresponds to elastic-plastic energy degradation due to the mutual work produced in the elastic-plastic field by stress dissymmetries due to the presence of residual stresses during the deviating deformation. Term γ takes into account the unstabililizing effect due to the different positions of plastic

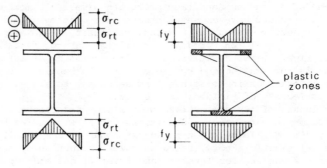

Fig. 9.71

zones in the flange.

All the above formulations are related to the case of pure bending, except for (9.58) and its derivations, which take into account variable moment conditions by means of the coefficient ϕ_1.

In order to generalize the problem, Faella and Mazzolani |70| proposed a procedure analysing flexural-torsional buckling of steel beams in the inelastic range and under different loading conditions, taking into consideration at the same time:

(a) Material strain-hardening

(b) Distribution of residual stresses

(c) Influence of transverse loads

They adopted the basic scheme of two forces symmetrically concentrated in a generic position, which made it possible to study the particular cases of the beam with the force concentrated at mid -span and the beam with two end couples. The critical moment was expressed in the following implicit form:

$$M_D = (M_{pl}/\bar{\lambda}^2)\left\{- B + \sqrt{A^2 - AC\rho_o\,(b/h)^2\,\bar{\lambda}^2/6}\right\} \tag{9.68}$$

where

λ = flange slenderness in the weaker plane of the beam

$\bar{\lambda}$ = λ/λ_e = dimensionless slenderness with refirence to the limit of proportionality slenderness ($\lambda_e = \pi\sqrt{E/f_y}$)

b,h= flange width and section depth rispectively

A,B,C= functions taking into account cross sectional shape, loading conditions, residual stress distribution, material strain hardening. They generalize functions α, β and γ of formula (9.65)

The most recent studies concerning beam buckling in the elastic-plastic range were carried out by Lindner |71|, Nethercot |72| and Vinnakota |73|.

In addition to studies still based on the bifurcation theory, a new approach has been successfully introduced according to which the problem is examined as a bi-axial bending phenomenon, in which geometrical imperfections - such as load eccentricity with reference to the web plane and initial torsional rotation - play a capital role.

The results of the rather numerous theories on flexural-torsional stability have an experimental basis consisting of few, but qualified researches among which are the tests carried out by Massey |74|, Dibley |75|, Fukumoto |76| , Kloppel and Unger |77|, Kitipornchai and Trahair |78|.

9.3.2 *Codification*

9.3.2.1 ECCS Recommendations

The method proposed in the ECCS Recommendations | 31 | can be applied within a field limited by the following assumptions:

(a) that the beam is a doubly symmetrical I-section produced by rolling or composed by welding

(b) that the loads act in the web plane and any eccentricity is accidental

(c) that cross sectional distortion and local buckling of the parts forming the beam are prevented

(d) that constraint conditions are those of a simple support; the contraint prevents displacements in the horizontal plane and torsional rotations, but does not absorb bending moments nor prevent warping (torsional support)

With reference to assumption (a), it must be pointed out that both available experimental results and theoretical solutions concerning plastic behaviour are limited to the case of doubly symmetrical sections. They can be considered as sections having a simple symmetry with reference to the vertical plane only if the elastic bifurcation theory is adopted. Therefore, as more precise information is lacking, it is advisable to treat unsymmetrical sections either assuming a reduced dimension for the wider flange, so as to render the section symmetrical with respect to the horizontal plane, or considering the compression flange as an isolated element.

Assumption (b) tends to exclude the application of the method to cases in which the load action deliberately introduces into the beam bi-axial bending and torsion.

Local buckling effects are excluded by assumption (c). They might take place when the width/thickness ratio of the various parts of the cross section is too great (see 9.6.2). Web buckling phenomenon (crippling)due to concentrated forces (see 9.6.3 must be prevented by means of appropriate stiffeners.

The method proposed in the ECCS Recommendations is based on studies that are all in respect of simply supported beams (according to assumption (d)). It is, however, possible to extend their results to other constraint conditions (continuous beams, cantilevers) by introducing coefficient ψ_1 (9.59). On the basis of such hypotheses, the verification method of the ECCS Recommendations requires that maximum stress σ due to the bending produced in the beam by the design loads be less than the limiting stress σ_D; i.e.:

$$\sigma \leq \sigma_D \qquad (9.69)$$

where

$$\sigma_D = \eta_r \alpha f_y \qquad (9.70)$$

where
η_r = a reduction factor
α = shape factor for bending about the strong axis
f_y = material yield stress
The reduction factor η_r is expressed by:

$$\eta_r = \frac{1}{(1 + \lambda_M^{2n})^{1/n}} \qquad (9.71)$$

where
n = the 'system factor'

$$\lambda_M = \sqrt{\alpha f_y / \sigma_{cr,D}} = \text{dimensionless slenderness for bending} \qquad (9.72)$$

$\sigma_{cr,D}$ = critical stress for flexural-torsional buckling in elastic field
The critical moment M_{cr} can be assessed according to formula (9.58) and coefficient ψ_1 makes it possible to take into reasonable account constraint conditions other than those of simple support (at least

454

for practical purposes). By combining equations (9.70) to (9.72),
the following expression is obtained:

$$\sigma_D = \alpha f_y \frac{\sigma_{cr,D}}{\sqrt[n]{\sigma_{cr,D}^n + (\alpha f_y)^n}} \qquad (9.73)$$

The system factor n should not be constant in principle, but
should vary depending on cross sectional shape, loading conditions,
load application position and its eccentricity, as well as on steel
quality and its production method.

As more precise data concerning such a relationship are not
available, and most of all for simplicity's sake, the proposal was
made to adopt the value: $n = 2.5$, a choice aimed at guaranteeing that
the corresponding dimensionless design curve (see Fig. 9.72) be a
mean value rather than a lower bound with reference to the results
of the 'exact' approach.

This decision is justified if one considers that some favourable
effects (such as material strain hardening, non-uniform bending moment
distribution and rotational restraints at loaded sections) have been
practically neglected.

The physical meaning of reduction factor η_r is to provide the ratio
between collapse moment for flexural-torsional buckling and fully
plastic moment:

$$\eta_r = \frac{M_o}{M_{pl}} \qquad (9.74)$$

As shown in Fig. 9.72, this ratio can be considered practically
equal to 1 for $\bar{\lambda}_M \leq 0.4$. In that area, therefore, the buckling
phenomenon is uninfluential and thus one can count on the full
yielding.

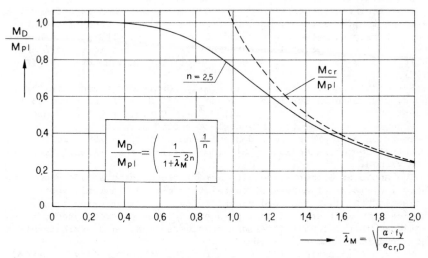

Fig. 9.72

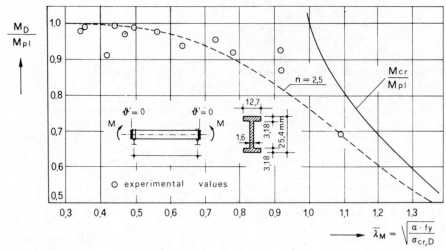

Fig. 9.73

The ECCS method - and particularly the choice of $n = 2.5$ - has been checked by a systematic comparison of the corresponding results with those of the most recent experimental researches and the most advanced theories.

The test results $|74|$ on simply supported steel beams under constant moment, with $f_y = 250$ N mm^{-2}, have been compared to curve $n = 2.5$ in Fig. 9.73. Except for a few discepancies for low values of $\bar{\lambda}_M$ (≤ 0.7) - that it would in any case be impossible to eliminate even by modifying the value of n - the agreement was satisfactory.

The steel cantilever with $f_y = 250$ N mm^{-2}, has been experimentally examined in $|77|$ with a load eccentricity e (force concentrated at the end) varying between $e = 0$ and 15 mm. The results showed that the value $n = 2.5$ gave close agreement for an eccentricity of $L/500$, while results for $e = L/1000$ were in correlation with $n = 2.91$ (Fig. 9.74).

Ref. $|75|$ concerns tests on laterally continuous beams with three spans, made of high strength steel ($f_y = 450$ N mm^{-2}) under constant moment over its central span (Fig. 9.75). Curve $n = 2.5$ provides a reasonable mean value for the complete series of tests.

Tests on a simply supported steel beam ($f_y = 250$ N mm^{-2}) with a central concentrated load $|78|$ also confirmed the reliability of the ECCS method.

The above experimental results were integrated by a wide numerical survey, by means of a general calculation method, including both inelastic behaviour and load eccentricity. The systematic survey examined the influence of many parameters (such as steel grade, cross sectional shape, longitudinal and transverse loading distribution) on the buckling phenomenon $|79,80|$.

From an examination of Lindner's results, the ECCS curve assessed

456

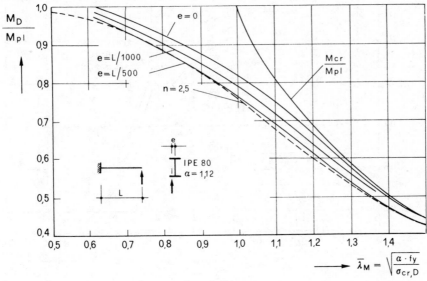

Fig. 9.74

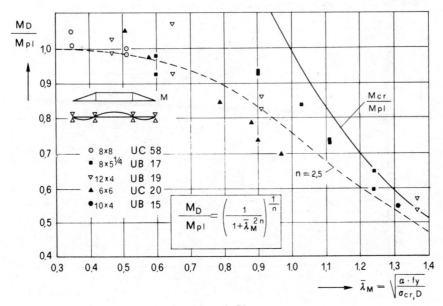

Fig. 9.75

457

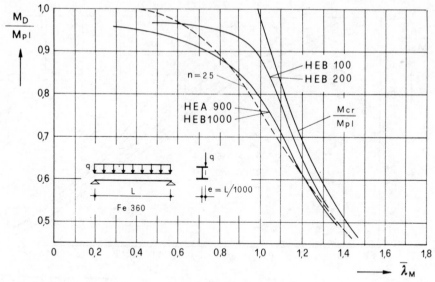

Fig. 9.76

with $n = 2.5$ was confirmed to be also theoretically valid. These results, furthermore, lead to the following conclusions:

The influence of steel quality is negligible: steel types Fe 360 and Fe 510 provide practically the same results

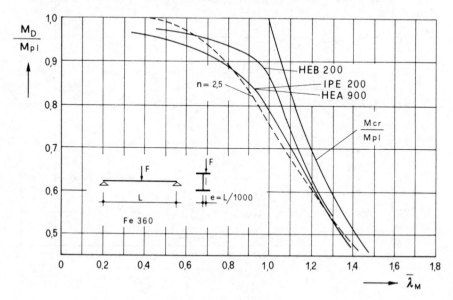

Fig. 9.77

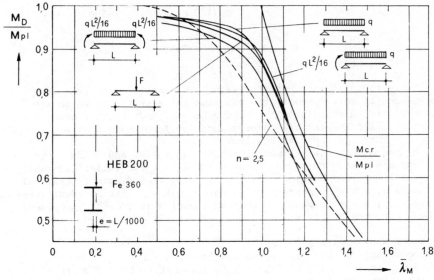

Fig. 9.78

Some very deep sections, such as HEB 1000, are less resistant than corresponding small sections, such as HEB 100 (Fig. 9.76) Less compact sections (IPE, HEA) are also less resistant than the more compact ones (HEB) (Fig. 9.77)
The most unfavourable loading condition is that of a uniformly distributed load together with concentrated end couples, even though, on the whole, the influence of load distribution is modest (Fig. 9.78)
On the contrary, the influence of load application point with reference to the centroid of the section is very important: obviously, if the load position is lowered, load bearing capacity is increased
At the same time as the above studies were being carried out, Fukumoto and Kubo |81| critically examined the use of the ECCS curve with $n = 2.5$ and its deterministic derivation. They considerably extended the experimental field, taking into consideration all studies available in the literature, including the numerous tests carried out in Japan. Altogether, 156 experiments on rolled beams (116 of which were Japanese) and 116 on welded beams (112 of which were Japanese) were considered. They carried out a probabilistic examination of the results, considering experimental ultimate moment characteristic value to be its mean value minus two standard deviations. The survey has shown that the formula adopted in the ECCS Recommendations is fit, in general, to interpret experimental results also statistically, provided the following values of factor n are adopted |81| :

459

	Rolled beams	Welded beams
For mean values	2.5	2.0
For characteristic values	1.5	1.0

It is thus cofirmed that the value 2.5 'meanly' interprets the behaviour of rolled beams, the experimental results for which are less scattered than those for welded beams. On the other hand, the statistic interpretation suggests adopting lower values (n = 1 to 1.5) proving more conservative results.

By comparing the experimental results on rolled and welded beams, it appears that, other conditions being equal, the latter generally have a lower load bearing capacity in respect of flexural-torsional stability. This is perfectly logical considering that residual stress intensity, which lowers load bearing capacity is greater in welded then in rolled profiles (see 4.2).

9.3.2.2 Approximate methods

By re-arranging and simplifying the formulations of the results obtained in the elastic field, it is possible to find simplified calculation methods for evaluating $\sigma_{cr,D}$ to be introduced into (9.72). From such formulations it is also possible to obtain some direct checking methods |3|, even though they are more conservative than that expressed by (9.69).

In particular, formula (9.58) can be written in terms of stress, taking $L = L_{c,h}$ for simplicity and introducing W_x the maximum section modulus, in either of the two following forms:

$$\sigma_{cr,D} = \psi_1 (1/W_x)(\pi/L) \sqrt{EI_y GI_T} \sqrt{1 + \pi^2 EI_\omega/(L^2 GI_T)} \qquad (9.75)$$

$$\sigma_{cr,D} = \psi_1 (1/W_x)(\pi^2/L^2) EI_y GI_\omega \sqrt{1 + L^2 GI_T/\pi^2 EI_\omega} \qquad (9.76)$$

It should be pointed out, first of all, that formula (9.75) contains under the root the ratio between warping torsional stiffness (EI_ω/L^2) and primary torsional stiffness (GI_T), while formula (9.76) contains the inverse of that ratio. Formula (9.76) may be written:

$$\sigma_{cr,D} = \psi_1 \, 0.975 \, \frac{i_y i_T}{\gamma i_x^2} \, \frac{h}{L} \sqrt{1 + 5.75 \alpha^2 (d^2/L^2)(i_y^2/i_T^2)} \qquad (9.77)$$

whereby (9.76) becomes (for $\alpha^2 \neq 0$):

$$\sigma_{cr,D} = \psi_1 \, 2.340 \, E \, \frac{\alpha^2}{\gamma} \, \frac{i_y^2}{i_x^2} \, \frac{hd}{L^2} \sqrt{1 + 0.175(L^2/d^2)\frac{i_T^2}{\alpha^2 i_y^2}} \qquad (9.78)$$

where the new symbols have the following meanings:
i_y and i_x = minimum and maximum radii of gyration respectively
γ = ratio between section depth and twice the distance of the farthest fibre from the neutral axis (γ = 1 for profiles symmetrical about the neutral axis)

460

h = section depth
d = distance between the centres of gravity of the flanges

$$\alpha \begin{cases} 0 \text{ for } \mathsf{T}, \mathsf{V}, \mathsf{L}, \text{ shaped sections} \\ 1 \text{ for sections with two equal flanges} \\ \dfrac{4I_1 I_2}{(I_1 + I_2)^2} \quad \text{for sections with two unequal flanges having moments} \\ \qquad\qquad\qquad \text{of inertia } I_1 \text{ and } I_2 \text{ about the web centroidal axis} \end{cases}$$

$i_{\mathrm{T}} = \sqrt{\dfrac{I_{\mathrm{T}}}{A}}$ with I_{T} given by (8.56) or (8.61)

Expressions (9.77) and (9.78) can be simplified as follows:
(a) In (9.77) to assume the root equal to one means neglecting the effect of warping on the primary torsional stiffness
(b) In (9.78), assuming the root equal to one means overlooking the primary torsional contribution with reference to the secondary one; in practice this corresponds to considering bending stiffness of the two flanges only

The two approaches indicated in (a) and (b) can lead to two different approximate methods, both on the safe side compared to the 'exact' solution. The former should be adopted whenever it is possible to depend largely on primary torsional stiffness and the latter whenever the warping effect is predominant.

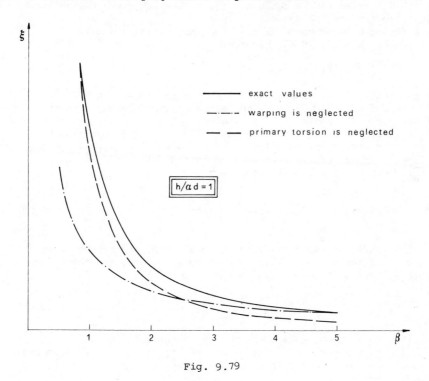

Fig. 9.79

Fig. 9.79 illustrates |82| the degree of accuracy of the two methods. In fact, it shows, as a function of ratio $\beta = L i_T / (h i_y)$ the values of the following ratio:

$$\xi = (\sigma_{cr,D}/E)\,(i_x/i_T)^2/\psi_1 \qquad (9.79)$$

obtained by applying the rigorous expression (9.77) and the values obtained by assuming a value of 1 for the root in (9.77) and (9.78).

If the secondary torsional stiffness EI_ω/L^2 is neglected and, consequently, the root of (9.77) is assumed equal to one, the critical stress is:

$$\sigma_{cr,D} = \psi_1\; 0.975\; E\; \frac{i_y i_T}{\gamma i_x^2}\; \frac{h}{L} \qquad (9.80)$$

By introducing the dimensionless ratios:

$$\rho_y = i_y/b \quad ; \quad \rho_x^2 = \gamma i_x^2/h^2 \quad ; \quad \rho_T = i_T/t_f \qquad (9.81)$$

one obtains:

$$\sigma_{cr.D} = \psi_1\; 0.975\; E\; \frac{\rho\,\rho_T}{\rho_x^2}\; \frac{bt_f}{hL} \qquad (9.82)$$

it is obvious that (9.82) can lead to a simple and useful design criterion, if the ratio $\rho_y\rho_T/\rho_x^2$ is more or less constant for a given group of profiles. This is so for the class of symmetrical I-shaped profiles listed in the table of Fig. 9.80 which gives the values of the ratio for the IPE and HEB series.

Assuming a lower bound of 0.60 for this ratio, and taking $E = 200000$ N mm^{-2}, (9.82) becomes:

$$\sigma_{cr,D} = \psi_1\; 117000\; \frac{bt_f}{hL} \qquad (9.83)$$

which practically coincides with the formula given in the US Recommendations |3,6,55| for the design of steel beams, whenever a safety coefficient of 1.5 is assumed.

In addition to IPE and HE profiles, the procedure can also be considered applicable to welded I-beams for which either of the two following groups of limitations is complied with, so that the condition $\rho_y\rho_T/\rho_x^2 \geq 0.6$ is satisfied.

$$\begin{cases} b/t_f \leq 20 \\ h/b \leq 4 \\ t_w/t_f \geq 0.5 \end{cases} \quad \text{or else} \quad \begin{cases} b/t_f \leq 20 \\ h/b \leq 3 \\ t_w/t_f \geq 0.3 \end{cases} \qquad (9.84)$$

If primary torsional stiffness is neglected, and consequently, the root of formula (9.78) is assumed equal to one, the following expression of critical stress is obtained:

$$\sigma_{cr,D} = \psi_1\; 2.340\; E\; \frac{\alpha^2 i_y^2}{\gamma i_x^2}\; \frac{hd}{L^2} \qquad (9.85)$$

With reference to I-shaped beams, if one assumes that:
Distance d between the centres of gravity of the two flanges and section depth h are related by $d = 0.98h$

 I	$\rho_x = \dfrac{i_x}{h}$	$\rho_y = \dfrac{i_y}{b_f}$	$\rho_T = \dfrac{i_T}{t_f}$	$\dfrac{\rho_y \rho_T}{\rho_x^2}$
IPE 100	0.407	0.225	0.52	0.70
200	0.413	0.224	0.50	0.65
300	0.416	0.223	0.50	0.65
400	0.414	0.220	0.50	0.64
500	0.408	0.216	0.50	0.65
600	0.405	0.212	0.50	0.65
HEB 100	0.416	0.253	0.53	0.77
200	0.427	0.253	0.52	0.72
300	0.433	0.253	0.52	0.70
400	0.428	0.246	0.52	0.70
500	0.424	0.242	0.50	0.68
600	0.420	0.239	0.50	0.68

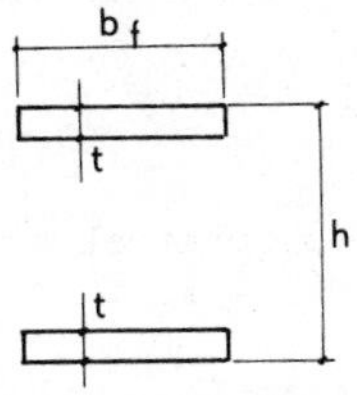

$$\frac{\rho_y \rho_T}{\rho_x^2} \geq 0.66$$

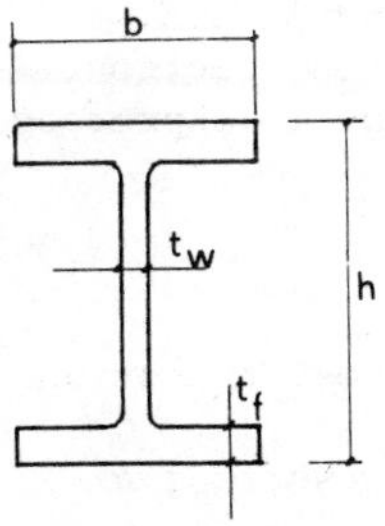

$$\frac{\rho_y \rho_T}{\rho_x^2} \geq 2.84 \; \frac{\sqrt{2 + 0.95 \left(\frac{t_w}{t_f}\right)^2 \frac{t_w}{t_f} \frac{h}{b_f}}}{b_f + \frac{t_w}{t_f} \frac{h}{b_f}}$$

$$\geq 0.57 \quad \text{for} \quad \begin{cases} h/b \leq 4 \\ t_w/t_f \geq 0.5 \end{cases}$$

$$> 0.58 \quad \text{for} \quad \begin{cases} h/b \leq 3 \\ t_w/t_f \geq 0.3 \end{cases}$$

Fig. 9.80

Factor γi_x^2 and the square of section depth are related by:

$$\gamma i_x^2 \cong 0.20 h^2 \qquad\qquad (9.86)$$

(for symmetrical I shaped profiles, this means that the maximum radius of gyration is almost equal to 45% of the depth)

Factor α^2 is of the order of one then for $E = 200000$ N mm^{-2}, the following expression can be obtained:

$$\sigma_{cr,D} = \psi_1 \frac{2\ 293\ 000}{\lambda^2} \qquad\qquad (9.87)$$

where
$\lambda = L/i_y$ = the slenderness in the weak plane

American Recommendations (AISC) apply a safety factor of 1.5 increased by 40% |3| to this check in order to take global account of the fact that the load might be applied to the upper flange, with an effect that is all the more unfavourable the more negligible primary torsional stiffness is compared to the secondary, as happens in the case at stake for assumption (b).

Fig. 9.79 shows that expression (9.87) offers a good approximation of the exact value of $\sigma_{cr,D}$ only in the field of its high values, usually exceeding yield stress. Therefore, it cannot be directly used in design, unless a correction is introduced in order to take elastic-plastic behaviour into account. A possible approach could be as follows. For $\psi_1 = 1$ (case of uniform bending) formula (9.87) can be written as follows:

$$\sigma_{cr,D} = \frac{11.4}{\lambda^2} E \qquad\qquad (9.88)$$

providing a value that is very near to that of Euler's critical stress for a simple strut, if the numerical value of $\pi^2 \cong 10$ is replaced by 11.4.

This results in the logical consequence of assumption (b), according to which beam stability is assured by the opposite bending of the two flanges. Loss of stability takes place when the critical stress is attained in the compression flange, causing lateral buckling of the flange and torsional rotation of the beam section.

The axial force undergone by the compression flange is expressed by:

$$N_f = M_{eq} S_x / I_x = (M_{Max}/\psi_1)(S_x/I_x) \qquad\qquad (9.89)$$

where
I_x = moment of inertia of the entire cross section about the strong axis
S_x = static moment of the compression flange about the strong axis
$M_{eq} = M_{max}/\psi_1$ = reference moment

Lateral buckling control can thus be performed approximately by checking the compression flange under the force N_f expressed by (9.89):

$$N_f/A_f \leq \sigma_c \qquad\qquad (9.90)$$

where
A_f = area of the compression flange

464

Maximum stresses σ_c to be introduced into (9.90) are those already defined in 9.2.2.3 for checking simple compression members, the values of which must be taken into relation to the slenderness λ of the compression flange between two successive torsional constraints, starting from curve 'c' for $t_f \leq 40$ mm and from curve 'd' for $t_f > 40$ mm.

Elastic-plastic behaviour has thus been indirectly introduced through critical stresses σ_c. They are, in fact derived starting from the ECCS buckling curves (see paragraphs 9.2.2.2 and 9.2.2.3) which take all geometrical and structural imperfections into account.

9.4 BEAM-COLUMNS

Besides axial forces, struts can also be loaded by bending moments due to:

Vertical load eccentricity

Transverse loads

Beam columns are thus the columns of steel framing for multi-storey buildings without bracing, industrial buildings and other structures (see 1.2 and 1.3). More generally the problem of bending concerns all frameworks - with or without bracing - whenever the frames co-operate in resisting horizontal and vertical forces. Even in pin-ended structures bending must be considered whenever the beam to column connection is not able to absorb the moment due to eccentricity between the column axis and the centre of gravity of the joint (see 1.4.1).

With reference to the shape of strut cross section, there are two types of buckling phenomena consistent with bending:

Plane buckling (see 9.4.1)

Flexural torsional buckling (see 9.4.2)

Plane buckling, which takes place by bending of the strut in the plane containing load eccentricity, is generally the determinant in the case of hot rolled double symmetrical profiles with open section (I-sections) and with hollow sections having a large torsional stiffness; i.e. for sections in which the shear centre practically coincides with the centre of gravity.

Flexural-torsional buckling is the determinant in the case of sections with a single symmetry ($T, C, \Pi, \ldots$) or with no symmetry at all ($L, l, \ldots$), in which the shear centre and centre of gravity do not coincide. Such shapes are common particularly in the field of cold formed profiles. Also in the case of centred load lack of coincidence between shear centre and centre of gravity may produce a torsional effect causing flexural-torsional buckling accompanied by twist. However, even in the case of such sections, it is important to study plane buckling, because often there are constraints preventing torsional rotation of the section.

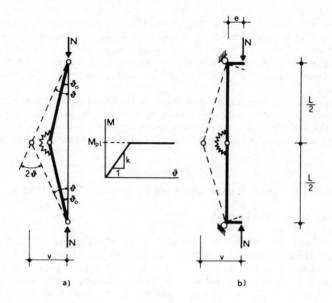

Fig. 9.81

9.4.1 *Plane buckling*

9.4.1.1 The kinematic model
In order qualitatively to analyse the behaviour of a beam column,
consider the kinematic model having one degree of freedom, formed by
two rigid parts hinged at their ends and mutually connected at
mid-span by a spring in which all deformability is concentrated (Fig.
9.81).

In the elastic field, if the spring stiffness is k, each elastic
rotation of the spring corresponds to a moment $M = k\theta$. If the model
is assumed to have an initial geometrical imperfection represented
by parameter θ_o, the spring rotation is nil for $\theta = \theta_o$.
Model equilibrium requires that

$$N \ \theta \ L/2 = M \tag{9.91}$$

where, in the elastic field:

$$M = 2(\theta - \theta_o)k \tag{9.92}$$

whence:

$$N = (4k/L)((\theta - \theta_o)/\theta) \tag{9.93}$$

For $\theta_o = 0$, formula (9.93) provides the Euler critical load for
the model without imperfections:

$$N_{cr} = 4k/L \tag{9.94}$$

and thus (9.93) can be written as

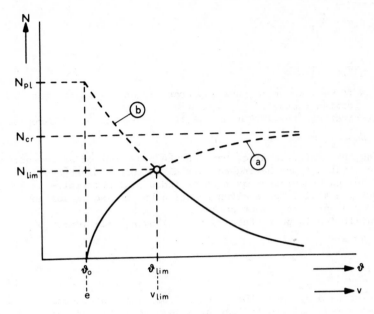

Fig. 9.82

$$N = N_{cr}(\theta - \theta_o)/\theta \qquad (9.95)$$

As shown by curve 'a' of Fig. 9.82, N_{cr} can be attained only for very large displacements ($\theta \to \infty$). Such a model was used to prove that, in the case of an initial geometrical imperfection θ_o, there is no equilibrium bifurcation, but a stable equilibrium is possible with a non-linear rotation (9.95) between load displacement parameter (end rotation), that may also be written as:

$$\theta = \frac{1}{1 - N/N_{cr}} \theta_o \qquad (9.96)$$

or, assuming: $\mu = N_{cr}/N$ $\qquad (9.97)$

$$\theta = \frac{\mu}{\mu - 1} \theta_o \qquad (9.98)$$

If the model is free from imperfection θ_o and has a straight initial configuration, but there is an eccentricity e in load application (Fig. 9.81), formulae (9.96) or (9.97) are still valid, provided the following replacement is introduced:

$$\theta_o = 2e/L \qquad (9.99)$$

In this case, it is convenient to assume a displacement v at mid-span, measured from the line of action of the force, related to rotation θ of the model of Fig. 9.81 as follows:

$$\theta = 2v/L \qquad (9.100)$$

Substituting from (9.99) and (9.100) in formula (9.97)

467

$$v = \frac{\mu}{\mu - 1} e \qquad (9.101)$$

Since $\mu = N_{cr}/N$,

$$\frac{\mu}{\mu - 1} = \frac{1}{1 - N/N_{cr}} \qquad (9.102)$$

This is well known in buckling problems and is called the 'amplification factor'.

Substituting from (9.99) and (9.100), formula (9.95) becomes:

$$N = N_{cr}(v - e)/v \qquad (9.103)$$

proving the perfect analogy between the behaviour of an eccentrically loaded strut and one having an initial out-of-straightness.

If spring behaviour – as that of industrial material – is not perfectly elastic, for a given rotation the spring yields and provides a constant moment of $M = M_{pl}$.

Equilibrium is guaranteed by the following relation:

$$Nv = M_{pl} \qquad (9.104)$$

or:

$$N\theta L/2 = M_{pl} \qquad (9.105)$$

both expressed by curve 'b' of Fig. 9.82. The intersection point between curves 'a' and 'b' has a value on the abscissa of:

$$v_{lim} = e + \frac{M_{pl}}{4k} L \qquad (9.106)$$

or

$$\theta_{lim} = \theta_o + \frac{M_{pl}}{2k} \qquad (9.107)$$

For higher displacement, equilibrium of the model (governed by curve 'b') is possible only for decreasing values of N, and thus it is unstable.

The limiting load N_{lim} for which there is transition between stable behaviour (curve 'a') and unstable behaviour (curve 'b') is expressed as:

$$N_{lim} = M_{pl}/(e + \frac{M_{pl}}{4k} L) \qquad (9.108)$$

or

$$N_{lim} = 2M_{pl}/(\theta_o L + \frac{M_{pl}}{2k} L) \qquad (9.109)$$

Assuming

$$N_{pl} = M_{pl}/e \quad ; \quad N_{cr} = 4k/L \qquad (9.110)$$

and, recalling (9.99), formulae (9.108-9.109) can be reduced to a single formula

$$1/N_{lim} = 1/N_{pl} + 1/N_{cr} \qquad (9.111)$$

known as the Merchant-Rankine formula |83|, which is often adopted to evaluate approximately the collapse load for framed structures in the unstable elastic-plastic range.

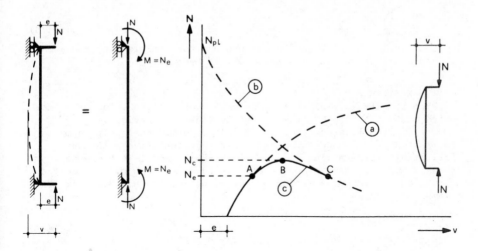

Fig. 9.83   Fig. 9.84

9.4.1.2 The perfect strut

The behaviour of an elastic strut free from imperfections and
eccentrically loaded (Fig. 9.83) is governed by curve 'c' of Fig.
9.84. The passage from the elastic to the plastic field is no longer
discontinuous, as for the kinematic model, but taken place gradually
along the curve $A - C$, demonstrating an elastic-plastic behaviour.

For low values of axial load ($N < N_e$), the strut remains in the
elastic field and the N/v relationship follows curve 'a'. Once load
N_e is attained (point A) the elastic field is left, as the
extreme fibre of the most loaded section attains yield stress. After
it has reached its maximum load bearing capacity N_c (point B), the
strut follows the unstable branch of the curve until point C, where
the most loaded section is completely yielded and the strut is
transformed into a mechanism, the behaviour of which is governed by
curve 'b'.

The condition allowing the strut to remain in the elastic field
is expressed by imposing the following limitation on its most loaded
section:

$$N/A + M/W \leq f_y \qquad (9.112)$$

As $M = Nv$ and according to (9.101):

$$M = N \frac{\mu}{\mu - 1} e$$

and formula (9.112) can be expressed as:

$$\frac{N}{A} + \frac{\mu}{\mu - 1} \frac{Ne}{W} \leq f_y \qquad (9.113)$$

commonly called the Perry-Robertson formula.

For I-sections, $i = 0.45h$ is a good approximation, whence:

$$W = \frac{2I}{h} = \frac{i^2 A}{0.5h} = 0.9 \, iA \tag{9.114}$$

Further, $\sigma_N = N/A$ being the stress due to axial loading and introducing slenderness $\lambda = L/i$, (9.113) becomes:

$$\sigma_N \, (1 + \frac{e}{L} \, \frac{\lambda}{0.9} \, \frac{\mu}{\mu - 1}) \leq f_y \tag{9.115}$$

From formula (9.115) it is possible to obtain the value of σ_N for which the strut departs from perfectly elastic behaviour.

However, the internal actions (M,N) at the mid-span section must be limited by the condition that their representative point must fall within its limiting domain (Fig. 9.85). For I-sections bent in the maximum stiffness plane, the limiting domain can be represented with reasonable approximation by relations (8.107)(8.113). For HE beams, for example:

$$M(N) = 1.11 M_{pl} (1 - N/N_{pl}) \quad \text{for} \quad N > 0.10 N_{pl}$$
$$M(N) = M_{pl} \quad\quad\quad\quad\quad \text{for} \quad N \leq 0.10 N_{pl} \tag{9.116}$$

According to (9.114)

$$M_{pl} = \alpha M_e = \alpha W f_y = 0.9 \alpha i A f_y \tag{9.117}$$

Therefore, the former formula under (9.116), with (9.117), becomes:

$$M(N) = 1.1 \times 0.9 \alpha i A f_y (1 - N/N_{pl}) = \alpha i N_{pl} (1 - N/N_{pl}) \tag{9.118}$$

Any combination $N - v$ for which the section can attain maximum moment $M(N)$ is thus expressed as:

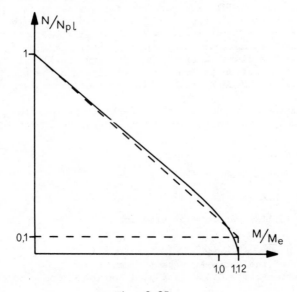

Fig. 9.85

$$M = Nv = \alpha i N_{pl} \{1 - N/N_{pl}\}$$

whence:

$$v = \alpha i \{1 - N/N_{pl}\} N_{pl}/N\}$$

or

$$v = \alpha \{-1 + N_{pl}/N\} L/h \tag{9.119}$$

Relation (9.119) shown in Fig. 9.84 (curve 'b') represents every combination of axial load N (for $N/N_{pl} \geq 0.10$) and of mid-span displacement v, for which the mid-span section is fully yielded, i.e. in plane $M - N$ of Fig. 9.85 the points representing stress fall on the boundary of the limiting domain. For $v = 0$, formula (9.119) gives value $N = N_{pl}$, whereas the expression asymptotically tends to $N = 0$ for v tending to infinity.

9.4.1.3 The industrial strut

Data. When perfect struts are replaced by industrial struts, the analysis of their actual behaviour and the evaluation of their load bearing capacity require, first of all, the definition of the parameters influencing that behaviour.

It has been shown in part from the study of the model (9.4.1.1) and of perfect struts (9.4.1.2) that, the structural scheme is characterized by the following parameters:

Eccentricity
Slenderness
Yield stress
Residual stress distribution
Cross sectional shape
Ultimate domain for the section
Initial geometrical imperfections (initial out-of-straightness or straightness defects)
To these must be added:
Load process
Bending moment distribution along the strut

The loading process influences the progressive yielding of the various fibres of the sections. In other words, for a strut with imperfections,it is not possible to determine a priori whether the concurrence of values $N - M$ characterizing maximum load bearing capacity is independent of the manner in which they are introduced into the strut by the applied loads.

Bending moment distribution in the strut, which is also influenced by initial geometrical imperfections, can modify behaviour at collapse. It is sufficient to imagine a strut loaded by two opposite end moments, to perceive that the mid-span section is influenced only by second order effects and not by the effects of bending due to loading. End sections, instead, are loaded only by applied loads M and N. Depending on bending moment distribution, the critical section can thus be a generic section located between the middle and the ends of the strut, not always easily identified a priori.

471

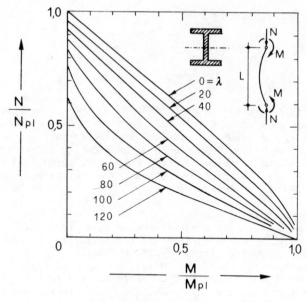

Fig. 9.86

Calculation method.   Load bearing capacity of a beam-column can be analysed using a computer simulation programme similar to those illustrate in 9.1.2, considering, for example, a discrete model such as that shown in Fig. 9.4.

Numerical calculation is thus able to provide concurrent values of M and N corresponding to maximum load bearing capacity of the strut. Such values are represented by a point in an $M - N$ curve.   By connecting the different points that can be obtained as N for a strut of given slenderness varies, a line, called the 'interaction domain' $M - N$ of the strut, is obtained, as shown schematically in Fig. 9.86.

The interaction domain intersects the N axis at the maximum value N_c bearable by a simply compressed imperfect strut $(M = 0)$, an intersection which thus depends on slenderness λ.   The interaction domain intersects axis M at the ultimate value of the bending moment M_{pl}, corresponding to complete yielding of a cross section of the strut.   In this case therefore, the intersection does not depend on strut slenderness.

Further, the interaction domain changes its shape as a function of λ : for $\lambda = 0$, it coincides with the ultimate domain $M - N$ of the cross section ($\lambda = 0$ means absence of geometrical effects due to axial load N) and therefore its concavity is always turned towards the origin.   For $\lambda \to \infty$, it must tend to reference axis M, thus representing the behaviour of a simply bent beam.   It follows that domain concavity is turned towards the origin for low slendernesses, whereas for high slendernesses it has a convexity towards the origin.

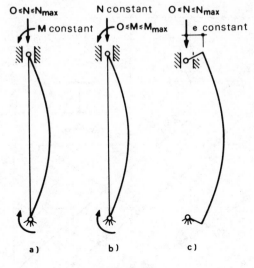

Fig. 9.87

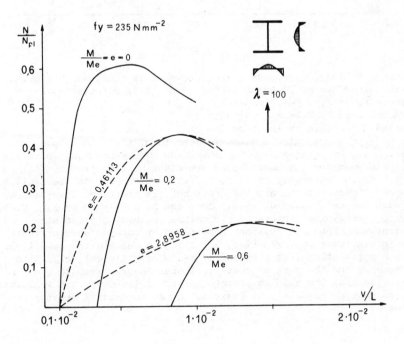

Fig. 9.88

473

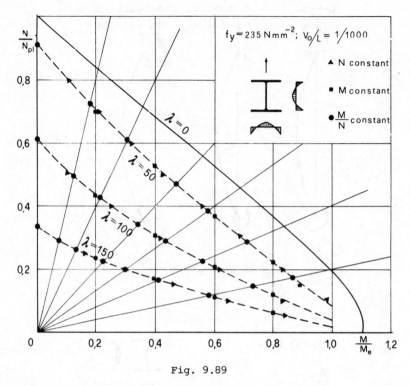

Fig. 9.89

Influence of loading process. The influence of the loading process
is emphasized by the diagrams of Fig. 9.87 |84|, concerning:
  Increase of N, M being constant
  Increase of M, N being constant
  N and M increase by the same amount (e = const)
  Fig. 9.88 shows, with a dimensionless form, the load-displacement
relationship for constant bending moment and increasing load (continous
line) and that for increasing load and constant eccentricity (dashed
line). Moment intensity is expressed with reference to the limiting
moment M_e. Curve trends are different but, in practice, they lead to
the same values of maximum load. This can be confirmed by Fig. 9.89,
showing some typical interaction domains. Values obtained at
constant axial load, at constant moment and at constant eccentricity
are, in excellent approximation, all lying on one and the same line.
  Such calculations, carried out by means of different calculation
programmes, for the various cross sectional shapes, lead to the
conclusion that the loading process has no influence on the evaluation
of load bearing capacity, at least as far as mono-axial bending is
concerned.

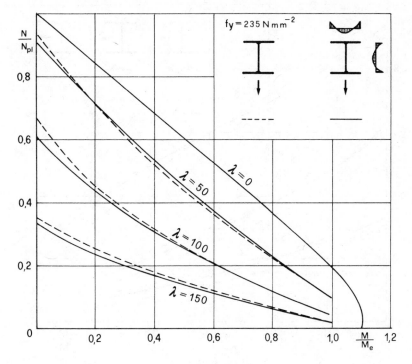

Fig. 9.90

Influence of residual stresses. Fig. 9.90 shows some interaction
domains obtained |84| with (continuous line) and without (dashed
line) the effect of thermal residual stresses.

It should be pointed out that their negative influence is reduced
as the bending moment acting on the strut increases and in some
cases it might even become favourable.

This is not so, however, for residual stresses due to cold
straightening (Fig. 9.91), which play a different role according
to the direction of bending, even though in practice the strut bends
in the weakest direction and thus their effect is always unfavourable.

Influence of straightness defects. The phenomenon of a sudden
passage from the initial asymmetrical configuration of a strut to
the symmetrical one at collapse has been often observed during tests
(Fig. 9.92).

Also, it has been shown that, at least in cases having a practical
interest -even with opposite end moment - initial asymmetrical
configurations lead to higher maximum load bearing capacities than
those corresponding to the same bending moment diagrams, with initial
symmetrical configurations.

This is illustrated in Fig. 9.93 |85|, comparing load-displacement
results for two different initial configurations of a very slender

475

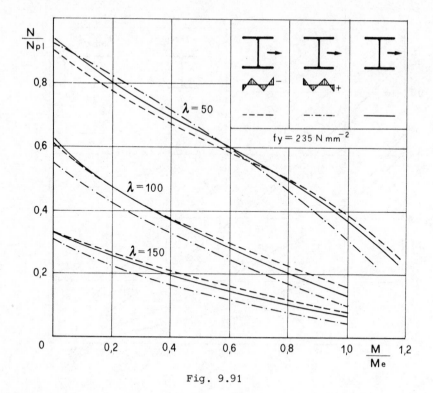

Fig. 9.91

strut ($\lambda = 150$) undergoing an asymmetrical moment distribution. Expressing initial deflection in the form:

$$v = v_1 \sin \pi x/L + c_2 \sin 2\pi x/L \qquad (9.120)$$

the two following cases can be defined:

Symmetrical case: $v_1/L = 1/1000 \qquad v_2/L = 0$

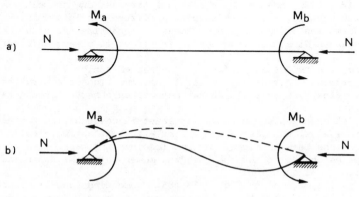

Fig. 9.92

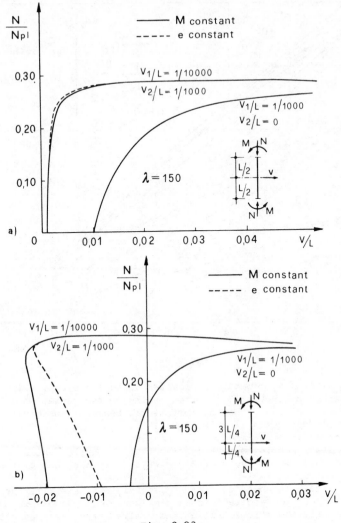

Fig. 9.93

Asymmetrical case: $v_1/L = 1/10000$ $v_2/L = 1/1000$
The two diagrams a) and b) of Fig. 9.93 show the evolution of transverse displacements of points at mid-span and at a quarter point of the strut, respectively. As to be expected, the two different loading systems (M = const, e = const) lead to practically the same results.

The asymmetrical case with an initial deflection having opposite curvatures evolves in the post-critical field towards a deflection all on one side, with a higher collapse load than that of the symmetrical case.

477

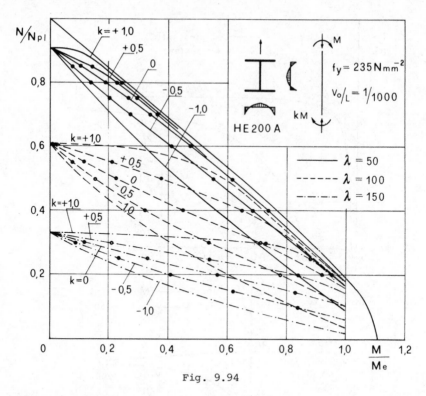

Fig. 9.94

Influence of bending moment distribution along the strut. Fig. 9.94 shows interaction curves |85| corresponding to the variation of ratio k between the values of the end moments for beam-columns having different slendernesses (λ = 50, 100, 150).

Some conclusions are immediately apparent:

For constant bending moment distribution (k = -1), maximum load bearing capacity of the strut is always less than that of the cross section

For bi-triangular bending moment distributions (k = +1), the resistance of the end section can influence the phenomenon in a way which is the more sensible, the smaller strut slenderness is.

Fig. 9.95 illustrates interaction curves for different bending moment distribution along the strut. In addition to the case of constant bending (a), the cases of two concentrated transverse loads at the quarter points (b) and one load at mid-span (c) are considered. A comparison of the corresponding results shows that, in general, behaviour is all the more favourable, the smaller is strut area undergoing moment values near maximum value.

9.4.1.4 Interaction domains

Many authors have proposed direct relations to express interaction domains $M - N$, based on an interpretation of experimental results

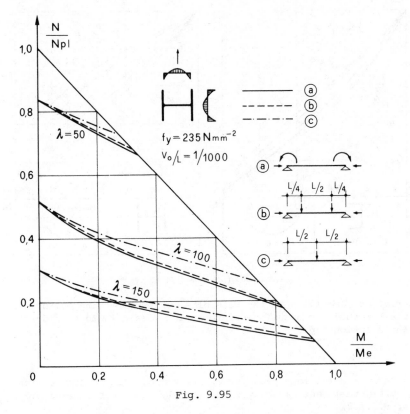

Fig. 9.95

or of numerical simulations in terms of strut slenderness, cross sectional shape, the presence of imperfections and bending moment distribution. The complexity of such formulations, in addition to the fact that most often their application is limited to particular cases, has made it preferable - practically for specifications - to adopt simplified criteria.

To adopt an approximate approach to the problem of beam-columns one must:

Produce interaction curves between moment and axial loads $(M - N)$ as slenderness λ varies, assuming a constant bending moment $M(x) = M_O$

Define an equivalent moment M_{eq}, to be used when the bending moment $M(x)$ varies along the beam. In general, one can assume $M_{eq} = \beta M_O$, M_O being a relevant value of the bending moment (e.g. maximum value, average value, maximum mid-span value, etc.)

It is obvious that these approximations often render the design too conservative. On the other hand, in order to obtain more realistic results, one can only resort to graphs or charts $|86,87|$. Graphs with axes N/N_c and M/M_u are perhaps the most useful (Fig. 9.96).

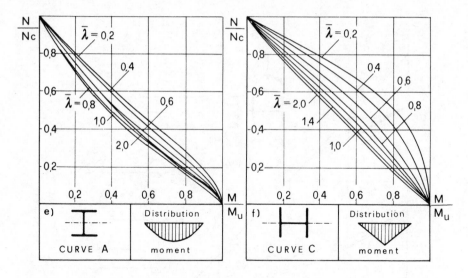

Fig. 9.96

Interaction formulae. In the case of a constant bending moment, the simplest method for assuming an interaction domain is to consider the linear relationship:

$$N/N_c + M/M_u = 1 \tag{9.121}$$

where

N_c = maximum axial load capacity of strut under simple compression
M_u = ultimate moment capacity of the simply bent section (see 8.2.3 or 8.3.1.2)

This method, however, is not on the safe side, because it has been observed that exact domains can have a convexity towards the origin (see Fig. 9.86). If corrected as follows, however:

$$\frac{N}{N_c} + \frac{M}{M_u(1 - N/N_{cr})} = 1 \tag{9.122}$$

where N_{cr} is the Euler critical load in the bending plane, the curves are convex towards the origin for any value of slenderness and therefore always on the safe side.

A different type of approach is possible by adding the external bending moment to the moment due to an imaginary eccentricity, taking second order effects into account. In this case, the formula becomes:

$$\frac{N}{N_{pl}} + \frac{M + Ne^*}{M_u(1 - N/N_{cr})} = 1 \tag{9.123}$$

When $M = 0$, N must assume the maximum axial bearing capacity N_c under pure compression and this leads to the expression:

$$e^* = \{N_{pl}/N_c - 1\}\{1 - N_c/N_{cr}\} M_u/M_{pl} \tag{9.124}$$

The interaction formula (9.123) with (9.124) always gives safe results and it is proposed in the ECCS Recommendations |31|.

Eccentricity $e*$ is not real eccentricity, as the bending effect is already included in the value of M. It is a 'parameter of imperfection', to be introduced in order to take global account of all imperfections: not only geometrical ones (out-of-straightness, initial curvature), but also structural ones (residual stresses).

The expression for $e*$ (9.124) is independent of the applied axial force N, but is a characteristic of the strut, being a function of its slenderness, through N_c, N_{cr}, and of the properties of its cross sections, through N_{pl} and M_u^c.

An equivalent but more expressive formulation of (9.123) may be found by introducing (9.124) into (9.123) and rearranging the results |7| giving:

$$M/M_u = (1 - N/N_c) \{1 - (N/N_{cr})(N_c/N_{pl})\}$$

which may be converted to:

$$\frac{N}{N_c} + \frac{M}{M_u \{1 - (N/N_{cr})(N_c/N_{pl})\}} = 1 \tag{9.125}$$

Equation (9.125) looks more familiar than (9.123). A comparison between (9.125) and (9.122) shows that the former gives more favourable results the smaller the ratio N_c/N_{pl}, i.e. the greater the slenderness λ.

Both equations (9.122) and (9.125) may lead to a unique design formula:

$$\frac{N}{N_c} + \frac{\mu}{k\mu - 1} \frac{M}{M_u} \leq 1 \tag{9.126}$$

where:

$$\mu = N_{cr}/N = \pi^2 EI/(NL^2) = (N_{pl}/N)(1/\bar{\lambda}^2) \tag{9.127}$$

$$k = 1 \qquad \text{for (9.122)}$$
$$= N_{pl}/N_c \quad \text{for (9.125)} \tag{9.128}$$

$$\bar{\lambda} = \lambda/\pi\sqrt{E/f_y} \text{ non-dimensional slenderness}$$

Equivalent moment. The definition of the equivalent moment:

$$M_{eq} = \beta M_o \tag{9.129}$$

does not appear to be rationally based in many cases. The problem varies depending on the following conditions:
(i)   Beam ends may or may not be free to move transversely
(ii) There may or may not be transverse loads along the beam

For beams which are part of sway frames and which do not carry transverse load, there is little information available. The US regulations require $\beta = 0.85$ and M_o to be the maximum end moment.

For beams which are part of frames braced against joint movement and which do not carry tranverse load various codes assume $M_o = M_a$ and β is given by one of the following formulae:

$$\beta = \sqrt{0.30(1 + (M_b/M_a)^2) - 0.40 \, M_b/M_a} \tag{9.130}$$

$$\beta = 0.60 - 0.40 \, M_b/M_a \quad \text{but} \quad \beta \geq 0.40 \tag{9.131}$$

These are proposed in references |88| and |89| respectively and

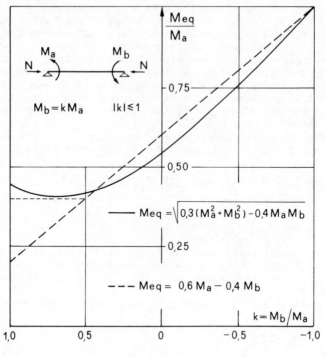

Fig. 9.97

differ only slightly from each other (Fig. 9.97). In these expressions M_b and M_a are end moment values ($M_b \leq M_a$) and M_b/M_a is positive whenever M_a and M_b are both clockwise or both anticlockwise.

The resulting equivalent moment $M_{eq} = \beta M_a$ is independent of the applied axial load N. This implies that the interaction given by (9.126) does not tend, as slenderness decreases, to the linear interaction

$$N/N_{pl} + M/M_u \leq 1$$

valid for the section. This discontinuity makes it necessary to check the resistance of the section at which the bending moment is highest.

The situation is even less satisfactory in the case of beams subject to transverse forces. US specifications |55| assume in equation (9.129) that M_o is equal to maximum end moment. In this case:

$$\beta = 1 + \psi N/N_{cr}$$

where:

$$\psi = \frac{\pi^2 \delta_o EI}{M_o L^2} - 1 \qquad (9.132)$$

δ_o is the maximum displacement due to the transverse load.

482

Case	$\psi\nu$	Case	$\psi\nu$
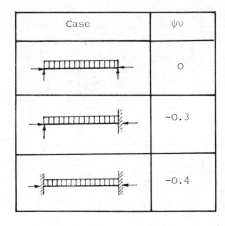	o		-0.2
	-0.3		-0.4
	-0.4		-0.6

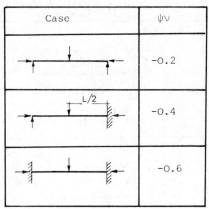

Fig. 9.98

For the allowable stress method used in the US Code (9.132) becomes:

$$\beta = 1 + (\psi\nu)\ N/N_{cr}$$

where ν is the safety coefficient and $(\psi\nu)$ is as given in Fig. 9.98.

Unlike the preceding cases, according to equation (9.132) the value of β depends upon beam slenderness and applied load whereas M_o is no longer the absolute maximum, but the maximum midspan moment. The ECCS Recommendations |31| are quite different. They assume:

$$M_{eq} = M_{eq,1} + M_1 \qquad\qquad (9.133)$$

thus expressing the equivalent moment as the sum of the equivalent moment $M_{eq,1}$ which would be attained if there were no transverse forces, and M_1 which is assumed equal to the bending moment caused by the transverse forces in a beam of the same length, but simply supported at its ends. In the cases when M_a and M_1 produce curvatures of a different sign and $|M_1| \leq |2M_a|$, one can assume in equation (9.133) that:

$$M_{eq} = M_{eq,1}$$

Using this approach (Fig. 9.99), the equivalent moment no longer depends on the applied axial load. Furthermore equation (9.133) produces sudden changes in the value of M_{eq} due to discontinuity in the value of M_o.

A different approach was used in |90| in order to define the equivalent moment for beams with no transverse end displacements loaded both by end moments and by intermediate transverse loads. About 1000 cases have been considered, simulating by computer program the behaviour of beams hinged at their ends, with different slendernesses, or section properties or with different loading.

It was shown that it is very difficult to establish better methods for design than those presently adopted by various codes. In fact, the results are so diverse that simple presentation is impossible,

483

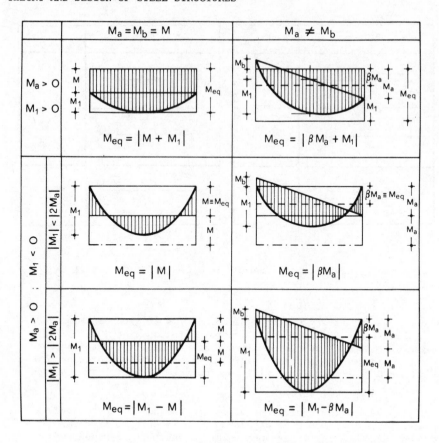

Fig. 9.99

unless radical approximations are made, resulting in loss of accuracy.
For beam-columns loaded only by end moments formula (9.131) may be
unconservative for low axial load values. If $N \rightarrow 0$, the value of β
must tend to unity. Thus equation (9.131) can be usefully replaced
by the following:

$$\beta = \begin{cases} \beta_1 & \text{for } N/N_c \geq 0.40 \\ 1 - \dfrac{1 - \beta}{0.4} \dfrac{N}{N_c} & \text{for } N/N_c \leq 0.40 \end{cases}$$

Where β_1 is given by equation (9.131) and N_c is the maximum axial
load carrying capacity of the strut in simple compression. For
beam-columns with transverse forces, the equivalent moment may be
related to the maximum moment in the span ($M_{eq} = M_o$) and it can
depend on the value of the ratio N/N_{cr}, N_{cr} being the Euler critical
load. It is also influenced by the load distribution. A good
approximation to true behaviour can be obtained by considering a
function of the ratio between the average value of the moments M_m

484

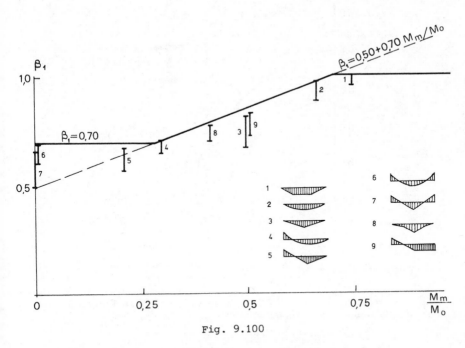

Fig. 9.100

and the maximum mid-span moment M_o. One can assume:

$$\beta = \begin{cases} \beta_1 & \text{for } N/N_{cr} \geq 0.50 \\ 1 - 2(1 - \beta_1)N/N_{cr} & \text{for } N/N_{cr} \leq 0.50 \end{cases}$$

where

$$\beta_1 = 0.50 + 0.70 \, M_m/M_o \quad \text{but } 0.70 \leq \beta_1 \leq 1.0$$

In Fig. 9.100 the results obtained by numerical simulation for different moment distributions, and for various shapes and slenderness are presented.

9.4.1.5 Bi-axial bending

The analysis of beam-column behaviour under bi-axial bending is a problem that is being studied by many authors. The aspects of the problem under investigation are similar - although much more complex - to those concerning mono-axial bending. Of particular interest are the influence of the loading process, which might in this case become significant, and that of residual stresses. These subjects have not yet been satisfactorly investigated. Corresponding results can be plotted by projecting on to a plane the space interaction surface $N - M_y - M_x$, shown in Fig. 9.101.

Approximate formulae describing such surfaces are still being defined. They may be found by generalizing design formula (9.122) to cover both directions, thus:

485

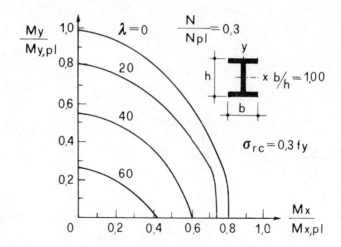

Fig. 9.101

$$\frac{N}{N_c} + \frac{M_{eq,x}}{M_{u,x}(1 - N/N_{cr,x})} + \frac{M_{eq,y}}{M_{u,y}(1 - N/N_{cr,y})} \le 1 \qquad (9.134)$$

The extension of formula (9.123) has not been sufficiently checked, but it has however been proposed in ECCS Recommendations

$$\frac{N}{N_{pl}} + \frac{\beta M_x + Ne^*_x}{M_{u,x}(1 - N/N_{cr,x})} + \frac{\beta_y M_y}{M_{u,y}(1 - N/N_{cr,y})} \le 1$$

$$\frac{N}{N_{pl}} + \frac{\beta_x M_x}{M_{u,x}(1 - N/N_{cr,x})} + \frac{\beta_y M_y + Ne^*_y}{M_{u,y}(1 - N/N_{cr,y})} \le 1 \qquad (9.135)$$

The symbols adopted in (9.134) and (9.135) have the same meaning as in the formulae for mono-axial bending. They have indices added to distinguish between the terms related to either the x or y axis. Both (9.134) and (9.135) may be summarized by the following formulations:

$$\frac{N}{N_c} + \frac{k\mu_x}{k\mu_x - 1}\left\{\frac{M_{eq,x}}{M_{u,x}} + \frac{\mu_y/(\mu_y - 1)}{\mu_x/(\mu_x - 1)}\frac{M_{eq,y}}{M_{u,y}}\right\} \le 1$$

$$\frac{N}{N_c} + \frac{k\mu_y}{k\mu_y - 1}\left\{\frac{\mu_x/(\mu_x - 1)}{\mu_y/(\mu_y - 1)}\frac{M_{eq,x}}{M_{u,x}} - \frac{M_{eq,y}}{M_{u,y}}\right\} \le 1 \qquad (9.136)$$

with

$$\mu_x = N_{cr,x}/N \qquad ; \qquad \mu_y = N_{cr,y}/N$$

$$k = \begin{cases} 1.0 & \text{for (9.134)} \\ N_{pl}/N_c & \text{for (9.135)} \end{cases}$$

Obviously for $k = 1$ the two formulae (9.136) are reduced to the one (9.134). Other, more precise design formulae, but specific only

486

for some column shapes, may be found in |92|.

9.4.2 *Flexural-torsional buckling*

9.4.2.1 Strut behaviour

In a strut not having a doubly symmetrical profile, bearing an axial
load applied at the centroid or an eccentric one, provided it does
not pass through the shear center, buckling takes place in a flexural
-torsional manner. The torsional effect due to the load not passing
through the shear center produces a twist of the cross section,
accompanying lateral bending (Fig. 9.102). The corresponding
unstable configuration is qualitatively similar to that of simply
bent beams (see 9.3.1).

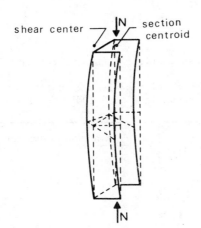

shear center — |N section centroid

Fig. 9.102

Flexural-torsional buckling is a phenomenon concerning in
particular cross sectional shapes typical of cold formed thin walled
profiles.
Consider the general case of a beam-column section having a simple
symmetry and imagine controlling its deflection v to remain within
the plane of symmetry containing the eccentricity as axial load
increases. One obtains the $N\text{-}v$ curve typical of plane buckling,
leading to collapse caused by yielding of the mid-span section (Fig.
9.103a).
If, instead, the strut can undergo displacements outside the plane
of symmetry as load increases, in addition to displacement v
corresponding to plane bending, a transverse displacement u and a
torsional rotation θ take place. The parameters u and v at first
increase linearly by small amounts and then suddenly become very
high, approaching the value of maximum flexural-torsional buckling
load (Fig. 9.103b).
Between the two types of behaviour (plane buckling and flexural
-torsional buckling), that corresponding to the lowest ultimate load

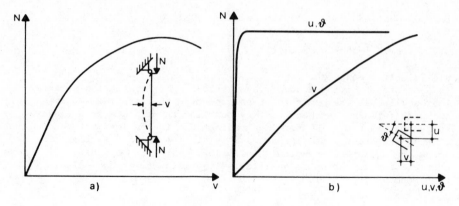

Fig. 9.103

is the determinant in stability checks. Whether the one or the other prevails depends on cross sectional shape, strut slenderness and load eccentricity.

In order to illustrate such a relation from a qualitative point of view, examine the example shown in Fig. 9.104. The strut profile

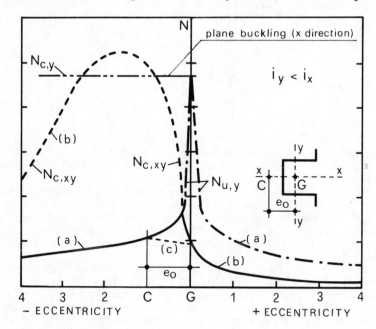

(a) plastic collapse

(b) flexural-torsional buckling

Fig. 9.104

is thin and omega-shaped, and load eccentricity varies between $\pm\ 4e_o$, where e_o is the distance between shear centre C and centre of gravity G. The $N - e$ curves plotted indicate:

Collapse load $N_{u,y}$ due to plane bending about axis $y - y$ (curve (a))

Flexural-torsional buckling load $N_{c,my}$ (curve (b))

It depends on the mutual position of curves (a) and (b) whether the one or the other type of buckling is responsible for the loss of load bearing capacity for each given eccentricity. The full line shows in which areas they provide the lowest collapse load.

It should,first of all,be pointed out that, if the load is centred ($e = 0$), the strut undergoes flexural-torsional buckling under a much lower load than $N_{c,y}$ corresponding to buckling in the $x-x$ plane of symmetry.

Further, whenever the axial load is applied towards the open side of the profile (positive eccentricity), curve (b) is always lower than curve (a) and thus there is flexural-torsional buckling. However, if the axial load is applied towards the closed side of the profile (negative eccentricity) beyond the shear centre, curve (a) is always the lower and therefore collapse takes place by lateral bending in plane $x - x$. Whenever the load is located in the intermediate area between the shear centre and the centre of gravity (interval $C\ G$), the two unstable modes coexist. Small eccentricity variations are related to large variations of collapse load, with a peak in connection with G. One can imagine, approximately, cutting the peak by connection curve (c) between curves (a) and (b).

Further, flexural-torsional behaviour in the negative eccentricity field is cut by collapse load $N_{c,y}$ for plane buckling due to bending about axis $y-y$.

The above behaviour is typical of all profiles having a single symmetry. What has been described for the omega-shaped profile can be qualitatively applied also to C , T and non-symmetrical I-sections, although differences between the two behaviours are less sensible. For such sections it is impossible to foresee a priori which buckling phenomenon will be the critical one so both possibilities must be checked.

9.4.2.2 Analysis of results

The problem of flexural-torsional buckling of struts and of beam-columns having a thin open profile is quite complex and even in the elastic field it is impossible to provide general solutions in a conclusive form.

For profiles having only one axis of symmetry $x - x$ in Fig. 9.105

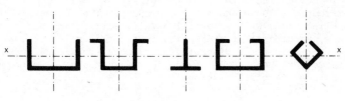

Fig. 9.105

the equation expressing the buckling condition for equilibrium bifurcation is of the following type:

$$(I_o/A)(N - N_{cr,x})(N - N_{cr,y})(N - N_{cr,\theta}) - N^2 e_o^2 (N - N_{cr,y}) = 0$$

where:

$$
\begin{aligned}
N_{cr,x} &= \pi^2 EI_x/L^2 \\
N_{cr,y} &= \pi^2 EI_y/L^2 \\
N_{cr,\theta} &= \{GI_T + E I_\omega \pi^2/L^2\} A/I_o
\end{aligned}
$$

(9.137)

are, respectively, the Euler critical loads for plane buckling about axes x and y and the critical load for pure torsional buckling (9.53), where:

A = cross sectional area
I_o = polar moment of inertia
e_o = distance between shear and gravity centres
L = buckling length

Formulae (9.137) show that equilibrium bifurcation can take place in the strut in two ways:

By simple bending about axis $y - y$ in the plane of symmetry, corresponding to the following critical load

$$N_{cr} = N_{cr,y}$$

By lateral bending, accompanied by twisting of the section, corresponding to a flexural-torsional critical load $N_{cr,xy}$ which is lower than both $N_{cr,x}$ and $N_{cr,\theta}$ and is expressed as:

$$N_{cr,xy} = 1/2\chi \left\{ (N_{cr,\theta} + N_{cr,x}) - \sqrt{(N_{cr,\theta} + N_{cr,x})^2 - 4\chi N_{cr,x} N_{cr,y}} \right\}$$

where $\chi = 1 - e_o^2 A/I_o$ is the bending versus torsion coupling factor. It is given in Fig. 9.104a as a function of some cross sectional shapes.

It might the simpler to evaluate the critical load for a pure torsional type, expressed as:

$$N_{cr,\theta} = EA \{k_1 (t/a)^2 + k_2 (a/L)^2\}$$

by applying quantities k_1 and k_2 given in Figs 9.106b,c,d, as functions of cross sectional shape, t being profile thickness and a being a characteristic dimension of the cross section, which can be directly determined in the figure.

In the elastic-plastic field, the results |93| of some theoretical solutions based on the bifurcation theory are shown in Fig. 9.107 for I-profiles, taking into account thermal residual stresses with a maximum compressive value of $\sigma_{rc} = 0.25 \, f_y$. The curves are for the three buckling cases with a central axial load: bending about the strong axis, bending about the weak axis and flexural-torsional buckling. For this class of profiles, flexural-torsional buckling is an intermediate behaviour between the two kinds of plane buckling.

The combined effects of residual stresses ($\sigma_{rc} = 0.25 f_y$) and of an initial curvature having $\theta_o = L/1,000$ were studied in |94| for

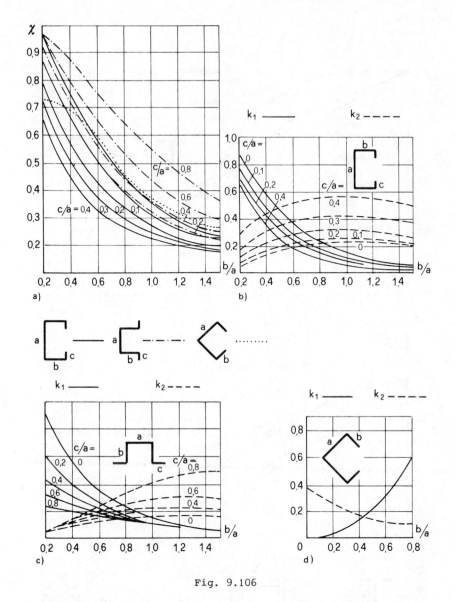

Fig. 9.106

T shaped profiles and for the following cases (Fig. 9.108):
  Plane buckling about axis x, without residual stresses (curve 1)
  Plane buckling about axis y, without residual stresses (curve 2)
  Plane buckling about axis y, with residual stresses (curve 3)
  Flexural-torsional buckling, without residual stresses (curve 4)
  Flexural-torsional buckling, with residual stresses (curve 5)

491

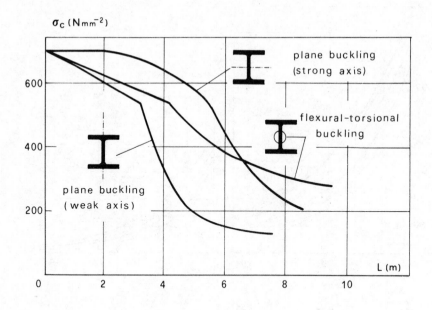

Fig. 9.107

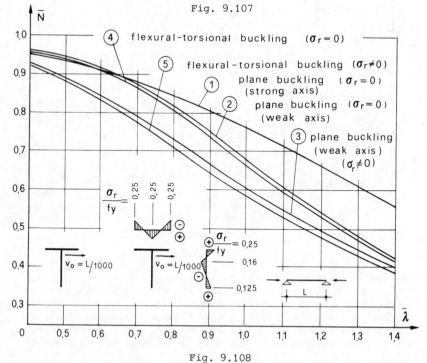

Fig. 9.108

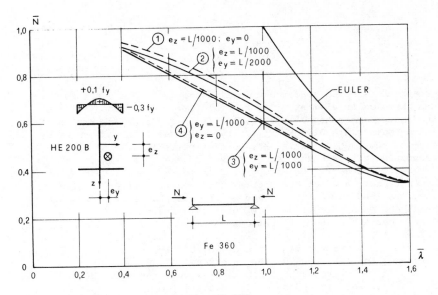

Fig. 9.109

Flexural-torsional buckling is always critical. In fact, whether there are any residual stresses or not, it leads to lower collapse load values (although only slightly lower ones) than those related to plane buckling about the weak axis (y)

The influence of smaller eccentricities ($e = L/1000$), together with residual stresses, on HEB 200 profiles was studied |95|. Corresponding results show that bi-axial behaviour provides slightly lower collapse load values than those of the most unfavourable mono--axial case (compare curves 3 and 4 of Fig. 9.109).

Solutions for other cases of practical interest do not, so far, seem to be available and the subject requires further systematic research.

9.4.2.3 Design methods

Design methods proposed for the purpose of establishing recommendations are substantially based on a generalization of the interaction domains for bi-axial bending (see 9.4.1.5).

In order to adapt them for the case of flexural-torsional buckling it is sufficient that, for $M_y = N = 0$, moment M_x be limited by value M_D for flexural-torsional buckling of a beam, rather than by ultimate plastic value $M_{u,x}$ of the bent cross section. Flexural-torsional buckling can thus be checked according to the following general formulae:

$$\frac{N}{N_c} + \frac{k\mu_x}{k\mu_x - 1}\left\{ \frac{M_{eq,x}}{M_D} + \frac{\mu_y/(\mu_y - 1)}{\mu_x/(\mu_x - 1)} \frac{M_{eq,y}}{M_{u,y}} \right\} \leq 1$$

$$\frac{N}{N_c} + \frac{k\mu_y}{k\mu_y - 1}\left\{ \frac{\mu_x/(\mu_x - 1)}{\mu_y/(\mu_y - 1)} \frac{M_{eq,x}}{M_D} - \frac{M_{eq,y}}{M_{u,y}} \right\} \leq 1$$

493

which, for $k = 1$, leads to:

$$\frac{N}{N_c} + \frac{\mu_x}{\mu_x - 1} \frac{M_{eq,x}}{M_D} + \frac{\mu_y}{\mu_y - 1} \frac{M_{eq,y}}{M_{u,y}} \leq 1$$

9.5 STRUTS AS STRUCTURAL ELEMENTS

The struts examined so far are ideal examples. If a strut is considered as an element of a structure, i.e. of a composition of other mutually connected struts, the following problems must be faced.

Constraints are different from hypothetical ones and are never perfect: there do not exist any perfect built-in joints. It is more reasonable to talk about the stiffness of the constraint, i.e. its fitness to oppose forces without undergoing any sensible displacement.

Constraints are not unvariable throughout the lifetime of a structure, and often they depend on the deformability of the other struts and on applied loads. A joint of a frame, for example, can offer a certain degree of fixity to a column, due to horizontal beam deformability. However, if a plastic hinge forms due to load that might act at any given time, the degree of fixity of the joint changes, increasing the free bending length of the column.

In order to solve such problems, it is necessary to seek for increasingly refined approaches. However, there are as yet no such sophisticated calculation instruments as to allow a complete simulation of entire structures undergoing buckling phenomena, and therefore able to render design options and related calculations unquestionable. It is thus an illusion to require a calculation instrument or a code to replace a good sense of an engineer which characterizes the quality of a design and, at the same time, gives a human dimension to designing, as it leaves a freedom of choice.

On the other hand, it is obvious that good sense alone does not prevent buckling. Approximate calculation methods and criteria, that have already proved to be satisfactory and are now established in practice, can be a guide and a means for producing, or at least stimulating, a sensitivity towards buckling problems which, in itself, leads to the recognition of the possibility of buckling, and therefore to considering how to prevent it.

The following illustrates some commonly adopted approaches to practical problems essentially concerning the determination of free bending length and the assessment of constraint stiffness and strength.

Free bending length - and therefore equivalent slenderness - can often be deduced either from an elastic analysis of the problem or from the collapse mechanism of the structure. In other cases - and typically, in the case of frames with sideway - interaction between column stability and overall stability of the structure can render results based on the equivalent slenderness criterion meaningless.

It is even more difficult to determine the necessary stiffness and strength of constaints. If a perfect structural system (i.e. free from any geometrical and structural imperfections) is analysed, it is sufficient to consider its stiffness without taking its strength into account. Consider, for example, the simple structure illustrated in Fig. 9.110a, which can be represented schematically as in Fig.

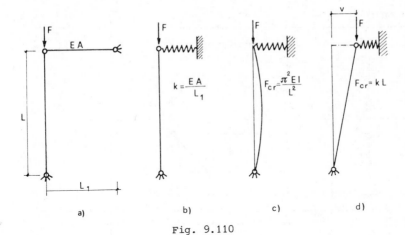

Fig. 9.110

9.110b |96|.

If the spring has constant $k > k_1 = \pi^2 EI/L^3$, the strut behaves as if the constraint were fixed (Fig. 9.110c); if the spring has a constant $k < k_1$, the constraint is not sufficiently rigid and thus gets involved in an overall buckling phenomenon (Fig. 9.110d). In actual practice, the problem is even more complex, as the system is not perfect and evolves in the elastic-plastic field. Load F causes a reaction in the constraint that depends on the extent of displacement v, which in turn is a function of initial imperfection and of strut and spring characteristics. Therefore, the value of such a reaction, required to design the constraint, is not known, nor can it be determined by conventional calculation methods. For these reasons, the various recommendations indicate semi-empirical and all embracing rules leading to designing constraints for a force equal to a fraction (1% to 2%) of the load acting on the strut to be stabilized. At the present state of knowledge in the field, such rules can however be assumed to be in favour of safety, at least for the most common design problems.

9.5.1 *Struts having different constraints*

9.5.1.1. Struts having end constraints
A strut of length L and provided with end constraints other than hinges can be designed on the basis of the following effective length:

$$L_c = \beta L \tag{9.138}$$

β being a coefficient depending on constraint conditions.

If the criterion illustrated in 9.1.3 with reference to equivalent slenderness is adopted, β values can be those determined by elastic analysis. Fig. 9.111 provides a list of the most frequent cases, together with some building examples. It is obvious that the hinge provides the least constraint and therefore to design a constraint

495

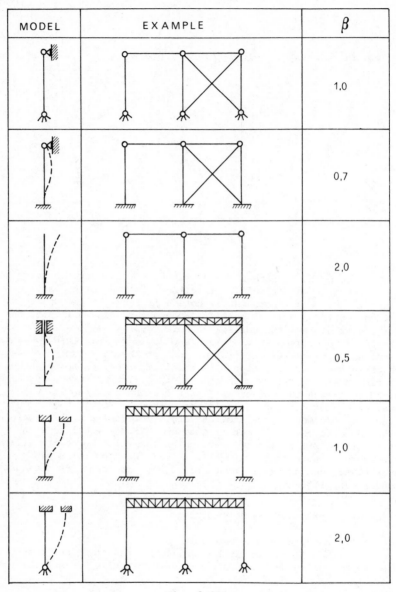

MODEL	EXAMPLE	β
		1,0
		0,7
		2,0
		0,5
		1,0
		2,0

Fig. 9.111

as a hinge is always in favour of safety. On the other hand, a
built-in joint can simulate either a rotationally efficient constraint
such as a foundation base designed for bending, or the constraint
provided to a column by a horizontal beam having a much greater stiffness. In order to take account of imperfect rotational fixity, it
496

might be realistic to increase by 10-20% values of β determined on the assumption of perfectly built-in constraints.

For the same reason, the last case illustrated in Fig. 9.111 should be considered cautiously: the horizontal beam deformabilty increases the value of β, which tends to infinity for an infinitely flexible girder (i.e. connected by a hinge to the columns, which corresponds to rendering the structure a mechanism).

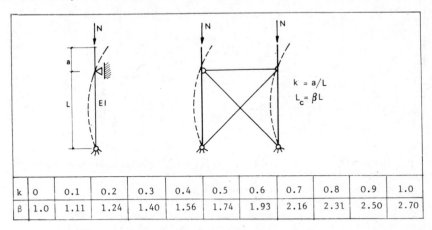

k	0	0.1	0.2	0.3	0.4	0.5	0.6	0.7	0.8	0.9	1.0
β	1.0	1.11	1.24	1.40	1.56	1.74	1.93	2.16	2.31	2.50	2.70

Fig. 9.112

9.5.1.2 Struts having intermediate constraints
A strut can be constrained at points other than its ends: a significant example is illustrated in Fig. 9.112, showing a bracing that does not reach the column top. The equation determining critical load in the elastic field is:

$$\alpha L(\tan \alpha kL + \tan \alpha L) - \tan \alpha L \cdot \tan \alpha kL = 0 \qquad (9.139)$$

where

$$\alpha^2 = N/EI \qquad ; \qquad k = a/L$$

If one assumes $x = \alpha L$, and with $\beta = \pi/x$, the equation is solved as follows:

$$N_{cr} = \alpha^2 EI = \frac{x^2 EI}{L^2} = \frac{\pi^2 EI}{(\beta L)^2} \qquad (9.140)$$

The table of Fig. 9.112 gives the values of β as k varies.

The same system of constraints, but with a strut for which the unsupported length L is greater than distance a between the two supports, is illustrated by the structure in Fig. 9.113. For a tending to zero, the strut behaves as if it were built in at ground level, and therefore $\beta = 2.0$. For $a \neq 0$ the equation determining the load is:

$$\alpha kL(\tan \alpha kL + \tan \alpha L) - \tan \alpha kL \cdot \tan \alpha L = 0 \qquad (9.141)$$

497

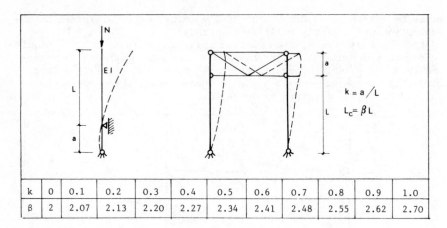

k	0	0.1	0.2	0.3	0.4	0.5	0.6	0.7	0.8	0.9	1.0
β	2	2.07	2.13	2.20	2.27	2.34	2.41	2.48	2.55	2.62	2.70

Fig. 9.113

The values of $\beta = \pi/x$ deduced therefrom are given in the table of Fig. 9.113.

Whenever a strut has more than two supports, it is hyperstatic.

Fig. 9.114 illustrates a typical structural example that can be represented schematically as a beam on three supports. The equation determining the critical load for this in the elastic field is:

$$\alpha kL(c\tan \alpha kL + c\tan \alpha L) - (1 + k) = 0 \qquad (9.142)$$

with $\alpha^2 = N/EI$ and $k = a/L$ as before.

The table of Fig. 9.114 gives the values of $\beta = \pi/x = \pi/\alpha L$ for $k < 1$. For $k > 1$, it is sufficient to exchange a and L in equation (9.142) and in the determination of effective length L_c.

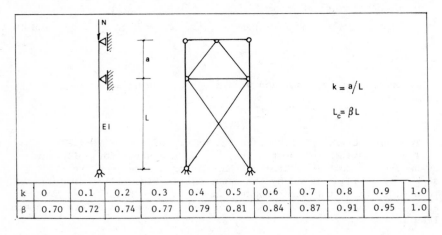

k	0	0.1	0.2	0.3	0.4	0.5	0.6	0.7	0.8	0.9	1.0
β	0.70	0.72	0.74	0.77	0.79	0.81	0.84	0.87	0.91	0.95	1.0

Fig. 9.114

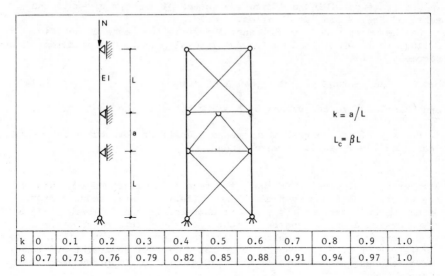

k	0	0.1	0.2	0.3	0.4	0.5	0.6	0.7	0.8	0.9	1.0
β	0.7	0.73	0.76	0.79	0.82	0.85	0.88	0.91	0.94	0.97	1.0

Fig. 9.115

Another rather interesting case is that of a strut on four supports, that can schematize the structure illustrated in Fig. 9.115. The central part - the shortest one - stabilizes the two lateral parts, reducing their effective length.

With the same conditions as the foregoing cases, the equation determining critical load in the elastic field is:

$$\alpha L - \tan \alpha L - \alpha L \tan \alpha L \tan \frac{\alpha k L}{2} = 0 \qquad (9.143)$$

k	0	0.1	0.2	0.3	0.4	0.5	0.6	0.7	0.8	0.9	1.0
β	0.5	0.53	0.57	0.61	0.65	0.70	0.75	0.81	0.87	0.93	1.0

Fig. 9.116

whence it is possible to deduce the values of $\beta = \pi/x$ given in the table of Fig. 9.115, as k varies.

On the other hand, if the central part is the longest one (Fig. 9.116), it is stabilized by the two lateral parts. The equation becomes:

$$\alpha k L - \tan \alpha\, kL - \alpha k L \ \tan \alpha\ kL \ \ \tan \frac{\alpha L}{2} = 0 \qquad (9.144)$$

and this leads to the values of $\beta = \pi/x$ given in the table of Fig. 9.116.

For $k = 0$, the central strut behaves as if it were built in at its ends; for $k = 1$, as if it were hinged.

9.5.1.3 Truss systems

Another typical problem is that of determining effective lengths (and therefore β) in truss systems. Some criteria, mostly deduced from German recommendations |54|, are illustrated in the following. In a truss designed optimizing its weight, all the struts reach maximum load bearing capacity simultaneously: therefore, at each joint, struts cannot offer each other a mutual degree of restraint and thus one must assume $\beta = 1$ for all the struts both for in plane and out of plane.

In actual practice, a truss is never optimal with reference to its weight: as profiles must be unified to achieve fabrication economy, it follows that maximum load bearing capacity is attained only in one member. The less loaded struts meeting at a joint thus provide a degree of rotational constraint to the most loaded ones

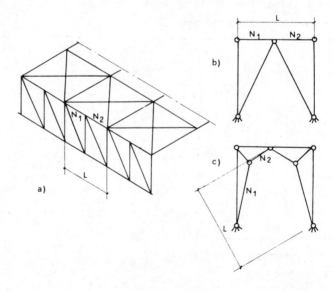

Fig. 9.117

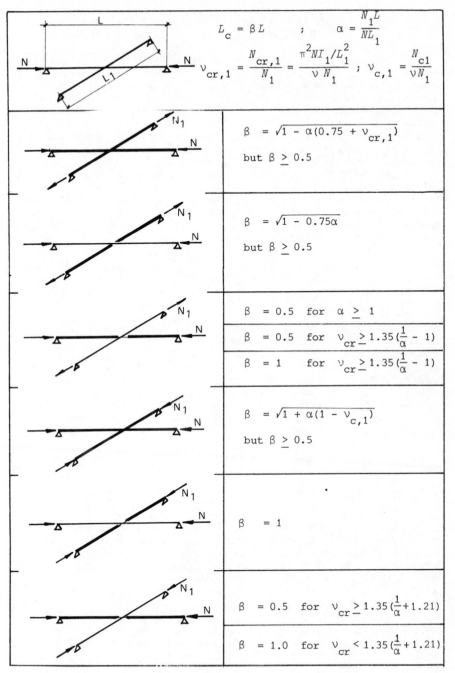

Fig. 9.118

and therefore reduce their effective length. If the compression chord section is constant, one can assume for it $\beta = 0.90$, both in plane and out of plane.

For wall struts (vertical bars and diagonals) one can instead assume $\beta = 0.80$ in the wall plane and $\beta = 1.0$ out of the plane. Such rules are applicable whenever the design ignores secondary bending moments and therefore the distance between the ideal hinges is assumed as length L of the various struts. Sometimes axial action varies within the same field: this might take place for an out-of--plane braced truss (Fig. 9.117a). Other examples are provided by out-of-plane buckling of the bracing girder shown in Fig. 9.117b or by the inclined struts shown in Fig. 9.117c. For $N_1 > N_2$, the strut can be checked for an axial load equal to N_1 and for an effective length $L_c = \beta L$ with $\beta = 0.75 \div 0.25\, N_2/N_1$.

If N_2 is a tensile force it must be given a minus sign and therefore the formula leads to a substantial reduction of effective length. However, the assumed value of β must not be less than 0.50, even if the formula leads lower values.

A strut, of length L and compressed by an axial load N, can be intersected by a second member of length L_1 and axially loaded by a tensile or compressive force N_1 (Fig. 9.118).

For in-plane bending, the crossing point must be considered fixed and therefore $\beta = 0.5$. For out-of-plane bending, two cases must be considered:

Members are jointed so that each passes through the other maintaining its stiffness unaltered

Only one member is integral, while the two parts of the other are joined to the first by means of a connection that must be considered as a hinge

However, the two members must be mutually connected efficiently, at least by a connection allowing the transmission of a force equal to one fourth of the axial force N acting in the strut. The following ratios are introduced:

$$\alpha = N_1 L/NL_1 ; \quad \nu_{cr,1} = \frac{N_{cr,1}}{\nu N_1} = \frac{\pi^2 EI/L_1^2}{\nu N_1} ;$$

$$\nu_{c,1} = \frac{N_{c,1}}{\nu N_1} = \sigma_c A_1/\nu N_1$$

in which $\nu = 1$, if the semi-probabilistic criterion is adopted, and $\nu = f_y/\sigma_{adm}$ if the allowable stress design is adopted. The effective length $L_c = \beta L$ can be deduced from the formulae of Fig. 9.118, for the cases considered therein. It is necessary, however, that $\beta \geq 0.5$, even if the formulae lead to lower values.

9.5.1.4 Angle bars
The determination of the effective length of the main struts for towers depends on the truss diagram and the shape of the section.

As more refined analyses are not available, the values of β given in the table of Fig. 9.119 can be adopted |54|. They are applicable whenever at least 50% of the axial force N in the most compressed vertical member is due to bending effects on the tower.

502

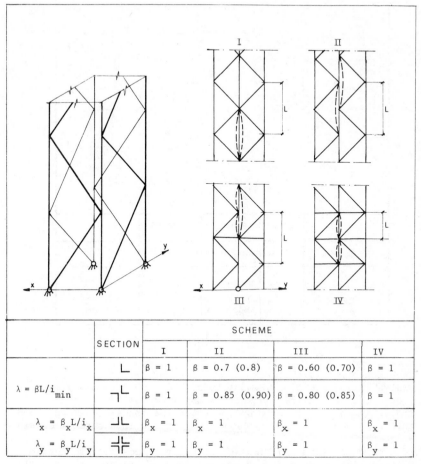

$\lambda = \beta L/i_{min}$	SECTION	SCHEME			
		I	II	III	IV
	L	$\beta = 1$	$\beta = 0.7 \ (0.8)$	$\beta = 0.60 \ (0.70)$	$\beta = 1$
	⌐L	$\beta = 1$	$\beta = 0.85 \ (0.90)$	$\beta = 0.80 \ (0.85)$	$\beta = 1$
$\lambda_x = \beta_x L/i_x$	⌐L	$\beta_x = 1$	$\beta_x = 1$	$\beta_x = 1$	$\beta_x = 1$
$\lambda_y = \beta_y L/i_y$	⫠⊩	$\beta_y = 1$	$\beta_y = 1$	$\beta_y = 1$	$\beta_y = 1$

Fig. 9.119

Otherwise, β values indicated in parentheses can be adopted.

Compressed single angle bars are mainly used in transmission towers and antennae, and as bracing diagonals. In these cases, the shear centre is located at the intersection of the centre lines of the two legs, and so does not coincide with the section centroid: it follows that a compressed angle bar buckles with an axial-torsional mode.

However, considering their vast use, many tests have been carried out, in quest of simpler design criteria, consistently with usual building details. They take into consideration |32|:

Angle bars connected on the two sides, and therefore simply compressed

Angle bars connected on one side only, and therefore eccentrically

503

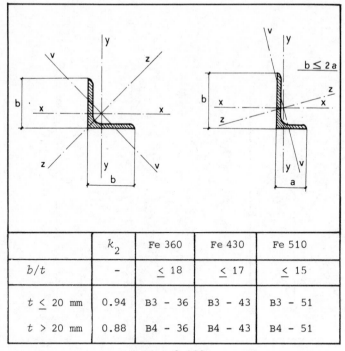

	k_2	Fe 360	Fe 430	Fe 510
b/t	-	$\leq$ 18	$\leq$ 17	$\leq$ 15
$t \leq 20$ mm	0.94	B3 - 36	B3 - 43	B3 - 51
$t > 20$ mm	0.88	B4 - 36	B4 - 43	B4 - 51

Fig. 9.120

compressed
They concern (Fig. 9.120):
Angles having equal or unequal legs, provided the longer leg
does not exceed twice the shorter one ($b \leq 2a$)
Angles in which the ratio b/t, between the longer leg length and
thickness does not exceed the limits set out in the table of Fig.
9.120.
If such conditions are complied with, it is possible to ignore
any interaction between flexural and torsional phenomena, as well
as any interaction between them and local stability. In fact,
experiments have shown that collapse of an angle bar usually takes
place due to local buckling of the outstanding leg, but for an axial
load practically identical to that determined on the basis of the
normal critical curves for struts, ignoring the penalty due to
torsion.
    Among all stability curves, the one most appropriate for
interpreting the behaviour of single angles is curve b |31|,
penalized by reduction in yield stress, depending on leg thickness,
according to a coefficient k_2, the values of which are indicated in
the table of Fig. 9.120. This leads to dimensional curves referred
to in the same figure for different types of steel and thicknesses.
    Therefore, in order to check a compressed single angle bar, it is
sufficient to calculate its slenderness, with reference to effective

length equal to the distance between intersection points of the centroidal axes of the truss without reduction ($\beta = 1$) and to the minimum radius of gyration of the section (about axis $v - v$, Fig. 9.120).

If both legs of the angle bar are connected, the load can be considered centred, ultimate stress being determined on the basis of the appropriate slenderness and of the dimensional curves indicated in Fig. 9.120, with reference to Fig. 9.40.

When only one leg of the angle bar is connected, eccentricity of loading and parasitic bending can be ignored provided reference is made to an equivalent slenderness equal to:

$$\lambda_{eq} = 0.60\lambda_c + 0.5757\lambda \quad , \quad \text{for } \lambda < \lambda_c\sqrt{2}$$

$$\lambda_{eq} = \lambda \quad , \quad \text{for } \lambda \geq \lambda_c\sqrt{2}$$

(9.145)

where

$$\lambda_c = \pi \sqrt{E/kf_y}$$

When the chords and the web members do not attain their maximum stress level for the same load conditions

$$\lambda_{eq} = 0.35\lambda_c + 0.75\lambda \quad , \quad \text{for } \lambda_c\sqrt{2} \leq \lambda \leq 3.5\lambda_c$$

may be used if the chords give good end restraints to the web members and at least two bolts in line are present at the end connections of the web members themselves.

9.5.2 Struts having a variable section

Tapered columns (i.e., struts having a gradually variable section) are no longer frequently adopted (Figs 9.121a,b): the expense of fabrication is not always rewarded by economy in weight. Their use is often justified by aesthetic or functional considerations (Fig. 9.121c).

Stepped columns (formed by two or more parts differing in geometry and loading) are more common: they are typical of industrial buildings with heavy cranes (Fig. 9.121d).

Tapered struts can be calculated assuming the following equivalent moment of inertia:

$$I_{eq} = kI_{max}$$

Coefficient k must be determined by applying the equivalent slenderness criterion: i.e., the elastic critical load is calculated for a strut having a variable section and the moment of inertia I_{eq} is found for a strut of constant section carrying the same critical load.

Given:

$$\gamma = \sqrt{I_{min}/I_{max}} \quad \text{for } \gamma \geq 0.10 \quad \text{and} \quad L_1/L \leq 0.5$$

one can assume, in good approximation |54|:
for the strut shown in Fig. 9.121a:

$$k = 0.17 + 0.33\gamma + 0.50\sqrt{\gamma} + (0.62 + \sqrt{\gamma} - 1.62\gamma)(L_1/L)$$

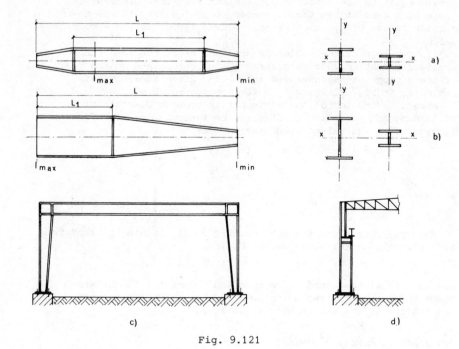

Fig. 9.121

for the strut shown in Fig. 9.121b:

$$k = 0.08 + 0.92\gamma + (0.32 + 4\sqrt{\gamma} - 4.32\gamma)(L_1/L)^2$$

For $L_1/L \geq 0.8$, one can assume $k = 1$ and for values of L_1/L from 0.5 to 0.8 inclusive a linear interpolation is possible between the values of k provided by the above formulae and one.

Also for stepped columns (Fig. 9.121d) there are as yet no numerical solutions obtained by simulation in the elastic-plastic field. The problem must therefore be studied in the elastic field, as those above.

If the following parameters are introduced (Fig. 9.122a)

$$\xi = I_1/I_2 \quad ; \quad \eta = L_2/L \quad ; \quad \alpha = N_2/(N_1 + N_2) = N_2/N_{tot}$$

$$\gamma_1^2 = N_1/EI_1 \quad ; \quad \gamma_2^2 = N_{tot}/EI_2 = \frac{\xi}{1 - \alpha}\,\gamma_1^2 \qquad (9.146)$$

elastic buckling (Fig. 9.122b) is governed by the following differential equations:

$$y_1'' + \gamma_1^2 y_1 = (M_1 + N_1 f_1)/EI_1$$

$$y_2'' + \gamma_2^2 y_2 = (M_1 + N_1 f_1 + N_2 f_2)/EI_2 \qquad (9.147)$$

where M_1 is the moment developed by the top constraint and f_1, f_2 are the transverse displacements at the top and at the step respectively.

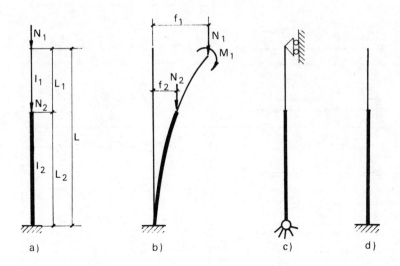

Fig. 9.122

Boundary conditions obviously depend on the type of end fixity. The solution of (9.147) and evaluation of the constants of integration to satisfy the boundary conditions lead to the characteristic transcendental equation:

$$f(\gamma_1 L_1, \gamma_2 L_2) = 0 \text{ with } \gamma_2 L_2 = (\gamma_1 L_1)(L_2/L_1)\sqrt{\xi/(1-\alpha)} \qquad (9.148)$$

Once the first root of (9.148) is found:

$$\bar{x} = \bar{\gamma}_1 L_1 = (\bar{\gamma}_2 L_2)(L_1/L_2)\sqrt{(1-\alpha)/\xi}$$

one gets:

$$\bar{\gamma}_1 = \bar{x}/L_1 \quad ; \quad \bar{\gamma}_2 = (\bar{x}/L_1)\sqrt{\xi/(1-\alpha)} \qquad (9.149)$$

From (9.149) and (9.146) at buckling:

$$N_{1,cr} = \bar{\gamma}_1^2 EI_1 = \bar{x}^2 EI_1/L_1^2 = \pi^2 EI_1/(\beta_1 L)^2$$

$$N_{tot,cr} = \bar{\gamma}_2^2 EI_2 = \frac{\bar{x}^2 EI_2}{L_2^2}(L_2/L_1)^2\frac{\xi}{1-\alpha} = \frac{\pi^2 EI_2}{(\beta_2 L)^2}$$

The effective legths are:

$$L_{c,1} = \beta_1 L = (\pi/\bar{x})L_1$$

$$L_{c,2} = \beta_2 L = (\pi/\bar{x})L_1\sqrt{(1-\alpha)/\xi} = \beta_1\sqrt{(1-\alpha)/\xi}$$

For the most common cases, equation (9.148) may be written as follows:

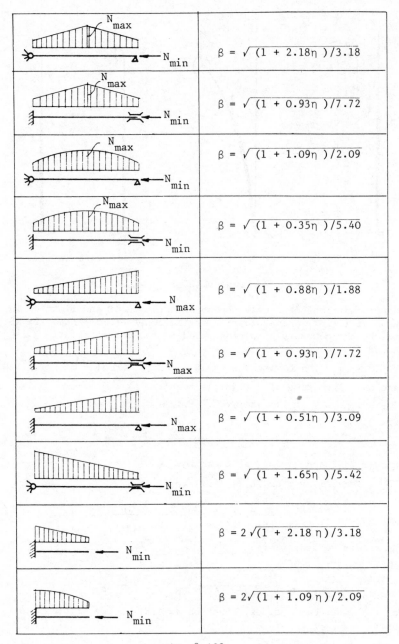

$$\beta = \sqrt{(1 + 2.18\eta)/3.18}$$

$$\beta = \sqrt{(1 + 0.93\eta)/7.72}$$

$$\beta = \sqrt{(1 + 1.09\eta)/2.09}$$

$$\beta = \sqrt{(1 + 0.35\eta)/5.40}$$

$$\beta = \sqrt{(1 + 0.88\eta)/1.88}$$

$$\beta = \sqrt{(1 + 0.93\eta)/7.72}$$

$$\beta = \sqrt{(1 + 0.51\eta)/3.09}$$

$$\beta = \sqrt{(1 + 1.65\eta)/5.42}$$

$$\beta = 2\sqrt{(1 + 2.18\eta)/3.18}$$

$$\beta = 2\sqrt{(1 + 1.09\eta)/2.09}$$

Fig. 9.123

Base and top pinned (Fig. 9.122c)

$$\{1 + (1 - \alpha)\ \frac{\eta}{1 - \eta}\}\gamma_1 L_1\ \cos\ \gamma_1 L_1\ \sin\ \gamma_2 L_2 +$$

$$(1 - \alpha)\{(1 - \alpha)/\eta\ + \alpha(1 - \eta)/\eta\}\ \gamma_2 L_2\ \sin\ \gamma_1 L_1\ \cos\ \gamma_2 L_2 +$$

$$- \alpha^2\ \sin\ \gamma_1 L_1\ \sin\ \gamma_2 L_2 = 0$$

Base fixed and top free (Fig. 9.122d)

$$\gamma_1 L_1\ \cos\ \gamma_1 L_1\ \cos\gamma_2 L_2 - \{(1 - \eta)/\eta\}(1 - \alpha)\gamma_2 L_2\ \sin\gamma_1 L_1\sin\gamma_2 L_2 = 0$$

Solution for other boundary conditions and a table of values for β_1 and β_2 may be found in $|97-98|$.

The effective length $L_c = \beta L$ for struts with variable axial load and different end constraints may be found from Fig. 9.123.

9.5.3 *Struts with elastic constraints*

9.5.3.1 Compressed struts

In actual practice, a constraint is sometimes formed by a deformable element that must be taken into account to avoid underestimating the danger of buckling.

Consider, for example, the strut shown in Fig. 9.124a: it is constrained at three supports. The material is assumed to be infinitely elastic and the strut perfectly straight, at critical load the buckled shape has a point of contraflexure are at the central support: the strut behaves as if it were formed of two struts, each having an effective length $L_c = L$.

If the intermediate constraint is elastic (Fig. 9.124b), there are two different possibilities. If the constraint is sufficiently rigid, i.e. if its stiffness exceeds a necessary minimum value , the strut behaves as that illustrated in Fig. 9.124a. If instead, the constraint is more deformable, it takes part in the buckling phenomenon and the effective length L_c is greater than L.

On the basis of the above intuitive considerations, it follows that - even when the phenomenon is bending examined from a purely elastic point of view and with reference to perfect struts - constraint stiffness must be checked to ensure that its deformability is contained within acceptable limits.

In practice the problem is even more complex, as structural elements are not perfect and are not made of infinitely elastic material. Hence:

The minimum stiffness the constraint must be provided with depends on the extent of initial imperfection

The constraint must be checked, also in terms of strength, with reference to the effect of the constraint reactions loading it, which are also functions of the displacements of the structural element being designed

The problem has not yet been solved satisfactily and therefore it is advisable to operate as follows:

A minimum stiffness of constraint is established, on the basis of calculations carried out in the elastic field on structures free

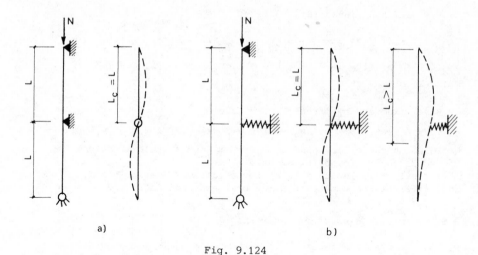

a)   b)

Fig. 9.124

from imperfections

Constraints are designed applying forces which are functions of the applied external loads and are determined a priori on the basis of experience, unless more sophisticated calculation criteria are available

It must, however, be pointed out that the subject discussed here does not usually concern struts the supports of which are formed by tension or compression members constrained to the ground (Fig. 9.125a) or to sufficiently rigid bracing elements (see 9.5.4): in fact, the axial stiffness of the constraining members does not penalize the compressed structural element.

On the other hand, the problem does concern structural elements in which vertical girders or stiffeners are entrusted with preventing out-of-plane buckling of the upper cord of a truss or of the compressed flange of a beam. Typical examples are provided by through bridges (Fig. 9.125b) or truss roof frames for industrial buildings, the upper chords of which are stabilized by appropriate bracing structures (Fig. 9.125c).

From a historical point of view, it is interesting that through bridges, with their truss beam walls and compression chords stabilized by the vertical frames, inspired the first studies of Engesser (1893) Chwalla (1929) and Bleich (1919), later improved by other authors, whose significant works are referred to in the SSRC Guidebook |6|.

The problem is substantially that illustrated in Fig. 9.126a, originally studied by Engesser on the basis of the following assumption |56|:

Straight upper chords, having a constant section and hinged ends, bearing a constant axial load

Many elastic supports, sufficiently near to one another that they can be considered as one continuous one

If the stiffness of each constraint (i.e. the ratio between applied force and displacement of its application point) is defined

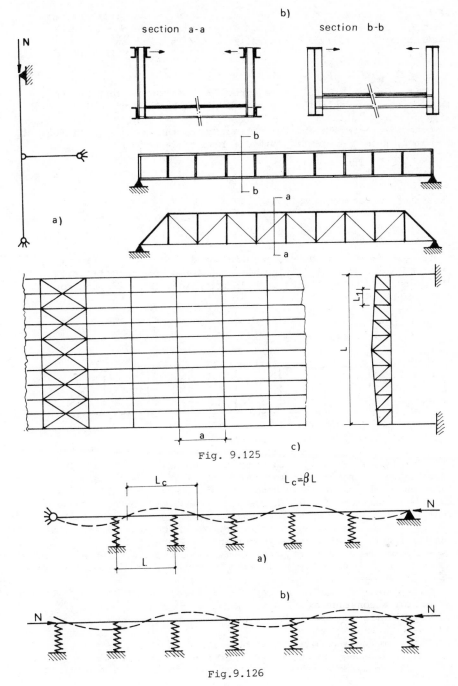

section a-a   section b-b

b)

a)

Fig. 9.125   c)

$L_c = \beta L$

a)

b)

Fig.9.126

511

as $k = F/v$, Engesser's solution contemplates that a strut, having bending stiffness EI, can attain its critical load N_{cr} in the elastic field, provided spring stiffness k is such that:

$$k \geq k_{min} = \frac{N_{cr}^2 L}{4\,EI} \qquad (9.150)$$

This solution is applicable, provided that the distance beween two consecutive contraflexure points of the buckled shape is $L_c = \beta L$, with $\beta \geq 1.2$.

Otherwise, formula (9.150) underestimates the necessary stiffness k. Since, from the elastic critical load for a strut:

$$EI = \frac{N_{cr}(\beta L)^2}{\pi^2} \qquad (9.151)$$

substituting for EI in (9.150) gives:

$$k \geq k_{min} = \frac{\pi^2 N_{cr}}{4\beta^2 L} \simeq 2.5\,\frac{N_{cr}}{\beta^2 L} \qquad (9.152)$$

Formula (9.152) can be extended in the elastic-plastic field, provided N_c replaces N_{cr}. In this connection, DIN Specifications |54| proposed the following control criterion, later adopted for many other codes.

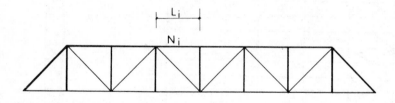

Fig. 9.127

Referring, for example, to the truss shown in Fig. 9.127, for each length of the compression chord the axial force is known, as a function of the external loading. Consider any length L_i of the chord with an axial force N_i. Its mean stress will be $\sigma_i = N_i/A_i$ and from the stability curves, the maximum slenderness λ_i compatible with this mean stress can be assessed. Then the maximum effective length compatible with strut performance will be $L_{c,i} = \lambda_i i_i$ (i_i being the radius of gyration of the section) and, hence, the $\beta_i = L_{c,i}/L_i$ ratio. Obviously, β_i increases together with λ_i, and therefore as σ_i decreases. It must be: $\beta_i \geq 1.0$

If the above operation is repeated for each length of chord, a mean value of coefficients

$$\beta_m = (\sum_i^n \beta_i)/n$$

can be determined and this should be ≥ 1.2. If $\beta_m < 1.2$, values of σ_i must be reduced, by increasing the chord section. If however $\beta_m \geq 1.2$, the minimum value of the constraints rigidity determined according to the limit state under consideration:

$$k_{min} = \frac{\pi^2}{4\beta_m^2} \frac{\nu N_{max}}{L_{min}}$$

(9.153)

where
N_{max} = maximum value of N_i
L_{min} = minimum value of L_i
$\nu = 1$, if the semi-probabilistic limit state method is adopted
ν = the safety coefficient, if the allowable stress method is adopted

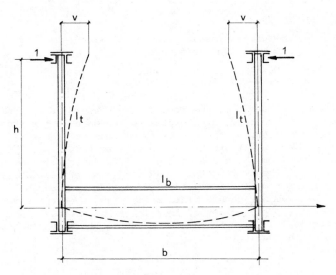

Fig. 9.128

Finally, constraint stiffness must be checked to see that it exceeds k_{min}. For this purpose (Fig. 9.128), it is sufficient to apply an average force to the constraints and to assess elastic displacement v. Stiffness is

$$k = 1/v = E / \left\{ \frac{h^3}{3I_t} + \frac{h^2 b}{2I_b} \right\}$$

(9.154)

The first term in the denominator is the vertical frame contribution, while the second term is the horizontal floor system contribution. If the contribution to flexural stiffness of the diagonals of the lateral beams can also be taken into consideration it follows that:

$$k = F / \left\{ \frac{h^3}{3I_t + 3I_d \left(\frac{h}{L_d}\right)^3} + \frac{h^2 b}{2I_b} \right\}$$

(9.155)

where I_d and L_d, respectively, are the moment of inertia and length of diagonals.

513

The above criterion considers the strut ends fixed.

If the strut ends, too, are connected by means of elastic constraints (Fig. 9.126b), the above formulae are no longer safe. In order to take account of end constraint deformability, still following DIN recommendations, the above procedure must still be applied to determine minimum stiffness k_{min} but it is also necessary to check that end and intermediate constraint stiffnesses k_e and k_i comply with the following limitations:

$$k_i \geq \eta k_{min} \quad ; \quad k_e \geq (\eta/\zeta) k_{min} \qquad (9.156)$$

where:

$$\zeta = \min(k_i)/k_e$$

is the ratio between the lowest value of intermediate constraint stiffness and end contraint stiffness

$$\text{and} \quad \eta = \frac{1 + 0.60\zeta\beta_m}{2} \left\{ 1 + \sqrt{1 - \frac{1.44\zeta\beta_m}{(1 + 0.6\zeta\beta_m)^2}} \right\} \qquad (9.157)$$

As already pointed out, the foregoing procedure is no longer safe if $\beta < 1.2$. In this case, by a different approach to the problem, it is possible to determine the necessary minimum stiffness k_{min}, to render the effective length L_c equal to L, so that $\beta = 1$. Such values can be deduced from the graph of Fig. 9.129, taken from the SSRC Guidebook |6|. They were determined on the assumption that, in the various lengths n of the beam, axial forces N_i vary parabolically between a maximum N_{max} and a minimum N_{min}, and that the same variation affects the moment of inertia of the various lengths of the chord. Obviously, it is on the safe side to use a constant moment of inertia over the whole length.

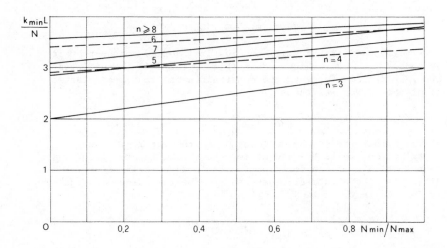

Fig. 9.129

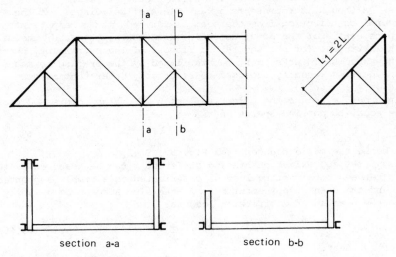

section a-a   section b-b

Fig. 9.130

The above checking procedure is not applicable whenever the strut is provided with only one intermediate constraint. This problem must be faced, for example, whenever the effective length of a diagonal strut has to be reduced (Fig. 9.130). In this case, according to DIN Specifications |54|, the constraint stiffness must be:

$$k \geq k_{min} = 16\,(N/L_1)\left\{(1/\beta) - (0.25 \; / \; \beta^2) - 0.75\right\} \qquad (9.158)$$

It appears that for $\beta = 1$ (i.e. for $L_c = L_1$) $k = 0$ and therefore no intermediate constraint is necessary. On the other hand, in order to obtain $\beta = 0.5$ and therefore, to halve the out-of-plane effective length:

$$k \geq 4N/L_1 = 2N/L \qquad (9.159)$$

Naturally, in this case, too, N must be replaced by νN, if the allowable stress method is being applied.

Constraints must not only be checked for stiffness: their strength must also be checked, particularly that of their connections. It is obvious that the forces loading the constraints depend on:

Axial force N in the strut that the constraints contribute to stabilizing

Amount of lateral displacement, which in turn is a function of axial force N and initial imperfection

It is not yet possible to provide a reliable quantification of such forces. Codes therefore adopt values confirmed by experience, which must be applied whenever more precise rules are not available. Following this criterion, constraint strength can be checked by applying, in addition to possible further effects of external loads, a force $F = N/100$, N being the axial force in the strut stabilized by the constraint.

515

The problem illustrated in Fig. 9.125 c has not yet been fully solved, although it represents a vast class of buildings. If the torsional contribution from the beam is ignored, it can be put into the perspective of the previously discussed cases: the upper chords of the n trusses (or the compressed flanges of the n principal beams, if they have solid webs) must be stabilized by the bracing by means of the purlins (usually supposed to react only under tension, see 1.4.4).

In order to assume the effective length L_c of the compressed chord to be equal to purlin spacing L_1 ($L_c = L_1$):

$$k \geq k_{min} \, Nn/L_1 \qquad (9.160)$$

k_{min} being the value deduced from Fig. 9.129 ($L_1 = L$).

As a first approximation, bracing stiffness k can be assessed as the ratio between total acting load qL distributed on a simply supported beam and the mean displacement v_m. As the displacement of a simply supported beam under distributed load is:

$$v(x) = \frac{1}{24} \frac{qL^4}{EI} \{(x/L) - 2(x/L)^3 + (x/L)^4\} \qquad (9.161)$$

it follows that:

$$v_m = \frac{1}{L} \int_0^L v(x)\,dx = 0.64 \frac{5}{384} \frac{(qL)L^3}{EI} \qquad (9.162)$$

and therefore:

$$k \simeq qL/v_m = 1/\{0.64 \frac{5}{384} \frac{L^3}{EI}\} \simeq 120 \frac{EI}{L^3} \qquad (9.163)$$

To sum up, the following relation must be complied with:

$$EI/L^3 \geq k_{min} \frac{N\,n}{120L_1} \qquad (9.164)$$

and therefore

$$EI/L^2 \geq k_{min} \frac{N\,m\,n}{120} \qquad (9.165)$$

where

$m = L/L_1$ = number of parts into which the chord is divided
n = number of trusses to be stabilized

When calculating the moment of inertia I the deformability due to shear must obviously be taken into account (see 8.1.4). It is apparent from this approach that bracing stiffness must be increased, the greater the number of points considered fixed.

It is more difficult to determine the forces loading the connections between the upper chord and the purlins and between the purlins and the rafter bracing. They are caused by two qualitatively different phenomena:

(a) The upper chord (Fig. 9.131a) is not perfectly straight and, therefore, the purlins can be loaded by forces $F_1 = 2N\sin \theta/2 \simeq N\theta$ where θ is out-of-straightness

(b) Trusses can be inclined relative to the direction of loading, showing a tendency towards flexural-torsional buckling (Fig.9.131b), that is opposed by the bracing by means of the purlins
Effect (a) can be considered as a local one, as it is unlikely

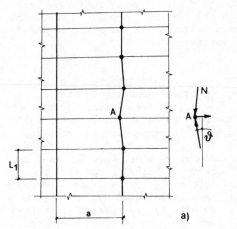
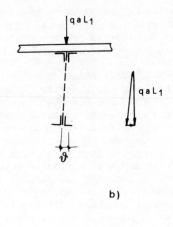

Fig. 9.131

that all the trusses have the same imperfection: force $N\theta$ corresponding to the ith truss is compensated by forces of the opposite sign
due to different imperfections in the other roof trusses. This is
why Dutch Regulations |99| advise dimensioning purlin connections
to the trusses for a force $F = 0.01N$, but not to accumulate such
forces when designing the bracing.

Effect (b), on the other hand, concerns the bracing, as it can be
summed up for all the n trusses that are stabilized by the same
bracing. Such an effect is a function of bracing deformability and
of the vertical load acting on the trusses, as well as of the number
of trusses. With reference to bracing design, Dutch Regulations
advise adding to external forces a uniformly distributed load equal
to:

$$q_b = 0.005 \; qna \qquad\qquad (9.166)$$

wherein:
q is load per surface unit acting on the roof
a is truss spacing
n is the number of trusses

A different approach was proposed in |100|. Bracing stiffness is
controlled verifying that its Euler critical load is at least 2.5
times the sum of maximum axial actions N in the compressed chords of
the trusses to be stabilized:

$$nN < \frac{N_{cr}}{2.5} = \frac{\pi^2 EI}{2.5L^2} \qquad ; \qquad EI/L^2 \geq 0.25nN \qquad (9.167)$$

This criterion leads to analogous values of (9.165) for $m = 8$ and
$k_{min} = 3.75$.

According to the same authors |100|, stength must be controlled by
imposing the following load on the bracing:

$$q_i = 0.025 \; nN/L \qquad\qquad (9.168)$$

517

as an alternative to the load due to other external actions. Such
a criterion differs from the Dutch one for two reasons: it considers
a load that is proportional to compression axial force and it does
not accumulate from the external forces.

9.5.3.2 Bent beams

Also with reference to flexural-torsional stability phenomena, the
effect of the constraints - formed by the other parts of the
structure - can be determinate. However, unlike the case of struts,
real conditions for beams are often more favourable than the ideal
ones assumed in calculations. In order to prove such a statement,
the typical structural arrangement in Fig. 9.132a will be analysed.
The example is a floor system: it is obvious that, in this case, one
cannot consider the stability of a beam in isolation, as the
stability of the entire structural system must be examined. Possible
lateral buckling of the principal beams is resisted not only by their
own flexural-torsional stiffness, but also by the flexural stiffness
of the secondary beams, which, although they are connected by means
of simple bolted joints, are quite suitable to provide a stabilizing
contribution.

The example illustrated in Fig. 9.132b shows a crane beam: as it
is necessary to oppose horizontal forces due to crane movement (see
3.3.2) it often ensues that a stiffener or a horizontal beam is
introduced, depending on crane size. Further, in such cases the
problem of fatigue often makes it advisable to limit stresses in the
struts, so that lateral stability problems do not arise.

In other cases, the loads applied to a beam form additional
constraints or might induce stabilizing effects. Examine, for example,
Fig. 9.132c showing beams bearing concrete or corrugated sheet slabs:
flexural-torsional buckling is opposed by the migration of the
support point of the overhanging slab, which causes a restoring
torsional moment and therefore a stabilizing effect.

These typical structural arrangements have not been illustrated
to diminish the importance of the problem at stake, but in order to
provide the opportunity of setting forth a few considerations.

The lateral stability of beams was studied - in this case as in
others - from a theoretical rather than a practical point of view,
possibly because in building practice the problem has always appeared
less conditioning than other forms of buckling: in actual building,
there is often some static reason that contributes to removing the
danger. However this cannot justify ignorance of the phenomenon, as
it is testified by the troubles that occur in structures designed
without the problem having been considered and in which loads are
applied so as not to form an additional constraint: e.g. plant
support beams or runway beams for assembly lines, not connected by
horizontal floor systems. The struts (compression flanges) in such
cases are those closest to the theoretical model and as it is often
unnecessary to give excessive importance to deformability limits
for such beams, it is not infrequent to notice collapse symptoms due
to flexural-torsional buckling.

In any case, it is evident that it is in favour of safety to
analyse the flexural-torsional buckling pehomenon:

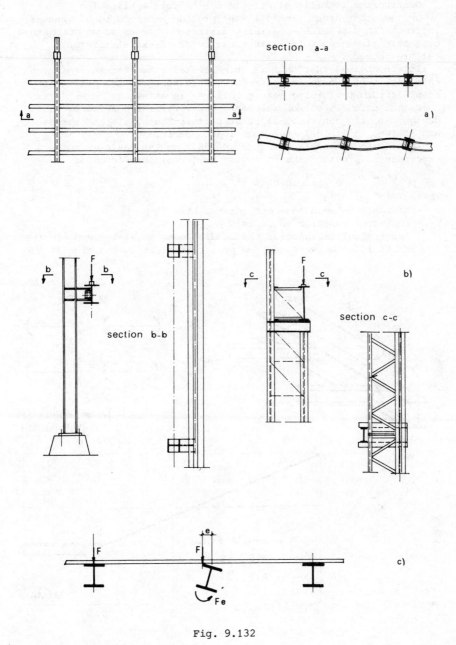

section a-a

a)

section b-b

b)

section c-c

c)

Fig. 9.132

Considering only lateral constraints formed by struts connected to fixed points or to bracing systems as effective

Overlooking stabilizing effects due to load application

On the other hand, the following problem seems to be of greater interest: to determine constraint distance L_c so as to guarantee that a plastic hinge can form in the beam before flexural-torsional failure occurs.

It is obvious that this is an essential datum for the collapse calculation of a beam in order to be able to admit the possibility of plastic hinges forming before collapse is attained.

Such a datum is of no interest, however, if the structure is designed to the conventional elastic limit state, i.e. if stresses are limited to yield in the most loaded fibre.

An AISC Recommendation issued in 1963, on the basis of some experiments, provided the following semi-empirical formula:

$$(60 - 40\, M/M_{pl})i_y > L_c \geq 35 i_y \qquad (9.169)$$

where (Fig. 9.133):

L_c = distance between adjacent constraints
i_y = radius of gyration about axis y-y
M = least absolute value of the bending moments at the end of the part included between the two constraints: it is positive if the

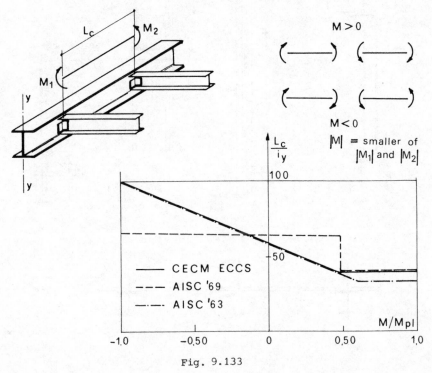

Fig. 9.133

part is bent with a single curvature, negative if there are opposite bending moments

M_{pl} = section plastic moment.

A few years later, Lay and Galambos |101| proved that that formula could be excessively conservative if $M > 0$ and not in favour of safety if $M < 0$. Therefore,and to extend the field to higher strength steels, AISC recommendations from 1969 adopt the following formulae (with f_y in N mm^{-2}):

$$L_c/i_y \leq 40 \times 235/f_y + 25; \quad \text{for} \quad -1 \leq M/M_{pl} < 0.5$$
$$L_c/i_y < 40 \times 235/f_y; \quad \text{for} \quad 0.5 \leq M/M_{pl} \leq 1.0$$
(9.170)

Fig. 9.133 compares the two formulations for f_y = 235 N mm^{-2}

The Recommendations of the European Convention, however, adopted formulae resembling the first American version and were not as strict for steel having a high yield. In fact, they propose:

$$L_c/f_y \leq (60 - 40 \, M/M_{pl})\sqrt{235/f_y}; \quad \text{for} \quad -1 \leq M/M_{pl} < 0.5$$
$$L_c/f_y \leq 40\sqrt{235/f_y}; \quad \text{for} \quad 0.5 \leq M/M_{pl} \leq 1.0$$
(9.171)

9.5.4 *Frame members*

The definition of effective length L_c = βL concerns axially compressed simple struts: it is equivalent to the length L of a strut having constant section, hinged at its ends, and carrying the same critical load N_{cr} in the elastic field as the strut in question (see 9.1.3).

This concept can be extrapolated for struts forming the members of a frame. In this case the effective length can be defined as the distance between two consecutive points of contraflexure of the critical shape in the elastic field (Fig. 9.134).

If sway is prevented, i.e. if the joints are constrained against transverse displacements (Fig. 9.134a) and if the frame is hinged at its base, it follows that $0.70 \leq \beta \leq 1.0$.

Extreme cases would result from the existence of an infinitely rigid or infinitely flexible girder.

In the case of a frame having its columns built in at their base:

$$0.50 \leq \beta \leq 0.70$$

the limits being once more suggested by the assumption of either infinite or zero flexural stiffness of the girder.

Obviously, a girder connected to the columns by joints that are not capable of transmitting bending moments must be considered as having zero stiffness: it is thus justified to assume β = 1 for pin-ended structures.

It is more complex to determine the effective length if sway is permitted (Fig. 9.134b). In fact, still assuming the limiting cases of an infinitely rigid or perfectly flexible girder, one obtains:

$2 \leq \beta \leq \infty$ if the base is hinged
$1 \leq \beta \leq 2$ if the base is built in

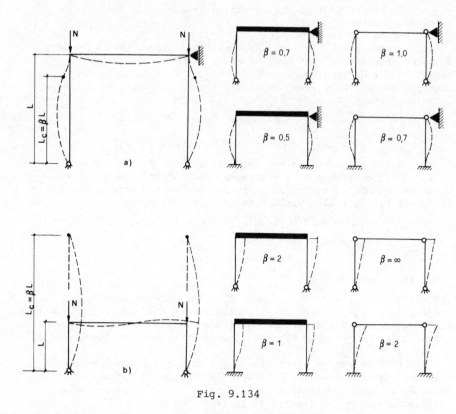

Fig. 9.134

In this case, the possible variation in β is much wider, so it is easier to assess effective length incorrectly.

The above considerations concerning single storey frames can be generalized, so as to be extended to frames of more than one storey (Fig. 9.135).

Here, too, effective length L_c is less than L when sway is prevented, whereas $L > L$ when sway is permitted. Effective length L_c depends on mutual stiffnesses of girder and columns and on the distribution of the load applied to the frame. In order to assess it correctly, the following data must be known:

Critical load factor, assuming an infinitely elastic behaviour

Distance between two consecutive points of contraflexure of the buckling shape

It is obvious that such a calculation cannot be proposed for practical purposes. Strictly speaking, it should be repeated for every loading condition, because effective length also depends on the mutual ratio of axial load in the columns belonging to the various floors. Furthermore, the equivalent slenderness criterion is not always applicable to the design of frame columns. It can be followed with a good degree of reliability to design frames in which sway is

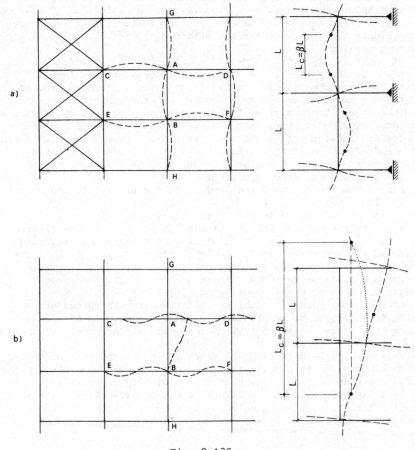

Fig. 9.135

prevented, but it might be grossly inappropriate - and not always in favour of safety - in the case of frames having movable joints.

For the latter frames, it is in any case preferable to determine the destabilizing effects of the vertical load, which offers at the same time a check on the overall stability of the frame and the additional bending moments to be included in column design.

For these reasons, methods currently suggested in the literature for assessing effective length in frames having either fixed (9.5.4.1) or movable (9.5.4.2) joints are discussed in the following separately from methods to be applied when determining destabilizing effects (9.5.4.3).

9.5.4.1 Frames with sway prevented
It is useful to define, first of all what 'frame with sway prevented'

523

means. In practice, the joints of frames can always move transversely: the extent of such a displacement might be great with respect to other diaplacements or so small that it can be neglected. To decide whether a frame can be defined as 'sway prevented' it is necessary to evaluate the ratio between the stiffnesses of the frame with and without bracing.

If the ratio is high (CECM-ECCS Recommendations provide that it must be greater than 5), the frame can be considered as 'sway prevented' and thus all transverse force effects can be entrusted to the bracings, without even assessing interaction between frame and bracings. Obviously, stiffnesses can be calculated by approximate methods, e.g. by applying unit transverse loads to the floor and assessing diplacements at the top: the ratio between the stiffnesses with and without bracing is the inverse of that between related displacements. Once it has been established that a frame may be considered as 'sway prevented', one must:

Assess the transverse force applied to the bracing caused by the vertical loads

Define the effective length of the column

Transverse force loading the bracing in addition to the external ones cannot be quantified by calculation unless the destabilizing effects of the vertical loads due to the deformabiliy of the system (see 9.5.4.3) are assessed. They are in any case slight in frames in which sway is prevented and therefore can usually be ignored compared to the external loads (wind, etc.).

On the other hand, it might be that the external forces are quantitatively slight or practically non-existent. In this case, transverse forces equal to 1/80 of the vertical forces may be applied |102|.

This is a semi-empirical rule which is equivalent to considering, as an alternative to the other loads, an earthquake producing in the structure static effects equivalent to a $0.0125g$ acceleration.

What has been illustrated before regarding effective length also applies here, i.e. effective length $L_c = L\beta$ of columns in frames having fixed joints is in any case such that:

$$0.5 \leq \beta \leq 1.0$$

It is a prudent design practice to assume β no less than 0.70: built in joints, obtained even by modern techniques, can never be considered perfect.

A practical and quick procedure for assessing β was developed by Julian and Lawrence |103| from the 'alignment charts' adopted in U.S. design practice and also referred to in the comments to AISC recommendations |55| and in the SSRG Guidebook |6|.

Consider the module of a multi-storey frame shown in Fig. 9.135a: it is formed by column AB, the effective length of which must be determined, the members constraining it at the ends, specifically the four beams AC, AD, BE, BF and the two columns AG and BH. These are all elastic constraints with reference to rotations at column ends A and B.

Let the following simplifying assumptions be made:

Perfectly elastic behaviour

524

All members with a constant section
Vertical loads applied only at the joints
Joints preventing relative rotation between the members
Identical ratio $L/\sqrt{N/EI}$ for all the columns
Moments distributed between the members meeting in a joint in
proportion to their respective stiffnesses k_i.
If it is further assumed that all the columns buckle
simultaneously, the equation determining $\beta = L_c/L$ is:

$$\frac{\alpha_1 \alpha_2}{4} (\frac{\pi}{\beta})^2 + \frac{\alpha_1 + \alpha_2}{2} (1 - \frac{\pi/\beta}{\tan \pi/\beta}) + \frac{2 \tan \pi/2\beta}{\pi\beta} - 1 = 0 \qquad (9.172)$$

where

$$\alpha_i = \Sigma k_c / \Sigma k_b \qquad (i = 1 \text{ or } 2)$$

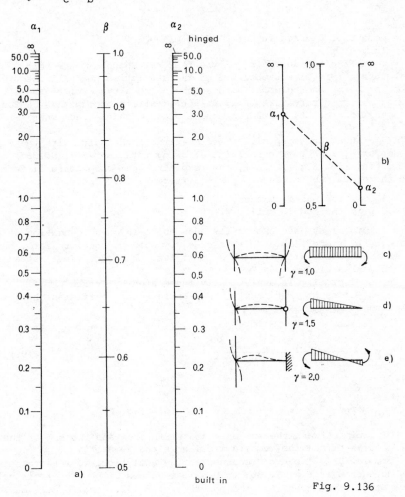

Fig. 9.136

is a parameter expressing the ratio between the sum of stiffnesses k_c of the columns meeting at a joint and the sum of beam stiffnesses k_b. The values of α_1 and α_2 refer respectively, to the upper and lower joint.

The solution to equation (9.172) can be obtained from the nomograph (alignment chart) of Fig. 9.136a. Values for α_1 and α_2 are given on two outside scales and for β on the middle scale. The line joining α_1 and α_2 intersects the middle scale at the required value of β Fig. 9.136b).

To determine α_1 and α_2, it is necessary to find column and beam stiffnesses. Stiffness k_i of the i-th member, disregarding elastic modulus E, can be expressed as:

$$k_i = \gamma_i \, I_i / L_i$$

where
I_i = moment of inertia in the bending plane
L_i = length
γ_i = a coefficient depending on the flexural shape of the beam

Equation (9.172) was obtained on the assumption that all the members (beams and columns) bend according to a symmetrical shape having a single curvature (Fig. 9.136c). Stiffness can thus be found by imposing a constant bending moment diagram on the beam when $\gamma_i = 1$.

If the beam ends opposite those meeting in the joint with the column are joined to a constraint allowing rotation (Fig. 9.136d) or preventing it (Fig. 9.136e), the resulting beam shapes are different and therefore the values of γ_i are different. They are:

$\gamma_i = 1.5$ if the end is hinged
$\gamma_i = 2.0$ if the end is built in.

A column might have one of its ends built-in on a foundation or a practically rigid slab. In this case, α would be equal to zero, but it is in any case wise to assume it no less than one. If, instead, the column is hinged, $\alpha = \infty$.

The same problem is analytically formulated by French Recommendations CM 1966 |104|. They express the coefficient by the following formula:

$$\beta = \frac{3 - 1.6(\xi_1 + \xi_2) + 0.84\xi_1\xi_2}{3 - (\xi_1 + \xi_2) + 0.28\xi_1\xi_2} \tag{9.173}$$

where:

$$\xi_i = \frac{\Sigma k_b}{\Sigma k_c + \Sigma k_b} = \frac{1}{\alpha_i + 1} \qquad (i = 1 \text{ or } 2)$$

is the ratio between the sum of beam stiffnesses and the sum of the stiffnesses of all the members meeting in the joint.

Formula (9.173) can be regarded as the analytical representation of the nomograph in Fig. 9.136: it provides values of β with errors not exceeding 2% compared to the solution of 9.172. The formula can

526

be simplified for some particular cases as follows:

If end '2' is hinged ($\xi_2 = 0$) $\beta = \dfrac{3 - 1.6\xi_1}{3 - \xi_1}$

If end '2' is built in ($\xi_2 = 1$) $\beta = \dfrac{0.7 - 0.38\xi_1}{1 - 0.36\xi_1}$

If ends '1' and '2' have the
same degree of constraint $\beta = \dfrac{1 - 0.6\xi}{1 - 0.2\xi}$
($\xi_1 = \xi_2 = \xi$)

Wood |105| suggested a different type of approach to the problem.
He considers the sub-element illustrated in Fig. 9.137a, which leads
to results that can be graphically illustrated |106| by the curves of
Fig. 9.137b.

In such curves, coefficient β is expressed as a function of the
two coefficients η_1 and η_2, which are ratios between column stiffness
k_c and the sum of the stiffnesses of all the members meeting in the
joint. That is:

$$\eta_i = \frac{k_c}{k_c + k_b} \qquad (i = 1 \text{ or } 2)$$

In order to apply this method to other frames than that shown in
Fig. 9.137a, the following considerations must be premised.

The curves were obtained on the assumption that beam ends are
built in and, therefore, that their flexural shape has a point of
contraflexure at 1/3 of the span (Fig. 9.137c). Beam stiffness is
thus expressed by the following relation:

$$k_{b,i} = \gamma_i \, I_i / L_i$$

where γ_i is a coefficient, for which the following values should be
assumed:
γ_i = 1 if the end is built in
γ_i = 0.75 if the end is hinged (Fig. 9.137d)
γ_i = 0.50 if the flexural shape has a single curvature (Fig. 9.137e).
It thus appears that coefficients k_x are equal to half of those
to be assumed when finding the stiffnesses of the beams concerned in
determining coefficients α_i for the U.S. 'alignment charts' or in
determining coefficients ξ_i of formula (9.173).

In general, if a column is also connected to other columns (Fig.
9.137f), coefficients η_i are modified as follows:

$$\eta_i = \frac{k_c + k_x}{k_c + k_x + \Sigma k_b}$$

where k_x is the stiffness of the column meeting in the joint in
question. It should be pointed out that in the case of a column
built in at its base, $\Sigma k_b = \infty$ and therefore $\eta = 0$. Conversely, for
a hinged column, $\Sigma k_b = 0$ and therefore $\eta = 1$.

In order to apply one of the above methods, it is necessary to
define which are the beams considered connected to the column: in fact
it is obvious that the manner in which the joint is formed influences

527

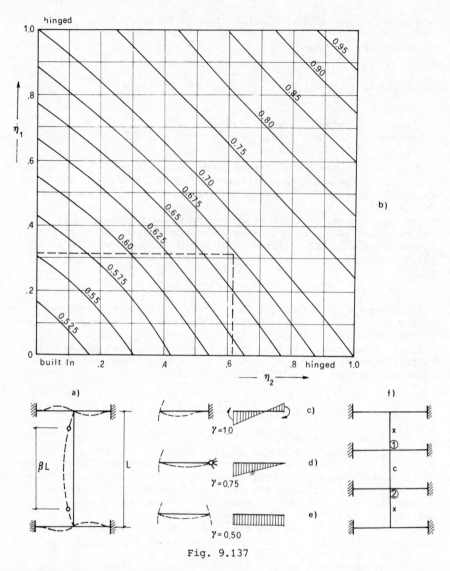

Fig. 9.137

the phenomenon. The French code expresses a criterion for assessing the efficiency of the constraint formed by the beams with reference to the calculation of term Σk_b, the beams considered connected must be joined to the column by the connection, the total height of which, measured as the distance between extreme bolt or weld axes, must be at least 3 times the radius of gyration assumed in the calculating column slendernesses.

In the case of joints with untorqued bolts, the loads acting on the beam can cause - even in service condition - joint settlements

528

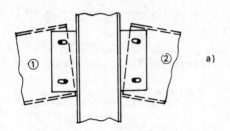
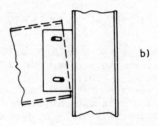

Fig. 9.138

due to hole-bolt clearance. It follows that, if the column tends to rotate (Fig. 9.138), e.g. clockwise, only bolt '1' is capable of opposing the rotation, because bolt '2' allows an inelastic settlement. For this reason, it is advisable to take into consideration only one of the two connected beams and, if they are different, the weaker one, i.e. that with the smaller stiffness k_b. An exterior column (Fig. 9.138b) connected by untorqued bolts must not be considered fixed by the beam and the joint must be regarded as a hinge.

Consider, for example, the frame shown in Fig. 9.139 and assess coefficient β for the columns between the third and fourth floors.

For line '1' the beams give no contribution at all, and thus β = 1.

For line '2' one beam has its other end connected to another column and thus must be regarded as having a single curvature, whereas the other one has its end connected to a hinge. Therefore:

$$\alpha_1 = \frac{50 + 50}{75 + 1.5 \times 75} = 0.53 \qquad \alpha_2 = \frac{50 + 50}{50 + 1.5 \times 50} = 0.80$$

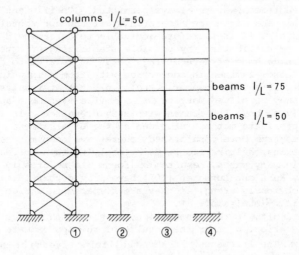

columns $I/L = 50$

beams $I/L = 75$

beams $I/L = 50$

Fig. 9.139

529

$$\xi_1 = \frac{1.5 \times 75 + 75}{50 + 50 + 1.5 \times 75 + 75} = 0.65 \qquad \xi_2 = \frac{1.5 \times 50 + 50}{50 + 50 + 1.5 \times 50 + 50} = 0.56$$

$$\eta_1 = \frac{50 + 50}{50+50+0.5 \times 75+0.75 \times 75} = 0.52 \qquad \eta_2 = \frac{50 + 50}{50+50+0.5 \times 50+0.75 \times 50} = 0.62$$

From Fig.9.136a or 9.137b or from formula (9.173), one always obtains $\beta \cong 0.72$.

For line '3' both the beams have their ends continuously connected to the adjacent columns and must therefore be considered as having a single curvature. It follows that:

$$\alpha_1 = \frac{50 + 50}{75 + 75} = 0.67 \qquad\qquad \alpha_2 = \frac{50 + 50}{50 + 50} = 1$$

$$\xi_1 = \frac{75 + 75}{50 + 50 + 75 + 75} = 0.60 \qquad \xi_2 = \frac{50 + 50}{50 + 50 + 50 + 50} = 0.50$$

$$\eta_1 = \frac{75 + 75}{50+50+0.50(75+75)} = 0.57 \qquad \eta_2 = \frac{50 + 50}{50+50+0.50(50+50)} = 0.67$$

In this case Fig. 9.136a, Fig. 9.137b and formula (9.137) all give $\beta \cong 0.75$.

Finally, for line '4' there is only one beam with an end connected to the adjacent column. Therefore:

$$\alpha_1 = \frac{50 + 50}{75} = 1.33 \qquad\qquad \alpha_2 = \frac{50 + 50}{50} = 2$$

$$\xi_1 = \frac{75}{50 + 50 + 75} = 0.43 \qquad \xi_2 = \frac{50}{50 + 50 + 50} = 0.33$$

$$\eta_1 = \frac{50 + 50}{50 + 50 + 0.5 \times 75} = 0.73 \qquad \eta_2 = \frac{50 + 50}{50 + 50 + 0.5 \times 50} = 0.80$$

and all three methods give $\beta \cong 0.83$

9.5.4.2 Frames with sway permitted

In an unbraced frame sway is always permitted. Coefficient β is then higher than one and might tend to infinity if the horizontal beams are very flexible: it is therefore advisable to determine β quite cautiously, underestimating any possible stabilizing effects.

Calculation methods of determining β can be related to the same criteria as those examined in connection with frames in which sway is prevented. It should however be pointed out that results for frames with sway permitted must be considered as even more approximate than those given above for frames with sway prevented. In fact, ever since the approximate methods provided by the US alignment charts or by Wood's nomograph were popularized, there have been many studies aimed at checking their reliability, by applying them to real structures: in many cases, results were found to be grossly unsafe. Even recently Kuhn and Lundgren |107| have proved that in many cases results are obviously wrong, if they are compared to a rigorous elastic stability analysis.

It is difficult, in this connection, to establish any criteria for determining a priori whether the results of the approximate methods are safe or not. The influence of the unstabilizing effects of the vertical load becomes a determinant factor and it is no longer possible to accept many of the assumptions underlying the approximate methods (see 9.5.4.1). In particular, the assumption that ratio $L/\sqrt{N/EI}$ is constant for all columns is unacceptable. In practice, this has never been confirmed (if it were so, the effective length would be the same for all columns), but non-compliance with the assumption entails only slight errors in frames in which sway is prevented. If sway is permitted, whenever ratio $L/\sqrt{N/EI}$ varies from one column to another, the approximate methods lead to errors that are often considerable.

Summing up, the approximate methods proposed by the alignment charts |103| or by Wood's nomograms |106| can be considered acceptable when sway is permitted only if frames are substantially regular, i.e.: whenever heights, moments of inertia and axial forces in the columns do not differ considerably between themselves.

For all these reasons of doubt, the approximate methods for determining effective length can be useful for a first design of frames having sway permitted; but in order to check their overall stability, it is preferable to adopt calculation methods allowing an assessment of the destabilizing effects of vertical loads (see 9.5.4.3).

Now consider a unit of the multi-storey frame shown in Fig. 9.135b, formed by column AB, the effective length of which must be determined, by the four beams AC, AD, BE, BF and the two columns AG and BH, forming the elastic contraints of column AB.

If the same simplifying assumptions set out in 9.5.4.1 are adopted, the equation determining β is:

$$\frac{\alpha_1 \alpha_2 (\pi/\beta)^2 - 36}{6(\alpha_1 + \alpha_2)} - \frac{\pi/\beta}{\tan \pi/\beta} = 0 \qquad (9.174)$$

where

$$\alpha_i = \Sigma k_c / \Sigma k_b \qquad (i = 1 \text{ or } 2)$$

The value of β can be obtained using the alignment chart of Fig. 9.140a: it is at the intersection of line $\alpha_1 - \alpha_2$ on the central scale. To find stiffnesses k_i, one must also assume:

$$k_i = \gamma_i I_i / L_i$$

where γ_i is a coefficient to be assumed equal to one, whenever the flexural shape of the member is bi-symmetrical (Fig. 9.140b). In the case of differently constrained beams, its value is:

$\gamma_i = 0.5$ if the end is hinged
$\gamma_i = 2/3$ if the end is prevented from rotating

Finally, if the column has one hinged end, $\alpha = \infty$. If the end is built-in, theoretically $\alpha = 0$, whereas in practice it is advisable to assume $\alpha = 1$, to take account of the fact that the built in joint is not perfectly stiff.

French recommendations CM66 |104| propose an approximate analytical formulation of the solution to (9.174), as follows:

$$\beta = \sqrt{\frac{1.6 + 2.4(\xi_1 + \xi_2) + 1.1\xi_1\xi_2}{\xi_1 + \xi_2 + 5.5\xi_1\xi_2}} \qquad (9.175)$$

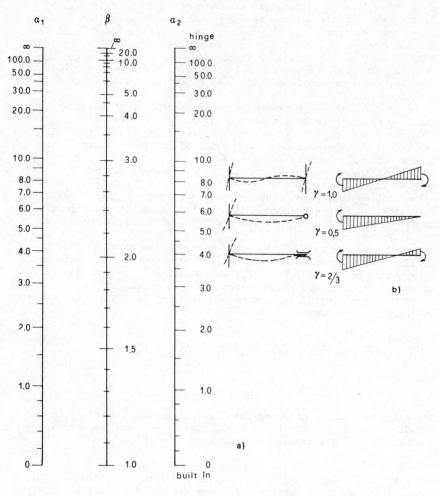

Fig. 1.140

as a function of coefficient ξ_i (i = 1 or 2), which is identical to that already defined for cases in which sway is prevented (see 9.5.4.1).

Formula (9.175) is reduced for some particular cases to:

If end '2' is hinged ($\xi_2 = 0$) $\beta = \sqrt{\dfrac{1.6 + 2.4\xi_1}{\xi_1}}$

If end '2' is built in ($\xi_2 = 1$) $\beta = \sqrt{\dfrac{4 + 3.5\xi_1}{1 + 6.5\xi_1}}$

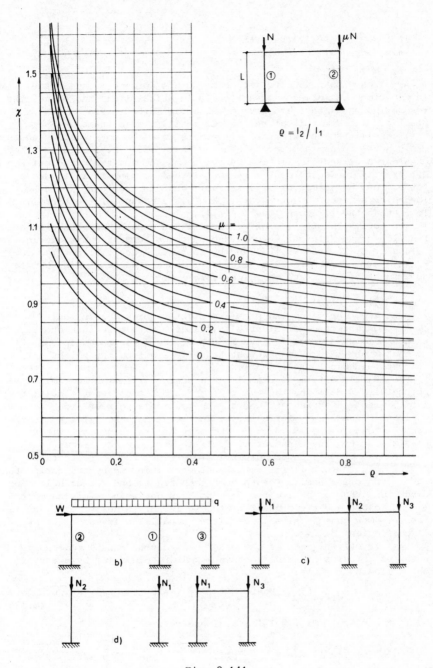

Fig. 9.141

533

If ends '1' and '2' have the same degree of constraint $(\xi_1 = \xi_2 = \xi)$

$$\beta = \sqrt{\frac{0.8 + 0.2\xi_1}{\xi_1}}$$

As already pointed out, such results are correct if the frames in question are formed by practically identical links and by columns resembling each other and having quite similar axial loads: in practice, for symmetrical and symmetrically loaded frames.

If one wishes to extend these results to other more complex cases, one can try to use a study by Chu and Chow |108| (Fig. 9.141). They consider a closed link formed by two beams and any two different and differently loaded columns: the most loaded column (1) is loaded by force N and has a moment of inertia I; the other column (2) is loaded by a force μN (with $\mu \leq 1$) and has a moment of inertia ρI, lower than the foregoing one ($\rho < 1$).

The effective length $L_{c,1}$ of column (1) can thus be expressed as a function of coefficient β_1 provided by the alignment chart of Fig. 9.140a or by formula (9.174) and by corrective coefficient χ, multiplying β_1, that can be deduced from the charts of Fig. 9.141a as a function of μ and ρ. Therefore:

$$L_{c,1} = \beta_{eff,1} \qquad \text{where} \qquad \beta_{eff,1} = \chi\beta_1$$

In practice, the following problems might require a solution:

How to correct coefficient β in frames having more than one bay

How to assess the effective length of the less loaded column (2), that might also have a height L_2 differing from the height of most loaded column (1)

Fig. 9.141b offers an example of a design criterion that might be useful in defining the terms of the problem: a generic multi-bay frame, with a distributed load on its horizontal girder and horizontal loads can, for each loading condition, be referred to the frame shown in Fig. 9.141c, loaded only at its joints. It is thus possible to examine a number of fictitious single bay frames, each formed by the most loaded column (also assumed to be the section having the greatest inertia), by another column and by the horizontal beam connecting them. For each of such frames and for each loading condition, the corrective coefficient χ is determined for the most loaded column and the highest of the series of values so obtained is chosen. One thus obtains an estimate, usually in favour of safety, of coefficient β (Fig. 9.141d).

The effective length of the other 'i' columns can be reasonably assessed in terms of the effective length of column (1) by means of the following formula:

$$L_{c,i} = \beta_{eff,i}L_i \qquad \text{where} \qquad \beta_{eff,i} = \beta_{eff,1}\, L/L_i\, \sqrt{\rho_i/\mu_i} \tag{9.176}$$

In the case of Fig. 9.141c, e.g. one gets:

$$\beta_{eff,1} = \chi\beta_1; \quad \beta_{eff,2} = \beta_{eff,1}\sqrt{\rho_2/\mu_2}; \quad \beta_{eff,3} = \beta_{eff,1}\sqrt{\rho_3/\mu_3}$$

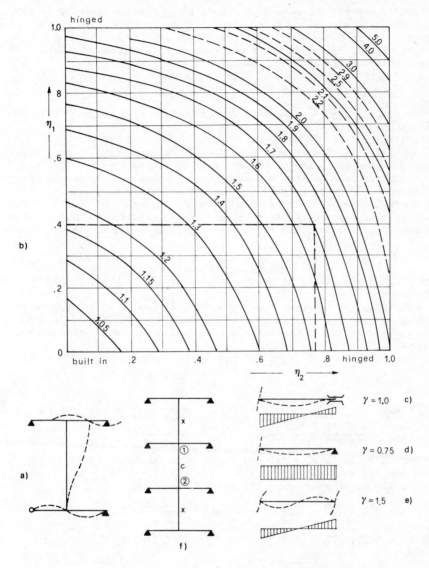

Fig. 9.142

The introduction of ratio L/L_i into (9.176) provides an approximate estimate of the effective bending length of columns having different heights. It should however be stressed that the above criteria can be properly adopted only for frames having links which differ only slightly from each other.

535

Wood's method, introduced in 9.5.4.1, can also be extended to deal with the case when sway is permitted. This method applied to the frame shown in Fig.9.142a   leads to the results illustrated by the curves of Fig. 9.142b in which coefficient β is expressed as a function of coefficients η_i (as defined in 9.5.4.1):

$$\eta_i = \frac{k_c}{k_c + \Sigma k_b} , \quad \text{with} \quad k_i = \gamma_i I_i / L$$

Coefficients γ_i - to be used in calculating stiffnesses - have the following values:

γ_i = 1 for columns
γ_i = 1 for beams having one end prevented from rotating (Fig. 9.142c)
γ_i = 0.75 for beams having one end free to rotate (Fig. 9.142d)
γ_i = 1.5 for beams with a bi-symmetrical flexural shape, and there-
    fore for beams connected to other columns (Fig. 9.142e).

In this case, too, in order to take the existence of other columns into account, and therefore to extend the calculation to multi-storey frames (Fig. 9.142f), the following coefficients can be referred to:

$$\eta_i = \frac{k_c + k_x}{k_c + k_x + \Sigma k_b}$$

where k_x is the stiffness of the column meeting in the joint in question.

Consider, for example, the frame shown in Fig. 9.143. Suppose the columns in the external lines have the same moments of inertia and axial loads equal to 60% of the corresponding values for the internal columns.

For the columns of lines '2' and '3' one has:

$$\alpha_1 = \frac{50 + 50}{75 + 75} = 0.67 \qquad \alpha_2 = \frac{50 + 50}{50 + 50} = 1$$

$$\xi_1 = \frac{75 + 75}{50 + 50 + 75 + 75} = 0.60 \quad \xi_2 = \frac{50 + 50}{50 + 50 + 50 + 50} = 0.50$$

$$\eta_1 = \frac{50 + 50}{50+50+1.5(75+75)} = 0.31 \quad \eta_2 = \frac{50 + 50}{50+50+1.5(50+50)} = 0.40$$

According to Figs. 9.140 and 9.142 or to formula (9.175), it follows that $\beta \cong 1.28$.

For the columns of lines '1' and '4', one has:

$$\alpha_1 = \frac{50 + 50}{75} = 1.33 \qquad \alpha_2 = \frac{50 + 50}{50} = 2$$

$$\xi_1 = \frac{75}{50 + 50 + 75} = 0.43 \qquad \xi_2 = \frac{50}{50 + 50 + 50} = 0.33$$

$$\eta_1 = \frac{50 + 50}{50 + 50 + 1.5 \times 75} = 0.47 \quad \eta_2 = \frac{50 + 50}{50 + 50 + 1.5 \times 50} = 0.75$$

and, from the same figures and formula, $\beta \cong 1.52$.

Coefficients β so determined are for equally loaded columns. If one wishes to take into account the fact that the extreme lines are less loaded ($\mu = N_4/N_3 = 0.60$), it is sufficient to introduce $\mu = 0.60$ and $\rho = 1$ into the curves of Fig. 9.141 (the columns are assumed to be identical). Therefore:

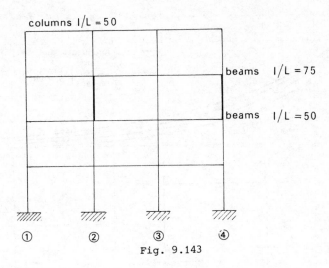

columns $l/L = 50$

beams $l/L = 75$

beams $l/L = 50$

①   ②   ③   ④

Fig. 9.143

$\chi = 0.90$

For the columns of the internal lines one can assume:

$\beta_{eff} = 0.90 \times 1.28 = 1.15$

For the columns of the external lines:

$\beta_{eff} = 1.15 \quad 1/0.60 = 1.49$

9.5.4.3 Destabilizing effect of vertical loads

It has already been pointed out that the methods discussed in 9.5.4.2
for determining the effective length of the columns of frames are
open to criticism: the simplifying assumptions needed in order
to formulate equation (9.174) are usually not complied with in
practice. The approximate methods reflect precisely enough the
behaviour of frames not subjected to sway: the range of β is limited
(from 0.5 to 1) and therefore the simplifying assumptions have a
limited influence on the results.

On the other hand, the difference between the simplified model and
the real structure is particularly sensible in the case of frames
subjected to sway, in which coefficient β has and unlimited range
(from 1 to ∞).

However, another more substantial criticism should be raised
against the methods discussed in 9.5.4.2: it has not yet been proved
that the equivalent slenderness approach is correct in the case of
frames subjected to sway. In the elastic-plastic range, transverse
displacements are much grater than those evaluated in the elastic
field and the increase in the destabilizing effects of vertical loads
due to plasticity cannot be disregarded.

From the theoretical point of view, the most rigorous approach
would be to analyse frames with the same numerical techniques as
those adopted for columns hinged at their ends. It would thus be

537

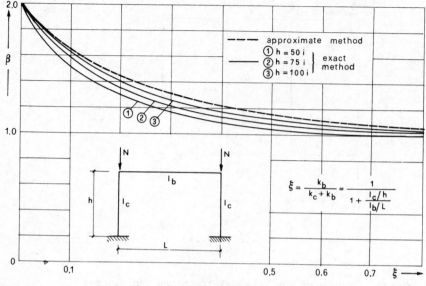

Fig. 9.144

possible currently to take into account structure geometrical imperfections and material non-linearities. However, it is obvious that such an approach is particularly difficult and needs a lot of computer time due to the great number of parameters affecting the problem |35|. On the other hand, only a similar approach would make it possible to decide whether the equivalent slenderness criterion is or is not reliable.

Such a check has been performed for simple frames |109|. In this case, the equivalent slenderness criterion is not only in favour of safety, but it is also sufficiently approximate: Fig. 9.144 shows the variation of coefficient β as a function of coefficient ξ defined in 9.5.4.2, coefficient β being expressed in the elastic field by formula (9.175) or by the curves of Fig. 9.140 and 9.142 and determined by a simulation programme in the elastic-plastic field (curves '1', '2', '3'). Numerical simulation calculations are in respect of HEA 200 sections with different slendernesses, bent in the maximum stiffness plane, having an initial imperfection $v_o/h = 0.002$ and the residual stress distribution shown in Fig. 9.8.

The formulae in the elastic field in this case always provide cautious and more precise values of β, as ratio h/i between column length and its radius of gyration increases.

Analogous calculations have not yet been carried out for multi-storey frames; researchers have concentrated their efforts towards aims that can be grouped into the two following types of approaches:

Plastic design

P - Δ methods

It is well known that classical plastic design provides a collapse

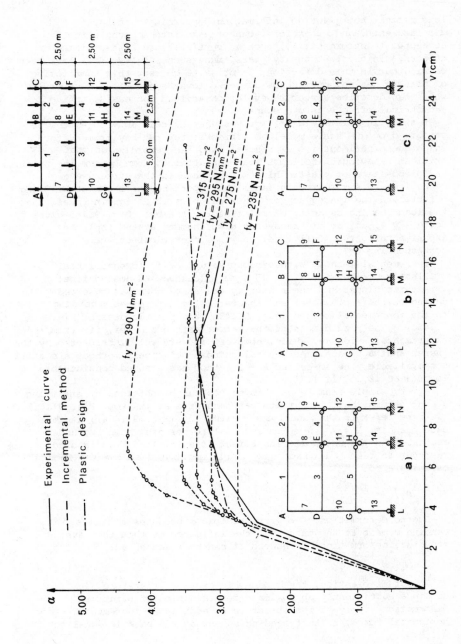

Fig. 9.145

load factor, but gives no information concerning structural displacements, and, therefore, cannot take into account effects connected to deformability, such as vertical load unstabilizing effects. In the last twenty years, many successful studies have been carried out in this direction, in order to correct this shortcoming of plastic design, one is now disposed towards numerical methods, which can describe structure evolution as load increases, taking geometrical effects into account.

A study in this field |110|proves what is said above. Tests were performed on real scale models of three-storey two-bay frames, for which the experimental equipment allowed the recording of post-critical behaviour. Collapse load was attained even before the number and disposition of plastic hinges had rendered the structure a mechanism: this result underlines the importance of the unstabilizing effects and the inadequacy of classical plastic design in foreseeing the load bearing capacity of multi-storey frames. Fig. 9.145 gives for $\alpha - v$ (load factor—transverse displacement at the top) the experimental curve (full line) and the theoretical curves corresponding to:

Incremental method, with various values of f_y (dashed lines)
Plastic design with f_y = 235 N mm^{-2} (dashed and dotted line)
Plastic hinges and their history are also shown, as observed during tests (Fig.9.145a) and theoretically determined according to the incremental method (Fig. 9.145b) and plastic design (Fig. 9.145c). The predictions of the latter method are very far from the experimental evidence, which, however, is very well interpreted by the incremental approach. As general purpose non-linear methods are still not available, the so-called $P - \Delta$ method has assumed considerable importance |6,35,111,112|.

A very simple example of this method can be obtained by analysing a bent column, built in at its base and free at its top (Fig. 9.146).

If we refer to the undeformed situation (Fig. 9.146a); at the base one has: $M = Vh$; $N = F$. Internal forces are, however, underestimated,as the unstabilizing effect of axial load F is ignored. In order to take it into account, displacement $v_{(1)}$ at the top should be calculated:

$$v_{(1)} = \frac{1}{3} \frac{Vh^3}{EI}$$

Due to deformation, the vertical load F produces an increase in bending moment at the base. If the influence of this increase on structure deformation is ignored it can be assessed as:

$$\Delta_{(1)}M = Fv_{(1)}$$

In actual practice, the increase is greater: therefore, in order to obtain a sufficiently approximate determination of structure deformation, apply to the structure an additional increment $\Delta_{(1)}V$ of horizontal force, so that bending moment at the base is equal to $\Delta_{(1)}M$:

$$\Delta_{(1)}V = \Delta_{(1)}M/h = Fv_{(1)}/h$$

Due to force $V + \Delta_{(1)}V$ (Fig. 9.146b), the strut displacement is:

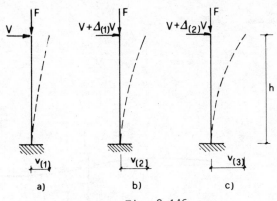

Fig. 9.146

$$V_{(2)} = \frac{1}{3} \; (V + \Delta_{(1)} V) \frac{h^3}{EI} = v_{(1)} (1 + \frac{\Delta_{(1)} V}{V})$$

It is thus possible to assess more nearly the unstabilizing effect of F. Its value is:

$$\Delta_{(2)} M = F v_{(2)} = \Delta_{(1)} M (1 + \frac{\Delta_{(1)} V}{V})$$

By applying at the top a force $V + \Delta_{(2)} V$ (Fig. 9.146) one gets:

$$\Delta_{(2)} V = \Delta_{(2)} M/h = F v_{(2)}/h$$

which gives a still better estimate of the unstabilizing effect of F. The operation can be repeated until the procedure converges. If the applied load F were near the critical value, the procedure would not converge and, in any case differences $v_{(i)} - v_{(1)}$ would be significant.

The $P - \Delta$ method can be applied in the elastic field, if the conventional elastic ultimate limit state is being checked, or in the elastic-plastic field, if the structure is being designed under collapse. If the allowable stress method is adopted, it must be applied by multiplying the vertical load by a safety coefficient.

The application of the $P - \Delta$ method to frames subjected to sway is illustrated in Fig. 9.147, with reference to a generic frame column included between levels i and $i + 1$. The following operations must be performed:

1. In connection with the generic ith floor, the sum of axial loads in the various columns is calculated: assume $R_i = \Sigma N_{ij}$ is the value of that sum ($j = 1, 2, \ldots n$ columns)

2. Displacement component $v_{i,(1)}$ is assessed for each floor, adding the effect of external transverse loads H_i to the estimated value of initial imperfections due to fabrication and erection tolerances (see 1.5)

3. In connection with the generic ith floor, additional imaginary shears are calculated as follows:

541

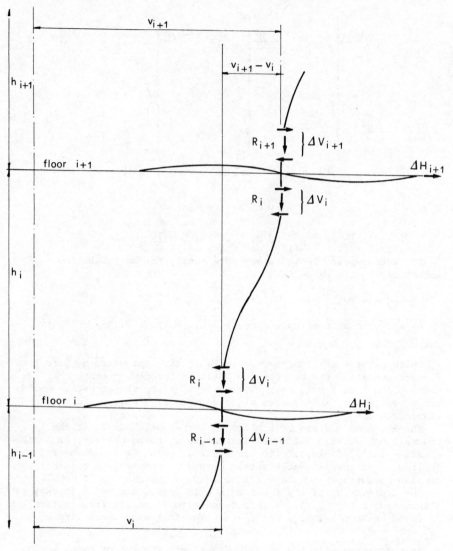

Fig. 9.147

$$\Delta_{(1)} V_i = \{v_{i+1,(1)} - v_{i(1)}\} R_i/h_i$$

and they are applied to the foot and the top of the column. At each floor, they are a function of relative displacement of the ith floor

4. The additional imaginary external force is determined:

$$\Delta_{(1)} H_i = \Delta_{(1)} v_i - \Delta_{(1)} i-1$$

and is applied at the ith level

5. The imaginary external force $\Delta_{(1)}H_i$ is combined with forces $H_{i,(1)}$ referred to in 2 above; the corresponding displacement components $v_{i,(2)}$, due to forces $H_{i,(2)} = H_{i,(1)} + \Delta_{(1)}H_i$, are calculated again

6. One must check that the difference $v_{i,(2)} - v_{i,(1)}$ is small compared to $v_{i,(1)}$. If it is not, operations 3 to 5 must be repeated

7. If the method converges quickly, it means that the structure is sufficiently rigid. If, instead, after 5 or 6 iterations no acceptable convergence has been obtained, it means that the structure is not sufficiently rigid to resist the applied vertical loads

8. Once convergence has been obtained at the nth iteration, strength and stability of the jth column is checked at each floor, in connection with the effect of the action of axial loads N_j and of bending moments M_j produced by vertical and horizontal $H_{i,(n)}^j$ loads. Usually, the strength checks must, however, be performed assuming floor distance h_i ($\beta = 1$) as the effective length: in fact the vertical load unstabilizing effect is already included in bending moment values, calculated with reference to the deformed structure

The procedure can be applied (exactly in the same way) also to braced frames: it is thus possible to determine the forces acting on the bracing in terms of structure deformability.

The $P - \Delta$ method, although it is more tedious than the approximate methods referred in 9.5.4.1 and 9.5.4.2, is certainly more realistic and is absolutely necessary in order to justify structural solutions that exceed the safety limitations set out in the various codes for structure deformability, whenever proper numerical studies are not carried out to guarantee overall stability.

In this connection, it is interesting to mention the proposals made by some researchers and given in the SSRC Guidebook |6| with reference to the conditions that must be complied with to avoid more sophisticated numerical analyses than first-degree ones (i.e. overlooking unstabilizing effects).

These conditions, that make it possible to set aside the empirical rule of limiting maximum displacement of a building to 1/500 of the height under the effect of transverse loads in service conditions, are as follows |6|.

A frame subjected to sway can be checked ignoring vertical load unstabilizing effects, and therefore with reference to the undeformed situation, provided:

All the columns are checked under bending with $\beta = 1$

The ratio between maximum axial force in each column and yield load under compression (assessed ingnoring stability phenomena) is less than 0.75

Column slenderness in the frame bending plane is less than 35

For all floors:

$$\Delta v_i / h_i < 0.15 \ \Sigma V / \Sigma N \qquad (9.177)$$

where
$\Delta_i = v_{i+1} - v_i$ = relative displacement of the ith floor
h_i = floor distance between levels i and $i + 1$
$\Sigma V, \Sigma N$ = the sum of shear and axial forces loading the column at the ith floor.

In inequality (9.177) is not complied with, a second-degree analysis is necessary (e.g. by the $P - \Delta$ method); if, however, it is complied with, but the other conditions are not (e.g. the slenderness limitation), then it is not certain that more sophisticated analyses are needed, but so far, their uselessness cannot be proved.

9.6 LOCAL BUCKLING

All the foregoing considerations do not include local buckling phenomena, i.e. the possibility that a part of the section might buckle out of its own plane (Fig. 9.148).

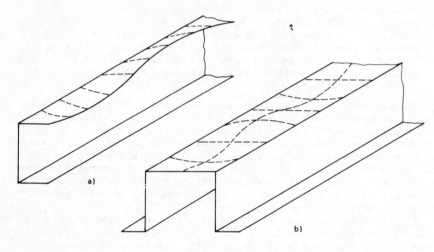

Fig. 9.148

As a general rule, local buckling might either reduce the ultimate strength or a section or, although not influencing its values, *Decreases* diminish the rotational capacity of the section in the plastic field, and therefore its ductility.

Consider, for example, the effect of buckling of a compressed part of the section of a structural element, the four types of behaviour illustated in Fig. 9.149 can take place.

For struts (Fig. 9.149a), the load corresponding to buckling of one part of the section might not be lower than that characterizing the stability of the strut: in this case, the local buckling phenomenon affects only the post-critical behaviour (curve '1') and not its ultimate strength (see 9.6.1.1). If, however, part of the section buckles prematurely, the ultimate load is less than the foregoing one (curve '2') and must be assessed according to the indications

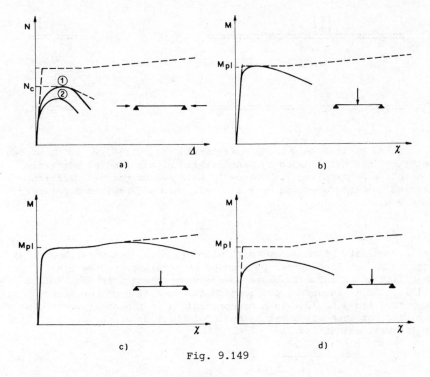

Fig. 9.149

contained in 9.6.1.2.

Compressed parts of the section can buckle once maximum load consistent with section bending performances has been attained. In this case (Fig. 9.149b), as curvature increases, bending moment decreases: maximum moment, when attained, cannot continue to be applied to the section. Such sections must be designed with reference to the conventional elastic limit state (see 9.6.2.1) but not under collapse, as they do not allow ductility and thus internal force redistribution.

If the width-thickness ratio in compressed parts of the section is lower than the foregoing one, local buckling takes place whenever elongations ε of the most stressed fibres are greater than ε_e corresponding to the material elastic limit (Fig. 9.149c).

In this case, the rotational capacity of the section and therefore ductility of the structural element, might allow a redistribution of internal forces and it is thus possible to design the structure at plastic limit state (see 9.6.2.2).

On the other hand, if there is a high ratio between compressed element length and thickness, local stability might take place before the section attains its maximum performance (Fig. 9.149d).

In this case, which is typical of so-called 'thin' profiles, it is actually local stability that conditions the design of the structural element (see 9.6.2.3).

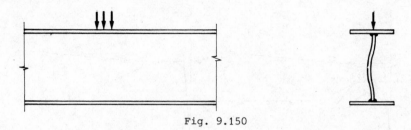

Fig. 9.150

Another typical case is illustrated in Fig. 9.150, which shows web buckling in a beam loaded by concentrated or distributed external forces ('web crippling'). Obviously this phenomenon has different origins from the foregoing ones and will thus be examined separately (see 9.6.3).

9.6.1 *Elements in compression*

Local stability of the compressed elements of a section without transverse stiffeners can be studied by reference to the elastic stability of an infinite plate having width b and thickness t (Fig. 9.151). It is loaded by compression forces acting on the simply supported sides having width b, constraints on the other two edges may vary but must allow extensional deformations of the plate.

The critical stress is:

$$\sigma_{cr} = k \frac{\pi^2 E}{12(1 - \nu^2)(b/t)^2} \qquad (9.178)$$

For E = 206 000 N mm^{-2} = 21 000 kg mm^{-2} and ν = 0.3, formula (9.178) can be expressed in the following dimensional form:

$$\sigma_{cr} = 186200 \, k/(b/t)^2 \text{ N mm}^{-2} = 18980 \, k/(b/t)^2 \text{ kg mm}^{-2} \qquad (9.179)$$

Coefficient k depends on the constraints along the non-loaded edges of the plate and therefore on the possible critical shape of the compressed element. Its values are set out in the table of Fig. 9.152.

9.6.1.1. Width-thickness ratios in struts

If there are no particular aesthetic or technological requirements, it is always economical to avoid the possibility of local buckling occurring before the load has attained the maximum carrying capacity of

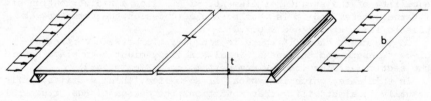

Fig. 9.151

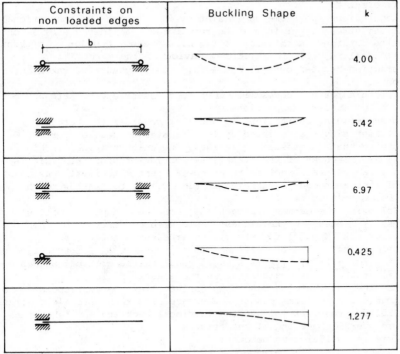

Constraints on non loaded edges	Buckling Shape	k
		4.00
		5.42
		6.97
		0.425
		1.277

Fig. 9.152

the strut.

In accordance with this criterion, it is possible to obtain from (9.178) the maximum value of the b/t ratio in the compressed element that is consistent with the possibility of attaining a conventional limiting stress σ_{lim}, before the element begins to buckle. From the following inequality:

$$\sigma_{cr} \leq \sigma_{lim} \tag{9.180}$$

it follows that:

$$\frac{b}{t} \leq \pi \sqrt{\frac{k}{12(1 - \nu^2)} \frac{E}{\sigma_{lim}}} \tag{9.181}$$

Assuming the values given in 9.6.1 for elastic constants E, ν, formula (9.181) can be expressed in the following non-dimensional form:

$$b/t \leq 0.951 \sqrt{k} \sqrt{E/\sigma_{lim}} \tag{9.182}$$

or in the dimensional form:

$$b/t \leq 28.148 \sqrt{k} \sqrt{235/\sigma_{lim}} \tag{9.183}$$

In order to render formulae (9.182) and (9.183) operative, it is necessary to define the values both of stress σ_{lim} and of constraint

547

coefficient k.

Regulations show some uncertainties in defining the conventional limiting stress σ_{lim}: it must be remembered that they are based on results obtained in the elastic field, whereas, in order to prevent local buckling, elastic-plastic behaviour of the strut and interaction between overall and local stability must be taken into account.

For this reason, codes generally adopt rather approximate and often extremely cautious assumptions.

When defining coefficient k, one must consider the actual constraint situation offered by the transverse section parts adjoining that in question. In fact, the unloaded sides in Fig. 9.151 are the edges of the transverse section of profile. This is why it is preferable not to refer coefficient k to ideal constraint conditions as those illustrated in Fig. 9.152, but rather to the transverse section type.

Following the German approach |54|, one can adopt a limiting stress σ_{lim} equal to the highest yield stress admitted for the material (355 N mm^{-2}) for strut slendernesses λ lower than 75, corresponding to the proportionality slenderness λ_e for that type of steel. For $\lambda > 75$, σ_{lim} is assumed equal to the critical stress $\sigma_{cr}(\lambda)$ of the strut, calculated in the elastic field.

Considering that the mean compressive stress σ_c corresponding to maximum strut performance is in any case less than both yield stress and Euler stress, it follows that no local buckling can take place before overall buckling of the strut.

Formula (9.182) thus becomes:

$$b/t \leq 0.951 \sqrt{k} \sqrt{21000/36} = 22.96\sqrt{k} \quad, \text{ for } \lambda < 75$$
$$b/t \leq 0.951 \sqrt{k}\sqrt{\frac{E}{\pi^2 E/\lambda^2}} = 0.303 \lambda\sqrt{k} \text{ , for } \lambda > 75 \tag{9.184}$$

With reference to constraint coefficients k, the various types of section are divided into the six classes illustrated in the table of Fig. 9.153. It shows the parts of the section (thick line) that provide a constraint with reference to the parts that have a greater tendency to buckling (fine line). The table also shows, for the various classes, the extreme cases of constraint corresponding to infinitely rigid or infinitely flexible contiguous stabilizing parts.

In order to define the real constraint conditions, one can introduce - approximately and depending on the case - either the ratio between the width thickness ratios of the stabilizing and stabilized parts:

$$\beta = \frac{b_1/t_1}{b/t} \tag{9.185}$$

or the coefficient:

$$\alpha = \sqrt[4]{b_2/b_1} \tag{9.186}$$

which is a function of the ratios between the widths of the parts constraining the two sides of the stabilizing part.

548

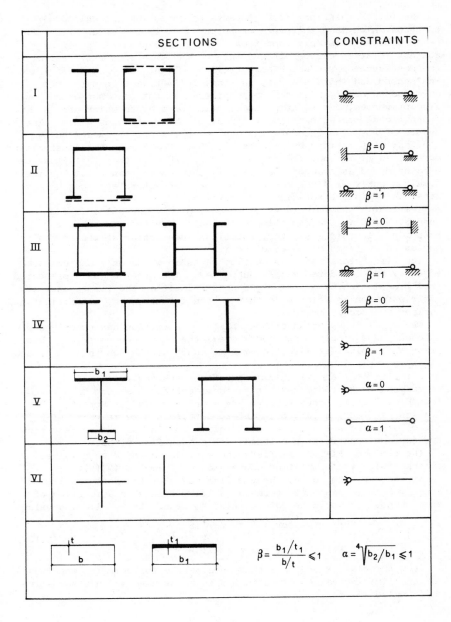

Fig. 9.153

Width b is generally assumed equal to the distance between the axes of the connections between section parts, i.e. between bolt or weld axes.

The first class includes the parts of a section the unloaded edges of which (and therefore those parallel to the strut axis) can be considered as simply supported. In fact, a possible constraint against rotation can be offered only by the torsional stiffness of the contiguous parts which however is so slight as to be negligible. The second and third classes, instead, include section parts the unloaded edges of which can be considered as not only supported, but also prevented from rotating by elastic constraints which are proportional to the bending stiffness of the continuous parts. The degree of rotational constraint can be approximately expressed by coefficient β defined by formula (9.185). For $\beta = 0$, the constraining part is so rigid compared to the constrained one, that the latter can be considered as clamped. For $\beta = 1$, the constraining part buckles simultaneously with the constrained part, and therefore does not offer an efficient rotational constraint.

The fourth class includes parts having only one unloaded edge supported and at the same time elastically constrained against rotation. In this case, too, rotational constraint stiffness can be expressed in terms of coefficient β.

The fifth class includes section parts in which only one unloaded edge may be considered simply supported. The other edge is prevented from translating by a less rigid constraint than the former one: its efficiency can be related to that of the former one by coefficient α defined by formula (9.186).

The sixth class includes those parts of a section having only one unloaded edge simply supported whereas the other one is free. This class thus coincides with the extreme cases of the fourth ($\beta = 1$) and fifth ($\alpha = 0$) classes.

For classes I and VI, German practice uses the values of k corresponding to their respective ideal constraints: one thus assumes (see Fig. 9.152), respectively, $k = 4$ and $k = 0.425$. For the other classes, instead, a parabolic interpolation is used in terms of β (or of α) between the values of k (given in Fig. 9.152) representing the extreme constraint conditions set out in Fig. 9.153.

The table of Fig. 9.154 lists the expressions of k so obtained, together with the simplified expression of formulae (9.148).

In the above approach, the yielding value of the best quality of steel was introduced into formula (9.181); safety is a function of steel type. It would be more correct to assume the following values:

$$\sigma_{lim} = 1.5f_y \qquad \text{for } \lambda < \pi \sqrt{E/1.5\,f_y}$$
$$\sigma_{lim} = \pi^2 E/\lambda^2 \qquad \text{for } \lambda > \pi \sqrt{E/1.5\,f_y} \tag{9.187}$$

From formulae (9.183) one thus obtains the following expressions for the limiting values of ratio b/t to be assumed as an alternative to formulae (9.184):

	k	b/t ≤	
		$\lambda < 75$	$\lambda > 75$
I	$\left\{\sqrt{k_1}\right\}^2$	45	$0.606\,\lambda$
II	$\left\{\sqrt{k_2} - (\sqrt{k_2} - \sqrt{k_i})\,\beta^2\right\}^2$	$52.5 - 7.5\beta^2$	$(0.70 - 0.10\,\beta^2)\lambda$
III	$\left\{\sqrt{k_3} - (\sqrt{k_3} - \sqrt{k_1})\,\beta^2\right\}^2$	$60 - 15\,\beta^2$	$(0.80 - 0.20\,\beta^2)\lambda$
IV	$\left\{\sqrt{k_4} - (\sqrt{k_4} - \sqrt{k_5})\,\beta^2\right\}^2$	$25.5 - 10.5\beta^2$	$(0.34 - 0.14\,\beta^2)\lambda$
V	$\left\{\sqrt{k_5} + (\sqrt{k_1} - \sqrt{k_5})\,\alpha^2\right\}^2$	$15 + 30\,\alpha^2$	$(0.20 + 0.40\,\alpha^2)\lambda$
VI	$\left\{\sqrt{k_5}\right\}^2$	15	$0.20\,\lambda$

$k_1 = 4.0$; $k_2 = 5.42$; $k_3 = 6.97$; $k_4 = 1.277$; $k_5 = 0.425$

Fig. 9.154

$$b/t \leq 22.98\,\sqrt{k}\,\sqrt{235/f_y} \qquad \text{for } \lambda < 76.7\,\sqrt{235/f_y}$$
$$b/t \leq 0.303\,\lambda\,\sqrt{k} \qquad \text{for } \lambda > 76.7\,\sqrt{235/f_y}$$
(9.188)

The US approach |55| is more pragmatic. It assumes the following values of the conventional limiting stress to be introduced into formulae (9.182):

$$\sigma_{lim} = 1.70\,f_y \qquad (9.189)$$

independeng of strut slenderness.

By introducing (9.189) into (9.182) and (9.183), the following expressions are obtained:

$$b/t \leq 0.729\,\sqrt{k}\sqrt{E/f_y}$$
$$b/t \leq 21.59\,\sqrt{k}\,\sqrt{235/f_y}$$
(9.190)

as an alternative to (9.188).

AISC Recommendations define the values of coefficient k to be adopted for the various types of section; the SSRC Guidebook indicates more correct values for that coefficent, according to the mutual stiffnesses of the parts forming the section.

The table of Fig. 9.155 indicates maximum values for the width-thickness ratio permitted by AISC recommendations for the various types of compressed elements, their comparison with the classes of Fig. 9.153 and coefficients k adopted to represent their degree of constraint.

A comparison of the results of adopting the US or the German method shows the former to be decidedly more in favour of safety: in fact, they do not take into account the level of stress meanly existing in a

551

Example	b/t	Category	k
web between two flanges flange between two webs	$b/t < 43.60\sqrt{\dfrac{235}{f_y}}$	I II III	$(2)^2$
web of T sections	$b/t < 21.80\sqrt{\dfrac{235}{f_y}}$	IV	$(1)^2$
unstiffened flanges stiffeners	$b/t < 16.35\sqrt{\dfrac{235}{f_y}}$	V	$(0.75)^2$
angles	$b/t < 13.08\sqrt{\dfrac{235}{f_y}}$	VI	$(0.60)^2$

Fig. 9.155

strut as a function of its slenderness, nor do they take advantage of the constraint against rotation which is offered by the bending stiffness of the elements adjoining the compressed part in question.

ECCS Recommendations |31| have adopted an intermediate approach between those referred to above. They consider the actually existing mean stress, but propose a more cautious interaction formula between overall and local stability than the German one, in order to take the elastic-plastic behaviour of the material into account. They assume:

$$\sigma_{lim} = \frac{\sqrt{\sigma_c}}{0.80}\frac{\sqrt{f_y}}{0.80} = \sqrt{(1.56\,\sigma_c)(1.56\,f_y)} \tag{9.191}$$

where

$\sigma_c = \dfrac{N_c}{A}$ is mean stress consistent with maximum performance of the strut (see 9.2.2).

It should be pointed out that the factor 0.80, by which both the contribution of σ_c and that of f_y are divided, has the function of a safety coefficient as a precaution against interaction between overall and local stability, and to guarantee that the latter does not influence the former.

If (9.191) is introduced into (9.182), the following formulae are obtained, in the non-dimensional or dimensional form:

$$b/t \leq 0.761\ \sqrt{k}\ E/\sqrt{\sigma_c f_y}$$
$$b/t \leq 22.52\ \sqrt{k}\ \sqrt{235/\sqrt{\sigma_c f_y}} \tag{9.192}$$

ECCS Recommendations, however, do not provide any typological criterion for the choice of coefficient k.

Finally, Fig. 9.156 shows, in terms of λ, maximum values of width-thickness ratio permitted by the German, U.S. and European approaches, for a constraint coefficient $k = 4$, applicable, for example, to the web of an I-section column. It can be seen that the data are

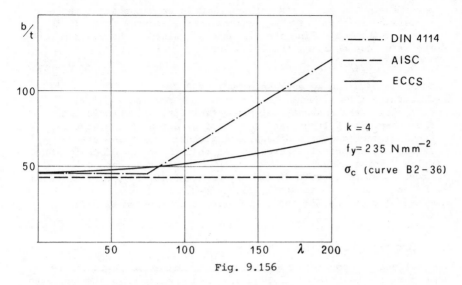

Fig. 9.156

identical for low slendernesses, while they differ substantially for high slendernesses.

9.6.1.2 Thin struts

For profiles with thin walls undergoing compression, it might some-times be useful to adopt greater-width thickness ratios than those set out in 9.6.1.1, in order to satisfy aesthetic, technological or functional requirements: thus the parts forming the section are not entirely exploited. In this case it is impossible to guarantee a priori that local buckling will not interact with overall buckling and lower the load bearing capacity of the strut as shown in Fig. 9.149a (curve '2').

There are at present no theoretical or numerical procedures quantifying interaction between overall and local stability, but the following experimental approach may be adopted:

A stub column test is performed (see 4.4.5) on a structural steel segment having low slenderness ($\lambda < 15$) and the value of mean strength against compression f_c is determined
Ratio $k = f_c/f_y$ is determined between the compressive strength obtained from the test and yield strength of the material. It does not in any case exceed one and, apart from the hardening effect due to cold forming, shows the strength reduction due to local buckling
Once compressive strength f_c is known, it is possible to define a conventional proportionality slenderness, by analogy with equation (9.3):

$$\sigma_c = \pi\sqrt{E/f_c} = \pi\sqrt{E/k \, f_y} \qquad (9.193)$$

Non-dimensional stability curve is referred to which must be applicable to sections in which there are no local instability

553

phenomena. The maximum non-dimensional force $\bar{N}$ the strut can bear is defined in terms of a non-dimensional slenderness $\bar{\lambda}$. It is then possible to define the dimensional value of the maximum stress the strut can bear, by the following:

$$\sigma_c = f_c \ \bar{N}(\lambda) = k \ f_y \ \bar{N}(\bar{\lambda}) \tag{9.194}$$

with $\bar{\lambda} = \lambda/\lambda_c$ and λ_c expressed by (9.193).

Coefficient k thus reduces strut performance for two reasons:

It reduces the conventional proportionality slenderness and, therefore, strut length and radius of gyration of the section being known, it increases the dimensional slenderness of the strut

It reduces compressive strength and therefore, non dimensional value $\bar{N}$ being known, it also reduces maximum stress σ_c.

In order to adopt formula (9.194) by applying (9.193), coefficients k must have been experimentally determined.

Some useful criteria for their evolution are reported in AISI Recommendations |113,114|.

9.6.2 *Compressed elements of a beam*

The influence of possible buckling of the compressed parts on the behaviour of a beam (Figs 9.149b,c,d) has already been demonstrated (at least from a qualitative point of view).

Such consideration leads to three criteria to be adopted when proportioning the section of a beam:

Checking that local instability does not alter section strength (Fig. 9.149b)

Checking that the section is suitable not only to attain its full plastic moment, but also to undergo significant plastic deformations (Fig. 9.149c)

Letting section performance be conditioned by local instability (Fig. 9.149d)

For the first two criteria, it must be checked that the width-thickness ratios of the compressed parts do not exceed the maximum ratios consistent with the conditions imposed by section strength, or by its fitness to deform plastically. For the third criterion, however, it is necessary to limit the stresses in the compressed parts of the sections that are due to the acting external loads. Obviously, if only strength is checked, one cannot count on possible important plastic deformations and, therefore, such a criterion is acceptable only when the design refers to the conventional elastic limit state. On the other hand, it is absolutely necessary to follow the second criterion in order to check the structure at collapse.

9.6.2.1 Elastic design

If check calculations do not rely on plastic hinges forming in the most loaded sections, it is not necessary for considerable ductility to be available: in order to guarantee adequate safety for the structure, it is sufficient that the limiting plastic moment can be attained in each section. A behaviour such as that illustrated in Fig. 9.149b is thus acceptable: in other words, a plastic deformation of from 2 to 4 times the elastic deformation can be attained before the descending branch begins.

Thus the problem does not differ from that examined in 9.6.1.1: width-thickness ratios of the compressed parts of the section must not exceed the maximum ratios defined for struts, obviously for $\lambda = 0$, because for beams the compressive stress should attain the yield limit. For $\sigma_c = f_y$, the three methods presented agree, as seen from a comparison of formulae (9.188), (9.190) and (9.192). It follows that the following expression can be adopted:

$$b/t \leq 0.761 \sqrt{k} \sqrt{E/f_y} \cong 22 \sqrt{k} \sqrt{235/f_y} \qquad (9.195)$$

with coefficients k as defined in Figs. 9.154 or 9.155.

9.6.2.2 Plastic design

The assumptions underlying plastic design require that all the plastic hinges needed for a collapse mechanism form before local buckling phenomena decrease the strength of the sections in which plastic hinges are present. In fact, the moment in a generic plastic hinge is assumed to remain constant and equal to the ultimate moment of collapse. It is therefore necessary that all the members are capable of allowing sufficiently great plastic rotations and, consequently, that their behaviour is of the type illustrated in Fig. 9.149c.

Various theories have been proposed in order to define the maximum width-thickness ratios of the compressed parts consistent with such requirements, and many experimental tests have been performed. The most interesting results have been given and discussed in |115|.

If it is assumed that strains can attain the value ε_1 (Fig. 9.157) representing the beginning of hardening ($\varepsilon_1 = 12$ to 15 ε_c), the limits given in Fig. 9.157 are obtained. These limits are set out in AISC recommendations and they have also adopted in the European (ECCS) Recommendations.

Fig. 9.158 shows some experimental results of bending tests on I-profiles |116|. The ratio between maximum strain ε_c attained in the compressed part before local buckling takes place, and the value ε_c at the elastic limit is plotted against the paramenter $b/t\sqrt{f_y/235}$. The continuous line shows the theoretical curve, as obtained by an approach based on elastic-plastic equilibrium bifurcation and assuming a $\sigma - \varepsilon$ diagram of the type shown in Fig. 9.157 with $\varepsilon_c/\varepsilon_2 = 12$. The experimental extrapolation of such a curve for $\varepsilon_c < \varepsilon_1$ is dashed: it was assumed from some French researches |117|, which are useful in assessing section ductility when b/t ratios exceed those indicated in Fig. 9.157.

9.6.2.3 Thin profiles

The behaviour qualitatively illustrated in Fig. 9.149a is characteristic of structural elements that are commonly called 'thin profiles'. In such elements, width-thickness ratios of the compressed parts exceed the maximum values set out in 9.6.2.1 and, therefore, local buckling phenomena condition section strength. In other words, in this case it is not possible to attain yield stress in the most loaded section.

'Thin profiles' have become very common since cold forming techniques by rolling or by press have made it possible to produce

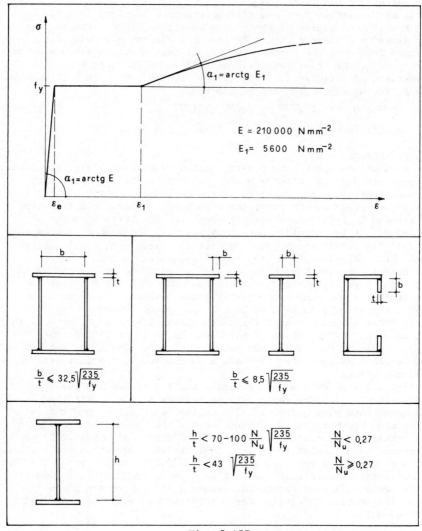

$$E = 210\,000 \ \text{N mm}^{-2}$$
$$E_1 = 5600 \ \text{N mm}^{-2}$$

$$\frac{b}{t} \leqslant 32.5 \sqrt{\frac{235}{f_y}}$$

$$\frac{b}{t} \leqslant 8.5 \sqrt{\frac{235}{f_y}}$$

$$\frac{h}{t} < 70 - 100 \frac{N}{N_u} \sqrt{\frac{235}{f_y}} \qquad \frac{N}{N_u} < 0.27$$

$$\frac{h}{t} < 43 \sqrt{\frac{235}{f_y}} \qquad \frac{N}{N_u} \geqslant 0.27$$

Fig. 9.157

profiles of various shapes, e.g. corrugated sheets (Fig.1.13) or profiles (Fig. 4.4). They were first studied in the fifties at Cornell University, and recommendations, comments and manuals were issued by AISI |113,114|. The results of those studies still form the basis of the recommendations of the various countries, although many studies and experiments have recently been performed both at Cornell and in England, Sweden and France |6,7|.

Compressed parts having a high width-thickness ratio can be divided into two different classes: stiffened and unstiffened elements. The

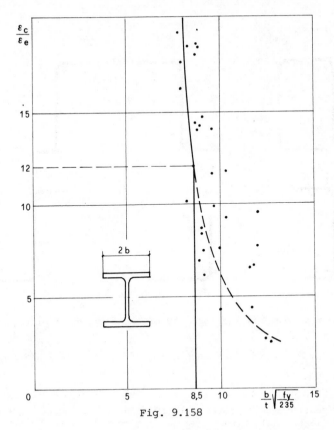

Fig. 9.158

former (Fig. 9.159a) can be considered as constrained on both the edges parallel to the direction of the force; the latter (Fig. 9.159b) have only one constrained edge, the other being free. Design criteria differ for the two cases: for stiffened elements, calculation is based on the definition of a reduced effective width b_{eff}, while for unstiffened elements, checks are performed to limit the compressive stress to a value that is a function of the b/t ratio

Stiffened elements. The calculation of stiffened compressed elements follows a concept that was first expressed by Von Karman (1932). He proposed interpreting the behaviour of an infinitely long compressed plate, simply supported on the edges parallel to the direction of the applied force, by ignoring the central buckled part (Fig. 9.148b) and assuming only the zone near the supports as effective (Fig. 9.159a).

Introduce into formula (9.182) the corresponding constraint coefficient $k = 4$ (Fig. 9.152) and assume $\sigma_{lim} = f_y$, in order to quantify the maximum load bearing capacity of the section, then

$$b_{eff}/t = 1.901 \ \sqrt{E/f_y}; \text{ provided } b_{eff}/t \le b/t \qquad (9.196)$$

557

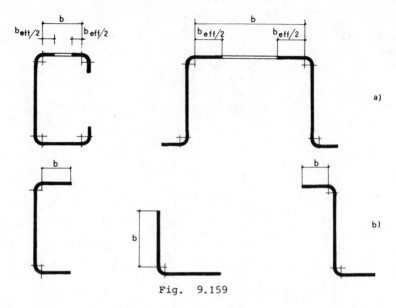

Fig. 9.159

The maximum load bearing capacity of the compressed plate, therefore, is:

$$F_u = f_y b_{eff} t = 1.901 \ f_y \ t^2 \ \sqrt{E/f_y} \ , \qquad \text{for} \quad b/t > 1.901 \ \sqrt{E/f_y}$$

or: (9.197)

$$F_u = f_y \ bt \qquad\qquad \text{for} \quad b/t < 1.901 \ \sqrt{E/f_y}$$

In the case of steel, formula (9.196) can be expressed in the following dimensional form:

$$b_{eff}/t = 56.30 \ \sqrt{235/f_y}, \ \text{provided:} \ b_{eff}/t \le b/t \qquad (9.198)$$

Formula (9.198) is a constant, as shown in Fig. 9.160a, when $(b_{eff}/t \sqrt{b_y/235})$ is plotted against $\sqrt{235/f_y} (b/t)$; it becomes a truncated hyperbola when $(b_{eff}/t)/(b/t)$ is plotted against $(b/t) \sqrt{235/f_y}$ (Fig. 9.160b).

Formula (9.182) and therefore (9.198), are applicable to ideal plates. Industrial plates, affected by imperfections, cannot attain the load bearing capacity defined in (9.197), as they buckle for lower load values. On the basis of tests, a linear interpolation of the experimental results was adopted in the plane of Fig. 9.160a:

$$\frac{b_{eff}}{t} = 1.901 \ \sqrt{E/f_y} \left\{ 1 - \frac{0.415}{b/t} \ \sqrt{E/f_y} \right\} \qquad (9.199)$$

by modifying formula (9.198) in favour of safety. Formula (9.199) is expressed in dimensional terms as follows:

$$\frac{b_{eff}}{t} = 56.26 \ \sqrt{\frac{235}{f_y}} \left\{ 1 - \frac{12.29}{b/t} \sqrt{\frac{235}{f_y}} \right\} \qquad (9.200)$$

558

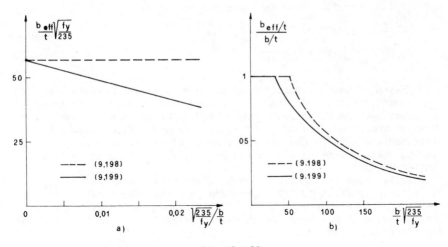

Fig. 9.160

and is indicated by the continuous lines in Figs. 9.160a,b.

Equation (9.199) provides the effective width at collapse. Experiments have shown that it could be considered correct also under different conditions, i.e. for values of $\sigma < f_y$ and that it could therefore also be written in the following form:

$$b_{eff}/t = 1.901\sqrt{\frac{E}{\sigma}} \left\{ 1 - \frac{0.415}{b/t} \sqrt{\frac{E}{\sigma}} \right\} \tag{9.201}$$

or, in dimensional terms:

$$\frac{b_{eff}}{t} = 56.26\sqrt{\frac{235}{\sigma}} \left\{ 1 - \frac{12.29}{b/t} \sqrt{\frac{235}{\sigma}} \right\} \tag{9.202}$$

Section calculation thus becomes an iterative operation: effective width b_{eff} is evaluated on the basis of a given value of σ and the value of σ is then calculated as a function of b_{eff}; and so on until the procedure converges.

Condition $b_{eff}/t = b/t$ in (9.199) expresses the limiting width-thickness ratio $(b/t)_{lim}$ for which width can be assumed to be entirely effective. It follows that:

$$(b/t)_{lim} = 1.29 \sqrt{E/\sigma} \tag{9.203}$$

and, therefore, in dimensional terms:

$$(b/t)_{lim} = 38.18 \sqrt{235/\sigma} \tag{9.204}$$

Obviously, the value of stress σ due to design loads must be introduced into formulae (9.201) to (9.204). If the allowable stress method is adopted, the value of σ to be introduced into the formulae is obtained on the basis of the service loads, multiplied by safety coefficient ν. As an alternative, formulae (9.201) to (9.204) can be adopted, introducing into them a stress σ obtained on the basis of the service loads, but dividing their coefficient by $\sqrt{\nu}$. For example

559

for $\nu = 1.5$, the following formulae are obtained:

$$\frac{b_{eff}}{t} = 45.94 \sqrt{\frac{235}{\sigma}} \left\{ 1 - \frac{10.03}{b/t} \sqrt{\frac{235}{\sigma}} \right\} \qquad (9.205)$$

$$(b/t)_{lim} = 31.17 \sqrt{235/\sigma} \qquad (9.206)$$

Unstiffened elements. Unstiffened elements have an intermediate behaviour between the 4th and 6th classes of Fig. 9.153. In fact, a cruciform section or an angle are formed by unstiffened elements only, whereas other sections (e.g. ⊏ - or ⌐-shapes) are formed by stiffened and unstiffened elements: buckling of the unstiffened element thus depends on the degree of constraint provided by the contiguous element.

Briefly, elastic buckling stress of an unstiffened element is expressed by formula (9.178), provided an intermediate value of coefficient k is adopted between $k = 1.277$ and $k = 0.425$, respectively corresponding to the cases of an edge prevented from rotating and one free to rotate (Fig. 9.152).

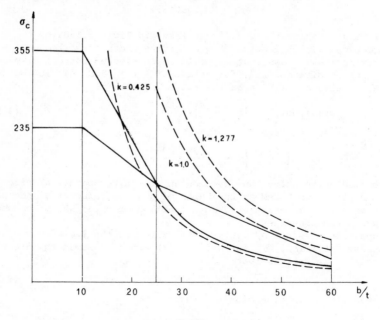

Fig. 9.161

Values of σ_{cr} as a function of b/t in both cases are given by dashed lines in Fig. 9.161, in addition to those determined for a coefficient $k = 1.0$, as adopted by AISC Recommendations for the webs of ⌐-shaped profiles (see Fig. 9.155).

The experimental researches underlying the AISI Recommendations |113,114| made it evident that at least for the most common type of sections:

For $b/t < 10$ local buckling is not a determinate factor; therefore yielding stress can be adopted for σ_c

For $b/t > 25$ local buckling takes place without producing any considerable plastic deformations.

Therefore, AISI recommendations proposed that, in the case of sections entirely formed by unstiffened profiles (cruciform shapes or angles) with $b/t > 25$, a constraint coefficient $k = 0.50$ be adopted: it is practically equal to that adopted for a simply supported edge. Thus, from formula (9.178) it follows that:

$$\sigma_c = \frac{0.50 \, \pi^2 E}{12(1 - \nu^2)(b/t)^2} \cong 0.452 \frac{E}{(b/t)^2} \qquad (9.207)$$

or, in dimensional terms:

$$\sigma_c = \frac{93100}{(b/t)^2} \quad (\text{Nmm}^{-2})$$

For the other types of section, the same recommendations proposed the line passing through the point defined by formula (9.207) for $b/t = 25$ and the tangent to the critical stress curve calculated for $k = 1$, as shown in Fig. 9.155.

$$\sigma_c = E (1 - 0.0141 \, b/t)/895 \qquad (9.208)$$

For intermediate width thickness ratios between 10 and 25, a linear interpolation is possible between the following values:

$$\sigma_c = f_y \quad \text{for} \quad b/t = 10; \qquad \sigma_c \cong E/1380 \quad \text{for} \quad b/t = 25$$

Therefore, for $10 \le b/t \le 25$:

$$\sigma_c = f_y \left\{ \left(\frac{5}{3} - \frac{2}{3} \frac{E}{1380 \, f_y} \right) - \left(1 - \frac{E}{1380 \, f_y} \right) \frac{b/t}{15} \right\} \qquad (9.209)$$

or, in non dimensional terms

$$\sigma_c = f_y \left\{ \left(1.667 - 0.432 \frac{235}{f_y} \right) - \left(1 - 0.648 \frac{235}{f_y} \right) \frac{b/t}{15} \right\} \qquad (9.210)$$

9.6.3 Web crippling

If static loads are applied to the upper flange of a beam, vertical stiffeners can be used to distribute their effects to the web (Fig. 9.162a). Otherwise they produce compression in the web (Fig. 9.162b). The various codes impose strength controls of the following type:

$$\sigma \le \alpha \, f_d \qquad \text{with} \qquad \alpha = 1.2 \text{ to } 1.3$$

σ being the local pressure on web section, allowing a 45° distribution of the concentrated effects. In addition to such controls, the problem of vertical stability must also be taken into consideration (Figs 9.162c,d).

A study by Basler |118| suggested the design criteria contained in the AISC recommendations and they in turn, were also the basis of provisions adopted by other codes. This approach considers the web

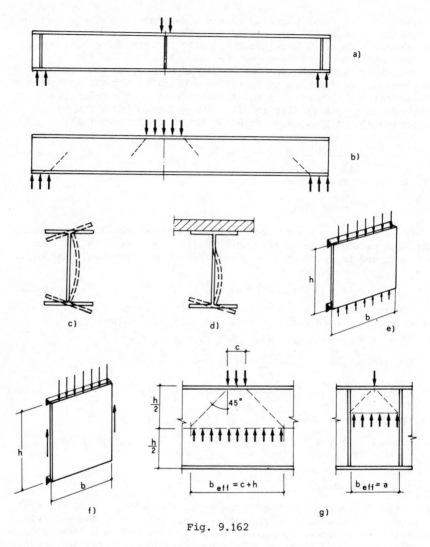

Fig. 9.162

as an elastic plate having its length h in the direction of the load and its width b equal to the beam span if there are no stiffeners, and equal to the stiffener spacing if they are present.

The critical stress in a plate supported on its four sides (Fig. 9.162) can be expressed by the following relation:

$$\sigma_{cr} = \frac{\pi^2 E}{12(1 - \nu^2)(h/t)^2} \xi = \frac{\pi^2 E}{12(1 - \nu^2)(b/t)^2} \xi_1 \qquad (9.211)$$

where ξ and ξ_1 are coefficients as follows:

$$\xi = n + h^2/nb^2; \qquad \xi_1 = \xi \ (b/h)^2 = n(b/h)^2 + 1/n \qquad (9.212)$$

In formulae (9.212), n is the number of semi-waves characterizing the critical shape.

Basler modified formula (9.211) on the basis of the following considerations. For $h/n \to 0$, i.e. for beams having very distant stiffeners, there is only one semi-wave and therefore the second part of the expression for coefficient ξ (9.212) is negligible. On the other hand, the two edges of the beam web are not compressed, but the web is supported by shear forces (Fig. 9.162f). It follows that the critical stress must be doubled and therefore $\xi = 2$. For $h/b \to \infty$, i.e. whenever stiffeners are very near each other, one returns to the case of the plate having its length in the direction of the force (Fig. 9.151) and therefore the critical stress as expressed in (9.211) must coincide with (9.178). It follows that $\xi = k$. If the contribution of contiguous panels is ignored (therefore, in favour of safety), the web can be considered as supported by the stiffeners. If also the flanges are free to rotate (Fig. 9.162c), then $\xi_1 = 4$.

By linearly interpolating between the above values for ξ and ξ_1, the following expressions can be adopted:

$$\xi = 2 + 4 \ h^2/b^2; \qquad\qquad \xi_1 = 4 + 2b^2/h^2 \qquad (9.213)$$

A similar approach can also be adopted if the web is constrained to a flange that is prevented from rotating (Fig. 9.162d). Then:

$$\xi = 5.5 + 4 \ b^2/h^2; \qquad\qquad \xi_1 = 4 + 5.5 \ b^2/h^2 \qquad (9.214)$$

Critical stress values expressed by formula (9.211) and calculated by introducing the above coefficients ξ or ξ_1 can provide a reasonable estimate in the elastic field. In order to take into account the influence of plastic deformation and interaction with the loading stresses in the web, Basler proposed introducing a strength reducing coefficient γ_m equal to 1.65. Therefore, for steel, the maximum stress is given by:

$$\sigma_c = 0.55 \ \xi \ E/(h/t)^2 \qquad (9.215)$$

or, in a dimensional form:

$$\sigma_c = 113000 \ \xi/(h/t)^2 \ (\text{Nmm}^{-2})$$

ξ being expressed by (9.213) or (9.214), respectively for flanges that are free to rotate or prevented from rotating.

Formula (9.215) defines maximum stress. To determine maximum concentrated load, it is necessary to adopt some reasonable assumptions concerning diffusion in the web. Basler suggested adopting an effective width equal to the smaller of that determined according to a 45° diffusion angle and the spacing of stiffeners.

Therefore:

$$F_c = t \ b_{eff} \ \sigma_c$$

b_{eff} = the smaller of $(h + c)$ and a (Fig. 9.162g), where
h = beam height
c = length of load
a = spacing of stiffeners (if any)

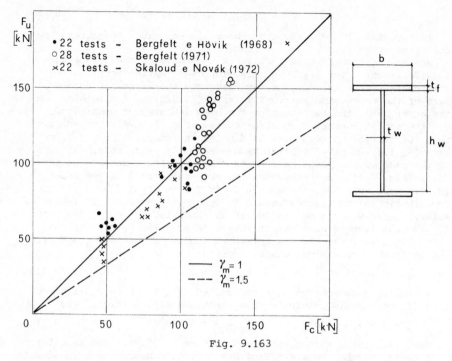

Fig. 9.163

Experimental tests have shown that Basler's simplified analysis sometimes leads to extremely cautious results. On the other hand, the interaction with bending stress in the beam is taken into account only by introducing a coefficient γ_m = 1.65, independent of stress values. Many authors undertook vast experimental studies seeking a more realistic formula. Unfortunately, as there is no theory suitable to guide one in the interpretation of the phenomenon and the choice of the most significant variables, each author tried to relate his experimental results by proposing formulae each of which was different from all the others. Herzog's attempt |119| seems to be the most interesting of all: he tried to interpret 72 experimental results obtained by various researchers (Fig. 9.163), for the purpose of providing a formula allowing the assessment of the maximum concentrated load F_c causing collapse due to web buckling.

With the following symbols:

b, t_f = flange width and thickness
h_w, t_w = web height and thickness
c = length along which the load is applied
a = stiffener spacing
σ = stress in the compressed flange at the load application point
f_y = yield stress
I_w = $h_w^3 t_w/$ 12 = moment of inertia of the web
I_f = $h_f^3 b/$ 12 = moment of inertia of the compressed flange

564

Herzog's proposed formula is:

$$F_c = \frac{\alpha t_w^2}{\gamma_m} \left\{ 1.2 + 1.25 \frac{I_f h_w}{I_w t_w} (1 + \frac{c}{h_w})^2 (0.85 + \frac{a}{100h_w}) \right\} \left\{ 1 - (\frac{\sigma}{f_y})^2 \right\}^{1/8}$$

(9.216)

In this formula (9.216), coefficient $\alpha \cong 0.04762 E$ should be assumed equal to $\alpha = 981$, F_c being in kN and t_w in cm; $\alpha = 100 F_c$ being in ton and t_w in cm; $\alpha = 1430 F_c$ being in kips and t_w in inches

Coefficient γ_m should be assumed equal to one if one wishes to obtain the mean value of experimental results.

As the standard deviation of test results was $\delta = 0.142$, is seems reasonable to assume a value of $\gamma = 1.4$ to 1.5 in order to evaluate design strength.

The results interpreted by formula (9.216) were obtained from experiments on models of beams made of steel having a yield strength f_y between 235 and 355 N mm^{-2} and with web thicknesses and heights as follows: $1 \leq t_w \leq 3$ mm; $300 \leq h_w \leq 700$ mm. The dimensional ratios involved in (9.216) varied as follows:

$$150 \leq h_w/t_w \leq 400$$
$$1.0 \leq a/h_w \leq 13.4$$
$$0 \leq c/h_w \leq 0.17$$
$$0.80 \leq F_c/\alpha \, t_w^2 \leq 2.88$$

In favour of safety and for the sake of simplicity, formula (9.216) can be replaced by the following:

$$F_c = \frac{1.2\alpha t_w^2}{\gamma_m} \left\{ 1 - (\sigma/f_y)^2 \right\}^{1/8}$$

(9.217)

which, for $\gamma_m = 1.43$, coincides with that proposed - still on an experimental basis - by Granholm |120| and modified according to |121|:

$$F_c = 0.040 \, E t_w^2 \{1 - (\sigma/f_y)^2\}^{1/8}$$

Formulae (9.216) or (9.217) directly provide the limiting value of a concentrated load; therefore, in order to apply them, it is not necessary to define the effective width b_{eff} according to the approximations illustrated in Fig. 9.162 g. However, rigorously speaking, they can be adopted only if the beam is loaded by the one external load. Where there is more than one applied load each having a value F and set at a spacing s, the field of application of the above formulae can be reasonably extended only provided s is considerably greater than the web height ($s > 4$ to 5 h_w). Otherwise, one should check that the mean stress $\sigma = F/(s t_w)$ does not exceed the limits expressed by (9.215).

Formulae (9.215) and (9.216) cannot be applied to cold formed profiles (Fig. 9.164), for which the influence of the folding radius is a determinate factor.

Winter |114| carried out many experiments concerning this problem and he suggested some empirical formulae, which are applicable to radius thickness ratios between 1 and 4. Such formulae were adopted

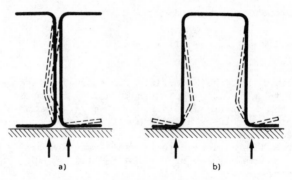

Fig. 9.164

in AISI recommendations; they make a distinction according to whether the load is applied at the ends of the beam or at an intermediate section and whether the section force is or is not sufficient effectively to prevent web rotation.

For sections of the type illustrated in Fig. 9.164a, such as I-shaped profiles formed by $[$-shaped elements or back-to-back angles, the following formulae are applicable:

If the load is applied at an intermediate section:

$$F_c = 100 \ t^2 \ f_y \ \{11.1 + 2.41 \ \sqrt{c/t}\} \tag{9.218}$$

If the load is applied at the ends of the beams:

$$F_c = 100 \ t^2 \ f_y \ \{7.4 + 0.93 \ \sqrt{c/t}\} \tag{9.219}$$

For sections of the type illustrated in Fig. 9.164b, such as L, $\mathbf{L}$ and Π-shaped profiles, the following formulae are applicable:

If the load is applied at an intermediate section:

$$F_c = 0.485t^2 f_y \{(3.050 + 23\,(c/t) - 0.09\,(c/t)(h/t) - 5\,h/t) \cdot \\ \cdot (1.06 - 0.06\,r/t)(1.22 - 0.22\,f_y/235)\} \tag{9.220}$$

If the load is applied at the ends of the beam:

$$F_c = 0.485t^2 f_y \{(980 + 42\,(c/t) - 0.22\,(c/t)(h/t) - 0.11\,h/t) \cdot \\ \cdot (1.15 - 0.15\,r/t)(1.33 - 0.33\,f_y/235)\} \tag{9.221}$$

In formulae (9.218) to (9.221)

r = internal radius
c = length along which the load is applied
h = inner web height
t = web thickness

9.7 STIFFENED PLANE PLATES

9.7.1 *Analysis of the problem*

Stiffened plates are components of more complex structural elements
(Fig.9.165). The structure of a modern deck system for a steel bridge
is formed by vertical and horizontal plates, which are stiffened in
two orthogonal directions (orthotropic plates). The web of a high
slender beam can be stiffened by means of vertical stiffeners and
possibly by horizontal ones, mainly located in the compressed area

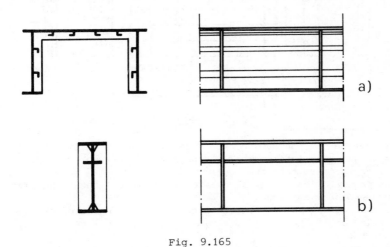

Fig. 9.165

The behaviour of a stiffened plate, particularly post buckling,
is severely influenced by stress distribution. There is a fundamental
distinction between:
Compressed plates
Bent plates
The first class includes, substantially, the compressed flanges
of the sections of deck systems for bridges, whereas the second class
includes beam webs. This distinction allows a first simplification
of the problem, as it can be divided into two parts that can be
studied separately: this is due to the fact that (as will be seen
later) strength reserves in the post critical field are scarce and
often negligible in compressed plates, whereas they are considerable
in bent plates.

9.7.1.1 Compressed plates

Compressed plates generally undergo the following stresses (Fig.
9.166):
(a) *Longitudinal stresses* (σ_1) related to the overall bending and
    compression effort produced on the entire structure by the
    applied loads. These stresses not only vary along the span,
    following bending moment variation, but can also vary along the

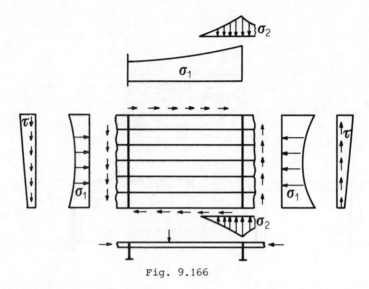

Fig. 9.166

transverse section due to the 'shear lag' effect. In addition,
there are also longitudinal stresses due to prevented warping
for non-uniform torsion
(b) *Tangential stresses* (τ) associated with overall shear and primary
and secondary torsion
(c) *Transverse stresses* (σ_2) due to transverse deformation of the
section or to the action of transverse diaphragms near the
supports
(d) *Bending stresses*, variable in the thickness of the plate, due to
the concentrated loads
Compressed plates are often idealized as a series of stiffened
plates, simply supported on the transverse diaphragms and loaded by
the combination of stresses listed above. Such a model entails an
approximation that sometimes cannot be accepted because of the
following complications.

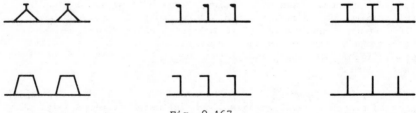

Fig. 9.167

Of a geometrical nature. Stiffeners are eccentric with reference to
the plane of the plate and often are not symmetrical with reference
to the vertical axis (Fig. 9.167). There are also inevitable
geometrical imperfections due to working tolerances. Furthermore

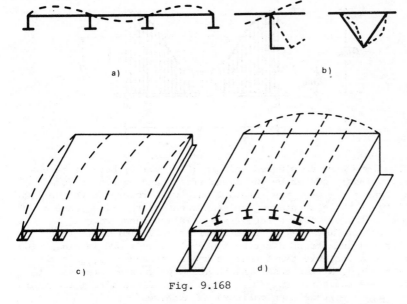

Fig. 9.168

there is, in practice, a continuity in both the longitudinal and transverse directions depending on the connection system that was adopted.

Of a mechanical nature. In addition to the existence of three-dimensional stresses due to the external loads, the welding process produces residual stress distributions in the various elements: at certain points, they may even attain yield. The degrading effects of residual compression stresses have an extremely important role with reference to the unstable behaviour of such structural elements.

As a consequence of this complex set of stresses, there are four independent possible types of buckling in compressed plates (Fig. 9.168):

(a) local buckling of individual panels between stiffeners
(b) local buckling of the compressed parts of stiffeners
(c) overall buckling of the stiffened plate with longitudinal bending between transverse diaphragms
(d) overall buckling of the plate with transverse bending between edge beams

It depends on the geometrical nature of the plate and on the stresses in it which of the above mechanisms will be responsible for the collapse of the compressed plate. It is thus impossible to make any general forecasts: one must realize that the problem is extremely complex and that, consequently, it is necessary to adopt approximate expedients.

With reference to behaviour in the post critical field, the following should be noted. Consider the phenomenon of longitudinal buckling of a compressed plate (case c): perpendicular stress

569

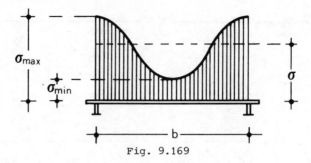

Fig. 9.169

distribution can be considered uniform (Fig. 9.169) because stress
concentrations produced by shear lag effect and by non uniform
torsion enear the webs are relevant only with reference to the local
stability of the elementary panels adjacent to the webs (case a).
As soon as the critical stress $\sigma = \sigma_{cr}$ is attained, however
redistribution takes place: it is qualitatively similar to that due
to shear lag and stress distribution evolves as shown in Fig. 9.169,
passing to a variable trend that is characterized by σ_{min} at the
centre and σ_{max} at the ends. As buckling progresses, the effect
increases as the ratio $\sigma_{max}/\sigma_{min}$ grows until first of all the yield
limit ($\sigma_{max} = f_y$) and then collapse is attained.

The structure therefore has a post-critical behaviour consisting
of an elastic and a successive elastic-plastic phase just before
collapse. On the other hand, it is obvious that the phenomenon has
a progressive and self exalting character, as no new type of static
behaviour is fit to oppose buckling. Strength reserve in the post
critical field for compressed plates is therefore scarce and it might
be advisable to overlook it.

9.7.1.2 Bent plates

Webs, or, better, web panels, bound by the flanges of a beam and by
vertical stiffeners generally simultaneously undergo bending and
shear (Fig. 9.170).

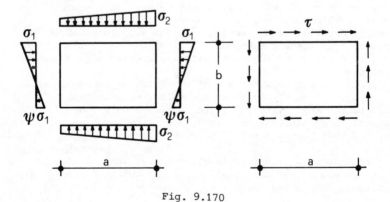

Fig. 9.170

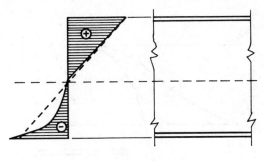

Fig. 9.171

In this case, the post critical phase is more pronounced than that of compressed plates. The corresponding strength reserve is far from being negligible, as the collapse load might be considerably higher than the elastic critical load.

Consider the two main loading conditions. First, from the point of view of bending, as the load increases the bi-triangular stress diagram tends to undergo the modifications shown in Fig. 9.171, due to the buckling of the compressed part of the web above the neutral axis.

If stability of the compressed part is guaranteed by flange stiffness (the flange being able to absorb the entire compression force), such behaviour might go beyond the elastic field in the post critical phase, until ultimate moment of the section is attained.

Secondly, from the point of view of shear, whenever the tangential stress (which can reasonably be considered constant throughout the entire depth of the web) attains its critical value, the plates buckle. Diagonal stress fields form in them, as has been proved by many experimental surveys with inactive compressed strips and strips in tension capable of resisting until they attain collapse due to material failure under tension.

If plate edges (flanges and vertical stiffeners) are sufficiently rigid, a new behaviour can thus take place in the post critical phase: it is of the truss type and is capable of resisting a much higher collapse load N_c than the critical load N_{cr} (Fig. 9.172).

Curve 'a' shows the theoretical behaviour there would be on the assumption of an ideal web without any initial imperfections, whereas curves 'b' and 'c' are for actual cases of webs affected, respectively, by slight and consistent deformations.

Collapse load N_c is practically independent of the initial deformation situation: web panels behave in the post critical field as struts constrained on four sides, undergoing great deformations.

The post critical behaviour of panels under shear can be examined according to the diagonal stress theory (Fig. 9.173). Suppose, first of all, that the web is very thin and divided by vertical stiffeners. The extreme deformability of the panels prevents them from bearing any compressive stress and they buckle forming diagonal waves. The shear force may be born by the vertical components of the diagonal tension forces which, while actually distributed over the whole web,

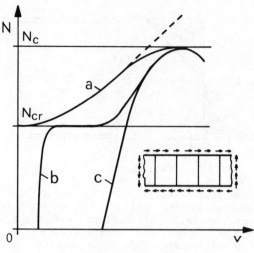

Fig. 9.172

can substantially be considered as concentrated on the diagonals of the individual panels, as in the truss diagram shown in Fig. 9.173b. This arrangement guarantees the strength of the structure until one of the following conditions takes place:

Diagonal tension stresses attain material yield

Vertical stiffeners are no longer able to resist the compression forces transmitted to them

Beam flanges are no longer able to bear the horizontal components of the diagonals in tension, in addition to the bending effects

The theory based on this model underestimates the ability of the web to resist compression: it is not zero, although it is modest. A more complete model (Fig. 9.173c), however, has the disadvantage of being statically indeterminate. Furthermore, in assessing the stiffness of the compressed strip, one must take into account the fact that the diagonals in tension provide an intermediate constraint against buckling of the compressed diagonals, the strength of which is thus greater (of the order of four times) than it would be if the diagonals in tension did not exist (Fig. 9.173d).

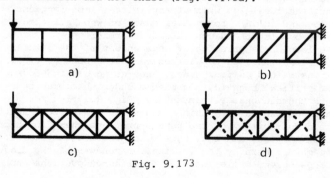

Fig. 9.173

In any case, structural collapse is attained in the post critical
field due to one of the following:
  Diagonal strips in tension collapse
  A kinematic mechanism takes place in the frame formed by the
  flanges and vertical stiffeners
  The first possibility is critical in the case of materials of
low ductility, whereas the second is mainly influenced by flange
and stiffener design.

9.7.1.3 Design criteria
Stability of plane stiffened plates can be analysed according to
two main approaches. They are:
  Linear theory, considering a perfectly elastic behaviour of the
  material in the field of small deformations and without any
  imperfections
  Non linear theory, introducing inelastic behaviour of the material
  in the field of large deformations, with both geometrical and
  mechanical imperfections.
  The following procedures belong to the linear theory.

Timoshenko's method |122,123|. This concerns the control of panels
provided with rigid stiffeners and therefore it studies buckling
conditions of the elementary panel in the elastic field.

Klöppel's method |124,125|. This considers plates provided with
both longitudinal and transverse stiffeners, of a rigid type
(buckling of the elementary panel) and of a flexible type (buckling
of the entire stiffened plate). Critical stresses are provided by
a series of graphs, in terms of panel dimensions and stiffener
rigidity. This method is a more sophisticated version of Timoshenko'
at its maximum degree of perfection.

Winter's method |114|. This controls flexural torsional stability
of a strut the section of which is composed by the longitudinal
stiffeners and by a plate having a reduced effective width b_{eff} (see
9.6.2.3).
  Timoshenko's and Klöppel's methods provided the basis for the
German regulations (DIN) and only indirectly allow post critical
strength reserves to be taken into account, by adopting lower safety
coefficients (1.25 to 1.35) than those adopted for yield (1.5).
  Methods based on the linear theory do not allow a correct estimate
of the ultimate load of the plate stiffener system, as they only
provide a critical load corresponding to elastic buckling and thus
do not allow assessment of the post critical reserves of the
structure. Therefore, they:
  Overestimate the strength of compressed plates because, on the
  one hand, they are characterized by scarce post ·critical strength
  reserves and, on the other, their behaviour is considerably
  influenced by geometrical and mechanical imperfections, which the
  linear theory does not take into account
  Underestimate the strength of bent plates, which have considerable
  reserves in the post critical field, as evidenced by the diagonal

573

stress field

It is thus obvious that design methods based on the linear theory provide a very poor model, with many unknown factors in interpreting the real problem, to the point that in 1968 the conclusions of the VIII IABSE Congress in New York included the following statements |7,126|:

"The linear theory of plate stability is not an adequate basis for the design of struts and girders consisting of thin-walled sections...

It is necessary to calculate the maximum membrane stresses and strains in the postbuckling range since average stresses, supposing plane sheet, are not a sufficient design criterion...

... We still lack a general mathematical approach to establish a design method taking into account all important parameters. New carefully planned test series and computer simulations, including the postbuckling behaviour (membrane actions) of the sheet, are therefore strongly recommended."

Such statements sound like a death sentence: in fact, immediately afterwards (1969-1971) four bridges with orthotropic decks collapsed during their erection |127-130| causing about fifty deaths in different countries (England, Austria, Australia, Germany).

Now, some ten years later, after an examination of the facts, some conclusions can be tentatively made |7,131|:

"The linear buckling theory, applied with inadapted factors of safety, was partly responsible for the collapse of four large box girder bridges during their erection.

Another, about equally important cause of these accidents, was the poor detailing rules and the absence of adequate fabrication tolerances."

In these last years numerous finite element calculation programmes have been developed: they are able to predict the ultimate strength of plate structures, taking large displacements and plastic effects duly into account.

Such methods - generally very sophisticated ones - can be considered as a reaction to the above mentioned events, which had explicitly condemned the traditional methods and their application. These new methods generally are a derivation of studies carried out from two separate points of view, examining:

Plates with predominant shear effects (see 9.7.1.4)

Plates with predominant compression effects (see 9.7.1.5)

Most of the calculation programmes proposed in these last years for the study of the stability of compressed plates are expensive research instruments, mainly used to calibrate simpler methods, while designers still need simplified engineering models. As far as such methods are concerned, a long time will probably pass before adequate calculation models will have been completely developed. Furthermore, even in cases for which such models partially exist, the new developments of structural engineering require their continuous

574

up dating and conforming to regulations in force.

'Standard models' with general characteristics should simulate the behaviour of plates with longitudinal and transverse stiffeners, including those with an asymmetrical section, generally loaded and with imperfections of various kinds. On the other hand, in the present state of knowledge, only the linear theory is able to provide general design cirteria, even though it often leads to unreliable results and must therefore be applied with proper precautions.

For these reasons, ECCS adopted provisional rules based on the linear stability theory, in order to help designers until the more refined methods become operative and acquire a general significance. Such provisional rules (see 9.7.2) are divided into two parts: the first part includes simple rules aimed at deriving collapse load from critical load, by multiplying the latter by appropriate correction factors that are defined in an extremely simple manner, although they are actually complex functions of panel slenderness, stiffener rigidity and geometrical and structural imperfections. The second part concerns the technological requirements connected with stiffener position, building details and manufacturing tolerances.

9.7.1.4 Web design

The post critital behaviour of beam webs was first studied by Wilson (1886). On the basis of noticing that railway bridge beams having very slender webs, but stiffened by vertical ribs, were not inconvenienced, he discovered (by operating on simple paper models) the diagonal stress field behaviour however without interpreting it analytically. This was done by Wagner (1930), who formulated the diagonal stress theory. Systematic theoretical and experimental studies in this direction were carried out by Basler and Thürlimann |132-134|, who proposed practical methods for taking into account the post critical behaviour of web-panels under both bending and shear: such methods were adopted in AISC recommendations.

The 'Basler and Thürlimann' method considers only the case of transverse ribs and, ignoring the flexural rigidity of the flange, supposes that the diagonal stress fields are anchored to the ribs only (Fig. 9.174a). This assumption corresponds to a truss beam behaviour.

Later, other authors moved from this assumption to that anchoring the tension diagonal strips also to the flanges of the beam (Fig. 9.174b). In this case, however, the mechanism of a frame formed by the flanges and the vertical ribs must be superposed on the truss beam mechanism. It can thus produce three possible modes of collapse: beam mechanism (Fig. 9.174c); panel mechanism (Fig. 9.174d); combined mechanism (Fig. 9.174e).

If the frame model is disregarded, collapse conditions must be imposed by the shear and tensile stresses in the web panels. If, however, it is taken into consideration, the stress induced in the flanges by the bending behaviour of the entire beam must be taken into account.

As in practice bending and shear stresses are both present, it is necessary to establish an interaction domain $V - M$. If, for

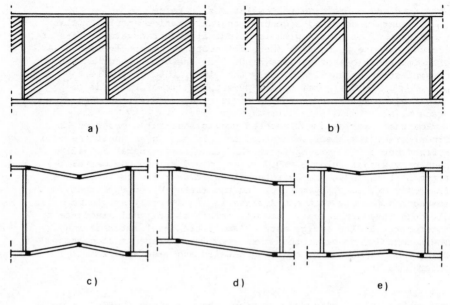

a)   b)

c)   d)   e)

Fig. 9.174

example, it is assumed that shear is entirely absorbed by the web and moment by the flanges only, the domain is defined by points A, B and C of Fig. 9.175a. If web bending contribution is admitted, point C, corresponding to $M_{f,pl}$, moves to C', corresponding to M_{pl}. If one furthermore considers flange shear contribution, point A, corresponding to $V_{w,pl}$ moves to A', corresponding to V_{pl}. Such a domain was proposed by Höglund |135|.

Basler |132| proposed a more cautious diagram $V - M$, shown in Fig. 9.175b in the non-dimensional form with reference to plastic shear V_{pl} and elastic limit moment M_e. The value of M_e increased by 10%,

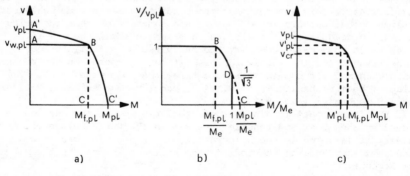

a)   b)   c)

Fig. 9.175

is assumed as plastic moment M_{pl}.

The collapse mechanisms corresponding to some of the principal design methods produced after that of Basler and Thürlimann are illustrated in Fig. 9.176. They are characterized by the notion of a diagonal tension field, plastic hinges in the flanges and panel constraint conditions.

The *Aarau method* (Herzog) takes into consideration a diagonal tension strip border at mid-panel height |136|. Distance c is fixed on the basis of many experimental results. The method provides approximate formulae for assessing ultimate shear as a function of panel characteristics.

According to the *Cardiff method* (Rockey and others) the stress field inclination does not necessarily coincide with the panel diagonal |137|. The collapse mechanism corresponds to four plastic hinges forming in the flanges. The *Karlsruhe method* (Steinhardt and others) assumes a diagonal field with a sinusoidal stress trend |138|. An analogous result is obtained by the step distribution contemplated by the *Lehigh method* (Chern and Ostapenko), that furthermore locates collapse in a panel mechanism with four hinges at the field corners |139|. The *Osaka method* (Komatsu) provides approximate formulae for determining distance c that identifies the mutual position of the hinges |140|. The *Prague – Cardiff method* (Skaloud and Rockey) is a simplification of the Cardiff method, obtained by assuming that direction of the tension stirp coincides with the panel diagonal |141|. The *Stockholm method* (Höqlund) is applicable to webs without vertical stiffeners: the web is assumed to be a system of orthogonal tension and compression members |135|. Plastic hinge positioning is based on empirical formulae. The *Tokyo method* (Fujii and others) considers a stress distribution resemblinc that of the Lehigh method, and assumes a collapse mechanism of the beam type in the flanges |142|.

Other methods not shown in Fig.9.176: include the *Göteborg Method* (Bergfelt) which considers the same hinge position as the Prague – – Cardiff method and optimizes stress field inclination according to Basler's method |143|; the *Zürich Method* (Dubas)which assumes a physical model based on collapse shear sub-division into three parts: critical shear, shear deriving from the corner effect of the diagonal strip and shear related to flange bending stiffness |144|; finally, the *Italian method* (Mele) which assumes a similar collapse mechanism to the Prague – Cardiff one and allow the existence of longitudinal stiffeners |145| to be taken into account.

A more detailed survey of these methods is contained in the ECCS Stability Manual and in the SSRC Guidebook |6,7|. A further comparison between them is made in the table of Fig. 9.177 on the basis of adopting the following requirements as parameters:

(a) Clearness
(b) Simplicity
(c) Computer need
(d) Existence of longitudinal stiffeners
(e) Applicability to hybrid girders
(f) Applicability to non-symmetrical girders
(g) Applicability to composite steel-concrete girders

577

METHOD	MECHANISM	BOUNDARY CONDITIONS
A ARAU (Herzog)	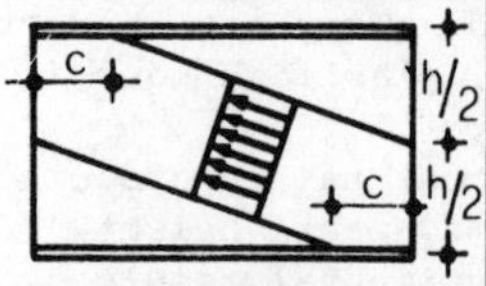	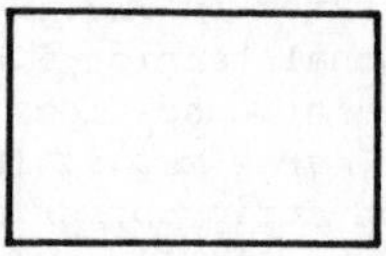
CARDIFF (Rockey & others)	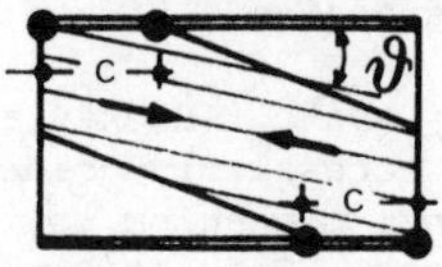	
KARLSRUHE (Steinhardt & others)		
LEHIGH (Chern & Ostapenko)	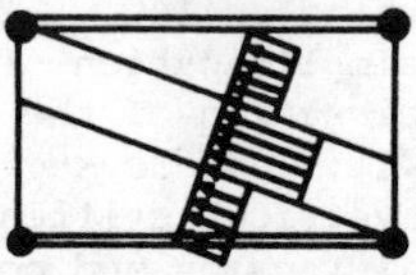	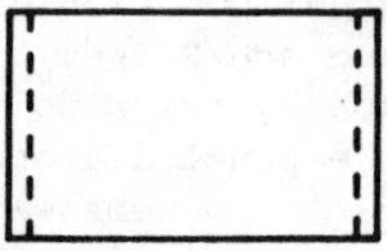
OSAKA (Komatsu)	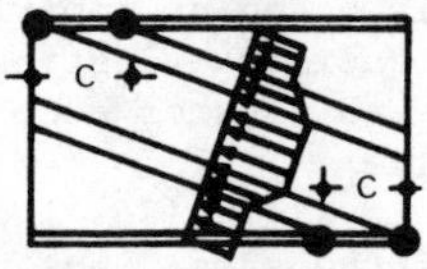	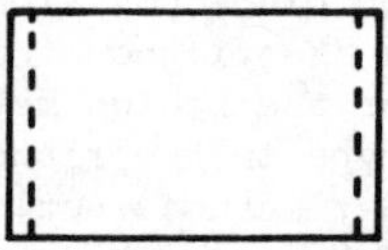
PRAHA-CARDIFF (Skaloud & Rockey)	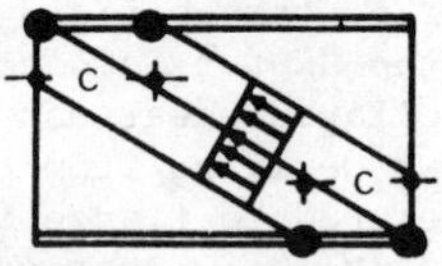	
STOCKHOLM (Höglund)	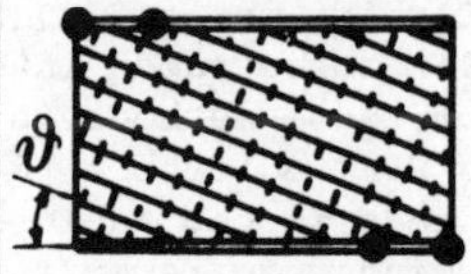	
TOKYO (Fujii &others	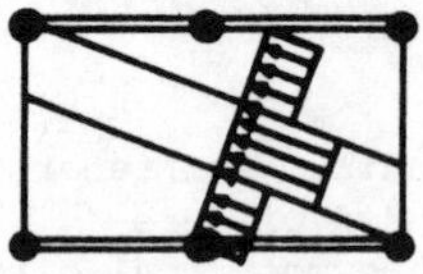	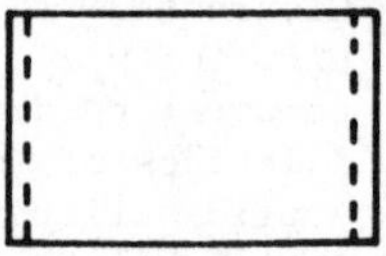

Fig. 9.176

Performance Method:	a	b	c	d	e	f	g	h	i	j	k	l	m
Araau	fair	no	no	yes	yes	yes	no	yes	no	yes	yes	yes	yes
Cardiff	very good	yes	charts	yes	yes	yes	-	yes	no	*	yes	yes	yes
Karlsruhe	very good	yes	no	yes	yes	yes	yes	yes	yes	*	yes	yes	yes
Lehigh	good	no	yes/no	yes/no	yes	yes	-	yes	no	yes	yes	yes	yes
Osaka	fair	no	yes	no	no	yes	no	no	no	no	yes	no	no
Prague-Cardiff	very good	no	no/yes	yes	yes	no	no	yes	yes	*	yes	yes	yes
Stockholm	very good	yes	no	no	yes	no	no	yes	yes	*	yes	yes	yes
Tokyo	good	no	no	no	yes	yes	-	yes	no	no	yes	yes	yes
Göteborg	very good	yes	no	no	yes	yes	-	yes	yes	no	yes	yes	yes
Zurich	very good	yes	no	no	yes	yes	-	yes	yes	no	yes	yes	yes
Italian	very good	yes	no	yes	yes	yes	-	no	no	no	yes	yes	yes

\- Nothing said

* mentioned

Fig. 9.177

(h) Useful specifications for evaluating ultimate bending capacity
(i) Taking into account axial forces
(j) Assessment of critical stress in compressed flanges
(k) Analysis of pure shear stresses
(l) Taking into account normal stresses due to bending
(m) Combination of shear and bending

9.7.1.5 Design methods for compressed plates

Among the methods based on the non-linear theory, that of Maquoi and Massonnet |146| was the first to face (although by an approximate approach) the problem of taking the post critical behaviour of orthotropic plates directly into account, considering both large deformations and inelastic material behaviour. In this method, the survey is carried out in the elastic phase, approximately taking into account material non-linearity by introducing a collapse criterion defined as that at which strut stresses attain the following value:

$$\sigma_c = \rho f_y \qquad \text{with} \qquad \rho = \rho_1 \rho_2$$

where:
f_y = material yield stress
ρ = overall efficiency , obtained as the product of the two partial efficiencies ρ_1 and ρ_2 of the stiffened plate

The primary efficiency ρ_1 of the stiffened plate, considered as an orthotropic plate having a uniformly distributed stiffness, (obtained by buttering the stiffeners in the longitudinal direction) is defined as the ratio between the mean value of the stresses on the loaded sides and the maximum stress attained at the connection to the web, which at collapse is equal to yield stress f_y.

Partial efficiency ρ_2 is a corrective term taking into account stiffener discontinuity which causes an actual stress distribution differing from that corresponding to infinitely close stiffeners (Fig. 9.178). It is expressed as:

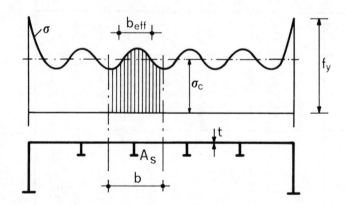

Fig. 9.178

$$\rho_2 = \frac{b_{eff}\, t + A_s}{b\, t + A_s} \qquad\qquad (9.222)$$

where:

b and b_{eff} = geometrical and effective width respectively

t = plate thickness

A_s = stiffener area

After this method, called the *Liège method*, many others were proposed. Their comparison |7| may be based on the opportunities offered by each method of taking into account the various aspects involved:

(a) Asymmetry of cross section of flange about the horizontal axis;

(b) Local buckling of the panels between stiffeners

(c) Stiffener flexural torsional buckling

(d) Existence of transverse stresses in the plate

(e) Existence of tangential stress in the plate

(f) Existence of concentrated forces

(g) Axial force variation along the span

(h) Curvature of the girder

(i) Overall post critical behaviour of the flange

(j) Residual stress influence

In the table of Fig. 9.179 these aspects are marked with an X against the various methods in which they are considered.

The *Merrison method* |147| complies with all the requirements referred to in the table. It is based on a semi-empirical criterion; it supposes that collapse occurs when the steel plate becomes plastic according to the prediction of an elastic analysis which takes into consideration large displacements and initial geometrical

Performance Methods	a	b	c	d	e	f	g	h	i	j
Liège	–	x	–	–	–	–	–	–	x	–
Merrison	x	x	x	x	x	x	x	x	x	x
Cambridge	x	x	–	–	x	–	x	–	–	x
Karlsruhe	x	x	–	–	–	–	–	–	x	–
Monash	x	x	–	–	–	–	–	–	–	–
Manchester	x	x	–	–	–	–	–	–	–	x
Imperial College	x	x	x	–	x	x	x	x	x	x
Zurich	–	x	–	–	–	–	–	–	–	–

Fig. 9.179

imperfections. These limitations do not allow a study of post elastic behaviour: they can be justified on the grounds that the method was composed in order to fill a gap in the regulations disclosed by the above mentioned bridge collapses.

The *Cambridge method* |148-149| is based on the both theoretical and experimental results: it studies the individual stiffener as an isolated element, applying the ECCS bending curves interpreted according to the Perry formula. The *Karlsruhe method* |138| analyses the stiffened plate in the field of large displacements with initial longitudinal imperfections under uniform compression.

The *Monash method* |150| also examines stiffeners as isolated struts and compares the steel plate critical stress to that of the stiffener in order to evaluate effective width to be used in the ultimate design. The *Manchester method* |151| too, considers compressed individual struts: it introduces a second degree elastic analysis to define effective width. The *Imperial College method* |152| considers practically all the aspects of the problem and has the advantage that it allows simple equations to the written in order to assess axial and bending stress distribution along the transverse section of the entire longitudinally stiffened plate having either open or closed section stiffeners. Corresponding results have provided satisfactory comparisons with the most commonly applied methods. Finally, the *Zürich method* |153| is a semi-empirical method, very simply applied; it does not, however, allow a determination of the importance of the various involved parameters. This is why it very often leads to inconsistent safety coefficient values with reference to collapse.

9.7.2 *ECCS Recommendations*

For the reasons explained in 9.7.1.3, the European Convention chose the linear theory approach as the basis for its Recommendations |31| concerning stiffened plates, until methods based on more refined models become operative. Such recommendations are for plane orthotropic plates, assumed free of any initial geometrical imperfection stiffened by symmetrical stiffeners and loaded in plane.

According to the linear theory, such plates remain plane until, at the critical load, equilibrium bifurcation occurs and out of plane displacements, the value of which cannot be determined by the linear approach, take place. Critical load does not coincide with the ultimate load of the plate; post critical strength reserves depend on edge conditions and the type of loading (see 9.7.1.1 and 9.7.1.2).

ECCS Recommendations indicate the values of an adequate corrective factor α_c transforming ultimate critical load which, according to existing stresses, may be either an amplifying factor if there are great post critical strength reserves, or a reducing one, if such reserves are scarce and, at the same time, imperfections have a degrading role (see 9.7.1.2 and 9.7.2.2).

This method is applicable provided very restrictive working tolerances are complied with: they are an essential part of the Recommendations in question (see 9.7.2.3).

9.7.2.1 Panel control

With reference to a rectangular plate having dimensions $a \times b$, loaded by perpendicular stresses σ_x, σ_y along two opposite side or by tangential stress τ, each one assumed to act separately, the following ideal critical stresses can be defined:

$$\sigma_{cr,x,o} = k_{\sigma,x}\ \sigma_{cr,o}$$
$$\sigma_{cr,y,o} = k_{\sigma,y}\ \sigma_{cr,o} \qquad (9.223)$$
$$\tau_{cr,o} = k_\tau\ \sigma_{cr,o}$$

In formulae (9.223), $\sigma_{cr,o}$ is a reference stress, corresponding to the Euler critical load for an infinite plate having width b and thickness t, hinged at its ends and having the following value:

$$\sigma_{cr,o} = \frac{\pi^2 E}{12(1 - \nu^2)}\ (t/b)^2 \qquad (9.224)$$

which for steel becomes (see also 9.6.1):

$$\sigma_{cr,c} = \frac{186200}{(b/t)^2}\ (\text{Nmm}^{-2}) = \frac{18980}{(b/t)^2}\ (\text{kgmm}^{-2}) \qquad (9.225)$$

Buckling coefficients $k_{\sigma,x}$, $k_{\sigma,y}$, k_τ depend on stress distribution, the ratio between panel sides ($\alpha = a/b$), edge conditions and rib geometry. For panels without intermediate stiffeners and hinged at their edge, the values of buckling coefficients can be determined using the table of Fig. 9.180.

For other constraint conditions, some values of k are given in the table of Fig. 9.181.

The case of panels with intermediate stiffeners depends on the above parameters, together with the flexural and extensional rigidity of the stiffeners. Relative flexural rigitity is expressed by the ratio between stiffeners and plate flexural rigidity:

$$\gamma = \frac{E\ I_s}{b\ D} = \frac{12(1 - \nu^2)\ I_s}{b\ t^3} \qquad (9.226)$$

or, for steel ($\nu = 0.3$):

$$\delta = 10.92\ I_s/bt^3 \qquad (9.227)$$

where I_s is the moment of inertia of the section formed by the stiffener outstand and the effective width b_{eff} of the plate assessed according to 9.6.

Relative extensional rigidity is expressed by the ratio between stiffener outstand area A_s and the transverse sectional area of the panel:

$$\delta = \frac{A_s}{bt} \qquad (9.228)$$

In the case of stiffened panels, buckling coefficients thus depend on the values of flexural rigidity and of extensional rigidity δ. Approximate expressions for k_σ and k_τ (applicable within the range of the above dimensional ratios) are given in the table

1	Compression $0 \le \psi \le 1$	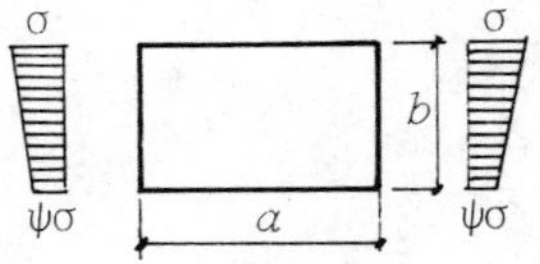	$\alpha \ge 1$	$k_\sigma = \dfrac{8.4}{1.1 + \psi}$
			$\alpha < 1$	$k_\sigma = \left(\alpha + \dfrac{1}{\alpha}\right)^2 \dfrac{2.1}{1.1 + \psi}$
2	Compression and bending $-1 < \psi < 0$	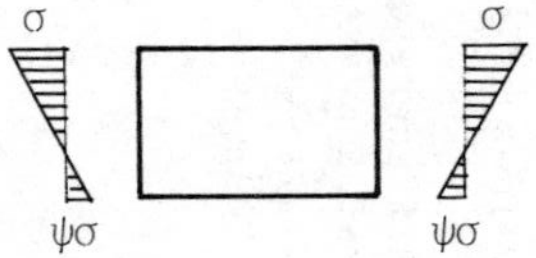		$k_\sigma = (1 + \psi)k_{\sigma,1} - \psi k_{\sigma,3} + 10\psi\,(1 + \psi)$ with: $k_{\sigma,1}$ from 1 for $\psi = 0$ $k_{\sigma,3}$ from 3
3	Simple bending $\psi = -1$	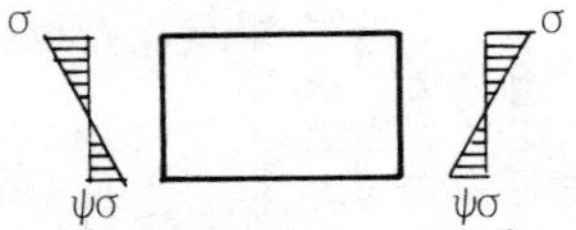	$\alpha \ge \dfrac{2}{3}$	$k_\sigma = 23.9$
4	Tension and bending $\psi < -1$	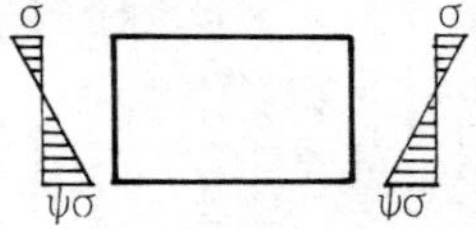	$\alpha < \dfrac{2}{3}$	$k_\sigma = 15.87 + 9.6\alpha^2 + \dfrac{1.87}{\alpha^2}$
5	Shear		$\alpha \ge 1$	$k_\tau = 5.34 + \dfrac{4.00}{\alpha^2}$
			$\alpha < 1$	$k_\tau = 4.00 + \dfrac{5.34}{\alpha^2}$

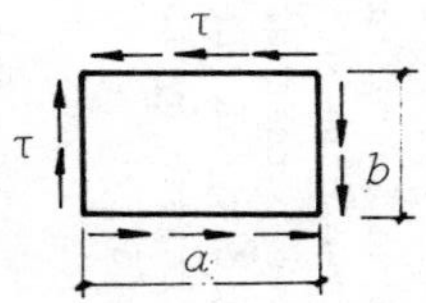

Fig. 9.180

buckling coefficients k_σ edge constraints	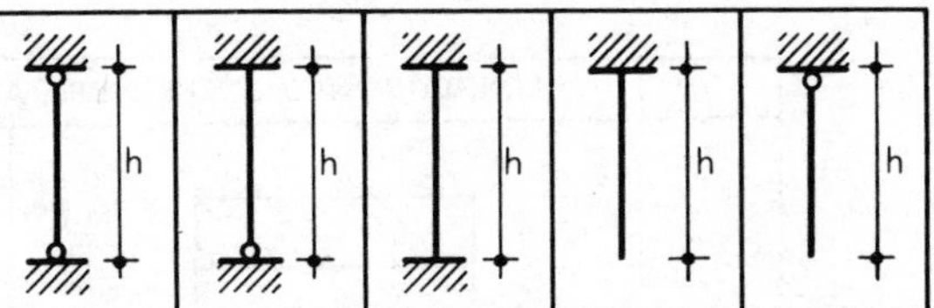				
load conditions	$\alpha \geq 1{,}0$	$\alpha \geq 0{,}8$	$\alpha \geq 0{,}7$	$\alpha \geq 1{,}6$	$\alpha \geq 1{,}5$
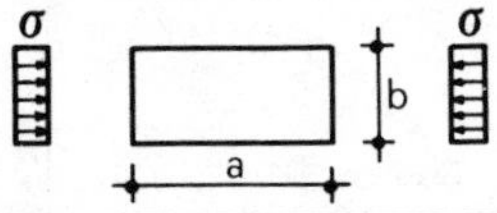	4,00	5,40	6,97	1,28	0,43
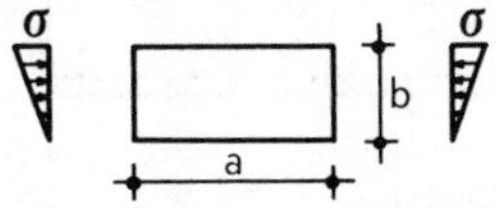	7,81	12,16	13,56	6,26	1,71
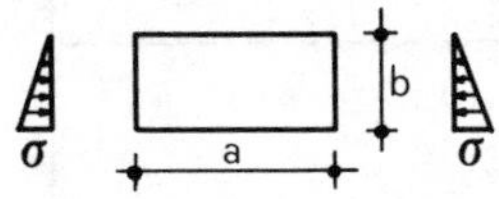	7,81	9,89	13,56	1,64	0,57

Fig. 9.181

of Fig. 9.182b for the cases illustrated in Fig. 9.182a: for one longitudinal stiffener alone (cases 1,2); for one transverse stiffener alone (cases 3,4) and for one stiffener in both directions. In all cases (cases 5,6) the loading conditions produce either perpendicular stresses only on two opposite sides or tangential stresses only. Indexes 'L' and 't' indicate that rigidities are related, respectively, to the longitudinal and transverse ribs.

Once the buckling coefficients are known for stiffened and unstiffened plates, formulae (9.223) enable one to determine the ideal values of the buckling stresses, imagined to act separately.

When, however, stress σ_x, σ_y and τ act simultaneously, the buckling condition is characterized by the following limiting buckling stresses:

$$\sigma_{cr,x}; \qquad \sigma_{cr,y}; \qquad \tau_{cr} \tag{9.229}$$

defining, case by case, a point in the stability domain of the plate (Fig. 9.183). This domain can be approximately defined by interaction formulae. For plates without stiffeners, equally constrained at their ends, clamped or simply supported, undergoing mono-axial stresses represented by σ_x and τ, the following formula has been proposed:

$$0.25(1 + \psi)\frac{\sigma_{cr,x}}{\sigma_{cr,x,o}} + \sqrt{0.25(3-\psi)\left\{\frac{\sigma_{cr,x}}{\sigma_{cr,x,o}}\right\}^2 + \left\{\frac{\tau_{cr}}{\tau_{cr,o}}\right\}^2} = 1 \tag{9.230}$$

585

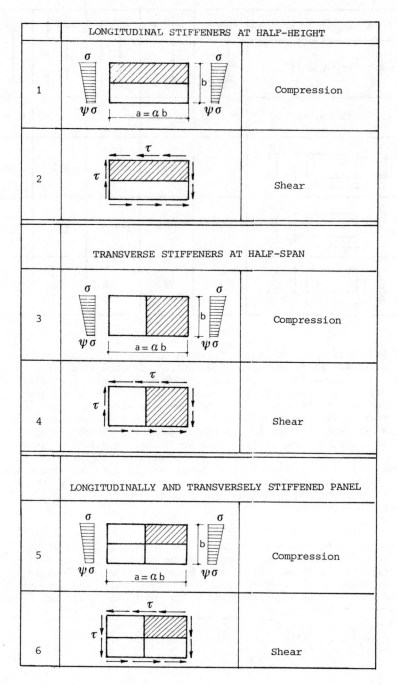

Fig. 9.182 a

	BUCKLING COEFFICIENTS
1	$\alpha \leq \sqrt[4]{1 + 2\gamma}$ $k_\sigma = \dfrac{2}{0.95(\psi + 1.1)} \dfrac{(1 + \alpha^2)^2 + 2\gamma}{\alpha^2 (1 + 2\delta)}$ $\alpha \geq \sqrt[4]{1 + 2\gamma}$ $k_\sigma = \dfrac{4}{0.95(\psi + 1.1)} \dfrac{1 + \sqrt{1 + 2\gamma}}{1 + 2\delta}$
2	$0.5 \leq \alpha \leq 2.0 \qquad k_\tau = 4.93(1+\alpha^2)\big/(\alpha^3\sqrt{\zeta})$ $\zeta = \dfrac{10.24(1 + \alpha^2)^2 + 3.16(1 + 9\alpha^2)^2 + 4.05\gamma}{(1 + \alpha^2)^2(1 + 9\alpha^2)^2 + 2\gamma(1 + \alpha^2)^2 + 2\gamma(1 + 9\alpha^2)^2}$ $+ \dfrac{10.24(1 + \alpha^2)^2 + 0.41(9 + \alpha^2)^2 + 13.11\gamma}{(1 + \alpha^2)^2(9 + \alpha^2)^2 + 2\gamma(9 + \alpha^2)^2 + 162\,\gamma(1 + \alpha^2)^2}$
3	$0.4 \leq \alpha \leq 1.0 \qquad k_\sigma = \dfrac{A - \sqrt{A^2 - B}}{1.43\alpha^2(\psi + 1.1)}$ $A = 1.5(1 + \alpha^2)^2 + 0.167(9 + \alpha^2)^2 + 3.33\alpha^2\gamma$ $B = (1 + \alpha^2)^2(9 + \alpha^2)^2 + 2\alpha^3\gamma\left\{(1 + \alpha^2)^2 + (9 + \alpha^2)^2\right\}$
4	$0.5 \leq \alpha \leq 2.0 \qquad k_\tau = 4.93(1 + \alpha^2)^2\big/(\alpha^3\sqrt{\zeta})$ $\zeta = \dfrac{10.24(1 + \alpha^2)^2 + 0.41(1 + 9\alpha^2)^2 + 13.11\alpha^3}{(1 + \alpha^2)^2(1 + 9\alpha^2)^2 + 162\,\gamma\alpha^3(1 + \alpha^2)^2 + 2\,\gamma\alpha^3(1 + \alpha^2)^2}$ $+ \dfrac{10.24(1 + \alpha^2)^2 + 3.16(9 + \alpha^2)^2 + 4.05\gamma\alpha^3}{(1 + \alpha^2)^2(9 + \alpha^2)^2 + 2\gamma\alpha^3(9 + \alpha^2)^2 + 2\gamma\alpha^3(1 + \alpha^2)^2}$
5	$0.9 \leq \alpha \leq 1.1 \qquad k_\sigma = \dfrac{(1 + \alpha^2)^2 + 2(\gamma_L + \gamma_t\alpha^3)}{\alpha^2(1 + 2\delta_L)}$
6	$0.5 \leq \alpha \leq 2$ $k_\tau = 2.60\,\dfrac{1+\alpha^2}{\alpha^3}\sqrt{(1 + \alpha^2)^2 + 2(\gamma_L + \alpha^3\gamma_t)}$

Fig. 9.182 b

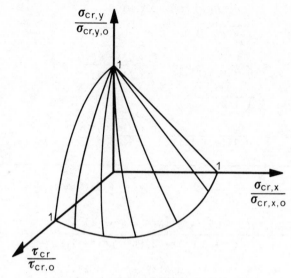

Fig. 9.183

In the case of $\psi = 1$ (uniform compression), formula (9.230) is applicable also to ribbed panels, whereas if $\psi < 1$, it is convenient to use the results given by Klöppel's curves. In this case, the following linear interaction formula can also be used in favour of safety, instead of (9.230):

$$\frac{\sigma_{cr,x}}{\sigma_{cr,x,o}} + \frac{\tau_{cr}}{\tau_{cr,x,o}} = 1 \qquad (9.231)$$

The limiting values of (9.229) can be combined according to the Von Mises criterion, to provide the ideal stress, which is equivalent to the considered stress state:

$$\sigma_{cr,id} = \sqrt{\sigma^2_{cr,x} + \sigma^2_{cr,y} - \sigma_{cr,x} \cdot \sigma_{cr,y} + 3\,\tau^2_{cr}} \qquad (9.232)$$

or, in the mono-axial case:

$$\sigma_{cr,id} = \sqrt{\sigma^2_{cr,x} + 3\,\tau^2_{cr}} \qquad (9.233)$$

In this case, σ_x and τ being the stresses acting on the panel, the ideal stress:

$$\sigma_{id} = \sqrt{\sigma^2_x + 3\,\tau^2} \qquad (9.234)$$

must be multiplied by a factor ν in order to obtain critical conditions (9.233):

$$\sigma_{cr,id} = \nu\,\sigma_{id} \qquad (9.235)$$

Further, assuming loads increase proportionally, it follows that:

$$\sigma_x/\sigma_{cr,} = \tau/\tau_{cr} = 1/\nu \qquad (9.236)$$

Substituting for $\sigma_{cr,x}$ and σ_{cr} from (9.236) in (9.230) and introducing the expression for ν so obtained into (9.235), one gets:

$$\sigma_{cr,id} = \frac{\sqrt{\sigma_x^2 + 3\,\tau^2}}{0.25(1 - \psi)\,\dfrac{\sigma_x}{\sigma_{cr,x,o}} + \sqrt{0.25(3 - \psi)\left\{\dfrac{\sigma_x}{\sigma_{cr,x,o}}\right\}^2 + \left\{\dfrac{\tau}{\tau_{cr,o}}\right\}^2}}$$

(9.237)

which is applicable in a mono-axial case. It provides:

$$\sigma_{cr,id} = \sigma_{cr,x,o} \quad \text{for} \quad \tau = 0$$

$$\sigma_{cr,id} = \sqrt{3}\,\tau_{cr,o} \quad \text{for} \quad \sigma_x = 0$$

(9.238)

In the bi-axial case, with σ_x, σ_y and τ, an analogous procedure leads to deducing the following formulae:

$$\sigma_{cr,id} = \frac{\sqrt{\sigma_x^2 + \sigma_y^2 - \sigma_x\sigma_y + 3\,\tau^2}}{0.25(4 - \rho)\dfrac{\sigma_y}{\sigma_{cr,y,o}} + \sqrt{0.25\left\{\dfrac{\sigma_y}{\sigma_{cr,y,o}}\right\}^2 + \left\{\dfrac{\sigma_x}{\sigma_{cr,x,o}}\right\}^2}}$$

(9.239)

for $\rho \leq 1$, or

$$\sigma_{cr,id} = \frac{\sqrt{\sigma_x^2 + \sigma_y^2 - \sigma_x\sigma_y + 3\,\tau^2}}{0.25(2 + \rho)\dfrac{\sigma_y}{\sigma_{cr,y,o}} + \sqrt{0.25(2 - \rho)\left\{\dfrac{\sigma_y}{\sigma_{cr,y,o}}\right\}^2 + \left\{\dfrac{\tau}{\tau_{cr,o}}\right\}^2}}$$

(9.240)

for $\rho > 1$, where

$$\rho = \frac{\tau\ k_{\sigma,x}}{\sigma_x \cdot k_\tau}$$

(9.241)

is a function of stress values and buckling coefficients.

The boundaries of the interaction domain (9.240) (see Fig. 9.183) are a straight line: for $\tau = 0$ ($\rho = 0$)

$$\frac{\sigma_{cr,x}}{\sigma_{cr,x,o}} + \frac{\sigma_{cr,y}}{\sigma_{cr,y,o}} = 1$$

(9.242)

and a parabola for $\sigma_x = 0$

Whenever $\sigma_{cr,id}$ exceeds the elastic limit of the material conventionally assumed equal to:

$$f_e = 0.8\,f_y$$

(9.243)

it is necessary to replace $\sigma_{cr,id}$ by the reduced critical stress:

$$\sigma_{cr,red} = \eta\sigma_{cr,id}$$

(9.244)

where the factor $\eta < 1$, expressed by ratio:

$$\eta = \sqrt{E_t/E}$$

(9.245)

takes plastic effects into account.

Reduction factor η (9.245) is a function of the stresses to which it is referred. In fact, if the following expression is assumed for tangent modulus E_t:

$$E_t = E \left\{ 1 - \frac{(\sigma - f_e)^2}{(f_y - f_e)^2} \right\}$$ (9.246)

and its value is calculated for stress level:

$$\sigma = \sigma_{cr,red}$$

one obtains, recalling (9.243):

$$E_t = E \left\{ 40 \frac{\sigma_{cr,red}}{f_y} - 25 \left(\frac{\sigma_{cr,red}}{f_y} \right)^2 - 15 \right\}$$ (9.247)

By introducing (9.247) into (9.244) and providing the solution with reference to $\sigma_{cr,red}$, the explicit expression of reduced critical stress is found:

$$\sigma_{cr,red} = f_y \frac{20 + \sqrt{25 - 15(f_y/\sigma_{cr,id})^2}}{25 + (f_y/\sigma_{cr,id})^2}$$ (9.248)

Once the critical ideal stress $\sigma_{cr,id}$ - or its possible reduced value $\sigma_{cr,red}$, has been found according to the linear theory, ECCS Recommendations allow a correcting factor α_c, to be applied provided it does not exceed an imposed limiting value α_c^*. Thus for panel stability:

$$\sigma_{id} < \alpha_c^* \eta \sigma_{cr,id} = \alpha_c^* \sigma_{cr,red}$$ (9.249)

where , in the most general case:

$$\sigma_{id} = \sqrt{\sigma_x^2 + \sigma_y^2 - \sigma_x \sigma_y + 3\tau^2}$$ (9.250)

The values of α_c^* are set out in 9.7.2.2, according to two procedures proposed in |31|.

9.7.2.2 Stiffener design

The possibility of adjusting the critical load value by means of corrective factors α_c^* is strictly related to the post critical strength reserves that can be expected from the structure.

So far, only experimental evidence can indicate the advisable value of α_c^*. In fact, the observation of the behaviour of stiffened plates has shown that the stiffener is very often involved in the buckling phenomenon for much higher values of its flexural rigidity (9.225) than those indicated by the linear theory. This is mainly due to stiffener asymmetries and to fabrication imperfections.

If the behaviour of a stiffened panel is observed according to the linear theory, a limiting value γ^* of its relative flexural rigidity can be defined. From the curve (Fig. 9.184) connecting critical stress σ_{cr} to rigidity γ (9.226) it can be seen that: σ_{cr} increases as γ increases, until a value σ_{cr}^*, corresponding to stiffness γ^* is reached above which critical stress remains constant. γ^* is

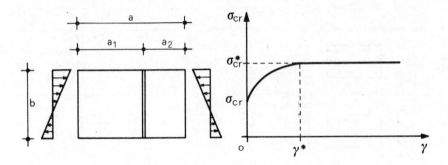

Fig. 9.184

called *optimal relative rigidity* depending on which stiffeners can be classed into two groups (Fig. 9.185):

Flexible stiffeners, if $\gamma < \gamma^*$

Rigid stiffeners, if $\gamma > \gamma^*$

In the first case (Fig. 9.185a), buckling concerns the whole of the stiffened panel, although the critical stress exceeds that which would apply if there were no stiffeners ($\gamma = 0$).

In the second case (Fig. 9.185b), buckling takes place separately in the two sub-panels divided by the stiffener which remains undeformed: it therefore seems useless to increase its rigidity any further, because the overall critical stress in any case corresponds to the lower of those of the two sub-panels.

For the most common types of stiffener and under various stress conditions, the expressions for γ^* have been assessed and they are listed ih the table of Fig. 9.186, in terms of the geometrical characteristics of the panel ($\alpha = a/b$) and of the ribs ($\delta = A_s/bt$, $\eta = \gamma_t/\gamma_L$). The cases examined, concerning bending loads with an entirely compressed panel, pure bending and shear action alone, were as follows:

Stresses due to compression or bending:

1. One longitudinal stiffener at mid-width
2. One transverse stiffener at mid-length

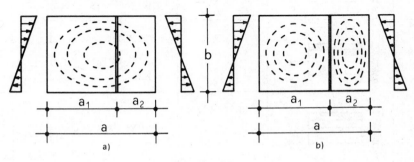

Fig. 9.185

		RANGE	γ^*	ξ
1	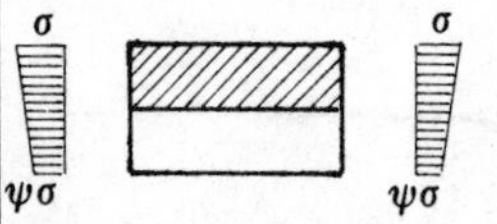	$\alpha < \sqrt{8(1+2\delta)} - 1$ $\alpha > \sqrt{8(1+2\delta)} - 1$	$(0.53+0.47\psi)\{\alpha^2/2[16(1+2\delta)-2] - \alpha^4/2 + (1+2\delta)/2\}$ $(0.53+0.47\psi)\{1/2[8(1+2\delta)-1]^2 + (1+2\delta)/2\}$	3
2	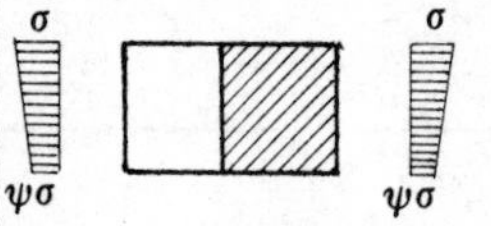	$0.4 \le \alpha \le 1.4$	$\dfrac{4(4/\alpha^2 - \alpha^2/4)}{\pi^2\alpha\left(1 - \dfrac{\pi^2\alpha^4}{12\alpha^4 - 48}\right)}$	3
3	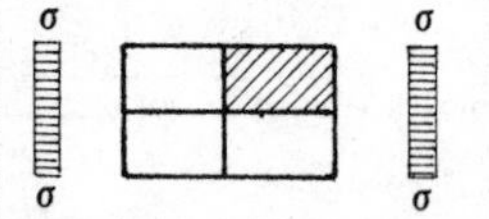	$0.9 \le \alpha \le 1.1$	$\gamma_L = (1+\alpha^2)^2 \dfrac{4(1+2\delta_L)-1}{2(1+\rho\alpha^3)}$ $\rho = \gamma_t/\gamma_L$	3
4	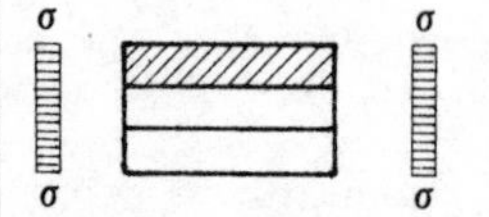	$\alpha < \sqrt{18(1+3)} - 1$ $\alpha > \sqrt{18(1+3)} - 1$	$\dfrac{\alpha^2}{3}\{36(1+3\delta)-2\} - \dfrac{\alpha^4}{3} + \dfrac{1+3\delta}{3}$ $\dfrac{1}{3}\{18(1+3\delta)-1\}^2 + \dfrac{1+3\delta}{3}$	4
5	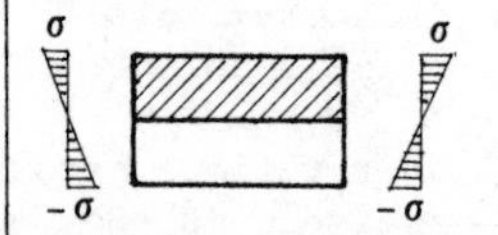		1.3	3
6	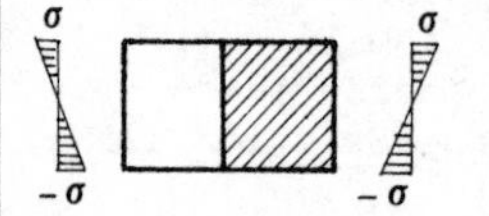	$0.6 \le \alpha \le 0.935$	$6.2 - 12.7\alpha + 6.5\alpha^2$	3
7	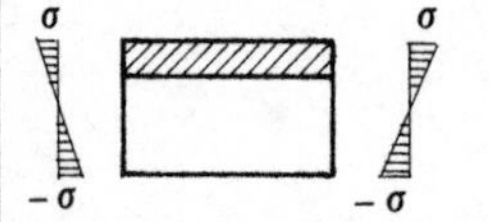	$\alpha \le 0.5$ $\alpha > 0.5$	$2.4 + 18.4\delta$ $(12+92\delta)(\alpha - 0.3)$ but $> 16 + 200\delta$	6
8	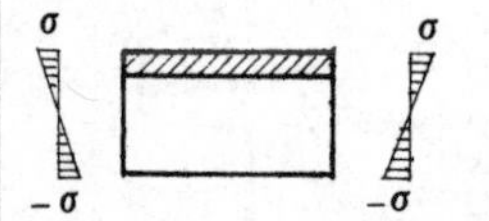	$0.5 \le \alpha \le 1$ $\alpha > 1$	$(21.3+112.6\delta)(\alpha-0.1)$ $(32-168.9)(\alpha - 0.4)$ but $> 50 + 200\delta$	7
9	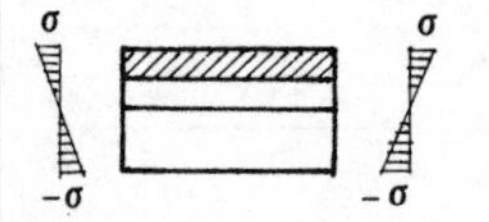	$\alpha \le 0.5$ $\alpha > 0.5$	$1.9 + 10.5\delta$ $(7-50\delta + 27\delta^2)\alpha - (1.6+14.5\delta + 13.5\delta^2)$ but $> 8+10\delta(7+20\delta)$	6

Fig. 9.186

		RANGE	γ^*	ξ
10	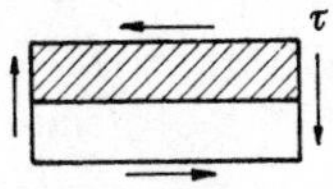	$0.5 \le \alpha \le 2$	$5.4\alpha^2(2\alpha + 2.5\alpha^2 - \alpha^3 - 1)$	3
11	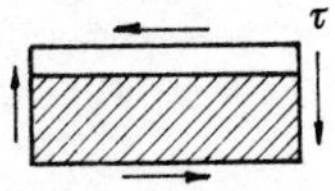	$0.5 \le \alpha \le 2$	$7.2\alpha^2(1 - 3.3\alpha + 3.9\alpha^2 - 1.1\alpha^3)$	6
12	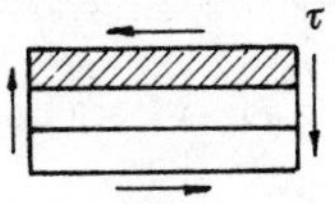	$0.3 \le \alpha \le 1$	$12.1\alpha^2(4.4\,\alpha - 1)$	4
13	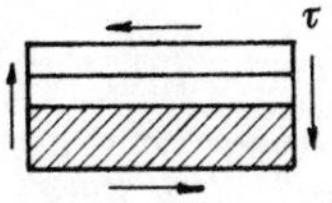	$0.5 \le \alpha \le 2.5$	$\dfrac{\alpha}{4.41}(2.71\alpha^4 - 21.06\alpha^3 + 56.97\alpha^2 - 15.56\alpha - 1)$	6
14	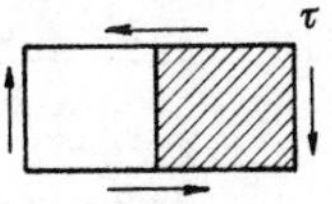	$0.5 \le \alpha \le 2$	$\dfrac{5.4}{\alpha}\left(\dfrac{2}{\alpha} + \dfrac{2.5}{\alpha^2} - \dfrac{1}{\alpha^3} - 1\right)$	3
15	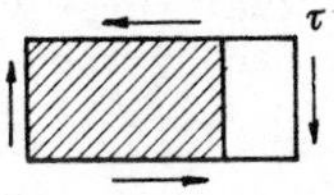	$0.5 \le \alpha \le 2$	$\dfrac{7.2}{\alpha}\left(1 - \dfrac{3.3}{\alpha} + \dfrac{3.9}{\alpha^2} - \dfrac{1.1}{\alpha^3}\right)$	3
16	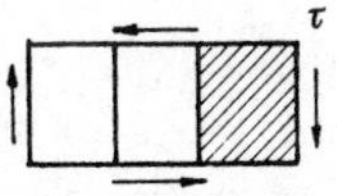	$1 \le \alpha \le 3.3$	$\dfrac{12.1}{\alpha}\left(\dfrac{4.4}{\alpha} - 1\right)$	3
17	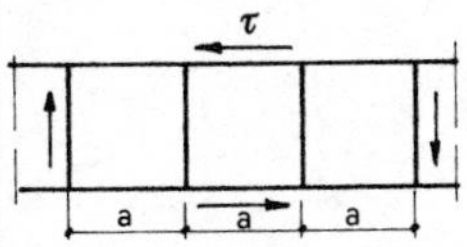	$0.2 \le \alpha \le 1$	$\dfrac{28}{\alpha} - 20\alpha$	3
18	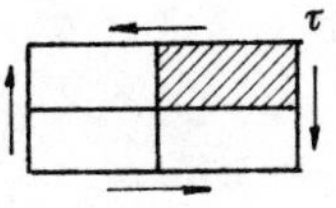	$0.5 \le \alpha \le 2$	$\gamma^*_t = \dfrac{6(1 - \alpha^2)^2}{\alpha^3 + 1/\eta}; \quad \eta = \dfrac{I_t}{I_L}$	3

Fig.9.186

Stresses due to simple compression:
   3. Two crossed stiffeners at mid-length and mid-width
   4. Two longitudinal stiffeners at width thirds
Stresses due to simple bending:
   5. One longitudinal stiffener at mid-width
   6. One transverse stiffener at mid-length
   7. One longitudinal stiffener at $b/4$ from compressed edge
   8. One longitudinal stiffener at $b/5$ from compressed edge
   9. Two longitudinal stiffeners at $b/4$ and $b/2$ from compressed
      edge.
Stresses due to shear:
   10. One longitudinal stiffener at mid-width
   11. One longitudinal stiffener at $b/4$ from compressed edge
   12. Two longitudinal stiffeners at width thirds
   13. Two longitudinal stiffeners at $b/4$ and $b/2$ from compressed
      edge
   14. One transverse stiffener at mid-length
   15. One transverse stiffener at 1/4 length
   16. Two transverse stiffeners at length thirds
   17. Equidistant transverse stiffeners
   18. Two crossed stiffeners at mid-length and mid-width

Contradicting the forecasts of the linear theory, the experiments carried out by Massonet showed that in reality stiffeners consigned according to the theoretical values of optimal rigidity γ^* do not remain perfectly undeformed until critical load is attained |54-155|.

If a designer wishes to profit from panel strength reserves in the post critical field, at least the transverse stiffeners, and preferably also the longitudinal ones, must remain undeformed until collapse. For this purpose, ECCS Recommendations suggest as a *'first procedure'* designing stiffeners with an actual relative flexural rigidity ξ times greater than that provided by the elastic theory:

$$\gamma^{**} = \xi \quad \gamma^* \tag{9.251}$$

The values of the increasing coefficient ξ, determined on the basis of the results of the above mentioned experiments in order to take account of the influence of geometrical and mechanical imperfections, are given in the last column of the table of Fig. 9.186, according to stiffener position and to load condition.

On the other hand, ECCS Recommendations suggest the following mean value of ξ:
   $\xi = 4$ for open section stiffeners
   $\xi = 2.5$ for closed section stiffeners
   If stiffeners are designed with

$$\gamma > \gamma^{**} \tag{9.252}$$

then it is guaranteed that the panel remains rigid, allowing a diagonal field to form in it which is typical of a good post critical behaviour.

If instead:

$$\gamma < \gamma^{**} \tag{9.253}$$

594

the stiffener is considered flexible and in this case, as far as the evaluation of buckling coefficients is concerned, the value of γ for all the stiffeners must be divided by ξ. The stiffener is thus considered as partially effective: this means that only its $1/\xi$ th part can be considered as effective.

If this approach is followed, the limiting values of the corrective factor α_c^* proposed in ECCS Recommendations are as listed in Fig. 9.187, in the 'first procedure' column.

Parameter ψ, defining stress diagram variation, cannot be less than -1.

When σ_x, σ_y and τ act simultaneously, one can assume:

$$\alpha_c^* = \left\{ \frac{\left\{\dfrac{\alpha_{c,x}^* \sigma_x}{\sigma_{cr,x,o}}\right\}^2 + \left\{\dfrac{\alpha_{c,y}^* \sigma_y}{\sigma_{cr,y,o}}\right\}^2 + \left\{\dfrac{\alpha_{c,xy}^* \tau}{\tau_{cr,o}}\right\}^2}{\left\{\dfrac{\sigma_x}{\sigma_{cr,x,o}}\right\}^2 + \left\{\dfrac{\sigma_y}{\sigma_{cr,y,o}}\right\}^2 + \left\{\dfrac{\tau}{\tau_{cr,o}}\right\}^2} \right\}^{1/2} \qquad (9.254)$$

and optimal stiffness γ to be introduced into (9.251) can be assessed by means of the following interaction formula:

$$\gamma^* = \frac{1}{\alpha_c^*} \frac{\gamma_x^*}{\nu_{c,y}} + \frac{\gamma_y^*}{\nu_{c,y}} + \frac{\gamma_{xy}^*}{\nu_{c,xy}} \qquad (9.255)$$

In (9.255), α_c^* was deduced from (9.254). Terms γ_x^*, γ_y^* and γ_{xy}^* indicate optimal stiffness values of the ribs, deduced from Fig. 9.186, assuming the panels loaded only by σ_x, σ_y, τ_{xy}, respectively, acting separately. Coefficients $\nu_{c,x}$, $\nu_{c,y}$ and $\nu_{c,xy}$ are the ratios between the values of reduced critical stress $\sigma_{cr,red}$ determined on the assumption that σ_x, σ_y and τ_{xy}, act on the panel separately and the corresponding values of the actual existing values of σ_x, σ_y and τ_{xy}.

As an alternative, if a designer does not need post critical strength it is possible unconditionally to follow the results of the linear theory, adopting the effective values of γ for assessing buckling coefficients. In this case, however, corrective factor limiting values must obviously be lower. The 'second procedure' suggested in ECCS Recommendations provides such values, given in the last column to the right of Fig.9.187. They, too, can be used by applying (9.254).

Correction factor	Stresses	Imposed values	
		1st procedure	2nd procedure
$\alpha_{c,y}$	Transverse compression (σ_y)	0.83	0.83
$\alpha_{c,x}$	Normal stresses (σ_x)	$1.05-0.11(1+\psi)$	$0.95-0.06(1+\psi)$
$\alpha_{c,x,y}$	shear (τ)	1.65	0.95

Fig. 9.187

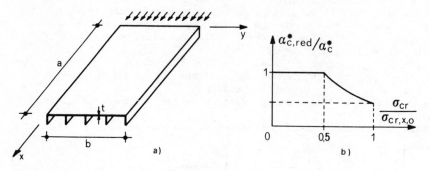

Fig. 9.188

It might, however, happen that such corrective factors must be further reduced, whenever panel stability (whether the panel is stiffened or not) is dominated by the 'column effect', maximizing imperfection influence and simultaneously reducing the post critical reserves to a negligible amount. This happens, for example, in the case of longitudinally stiffened plates with a prevailing longitudinal compression, whenever the Euler critical stress σ_{cr} of the plate, imagined as a column simply supported at its ends and having span 'a' and its transverse section formed by the plate and the longitudinal stiffeners (Fig. 9.188a), exceeds half the buckling stress (9.223), i.e. whenever:

$$1 > \sigma_{cr}/\sigma_{cr,x,o} > 0.5$$

independently of compression stress variability.

As:

$$\sigma_{cr} = \frac{\pi^2 EI}{a^2 A} \qquad (9.256)$$

$$\sigma_{cr,x,o} = k_{\sigma,x} \frac{\pi D}{b^2 t} \qquad (9.257)$$

where:

$$I = \frac{bt^3}{12} + \Sigma \ I_{s,L}$$

$$A = bt + \Sigma A_s \qquad (9.258)$$

$$D = \frac{Et^3}{12(1 - \nu^2)}$$

Then with the following conditions:

$$\alpha = a/b \quad ; \quad \gamma_L = \frac{EI_{s,L}}{bD} \quad ; \quad \delta_L = \frac{A_{s,L}}{bt} \qquad (9.259)$$

one obtains:

$$\frac{\sigma_{cr}}{\sigma_{cr,x,o}} = \frac{(1 - \nu^2) + \Sigma\gamma_L}{\alpha^2 \, k_{\sigma,x}(1 + \Sigma\delta_L)} \qquad (9.260)$$

where $\Sigma\gamma_L$ and $\Sigma\delta_L$, respectively, indicate relative flexural and extensional rigidity of all the longitudinal stiffeners.

In the interval defined by formula (9.255), the reduced corrective factor is expressed as follows:

$$\sigma^*_{c,red} = \alpha^*_c \frac{1}{1 + \{(\sigma_{cr,red}/\sigma_c) - 1\}\{2(\sigma_{cr}/\sigma_{cr,x,o}) - 1\}} \qquad (9.261)$$

which provides the trend referred in Fig. 9.188b, where:

$\sigma_{cr,red}$ = reduced critical stress (9.244);

σ_c = maximum compressive stress for the column of Fig.9.188a, deduced from curve 'b' (see 9.2.2.2).

Corrective factors α^*_c are thus strictly related to rib design, for which ECCS Recommendations provide special guidance.

In the case of transverse web ribs, the condition $\gamma_t \geq \gamma^*_t$ leads to:

$$I_{s,t} \geq \xi\{2.5(h_w/a)^2 - 2\} \, at_w^3 \qquad (9.262)$$

provided $I_{s,t} > 0.5\xi \, at_w^3$

where:

b_w, t_w = web height and thickness, respectively;

a = transverse stiffener distance

Another possible approach for designing transverse stiffeners is to imagine that the diagonal stress fields have formed in the panels. In this case, the transverse stiffeners can be checked as struts undergoing an axial force equal to 0.3 times shear V. The minimum stiffener area can then be expressed as:

$$A_s = \chi \, \frac{0.3 \, V}{t_y} - 12 \, t_w^2$$

where:

$$\chi = \begin{cases} 1 \text{ for stiffeners that are symmetrical with reference to the plate plane} \\ 2 \text{ for stiffeners on one side only provided their slenderness is below 25} \end{cases}$$

For longitudinal stiffeners, condition $\gamma^*_L \geq \gamma^{**}_L$ indicates that the optimal value of ratio $I_{s,L}/\xi$ is that which equalizes the ideal critical stresses corresponding to the individual panel and to the entire stiffened plate.

9.7.2.3 Detailing and tolerances

The aim of the ECCS Recommendations is to contain within acceptable limits those causes (geometrical and mechanical imperfections, working tolerances, errors in the design of building details) that might render it unsafe to apply the linear theory according to the method of the corrective factors |31,156|.

In order to rely on strength reserves in the post critical field, all the transverse and longitudinal stiffeners must be welded.

Wherever they cross each other, continuity must be guaranteed: the

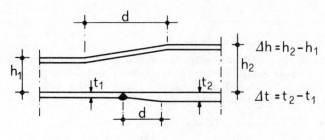

Fig. 9.189

longitudinal stiffener must pass through openings in the transverse stiffener, or it must be welded to it whenever no important shrinkage phenomena are foreseen. The transverse stiffener must be connected to the compression flange of the beam and to the tension flange if there are any concentrated loads.

Some fundamental principles must be complied with in detailing:
The connections in the stiffeners must be rigid
Variation in height Δh or thickness Δt of the transverse section of the stiffeners must be properly tapered along a part d for which (Fig. 9.189):

$$\Delta h/d < 1/6 \qquad \Delta t/d < 1/5$$

The longitudinal stiffeners of the compressed plates must not have any opening across welded joints exceeding $6\,t$ (Fig. 9.190). The webs of the transverse stiffeners, in which there are openings to allow the continuity of the longitudinal stiffeners, must have a minimum height:

$$h_w \geq 1.75\ c$$

if c is hole height (Fig. 9.191).
In the compressed areas, stiffener connections must be symmetrical, in order to avoid unfavourable effects due to eccentricity and bolted joints must have a double cover plate
In the case of welded connections between two plates having

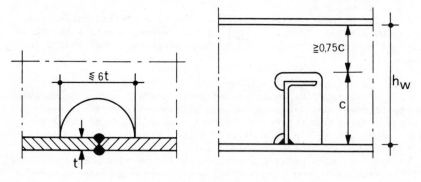

Fig. 9.190

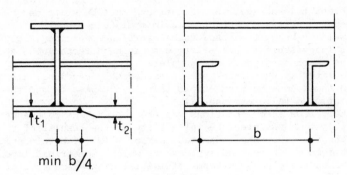

t_1 t_2

min $b/4$

Fig. 9.191

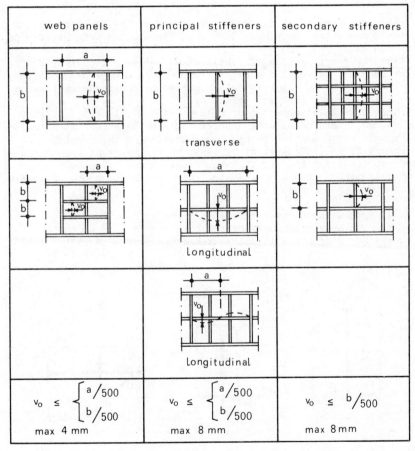

web panels	principal stiffeners	secondary stiffeners
	transverse	
	Longitudinal	
	Longitudinal	
$v_o \leq \begin{cases} a/500 \\ b/500 \end{cases}$ max 4 mm	$v_o \leq \begin{cases} a/500 \\ b/500 \end{cases}$ max 8 mm	$v_o \leq b/500$ max 8 mm

Fig. 9.192

different thicknesses, the thinner one must be stiffened by a transverse stiffener set at a maximum distance equal to $b/4$ from the weld, b being the longitudinal stiffener spacing (Fig. 9.192) It is inadvisable to use stiffeners having a rectangular section in the compressed parts of the plate

Such criteria, which are perhaps exceedingly cautious, aim at avoiding those construction errors that in the past caused the above mentioned failures (see 9.7.1.3).

Further, ECCS Recommendations set precise working tolerances that are also able to guarantee the reliability of the results of the linear theory. They are indicated in Fig. 9.192 with reference to the most common fabrication systems for which geometrical imperfection limit values are given, in terms of initial displacement v_0 of the individual parts forming the stiffened plane plates.

9.7.3 *Beam webs*

The vast interest of this subject is proved by the great number of design methods that have been developed, some which were briefly illustrated in 9.7.1.4.

From the point of view of fabrication, the traditional system of stiffening the web by means of transverse and possibly longitudinal stiffeners is being progressively replaced by the tendency towards abolishing any intermediate stiffener.

Studies in this respect have been carried out largely in the US and Sweden, where beams with unstiffened webs have been used for over ten years. Such a solution has been codified in the Swedish Recommendations |157| ever since 1966, on the basis of Nylander's studies, |158,159| and it will undoubtedly spread very soon to other Countries |160|,because of its simplicity considering the increasing incidence of handwork.

9.7.3.1 Webs without intermediate stiffeners
Design. The only stiffeners of the beam are located at the end supports or under possible concentrated loads (Fig. 9.193). They

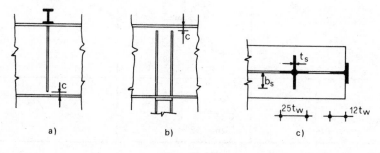

a)   b)   c)

Fig. 9.193

must be symmetrical with reference to the web plane and set at the following distance from the flange opposite the loaded side:

$$c < 4 \, t_w$$

Such stiffeners must be designed to bear the concentrated load (Fig. 9.193a) or the support reaction (Fig. 9.193b).

According to AISC recommendations |55|, the resistant section to be taken into consideration is that formed by the two coupled stiffeners plus a part of web equal to $25t_w$ for intermediate stiffeners and $12t_w$ for end stiffeners (Fig. 9.193c). The effective length of the stiffener is assumed as 3/4 of beam height, whereas transverse dimension ratios must be limited as follows:

$$b_s/t_s \leq 0.55 \, \sqrt{E/f_y} = 16.27 \, \sqrt{235/f_y} \qquad (9.263)$$

The stiffeners are considered rigid whenever they not only bear the concentrated load, but provide also a strong anchorage to the diagonal tension bands (Fig. 9.194). According to Höglund |135,161| if: $\lambda_w > 3.4$, λ_w being expressed as

$$\lambda_w = h_w/t_w \, \sqrt{f_y/E} \cong h_w \, \sqrt{f_y/235} \, /(29.6t_w) \qquad (9.264)$$

the stiffeners must bear a horizontal component equal to:

$$q = 2.5 \, f_y \, t_w/\lambda_w$$

If instead $\lambda_w \leq 3.4$ a flexible stiffener is sufficient: in other words, the stiffener need not comply with this particular requirement.

From the geometrical point of view, the stiffener can be considered rigid if - with reference to Fig. 9.195a - the detail is designed with two stiffeners so that:

$$d > 0.18 \, h_w; \qquad A_{s,1} > 0.1 \, A_w$$

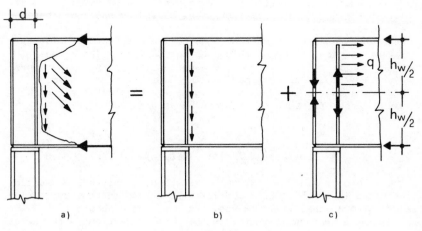

a)   b)   c)

Fig. 9.194

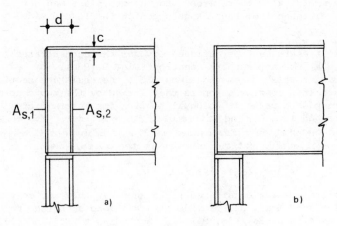

Fig. 9.195

whereas $A_{s,2}$ must be designed to bear the support reaction.

Where there is only one end stiffener (Fig. 9.195b), it must be considered flexible.

For the rest of the beam, where the web is not stiffened, its slenderness must be limited by the following relation:

$$h_w/t_w < 0.4 \ E/f_{y,f} = 350 \ (235/f_{y,f})$$ (9.266)

In order to avoid buckling phenomena, the flange dimensions must comply with the following relation:

$$h_w/t_w \leq 0.88 \ \sqrt{E/f_{y,f}} \cong 26 \ \sqrt{235/f_{y,f}}$$ (9.267)

Conventional elastic limit state. In a beam so designed the web is protected against local effects due to concentrated vertical loads, but its buckling due to normal stresses caused by shear and bending is accepted.

This is taken into account by considering a reduced or effective resistant section, although this does not mean that the stress distribution in the web is exactly assessed: the purpose is to provide an overall estimate of the load bearing capacity of the beam.

Depending on web slenderness λ_w (9.264), the elastic strength modulus of the effective section can be assumed equal to:

$$
\begin{aligned}
W_{red} &= W_x & &\text{if } \lambda_w \geq 4.8 \\
W_{red} &= \{1 - 0.15 \, (A_w/A_f)(1 - 4.8/\lambda_w)\}W_w & &\text{if } \lambda_w < 4.8
\end{aligned}
$$ (9.268)

The values provided by (9.268) correspond to the reduced section illustrated in Fig. 9.196a. The more the stress in the compression flange is lower than yield stress, the more these values underestimate section strength. In the case of sections with unequal flanges, the resistant section can be identified as illustrated in

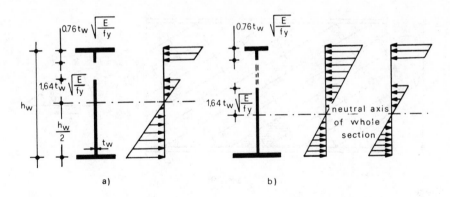

Fig. 9.196

Fig. 9.196b. For hybrid beams, Höglund |135| suggested applying a reduction to formulae (9.268) as follows:

$$\Delta W = \frac{t_w \, h_w^3}{12} \left\{ 1 - \frac{f_{y,w}}{f_{y,f}} \right\}^2 \left\{ 2 + \frac{f_{y,w}}{f_{y,f}} \right\} \qquad (9.269)$$

in order to take account of the fact that, as usually $f_{y,f} > f_{y,w}$, the web becomes plastic while the flange is still in the elastic field.

Once the reduced section modulus has been assessed, control at the conventional elastic limit state requires:

$$\sigma = M/W_{red} \leq f_y \qquad (9.270)$$

Serviceabiliy limit state. In order to determine displacements, one must use reduced moments of inertia given by:

$$I_{red} = I_x \qquad\qquad\qquad \text{if } \lambda_w \leq 4.8$$
$$\qquad\qquad\qquad\qquad\qquad\qquad\qquad\qquad\qquad (9.271)$$
$$I_{red} = \{1 - 0.08 \; A_w/A_f \; (1 - 4.8/\lambda_w)\} \, I_x \quad \text{if } \lambda_w > 4.8$$

where λ_w is determined by (9.264), replacing the yield stress f_y in it by the actual stress under service loads.

Whenever (as in the case of deep and short beams) shear deformability is important, a good approximation may be obtained by increasing the bending displacements assessed by means of formulae (9.271), by the quantity:

$$1 + 15 \, (h_w/L)^2 \, A_f/A_w \qquad\qquad \text{for simply supported beams}$$

$$1 + 75 \, (h_w/L)^2 \, A_f/A_w \qquad\qquad \text{for intermediate spans of continuous beams, wherever loaded.}$$

Ultimate limit state. The ultimate value of shear force can generally be expressed as:

$$V_u = V_{u,w} + V_{u,f} \qquad (9.272)$$

603

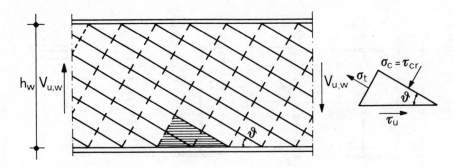

Fig. 9.197

where:

$V_{u,w}$ = ultimate value of shear due to the in plane membrane behaviour of the web

$V_{u,f}$ = ultimate value of shear supported by the flanges, due to their flexural rigidity

If there are no intermediate stiffeners, it is not possible to foresee the forming of a frame-like mechanism, as indicated in Fig. 9.174, and therefore one can assume $V_{u,f} = 0$.

When assessing $V_{u,w}$ reference must be made to the post critical behaviour that takes place due to the stress redistribution following web buckling. At this point, according to the Stockholm method, the web can be conceived as if it were replaced by a system of orthogonally intersecting bars, having and inclination θ that changes as the load increases (Fig. 9.197). The compression bars bear a constant stress, corresponding to the critical tangential stress τ_{cr}, whereas in the tension bars the stress τ_t increases as their inclination θ decreases.

Value $V_{u,w}$ is attained whenever the plastic level according to the Von Mises criterion is reached at the intersection between the bars. It is given by:

$$V_{u,w} = A_w \tau_u$$

The design values of τ_u are provided in the table of Fig. 9.198, as a function of slenderness λ_w (9.264).

$\lambda_w = \dfrac{h_w}{t_w} \sqrt{\dfrac{f_y}{E}}$	DESIGN VALUES τ_u	
	END STIFFENERS	
	FLEXIBLE (Fig. 9.195 b)	RIGID (Fig. 9.195 a)
$\lambda_w \leq 2.4$	$0.58\, f_y$	$0.58\, f_y$
$2.4 \leq \lambda_w \leq 3.4$	$1.4\, f_y / \lambda_w$	$1.4\, f_y / \lambda_w$
$3.4 \leq \lambda_w$	$1.4\, f_y / \lambda_w$	$2.22\, f_y / (\lambda_w + 2)$

Fig. 9.198

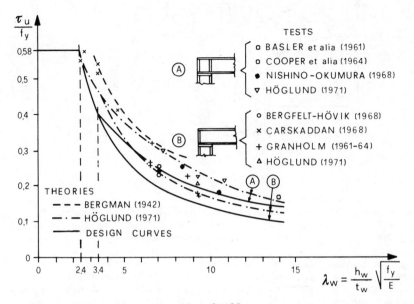

Fig. 9.199

These values are so applicable to stiffeners over the intermediate supports of continuous beams. The definition of a rigid or flexible stiffener was provided above. It is intuitive that, if the stiffener is not sufficiently rigid as to guarantee the anchorage of the diagonal tension bands, the ultimate shear value $V_{u,w}$ must be reduced.

These results are based on Höglund's experiments |161| and were adopted by the Swedish code |162|. Fig. 9.199 provides a comparison |163| between the experimental results obtained by various authors, the results of Bergman's theories |164| and those of Höglund |161|, as well as the design curves corresponding to the expressions given in Fig. 9.198. The latter are the lower bounds to both the theoretical and experimental values, with reference to rigid and flexible end stiffeners.

The influence of bending moment is analysed by comparing its value with the value of the full plastic moment of the flanges only

$$M_{f,pl} = A_f(h - t_f)f_{y,f} \qquad (9.273)$$

If $M < M_{pl}$, the bending moment can be considered to have no influence on the ultimate shear value $M_{u,w}$.

If, however, $M > M_{f,pl}$, Basler's formula |132| provides the interaction bond between flexural and shear load bearing capacities (Fig. 9.200):

$$M_u = M_{f,pl} + (M_{pl} - M_{f,pl}) \left\{ 1 - (V/V_{u,w})^2 \right\} \qquad (9.274)$$

If $V_{u,f} = 0$, curve A-B of the domain shown in Fig. 9.200 coincides

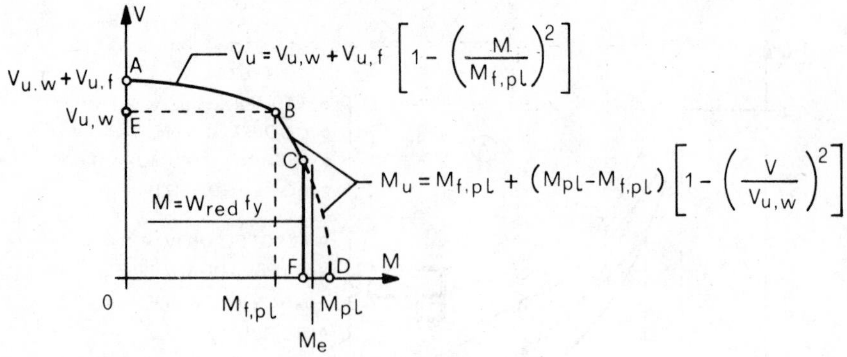

Fig. 9.200

with segment EB. This domain is further limited by segment CF, expressing the reduced value of the flexural load bearing capacity M_{red} due to web buckling in the compressed areas:

$$M_{red} = M_e$$

where $M_{red} = W_{red} \, \hat{f}_y$, and W_{red} is given by formulae (9.268).

According to $|160|$, the compatibility condition between normal stresses due to bending σ_M and tangential stresses τ due to shear is expressed as:

$$\sigma_M/f_y + 0.63 \ \tau/\tau_u \le 1.38 \qquad (9.275)$$

where the values of τ_u are those given in the table of Fig. 9.198.

9.7.3.2 Webs with intermediate stiffeners

From 9.7.1.4 it appears that there exist many design methods aimed at interpreting (more or less approximately) the real behaviour of webs divided into panels by transverse stiffeners. After having examined such methods, an attempt will now be made to deduce some general conclusions indicating which is the best approach as far as design is concerned.

It has been seen that there are two substantially different points of view. The first (that followed in the ECCS Recommendations - see 9.7.2) is based on the results of the linear stability theory. In this context, the critical state, possibly rectified by appropriate factors taking post critical reserves into account (see 9.7.2.1), is chosen as the limit against which the web panel must be guaranteed. With reference to their relative flexural rigidity, the stiffeners can be considered rigid if $\gamma > \gamma^{**}$ and therefore buckling takes place in the individual web panels (see Fig. 9.185b), or else flexible if $\gamma < \gamma^{**}$ i.e. if the stiffeners are involved in web buckling (see Fig. 9.185a) where

$$\gamma^{**} = \xi \ \gamma^* \qquad (9.276)$$

is the effective flexural rigidity rectifying optimal (flexural) stiffness γ^* provided by the linear theory by means of the empirical

606

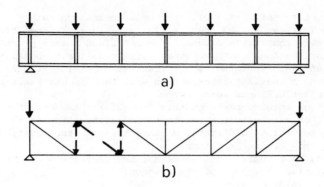

Fig. 9.201

coefficient $\xi > 1$ assumed from Massonnet's experiments $|154,155|$ for the purpose of guaranteeing stiffener undeformability until collapse (see 9.7.2.2).

The second point of view, which has been followed by many methods, consists in adopting proper models for interpreting the behaviour of beams with stiffened webs in the post critical phase. The behaviour of a beam beyond the critical state can thus be considered equivalent to that of an ideal truss (Fig. 9.201).

In order that such a mechanism of behaviour may take place, in addition to an adequate flexural stiffness, an adequate extensional stiffness of the vertical stiffeners is also required:

$$\delta > \delta* \qquad (9.277)$$

From the survey of such collapse methods (see Fig. 9.176) it clearly appears, given from present knowledge in this field, that a design procedure, although approximate, must be simple to apply, and its static model must be clear.

Among the collapse mechanisms adopted by the various models, the most frequent one (no doubt because it interprets experimental results best) is that illustrated in Fig. 9.202.

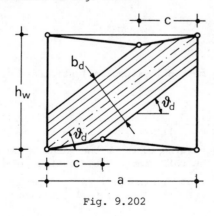

Fig. 9.202

In fact, the vast series of experiments performed mainly in Cardiff, Lehigh and Trieste, showed that, at failure, in each panel between the stiffeners there is a diagonal tension field having practically the same inclination θ_d as, and symmetrically disposed about, the panel diagonal, and anchored also to the flanges. The width of the stress band b_d is such that its intersection with the flanges takes place where the plastic hinges form.

The various methods based on this type of mechanism differ, although often very slightly, as to the quantification of the model of Fig. 9.202 according to distance c characterizing the position of the plastic hinges in the flange.

In the case of symmetrical I-beams, two parameters can be defined for the relative rigidity of the flanges:

Flexural rigidity

$$\gamma_o = \frac{I_f}{a^3 t_w} = \frac{b_f t_f^3}{12 a^3 t_w} \tag{9.278}$$

Extensional rigidity

$$\delta_o = \frac{t_f b_f}{t_w h_w} \tag{9.279}$$

The various methods generally have an expression for the ratio c/a depending on one of the above rigidities.

Flexural stiffness γ_o is assumed in the following methods:
Aarau $|136|$:

$$c/a = \frac{\gamma_o \cdot 10^6}{2 \times 10^6 \cdot \gamma_o + 16} \tag{9.280}$$

Göteborg $|143|$:

$$c/a = 5 \sqrt[4]{\gamma_o} \tag{9.281}$$

Trieste $|144|$:

$$c/a = \frac{\gamma_o \cdot 10^6}{2 \times 10^6 \cdot \gamma_o + 12} \tag{9.282}$$

In the following methods, extensional stiffness δ_o is assumed instead:
Osaka $|140|$:

$$c/a = \frac{1.782 \, \delta_o + 0.38}{2(1.782 \, \delta_o + 1)} \tag{9.283}$$

Cardiff - Prague, according to which the value of c/a are determined by the solution to the following cubic equation $|165|$:

$$\{c/a\}^3 - \{c/a\}^2 + 4\delta_o \, (1/\sin \theta_d) \, f_y/\sigma_{ref} = 0 \tag{9.284}$$

where θ_d is the geometrical inclination of the panel diagonal, given by:

$$\sin \theta_d = \frac{1}{\sqrt{1 + \alpha^2}} \qquad \text{where} \quad \alpha = a/h_w \qquad (9.285)$$

and

$$\sigma_{ref} = -\frac{3}{2} \tau_{cr,o} \sin 2\theta_d + \sqrt{f_y^2 + \tau_{cr,o}^2 \{(\frac{3}{2} \sin 2\theta_d)^2 - 3\}} \qquad (9.286)$$

where

$$\tau_{cr,o} = k_\tau \, \sigma_{cr,o} \quad \text{given by (9.223) for} \quad \tau_{cr,o} < (0.8/\sqrt{3})f_y \quad \text{or:}$$

$$\tau_{cr,o} = \frac{f_y}{\sqrt{3}} \left\{ 1 - \frac{0.16 \, f_y}{\sqrt{3} \, \tau_{cr,o}} \right\}, \quad \text{for} \quad \tau_{cr,o} > \frac{0.8}{\sqrt{3}} f_y$$

Stockholm |134|

$$c/a = 0.25 \, \delta_o \, t_f/h_w = 0.25 \, \delta_o^2 \, t_w/b_f \qquad (9.287)$$

The Tokyo method |142| instead, is dependent on flange rigidity and directly assumes:

$$c/a = 0.5$$

an asymptotic value of (9.280) and (9.282) as γ_o increases (see Fig. 9.203).

Fig. 9.203 compares the values of c/a provided by applying (9.280), (9.281) and (9.282) as a function of γ_o. It can be seen that (9.280)

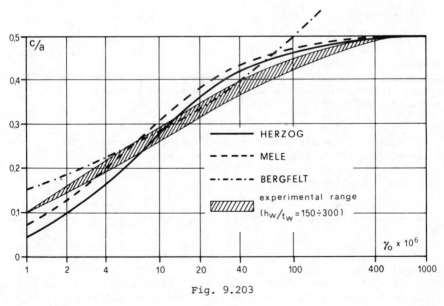

Fig. 9.203

and (9.282) agree very well with the experimental results, whereas (9.281) provides reliable results within the values of $\gamma \leq 10^{-4}$.

The Cardiff-Prague method depends not only on extensional stiffness δ_o, but also on the tensile stress σ_{ref} (expressed by (9.286) in the diagonal band, considered acting in the direction of the geometrical diagonal of the panel. This expression is based on the observation that the stresses in the various points of the diagonal band are formed by the following three components:

$$\sigma_x = \sigma_t + \tau_{cr,o} \sin 2\,\theta_d$$
$$\sigma_y = -\,\tau_{cr,o} \sin 2\,\theta_d \qquad\qquad (9.288)$$
$$\tau = \tau_{cr,o} \cos 2\,\theta_d$$

where

σ_t = tensile stress due to post critical behaviour according to the model of Fig. 9.202

$\tau_{cr,o}$ = stress corresponding to panel critical state, axes x and y being assumed as references, x coinciding with the diagonal.

Plasticity, following Von Mises, is attained when:

$$f_y = \sqrt{\sigma_x^2 + \sigma_y^2 - \sigma_x \sigma_y + 3\tau^2} \qquad\qquad (9.289)$$

Substituting the expression of (9.288) in (9.289) and assuming $\sigma_t = \sigma_{ref}$, one obtains a second degree equation from which expression (9.286) for σ_{ref} can be obtained.

It should be pointed out that the stresses expressed by (9.288) are included between pure traction:

$$\sigma_x = \sigma_t = f_y \ ; \quad \sigma_y = 0 \ ; \quad \tau = 0$$

and pure shear:

$$\sigma_x = -\sigma_y = f_y/\sqrt{3}$$

On the Von Mises ellipse (Fig. 9.204) these limits are represented, respectively, by points B and A.

In order to obtain a simpler expression of σ_{ref}, Mele proposed a linear interpolation of the Von Mises domain in its part A - B by means of the following linear equation:

$$\sigma_1 = f_y + (\sqrt{3} - 1)\sigma_2 \qquad\qquad (9.290)$$

If $\theta_d = 45°$, the stresses expressed in (9.288) become the following principal stresses:

$$\sigma_1 = \sigma_t + \tau_{cr,o}$$
$$\sigma_2 = -\tau_{cr,o} \qquad\qquad (9.291)$$

By substituting these in (9.290) and putting $\sigma_{ref} = \sigma_t$ one obtains:

$$\sigma_{ref} = f_y \{1 - \sqrt{3}\ \tau_{cr,o}/f_y\} \qquad\qquad (9.292)$$

Compared to the σ_{ref} proposed by Rockey (9.286), formula (9.292) provides an acceptable approximation (deviation < 10%) for the evaluation of ultimate shear. However, if it is adopted to design transverse stiffeners, it must be rectified by multiplying it by

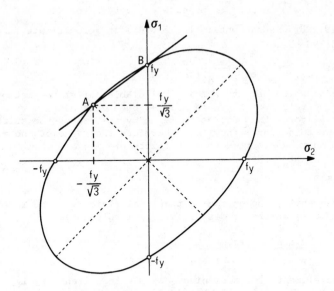

Fig. 9.204

the appropriate coefficient η as follows:

$$\eta = 1.6 - \alpha, \text{ for } 0.2 \leq \alpha \leq 0.6 \tag{9.293}$$

$$\eta = 1 \qquad , \text{ for } \alpha > 0.6$$

Reference stress σ_{ref} allows the assessment of the ultimate shear strength $V_{u,f}$ due to post critical behaviour according to the model of Fig. 9.202 from which:

$$b_d = 2 c \sin \theta_d \tag{9.294}$$

Since

$$V_{u,f} = \sigma_{ref} b_d t_w \sin \theta \tag{9.295}$$

and:

$$\sin \theta_d = \frac{1}{\sqrt{1 + \sigma^2}}$$

(9.294) can also be expressed as follows:

$$V_{u,f} = \frac{2 c t_w}{1 + \alpha^2} \sigma_{ref} \tag{9.296}$$

As ultimate shear contribution to panel critical state is given by:

$$V_{u,w} = \tau_{cr,o} t_w h_w \tag{9.297}$$

the overall ultimate shear capacity is:

611

$$V_u = V_{u,w} + V_{u,f} = \tau_{cr,o} \, t_w \, h_w + \frac{2 \, c \, t_w}{1 + \alpha^2} \, \sigma_{ref} \qquad (9.298)$$

From (9.298) one can obtain the expression for ultimate tangential stress, given by:

$$\tau_u = \tau_{cr,o} + \sigma_{ref} \frac{2\alpha}{1 + \alpha^2} \frac{c}{a} \qquad (9.299)$$

the value of which becomes readily calculable, after the method defining the values of σ_{ref} and c/a has been chosen.

The axial force absorbed at collapse by the transverse stiffeners in practice coincides with $V_{u,f}$. Therefore the value of the minimum area that must be allotted to such stiffeners is:

$$A_{s,min} = \frac{V_{u,f}}{f_y} = \frac{2 \, c \, t_w}{1 + \alpha^2} \frac{\sigma_{ref}}{f_y} \qquad (9.300)$$

and, consequently, the optimal extensional rigidity to be introduced into transverse stiffener design (9.277) is given by:

$$\delta^* = \frac{A_{s,min}}{t_w \, h_w} = \frac{\sigma_{ref}}{f_y} \frac{2\alpha}{1 + \alpha^2} \frac{c}{a} \qquad (9.301)$$

All the elements for designing a beam with a transversely stiffened web are now available and design can be done with reference to two limit states, defined as follows:

Serviceability limit state (at critical state)

Collapse limit state

Control at serviceability limit state consists in checking that the panel does not buckle under service loads, amplified by a suitable coefficient ν_1. This criterion is based on the observation that the critical load does not express the maximum load bearing capacity of the structure and therefore does not characterize the ultimate limit state, but only indicates the loads at which large transverse deformations of the panel occur. It can be considered a serviceability limit state, as web panel buckling can create psychological and aesthetic problems to the users.

The required check can thus be expressed as:

$$\sigma_{id} < \sigma_{cr,red}/\nu_1 \qquad (9.302)$$

$\sigma_{cr,red}$ being given in 9.7.2.1 according to the linear theory, whereas σ_{id} is calculated on the basis of the service loads.

In order to guarantee post critical resources, stiffeners must keep their rigidity. Thus their relative flexural rigidity must be at least equal to their optimal one γ^* (see the table of Fig. 9.186), and their extensional rigidity δ must be adequate to allow the diagonal behaviour in the post critical field (Fig. 9.201).

For the control at the ultimate limit state, stresses must be evaluated on the basis of loads multiplied by load factors γ_f.

By generalizing (9.299) for a condition of bending and shear stresses, an expression of the following type can be assumed as a reference tangential stress:

612

$\gamma_o \, 10^4$	c/a	$\gamma_o \, 10^4$	c/a	$\gamma_o \, 10^4$	c/a
1	0.07	11	0.32	40	0.43
2	0.12	12	0.33	45	0.44
3	0.17	13	0.34	50	0.45
4	0.20	14	0.35	70	0.46
5	0.23	15	0.36	100	0.47
6	0.25	17	0.37	200	0.48
7	0.27	20	0.38	500	0.49
8	0.28	22	0.39	1000	0.50
9	0.30	25	0.40	2000	0.50
10	0.31	30	0.42	5000	0.50

Fig. 9.205

$$\tau_{ref} = \frac{\sigma_{cr,red}}{\sqrt{3}} + \sigma_{ref} \frac{2\alpha}{1 + \alpha^2} \frac{c}{a} \qquad (9.303)$$

It is the sum of two contributions: one due to the strength of the structure prior to panel buckling and the other corresponding to the post critical reserves.

In particular, if the procedure suggested by Mele is adopted, the σ_{ref} given by (9.292) and the ratio c/a given by (9.282) must be introduced into (9.303). The ratio c/a in question is set out in Fig. 9.205 as a function of the flange relative flexural rigidity γ_0 (9.278).

Formula (9.303) thus becomes:

$$\tau_{ref} = \frac{\sigma_{cr,red}}{\sqrt{3}} + (f_y - \sigma_{cr,red}) \frac{\alpha}{1 + \alpha^2} \frac{c}{a} \qquad (9.304)$$

Formula (9.304) is the basis of the design method proposed in the Italian recommendations |166|.

The tangential stress actually existing in the web must be:

$$\tau_m \leq \tau_{ref}/\gamma_m \qquad (9.305)$$

If this condition is not complied with γ_m is an appropriate resistance factor (in |166|, $\gamma_m = 1.2$ is assumed).

If, furthermore:

$$\tau_m \leq \frac{\sigma_{cr,red}/\sqrt{3}}{\gamma_m} \qquad (9.306)$$

it is not necessary to require any post critical resources from the panel, but it is sufficient to guarantee an adequate flexural stiffness of the transverse stiffeners. For this purpose, it is sufficient to design the stiffener so that its moment of inertia is (recalling (9.227)):

613

$$I_s \geq 0.092 \; \gamma^* \; h_w t_w^3 \tag{9.307}$$

Where stiffeners are symmetrical with reference to the web the moment of inertia I_s is assumed to be calculated with reference to the web mean plane. If stiffeners are fixed on one side, I_s is calculated with reference to the plane of their connection to the web (Fig. 9.206):

$$I_s = I_x + Ae^2 \tag{9.308}$$

Optimal flexural rigidity can be assumed approximately equal to:

$$\gamma^* = (28/\alpha) - 20 \; \alpha, \quad \text{for } 0.4 \leq \alpha \leq 1 \tag{9.309}$$

$$\gamma^* = 8 \qquad , \quad \text{for } \quad \alpha > 1$$

according to the extremely simple solution proposed by Stein and Fralich |167| and, which leads to either values of γ^* than the rigorous ones obtained by Klöppel |124-125|. Formula (9.309) therefore is on the safe side and in practice is equivalent to assuming an increase coefficient ξ (9.276) equal to about 1.5 to 2.

If (9.306) is not complied with, not only the flexural rigidity corresponding to (9.307), but also an adequate extensional rigidity of the stiffeners must be assured, in order that the post critical behaviour of the panel be such as to withstand at least up to ν_c times the service loads. For this purpose, recalling (9.228), it must be ensured that:

$$A_s \geq \delta^* \; h_w \; t_w \tag{9.310}$$

Area A_s of the stiffener is assumed to be an effective area equal to the geometrical one increased by a portion of web, having a width extending to not more than 5 times web thickness on either side of the stiffener (Fig. 9.206b).

Relative rigidity δ^* can be calculated according to (9.301), in which, in conformity with what has been said the following

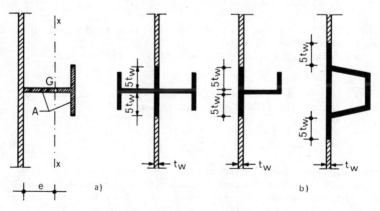

Fig. 9.206

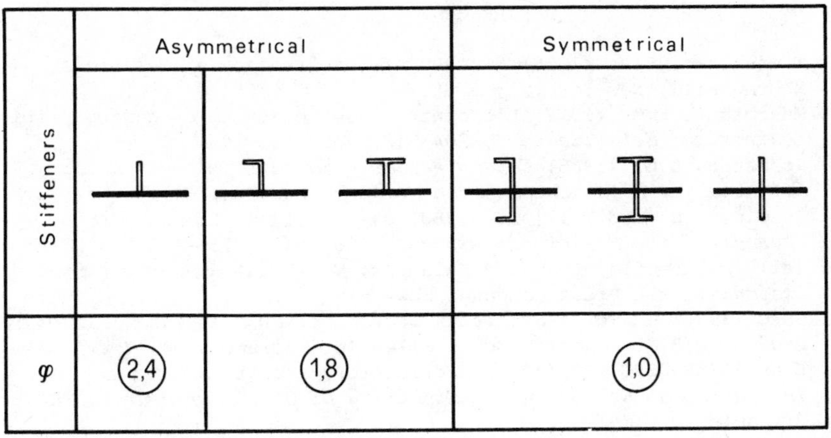

	Asymmetrical			Symmetrical		
Stiffeners						
φ	$\textcircled{2,4}$		$\textcircled{1,8}$		$\textcircled{1,0}$	

Fig. 9.207

expression for the reference stress is assumed:

$$\sigma_{ref} = (f_y - \sigma_{cr,red})\eta \qquad (9.311)$$

η being the appropriate rectifying coefficient (9.293).

With the following situation:

$$\rho = \frac{2\alpha}{1 + \alpha^2}\,\eta \qquad (9.312)$$

and introducing the rectifying coefficient ϕ, suggested by Basler to take stiffener form and position (see Fig. 9.207) into account, one obtains:

$$\delta^* = \{1 - \sigma_{cr,red}/f_y\}\rho\phi \quad (c/a)_{min} \qquad (9.313)$$

The value $(c/a)_{min}$ is found my imposing (9.305) by equality

$$(c/a)_{min} = \frac{\nu_c\,\tau_m - \sigma_{cr,red}/\sqrt{3}}{\dfrac{2\alpha}{1+\alpha^2}\,(f_y - \sigma_{cr,red})} \qquad (9.314)$$

It is interesting to point out that this approach may be considered as an intermediate one between the linear design method and the US recommendations (AISC), based on Basler's and Thürlimann's model

615

REFERENCES

1. Timoshenko, S.P. (1953) *History of the Strength of Materials*, McGraw Hill, New York
2. Mazzolani, F.M. (1977) L'evoluzione del concetto di carico critico, *Costruzioni Metalliche*, 4, 204-210; see also |48|
3. Johnston, B.G. (1966) *Column Research Council Guide to Design Criteria for Metal Compression Members*, 2nd Edn, Wiley, New York
4. Beedle, L.S. and Tall, L. (1980) Basic Column Strength, *ASCE Journal of the Structural Division*, 86, ST 7, 139-173
5. Tall, L., Beedle, L.S. and Galambos, V. (1964) *Structural Steel Design*, Ronald Press Company, New York
6. SSRC (1976) *Guide to Stability Design Criteria for Metal Structures*, 3rd Edn, (B.J. Johnston Edr) Wiley Interscience, New York
7. ECCS-IABSE-SSRC-CRC (1976) Introductory report, *Second International Colloquium on Stability of Steel Structures*, ECCS, Brussels
8. ECCS-IABSE-SSRC-CRC (1976) *Second International Colloquium on Stability of Steel Structures, Preliminary Report*, ECCS, Brussels
9. ECCS-IABSE-SSRC-CRC (1977) *Second International Colloquium on Stability of Steel Structures, Final Report*, ECCS, Brussels
10. ECCS-IABSE-CRC (1977) *Regional International Colloquium on Stability of Steel Structures*, Final Report, Budapest
11. ASCE (1977) *International Colloquium on Stability of Structures under Static and Dynamic Loads*, Washington D.C., ASCE, New York
12. AISC (1981,82) Stability of metal structures. A world vew, *Engineering Journal*, AISC, 3/81, 90-120, 4/81, 154-196, 1/82, 27-62, 2/82, 101-138
13. Ketter, R.L., Kaminsky, E.L. and Beedle, L.S. (1955) Plastic deformation of wide flange beam-columns, *Transactions ASCE*, 120, 1028-1069
14. Ketter, R.L. (1955) Stability of beam-column above the elastic limit, *Proceedings ASCE*, 81, Separate No. 692, 1-27
15. Galambos, T.V. and Ketter, R.L. (1969) Columns under combined bending and thrust, *ASCE Journal of the Mechanical Division*, 85, EM2, 1-30
16. Vinnakota, S. and Badoux, J.C. (1970), Flambage élastoplastique des poutres-colonnes appuyées sur des ressorts, *Construction Métallique*, 2, 5-16
17. Vinnakota, S. and Aoshima, Y. (1974) Inelastic behaviour of rotationally restrained columns under biaxial bending, *The Structural Engineer*, 52, 245-255
18. Ojalvo, M. (1960) Restrained columns, *ASCE Journal of the Mechanical Division*, 86, EM 5, 1-11
19. Levi, V., Driscoll, G.C. and Lu, L.W. (1965) Structural subassemblages prevented from sway, *ASCE Journal of the Structural Division*, 91, ST 5, 103-123
20. Batterman, R.H. and Johnston, B.C. (1967) Behaviour and maximum strength of metal columns, *ASCE Journal of the Structural Division*, 93, ST 2, 205-230
21. Frey, F. (1971) Calcul de flamblement des barres industrielles, *Bulletin Technique de la Suisse Romande*, 11, 239-258

22. Beer, H. (1966) Beitrag zur Stabilitätsuntersuchung von Stabwerken mit Imperfektionen, *IABSE Publications*, 26, 43-60
23. Beer, H. and Schulz, G. (1969) Die Traglast des planning mittung gedrückten Stabs mit Imperfektionen, *VDI-Zeitschrift*, 111, 1537-1541, 1683-1687, 1767-1772
24. Beer, H. and Schulz, G. (1970) Bases théoretiques des courbes européennes de flambement, *Construction Métallique*, 3, 37-57
25. Ballio, G., Petrini, V. and Urbano C. (1973) Simulazione numerica del comportamento di elementi strutturali compressi per incrementi finiti del carico assiale, *Costruzioni Metalliche*, 1, 27-37
26. Faella, C. and Mazzolani, F.M. (1974) Simulazione del comportamento di aste industriali inelastiche sotto carico assiale, *Costruzioni Metalliche*, 4, 235-248
27. Sfintesco, D. (1970) Fondement expérimental des courbes européennes de flambement, *Construction Métallique*, 3, 5-12
28. Jacquet, J. (1970) Essais de flambement et exploitation statistique, *Construction Métallique*, 3, 13-16
29. Carpena, A. (1970) Détermination des limites élastiques pour l'analyse du flambement, *Construction Métallique*, 3, 58-63
30. Ballio, G. (1977) Problemi di instabilità delle strutture in acciaio: situazione attuale delle conoscenze ed aspetti progettuali, *Costruzioni Metalliche*, 5, 231-240
31. CECM ECCS (1978) *European Recommendations for Steel Construction*, ECCS, Brussels
32. ASCE (1971) *Guide for Design of Steel Transmission Towers*, ASCE, New York, See Also 1967, *ASCE Journal of Structural Division*, 93, ST 4, 245-282
33. Strating, J. and Vos, H.J. (1972) *Computer Simulation of the ECCS Buckling Curve Using a Monte Carlo Method*, Proceedings, International Colloquium on Column Strength, IABSE Reports of the Working Commissions, 23, Zurich, 334-358
34. Ballio, G., Finzi, L., Urbano, C. and Zandonini, R. (1975) Flambement des tubes en acier à haute limite élastique, *Construction Métallique*, 2, 5-14
35. Council on Tall Buildings, Group SB (1979) *Structural Design of Tall Steel Buildings*, Volume SB of Monograph on Planning and Design of Tall Buildings, ASCE, New York
36. Mazzolani, F.M., Di Carlo, A. and Pignataro, M. (1977) *Post Buckling behaviour of Multistory Steel Frames*, Proceedings of the International Conference on Stability of Structures under Static and Dynamic Loads, Washington, 194-211, see |11|
37. Mazzolani, F.M. (1973) Influenza delle imperfezioni sulla stabilità delle colonne in acciaio, *Costruzioni Metalliche*, 6, 380-390
38. Mazzolani, F.M. (1972) *Buckling Curves of Hot Rolled Steel Shapes with Structural Imperfections*, Proceedings, International Colloquium on Column Strength, Paris, IABSE Reports of the Working Commissions, 23, Zurich, 152-161
39. Yang, C.H., Beedle, L.S. and Johnston, B.S. (1952) Residual stress and the yield strength of steel beams, *Welding Journal*, 31, Research Supplement, 205S-230S
40. Huber, A.W. and Beedle, L.S. (1954) Residual stress and the compressive strength of steel, *Welding Journal*, 33, No. 12

617

41. VanKaren, R.C. and Galambos, T.V. (1964) Beam-column experiments, *ASCE Journal of the Structural Division*, 90, ST 2, 224-256
42. Nitta, A. and Thürlimann, B. (1962) Ultimate strength of high yield strength constructional alloy circular columns. Effect of cold straightening, *IABSE Publications*, 22, 265-288
43. Tall, L. (1966) *Welded Built up Columns*, Fritz Engineering Laboratory, Report No. 249.29, Lehigh University, Bethlehem, Pennsylvania
44. Osgood, W.R. (1957) *The Effect of Residual Stress in Column Strength*, Proceedings, 1st U.S. National Congress on Applied Mechanics, 415-418
45. Thürlimann, B. (1960) New aspects concerning inelastic instability of steel structures, *ASCE Journal of Structural Division*, 86, ST 1, 99-120
46. Godfrey, G.B. (1962) The allowable stresses in axially loaded steel struts, *The Structural Engineer*, 40, 97-112
47. Beer, H. and Schulz, G. (1972) *The European column curves*, Proceedings, International Colloquium on Column Strength, Paris, IABSE Reports of the Working Commissions, 23, Zurich, 385-398
48. Maquoi, R. and Rondal, J. (1978) Mise en équation des nouvelles courbes européennes de flambement, *Construction Métallique*, 1, 17-29
49. Rondal, J. and Maquoi, R. (1979) Formulation d'Ayrton-Perry pour le flambement des barres métallique, *Construction Métallique*, 4, 41-53
50. Dwight, J.B. (1972) *Use of Perry Formula to Represent the New European Strut Curves*, Proceedings, International Colloquium on Column Strength, Paris, IABSE Reports of the Working Commissions, 23, Zurich, 399-412
51. Finzi, L. and Urbano, C. (1977) *Minimizing the Number of Buckling Curves in Codes*, Proceedings of the International Conference on Stability of Structures under Static and Dynamic Loads, Washington, 688-697, see |11|
52. AASHO (1957, 65, 69, 73) *Standard Specifications for Highway Bridges*, American Association of State Highway and Transportation Officials, Washington
53. AREA (1970) *Specifications for Steel Railway Bridges*, American Railway Engineering Association, Chicago
54. DIN 4114 (1952) *Stahlbau, Stabilitätsfalle (Knickung, Kippung, Beulung) Berechnungsgrundlagen vorschriften und Richtlinienen*; see also VDE (1953) *Stahl im Hochbau*, Stahleisen, Düsseldorf; see also Galambos, T.V. and Jones, J. *German Buckling Specification DIN 4114 English Translation*, Column Research Council, New York
55. AISC (1978) *Specification for the Design, Fabrication and Erection of Structural Steel for Buildings*, American Institute of Steel Construction, Chicago
56. Bleich, F. (1952) *Buckling Strength of Metal Structures*, McGraw Hill, New York
57. Lin, F.J., Glauser, E.C. and Johnston, B.G. (1970) Behaviour of laced and battened structural members, *ASCE Journal of the Structural Division*, 96, ST 7, 1377-1402

58. Ballio, G., Finzi, L. and Zandonini, R. (1977) *A Theoretical Approach to the Behaviour of Centrally Compressed Built up Struts*, Preliminary report of Second International Colloquium on Stability of Steel Structures, Liège, 77-84, see |8|

59. Petrini, V. and Zandonini, R. (1977) Simulazione numerica del comportamento di aste composte con elementi ravvicinati, *Costruzioni Metalliche*, 3, 128-140

60. Timoshenko, S. and Gere, J. (1964) *Theory of Elastic Stability*, 2nd Edn., McGraw Hill, New York

61. Salvadori, M.G. (1956) Lateral buckling of eccentrically loaded I Columns, *Transaction ASCE*, 121, 1163-1178

62. Clark, J.W. and Hill, H.N. (1960) Lateral buckling of beams and girders, *ASCE Journal of Structural Division*, 86, ST 7, 175-196

63. Kerensky, O.A., Flint, A.R. and Brown, W.C. (1956) The basis for design of beams and plate girders in the revised british standard 153, *Proceedings, The Institution of Civil Engineers*, Part III, 5, 396

64. Flint, A.R. (1950) The stability and strength of slender beams, *Engineering*, 170,545

65. Neal, B.G. (1950) The lateral instability of yielded mild steel beams of rectangular cross section, *Philosophical Transaction, Royal Society*, London, 242 (A), 848

66. Horne, M.R. (1952) *The Lateral Instability of I Beams Stressed Beyond the Elastic Limit*, Report of British Welding Research Association, London

67. Galambos, T.V. (1963) Inelastic lateral buckling of beams, *ASCE Journal of the Structural Division*, 89, ST 5, 217-242

68. Como, M. and Mazzolani, F.M. (1969) Ricerca teorico sperimentale sullo svergolamento nel piano e fuori del piano dei profilati in presenza di tensioni residue, *Costruzioni Metalliche*, 3, 212-243

69. Mazzolani, F.M. (1970) *Plane, Torsional and Lateral Buckling of I Shapes Taking Into Account Residual Stresses*, Proceeding of the IV Conference on Metal Constructions, Warsaw

70. Faella, C. and Mazzolani, F.M. (1972) Instabilità flessotorsionale inelastica di travi metalliche sotto carichi trasversali, *Costruzioni Metalliche*, 6, 433-450

71. Lindner, J. (1971) Näherungsweise Ermittlung der Traglasten von auf Biegung und Torsion beanspruchten Stäben, *Die Bautechnik*, 48, 160-170

72. Nethercot, D.A. (1974) Buckling of welded beams and girders, *IABSE Publications*, 34, 107-121

73. Vinnakota, S. (1976) Inelastic stability of laterally unsupported I beams, *Proceedings, 2nd National Symposium on Computerised Structural Analysis*, Washington University

74. Massey, C. (1964) The lateral stability of steel I beams in the plastic range, *Civil Engineering Transactions, Institution of Engineers Australia*, CE6, No. 2, 119

75. Dibley, J.E. (1969) Lateral torsional buckling of I sections in grade 55 steel, *Proceedings, Institution Civil Engineers*, 43, 599-627

76. Fukumoto, Y., Fujiwara and Watanabe, N. (1971) Inelastic lateral buckling tests on welded beams and girders, *Proceedings, Japanese Society of Civil Engineers*, Paper 189

619

77. Klöppel, K. and Unger, B. (1971) Eine experimentelle Untersuchung des Kippverhaltens von Kragträgern im elastischen und Plastischen Bereich im Hinblick auf eine Neufassung des Kippsicherheitsnachweises der DIN 4114, *Der Stahlbau*, 40, 375-384

78. Kitipornchai, S. and Trahair, N.S. (1975) Inelastic buckling of simply supported steel I beams, *ASCE Journal of the Structural Division*, 101, ST 7, 1333-1348

79. Lindner, J. (1974) Der Einfluss von Eigenspannungen auf die Traglast von I Trägern, *Der Stahlbau*, 2, 39-45, 3, 86-91

80. Lindner, J., and Bamm, D. (1977) *Influence of Realistic Yield Stress Distributions on Lateral Torsional Buckling Loads*, Preliminary Report of Second International Colloquium on Stability of Steel Structures, Liège, 213-216, see |8|

81. Fukumoto, Y. and Kubo, M. (1977) *An Experimental Revew of Lateral Buckling of Beams and Girders*, Stability of Structures under Static and Dynamic Load, Washington, 541-555, see |11|

82. Ballio, G. (1973) Stabilità laterale delle travi inflesse, *Costruzioni Metalliche*, 3, 149-156

83. Merchant, W. (1954) The failure load of rigidly jointed framework as influenced by stability, *The Structural Engineer*, 32, 185-190

84. Ballio, G., Petrini, V. and Urbano, C. (1973) The effect of loading process and imperfections on the load bearing capacity of beam columns, *Meccanica*, 8, 56-66

85. Ballio, G., Petrini, V. and Urbano, C. (1974) Loading effects in beam-columns, *Meccanica*, 9, 256-273

86. Roik, K. and Wagenknecht, G. (1976) *Traglastdiagramme zur Bemessung von Druckstäben mit doppeltsymmetrischem Querschnitt aus Baustahl*, KIB, Berichte, 27, Vulkan Verlag, Essen

87. Roik, K. and Bergmann, R. (1977) *Steel Column Design*, Preliminary Report of Second International Colloquium on Stability of Steel Structures, Liège, 339-348, see |8|

88. Austin, W.J. (1961) Strength and design of metal beam-columns, *ASCE Journal of the Structural Division*, 87, ST 4, 1-32

89. Campus, F. and Massonnet C. (1955) *Recherches sur le flambement de colonnes en acier A 32, à profil en double té, sollicitées obliquement*, Bulletin du Centre d'études de recherches et d'essais, 7, 119-338, Tome VII, Liège, see also *ASCE Journal of the Structural Division*, 85, 75-111

90. Ballio, G. and Campanini, G. (1981) Equivalent bending moment for beam-columns, *The Journal of Constructional Steel Research*, 1, No. 3, 13-23

91. Birnstiel, C. and Michalos, I. (1962) Ultimate loads of H columns under biaxial bending, *ASCE Journal of the Structural Division*, 89, ST 4, 161-197

92. Chen, W.F. and Atsuta, T. (1977) *Theory of Beam Columns, Vol. 1, In Plane Behaviour and Design, Vol. 2, Space Behaviour and Design*, McGraw Hill, New York

93. Nishino, F., Tall, L. and Okumura, T. (1968) Residual stress and torsional buckling strength of H and cruciform columns, *Transaction Japan Society of Civil Engineers*, 160, 75

94. Randl, E. (1972) *Beitrag zur Berechnung von Stäben aus Baustahl mit strukturellen und geometrischen Imperfektionen unter Berücksichtigung der räumlichen Verformungsgeometric*,Dissertation in Graz

95. Lindner, J. (1972) *Theoretical Investigations of Columns under Biaxial Loading*, International Colloquium on Column Strength, Paris, IABSE Reports of Working Commissions, Vol. 23, Zurich, 182-190

96. Winter, G. (1960) Lateral Bracing of Columns and Beams, *Transactions ASCE*, 125, 807-845

97. Anderson, J.P. and Woodward, J.H. (1972) Calculation of effective lengths and effective slenderness ratios of stepped columns, *Engineering Journal*, AISC, 4, 157-166

98. Agraval, K.M. and Stafiej, A.P. (1980) Calculation of effective length of stepped columns, *Engineering Journal*, AISC, 4, 96-105

99. NEN 3851 (1972) *Technical Principles for the Design and Calculation of Building Structures*, Netherlands Standards Institution, Unofficial Translation

100. Finzi, L. and Nova, E. (1969) *Elementi Strutturali*, CISIA, Milan

101. Lay, M.G. and Galambos, T.V. (1967) Inelastic beams under moment gradient, *ASCE Journal of the Structural Division*, 93, ST 1, 381-399

102. CNR UNI (1980) *Costruzioni in acciaio. Istruzioni per il calcolo, l'esecuzione e la manutenzione*, Consiglio Nazionale delle Ricerche, Rome

103. Julian, O.G. and Lawrence, L.S. (1959) *Notes on J and L Nomograms for Determination of Effective Lengths*, unpublished, see |55|, Commentary

104. CM (1966) *Régles de calcul des constructions en acier*, Eyrolles, Paris

105. Wood, R.H. (1974) Effective lengths of columns in multi-storey buildings, in three parts, *The Structural Engineer*, 52, 235-244, 295-392, 341-346

106. Wood, R.H. and Roberts, E.H. (1975) A graphical method of preventing sidesway in the design of multi-storey buildings, *Proceedings of Institution of Civil Engineers*, London, 59, 353-372

107. Kuhn, G. and Lundgren, H.R. (1977) *An Appraisal of the Effective Length Alignment Charts*, Stability of Structures under Static and Dynamic Load, Washington, 212-242, see |11|

108. Chu, K.H. and Chow, H.L. (1969) Effective column length in unsymmetrical frames, *IABSE Publications*, 29, 1-15

109. Setti, P. and Zandonini, R. (1982) Stability of single storey unbraced frames, *Costruzioni Metalliche*, 2, 79-98

110. Pagano, M. and Mazzolani, F.M. (1966) Indagine sperimentale al vero sulla instabilità elastoplastica di due telai multipiano in acciaio, *Costruzioni Metalliche*, 4, 213-225, 5, 297-320, 6, 387-398

111. Rosenblueth, E. (1965) Slenderness effects in buildings, *ASCE Journal of the Structural Division*, 91, ST 1, 229-252

112. Adams, P.F. (197..) *Design of Steel Beam-Columns*, Proceedings, Canadian Structural Engineering Conference, Montreal, Canada

113. AISI (1968) *Specification for the Design of Cold Formed Steel Structural Members*, American Iron and Steel Institute, New York

114. Winter, G. (1968) *Commentary on the 1968 Edition of Light Gage Cold Formed Steel Design Manual*, American Iron and Steel Institute New York

115. Peköz, T. and Winter, G. (1980) Cold formed steel construction, *IABSE Periodica*, 1

116. Massonnet, C. and Save, M. (1976) *Calcul plastique des constructions*, Vol. I, 3rd Edition, Editor Nelissen, Angleur, Liège

117. CTICM(1978) *Manuel pour le calcul en plasticité des constructions en acier*, Centre Technique Industriel de la Construction Métallique, Puteaux

118. Basler, K. (1961) *New Provisions for Plate Girder Designs*, Proceedings National Engineering Conference, AISC

119. Herzog, M. (1974) Ultimate strength of plate girders from tests, *ASCE Journal of the Structural Division*, 100, ST 5, 349-863

120. Granholm, C.A. (1960) *Tests on Girders with Thin Web Plates*, Rapport 202, Institutionen för Byggnadsteknik, Chalmers Tekniska, Högskola, Goteborg

121. Bergfelt, A. (1971) *Studies and Tests on Slender Plate Girders without Stiffeners*, Proceedings of Colloquium on Design of Plate and Box Girders for Ultimate Strength, IABSE

122. Timoshenko, S. (1915) Stability of rectangular plates with stiffeners, *Mem. Inst. Engrs. Ways of Commun*, 89

123. Timoshenko, S. (1921) Uber die Stabilität verstefter Platten, *Der Eisenbau*, 12

124. Klöppel, E.K. and Scheer, K.H. (1960) *Beulwerte ausgesteifter Rechteckplatten, Band I*, Ernst & Sohn, Berlin

125. Klöppel, E.K. and Müller, K.H. (1968) *Beulwerte ausgesteifter Rechteckplatten, Band II*, Ernst & Sohn, Berlin

126. Massonnet, C. (1968) *Poutres de grandes dimensions à âme pleine*, Preliminary Report of the 8th Congress of IABSE, New York, 157-177

127. Sattler, K. (1970) Nochmals: Betrachtungen über die Brachursachen der neuen Wiener Donaubrücke, *Der Tiefbau, Ingenieurbau und Strassenbau*, 948-950

128. (1970) Collapse of Cheddar Bridge ouver Milford Haven, *Civil Engineering and Public Works Revue*, 605-653

129. (1970) West Gate Bridge collapse in Melbourne, *Civil Engineering and Public Works Review*, 1328-1329

130. (1971) Koblenz November 1971 Fourth major steel bowgirder bridge collapse, *The Consulting Engineer*, 23-24

131. Maquoi, R. and Massonnet. C. (1972) Leçons à tirer des accidents survenus à quatre grands ponts métalliques en caisson, *Annales des Travaux Publics de Belgique*, 2, 69-84

132. Basler, K. and Thürlimann, B. (1961) Strength of plate girders in bending, *ASCE Journal of the Structural Division*, 87, ST 7, 153-180

133. Basler, K. (1961) Strength of plate girders in shear, *ASCE Journal of the Structural Division*, 87, ST 8, 151-180

134. Basler, K. (1961) Strength of plate girders under combined bending and shear, *ASCE Journal of the Structural Division*, 87, ST 7, 181-197

135. Hoglund, T. (1973) *Design of Thin Plate I Girders in Shear and Bending with Special Reference to Web Buckling*, Institutionen för Byggnadsstatik Kungl. Tekniska Högskolan, Stockholm, 94

136. Herzog, M (1974) Die Traglast unversteifter und versteifter, dünnwandiger Blechträger unter reinem Schub und Schub mit Biegung nach Versuchen, *Der Bauingenieur*, <u>49</u>, No. 10, 382-389
137. Porter, D.M., Rockey, K.C. and Evans, H.R. (1975) The collapse behaviour of plate girders loaded in shear, *The Structural Engineer*, <u>53</u>, No. 8, 313-325
138. Steinhardt, O. and Valtinat, G. (1972) *Theorie und Berechnung Beulgefährdeter vollwand und kastenträger*, Schriftenreihe, Helf 9, Technische Universität Hannover, Lehrstuhl für Stahlbau
139. Ostapenko, A. and Chern, C. (1971) *Ultimate Strength of Longitudinally Stiffened Plate Girders under Combined Loads*, Colloquium on Design of Plate and Box Girders for Ultimate Strength, London, IABSE Proceedings, <u>11</u>, 301-313
140. Komatsu, S. (1971) *Ultimate Strength of Stiffened Plate Girders Subjected to Shear*, Colloqium on Design of Plate and Box Girders for Ultimate Strength, London, IABSE Proceedings, <u>11</u>, 49-65
141. Skalood, M. and Rockey, K. (1971) *The Ultimate Strength and behaviour of Plate Girders Loaded in Shear*, Colloquium on Design of Plate and Box Girders for Ultimate Strength, London, IABSE Proceedings, <u>11</u>, 1-19
142. Fujii, T., Fukumoto, Y., Nishino, F. and Okumura, T., *Research Works on Ultimate Strength of Plate Girders and Japanese Provisions on Plate Girder Design*, Colloquium on Design of Plate and Box Girders for Ultimate Strength, London, IABSE Proceedings, <u>11</u>, 22-48
143. Bergfelt, A. (1973) *Plate Girders with Slender Webs, Survey and a Modified Calculation Method*, Summary in English of the Report No. 11.2.2 to "Nordiske Borskningsdager for Stalkonstruksjoner", Oslo
144. Dubas, P. (1974) *Zur Erschöpfungslast schubbeanspruchter Stehbleche*, Professor Steinhardt Festschrift, Karlsruhe
145. Mele, M. (1973) Sul dimensionamento delle nervature di irrigidimento d'anima nelle travi in acciaio, *Costruzioni Metalliche*, <u>4</u>, 258-270, <u>5</u>, 315-332
146. Maquoi, R. and Massonnet, C. (1971) Non-linear theory of postbuckling resistance of large stiffened box girders, *IABSE Publications*, <u>31</u>-II, 91-140
147. Merrison Committee (1974) *Inquiry into the Basis of Design and Method of Erection of Steel Box Girder Briges: Interim Design and Workmanship Rules*, Her Majesty's Stationary Office, Department of the Environment, London
148. Dwight, J.B. and Little, G.H. (1974) *Stiffened Compression Panels. A Design Approach*, Cambridge University Technical Report No. CUED/C Struct/TR 38
149. Dwight, J.B. and Little, G.H. (1976) Stiffened steel compression flanges. A simpler approach, *The Structural Engineer*, <u>12</u>, 501-509
150. Murray, N.W. (1975) Analysis and design of stiffened plates for collapse load, *The Structural Engineer*, 153-158
151. Horne, M.R. and Narayanan, R. (1975) *An Approximate Method for the Design of Stiffened Steel Compression Panels*, Proceedings of The Institution of Civil Engineers: Research & Theory, <u>9</u>

152. Chatterjee, S. and Dowling, P.J. (1975) *Proposed Design Rules for Longitudinal Stiffeners in Compression Flanges of Box Girders*, CESLIC Report BG40, Imperial College, London

153. Dubas, P. (1971) Essais sur le comportement post critique de poutres en caisson raidies, *IABSE Report of Working Commission*, 11, 367-379

154. Massonnet, C. (1951) Recherches experimentals sur le voilement de l'ame de poutres à ame pleine, *Bulletin du CERES*, Liege, 5, 67-240 and (1952) Preliminary Report of the 4th Congress of IABSE, Cambridge, 539-555

155. Massonnet (1960) Stability considerations in the design of steel plate girders, *ASCE Journal of the Structural Division*, 86, ST 1, 71-97

156. Massonnet, C. and Janss, J. (1981) *State of Art Report on Tolerances in Steel Plated Structures*, The Design of Steel Bridges edited by Rockey and Evans, Granada, London, 83-118

157. HSI balken (1966) *Provisoriska normer för svetsade stalbalkar, typ HSI (Provisional Specifications for Welded Steel Girders, Type HSI)*, AB Gränges Hedlund, Stockholm

158. Nylander, H. (1963) *Stabilited av tryckt fläns vid tunväggig I-balk (Stability of Compression Flange of Thin Plate I Girders)* Bulletin No. 45 Division of Building Statics and Structural Engineering, Royal Institute of Technology, Stockholm

159. Nylander, H. (1964) *Stabilited av kontinuerling I balk med sidostagad överfläns (Stability of Continuous I Girders with Braced Top Flange)* Bulletin No. 50, Division of Building Statics and Structural Engineering, Royal Institute of Technology, Stockholm

160. CECM ECCS EKS (1977) *Steifenlose Stahlskelettragwerke und dünnwandige Vollwandträger*, Ernst & Sohn, Berlin, Munchen, Dusseldorf

161. Höglund, T. (1971) *Livets verkningssätt och bärförmaga hos tunnväggig I balk*, Bulletin No. 93, Division of Building Statics and Structural Engineering, Royal Institute of Technology, Stockholm

162. StBK K2 (1973) *Kommentarer till Stalbyggnadsnorm 70 Knäckning, vippning och buckling (Comments on the Regulations for Steel Structures, Buckling, Lateral Buckling, Local Buckling)*, Worked out by Nylander, H. in collaboration with Höglund, T. and Johansson, B.. The National Swedish Committee on Regulations for Steel Structures, Stockholm

163. Frey, F. and Anslijn, R. (1971) *Dimensionnement des poutres a âme plein sans raidisseurs*, CRIF, Centre des Recherches Scientifique et Technique de l'Industrie des Fabrications Metalliquès, Bruxelles MT 114

164. Bergman, S.G.A. (1948) *Behaviour of Buckled Rectangular Plates under the Action of Shearing Forces*, Institute of Structural Engineers and Building Representatives, Stockholm

165. Rockey, K.C. (1971) *An Ultimate Load Method of Design for Plate Girders*, Development in Bridge Design and Construction, Crosby Lockwood, London

624

166. Ballio, G., Caramelli, S., De Miranda, F. and Mele, M. (1975) Commenti alle nuove istruzioni CNR: nervature di irridimento delle anime di travi a parete piena, *Costruzioni Metalliche*, 5, 229-241
167. Stein, M. and Fralich, R.W. (1949) *Critical Shear Stress of Infinitely Long, Simply Supported Plates with Transverse Stiffeners*, NACA, Tech. Note 1851

Allen H.S. & Bukson PS 1980
BACKGROUND TO BUCKLING
McGRAW-HILL